Theory and Treatment Planning in Family Therapy: A Competency-Based Approach

DIANE R. Gehart, Ph.D.

California State University, Northridge

Australia • Brazil • Japan • Korea • Mexico • Singapore • Spain • United Kingdom • United States

Theory and Treatment Planning in Family Therapy: A Competency-Based Approach
Diane R. Gehart

Product Director: Jon-David Hague
Product Manager: Julie Martinez
Product Assistant: Nicole Richards
Media Developer: Sean Cronin
Marketing Manager: Shanna Shelton
Art and Cover Direction, Production Management, and Composition: Carolyn Deacy and Jitendra Kumar, MPS Limited
Manufacturing Planner: Judy Inouye
Photo Researcher: Saranya Sarada
Text Researcher: Pinky Subi
Text and Cover Designer: Carolyn Deacy
Cover Image: © Didier Marti/Getty Images

© 2016 Cengage Learning

WCN: 01-100-101

ALL RIGHTS RESERVED. No part of this work covered by the copyright herein may be reproduced, transmitted, stored or used in any form or by any means graphic, electronic, or mechanical, including but not limited to photocopying, recording, scanning, digitizing, taping, Web distribution, information networks, or information storage and retrieval systems, except as permitted under Section 107 or 108 of the 1976 United States Copyright Act, without the prior written permission of the publisher.

> For product information and technology assistance, contact us at
> **Cengage Learning Customer & Sales Support, 1-800-354-9706**
> For permission to use material from this text or product,
> submit all requests online at **www.cengage.com/permissions**
> Further permissions questions can be emailed to
> **permissionrequest@cengage.com**

Library of Congress Control Number: 2014939779

ISBN: 978-1-285-45643-0

Cengage Learning
20 Channel Center Street
Boston, MA 02210
USA

Cengage Learning is a leading provider of customized learning solutions with office locations around the globe, including Singapore, the United Kingdom, Australia, Mexico, Brazil, and Japan. Locate your local office at **www.cengage.com/global**

Cengage Learning products are represented in Canada by Nelson Education, Ltd.

To learn more about Cengage Learning, visit **www.cengage.com**

Purchase any of our products at your local college store or at our preferred online store **www.cengagebrain.com**

Printed in the United States of America
2 3 4 5 6   22 21 20 19 18

Dedication

In the past few years, the field of family therapy has lost many whose contributions are our mainstay. This book is dedicated to those who have paved the way for the next generation. We are forever in their debt.

Gianfranco Cecchin
Whose laughter, humility, and acceptance transformed me

Tom Andersen
Whose presence was angelic: the most "gentle" man I have ever met

Paul Watzlawick
Whose courage and kind words I shall never forget

Steve de Shazer
Whose brilliance dazzled me

Insoo Kim Berg
Whose energy and enthusiasm inspired the best in me

Michael White
Whose ideas opened new worlds for me

Jay Haley
Who taught me the logic of paradox

Ivan Boszormenyi-Nagy
Who reminded me to focus on what really matters

Brief Contents

Preface xvi
Acknowledgments xx
About the Author xxii
Author's Introduction xxiii

PART I Introduction to Competency, Theory, and Treatment Planning 1

1. Competency and Theory in Family Therapy 3
2. Research and Ethical Foundations of Family Therapy Theories 15
3. Theory-Specific Case Conceptualization and Treatment Planning 35
4. Philosophical Foundations of Family Therapy Theories 47

PART II Family Therapy Theories 67

5. Systemic Therapies: MRI and Milan 69
6. Strategic Therapy 103
7. Structural Family Therapy 125
8. Satir's Human Growth Model 147
9. Symbolic-Experiential and Internal Family Systems 173
10. Intergenerational and Psychoanalytic Family Therapies 203
11. Cognitive-Behavioral and Mindfulness-Based Couple and Family Therapies 233
12. Solution-Based Therapies 271
13. Narrative Therapy 301
14. Collaborative Therapy and Reflecting Teams 329
15. Evidence-Based Treatments in Couple and Family Therapy: Emotionally Focused Therapy and Functional Family Therapy 357
16. Evidence-Based Group Treatments for Couples and Families 395

PART III Cross-Theoretical Case Conceptualization and Integration 407

17. Cross-Theoretical Case Conceptualization and Integration 409

Name Index 445
Subject Index 447

Contents

Preface xvi
Acknowledgments xx
About the Author xxii
Author's Introduction xxiii

PART I Introduction to Competency, Theory, and Treatment Planning 1

1 Competency and Theory in Family Therapy 3

Lights, Camera, Action! 3
Competency and Theory: Why Theory Matters 4
  Why All the Talk About Competency? 4
  Competency and (Not) You 4
  Common Threads of Competencies 5
  Diversity and Competency 5
  Research and Competency 6
  Law, Ethics, and Competency 6
  Person-of-the-Therapist and Competency 7
How This Book Is Different and What It Means to You 7
  Lay of the Land 8
  Anatomy of a Theory 8
  Voice and Tone 10
Suggested Uses for This Text 10
  Suggestions for Thinking About Family Therapy Theories 10
  Suggestions for Using This Book to Learn Theories 11
  Suggestions for Using This Book to Write Treatment Plans 11
  Suggestions for Use in Internships and Clinical Practice 12
  Suggestions for Studying for Licensing Exams 12
  Suggestions for Faculty to Measure Competencies and Student Learning 12
Online Resources for Students 13
Online Resources for Instructors 13
References 14

2 Research and Ethical Foundations of Family Therapy Theories 15

Lay of the Land 15
Research and the Evidence Base 15
Research-Informed Clinician Model 16
The Minimum Standard of Practice: Evidence-Based Practice (EBP) 17

Heart of the Matter: Common Factors Research 17
  Lambert's Common Factors Model 18
  Wampold's Common Factors Model 19
  Client Factors 19
  Therapeutic Relationship 19
  Therapeutic Model: Theory-Specific Factors 20
  Hope and the Placebo Effect: Expectancy 20
  Diversity and the Common Factors 20
  Do We Still Need Theory? 21
Show Me Proof: Evidence-Based Treatments 21
  Empirically Supported Treatments and Their Kin 21
  Research in Perspective: Limitations of the Evidence-Based Movement 23
Review of the MFT Evidence Base 23
Legal and Ethical Issues in Couple and Family Therapy 24
  Lay of the Land: More than Just Rules 24
  The Big Picture: Standards of Professional Practice 25
  Specific Legal and Ethical Concerns in Couple and Family Work 26
  Current Legal and Ethical Issues in Couple and Family Work 30
  Conclusion 31
Online Resources for Research 32
Online Resources for Law and Ethics 32
References 32

3 Theory-Specific Case Conceptualization and Treatment Planning 35

Treatment + Plan = ??? 35
Distinguishing a Therapist from Hairdresser: Theory-Specific Case Conceptualization 35
  Writing a Theory-Specific Case Conceptualization 36
  Identifying a Path: Treatment Planning 36
A Brief History of Mental Health Treatment Planning 37
  Symptom-Based Treatment Plans 37
  Theory-Based Treatment Plans 37
Clinical Treatment Plans 38
Writing Useful Therapeutic Tasks 40
  Initial Phase 41
  Working Phase 41
  Closing Phase 41
  Diversity and Treatment Tasks 41
Writing Useful Client Goals 42
  The Goal Writing Process 43
  Initial Phase Client Goals 44
  Working-Phase Client Goals 44
  Closing-Phase Client Goals 45
Writing Useful Interventions 45
Do Plans Make a Difference? 46
Online Resources 46
References 46

Contents vii

4 Philosophical Foundations of Family Therapy Theories 47

Lay of the Land 47
Systemic Foundations 48
    Rumor Has It: The People and Their Stories 48
    The Spirit of Systemic Therapy 49
    Introduction to Key Concepts in Systemic Theory 49
    Key Concepts Related to the Structure of Systems 49
    Key Concepts Related to Communication Patterns in Systems 51
    Translating Systemic Theoretical Concepts Into Action 54
Social Constructionist Foundations of Family 54
    Systems Theory and Social Constructionism: Similarities and Differences 54
    Rumor Has It: The People and Their Stories 55
    The Postmodern Spirit 56
    Postmodern Assumptions 56
    Social Constructionism, Postmodernism, and Diversity 58
Philosophical Overview 58
Rock-Paper-Scissors and Other Strategies for Choosing a Theory 59
    How to Choose: Dating Versus Marrying 59
    Defining Your Philosophy 60
    Modernism 60
    Humanism 61
    Systemic Therapy 62
    Postmodern Therapy 63
    Dancing with Others Once You Marry 64
Online Resources 65
References 65

PART II Family Therapy Theories 67

5 Systemic Therapies: MRI and Milan 69

Lay of the Land 69
MRI Systemic Therapy 70
    In a Nutshell: The Least You Need to Know 70
    The Juice: Significant Contributions to the Field 71
    Rumor Has It: The People and Their Stories 71
    The Big Picture: Overview of Treatment 72
    Making Connection: The Therapeutic Relationship 73
    The Viewing: Case Conceptualization and Assessment 74
    Targeting Change: Goal Setting 78
    The Doing: Interventions 79
Putting It All Together 80
    Case Conceptualization Template 80
    Treatment Plan Template for Individual 81
    Treatment Plan Template for Couple/Family 82
Milan Systemic Therapy 83
    In a Nutshell: The Least You Need to Know 83
    The Juice: Significant Contributions to the Field 84
    Rumor Has It: The People and Their Stories 84
    The Big Picture: Overview of Treatment 85

　　　　Making Connection: The Therapeutic Relationship　85
　　　　The Viewing: Case Conceptualization and Assessment　87
　　　　Targeting Change: Goal Setting　88
　　　　The Doing: Interventions　88
　　Putting It All Together　90
　　　　Case Conceptualization Template　90
　　　　Treatment Plan Template for Individual　90
　　　　Treatment Plan Template for Couple/Family　92
　　Research and the Evidence Base　93
　　Clinical Spotlight: Multisystemic Therapy　94
　　　　Goals　94
　　　　Case Conceptualization　94
　　　　Principles of Intervention　94
　　Tapestry Weaving: Diversity Considerations　95
　　　　Ethnic, Racial, and Cultural Diversity　95
　　　　Sexual Identity Diversity　95
　Online Resources　96
　References　97
　Systemic Case Study: Blended Family Issues　98
　　　　Systemic Case Conceptualization　98
　　　　Systemic Treatment Plan　99

6　Strategic Therapy　103

Strategic Therapy　103
　　In a Nutshell: The Least You Need to Know　103
　　The Juice: Significant Contributions to the Field　103
　　Rumor Has It: The People and Their Stories　104
　　The Big Picture: Overview of Treatment　104
　　Making Connection: The Therapeutic Relationship　107
　　The Viewing: Case Conceptualization and Assessment　107
　　Targeting Change: Goal Setting　109
　　The Doing: Interventions　109
Putting It All Together　112
　　Case Conceptualization Template　112
　　Treatment Plan Template for Individual　113
　　Treatment Plan Template for Couple/Family　114
Research and the Evidence Base　116
Clinical Spotlight: Brief Strategic Family Therapy　116
　　Goals　116
　　Case Conceptualization　116
　　Principles of Intervention　117
Tapestry Weaving: Diversity Considerations　117
　　Ethnic, Racial, and Cultural Diversity　117
　　Sexual Identity Diversity　118
Online Resources　119
References　120
Strategic Case Study: Intergenerational Conflict in Chinese American Family　120
　　Strategic Case Conceptualization　121
　　Strategic Treatment Plan　122

7 Structural Family Therapy 125

Lay of the Land 125
    In a Nutshell: The Least You Need to Know 125
    The Juice: Significant Contributions to the Field 125
    Rumor Has It: The People and Their Stories 128
    The Big Picture: Overview of Treatment 128
    Making Connection: The Therapeutic Relationship 129
    The Viewing: Case Conceptualization and Assessment 130
    Targeting Change: Goal Setting 132
    The Doing: Interventions 132

Putting It All Together 135
    Case Conceptualization Template 135
    Treatment Plan Template for Individual 135
    Treatment Plan Template for Couple/Family 137

Research and the Evidence Base 138

Clinical Spotlight: Ecosystemic Structural Family Therapy 139
    Case Conceptualization 139
    Goals 139
    Interventions 139

Tapestry Weaving: Working with Diverse Populations 140
    Cultural, Ethnic, and Socioeconomic Diversity 140
    Sexual Identity Diversity 141

Online Resource 141

References 142

Structural Case Study: African American Family Adjusts to Divorce 142
    Structural Case Conceptualization 143
    Structural Treatment Plan 144

8 Satir's Human Growth Model 147

Lay of the Land 147
    Common Assumptions and Practices 147

The Satir Growth Model 148
    In a Nutshell: The Least You Need to Know 148
    The Juice: Significant Contributions to the Field 148
    Rumor Has It: The People and Their Stories 151
    The Big Picture: Overview of Treatment 152
    Making Connection: The Therapeutic Relationship 153
    The Viewing: Case Conceptualization and Assessment 154
    Targeting Change: Goal Setting 157
    The Doing: Interventions 158
    Interventions for Special Populations 160

Putting It All Together 161
    Case Conceptualization Template 161
    Treatment Plan Template for Individual 162
    Treatment Plan Template for Couple/Family 163

Research and the Evidence Base 164
    Research on Humanistic Principles 164

Tapestry Weaving: Working with Diverse Populations 165
    Cultural, Ethnic, and Gender Diversity 165
    Sexual Identity Diversity 165

Online Resources 166

References 166

Experiential Case Study: Son Comes Out in Immigrant Persian Family 167
    Satir Case Conceptualization 168
    Satir Treatment Plan 169

9 Symbolic-Experiential and Internal Family Systems 173

Lay of the Land 173
Symbolic-Experiential Therapy 173
    In a Nutshell: The Least You Need to Know 173
    The Juice: Significant Contributions to the Field 174
    Rumor Has It: The People and Their Stories 174
    The Big Picture: Overview of Treatment 175
    Making Connection: The Therapeutic Relationship 175
    The Viewing: Case Conceptualization and Assessment 176
    Targeting Change: Goal Setting 178
    The Doing: Interventions 179
Putting It All Together 181
    Case Conceptualization Template 181
    Treatment Plan Template for Individual 182
    Treatment Plan Template for Couple/Family 184
Internal Family Systems Therapy 185
    In a Nutshell: The Least You Need to Know 185
    The Juice: Significant Contributions to the Field 185
    Rumor Has It: The People and Their Stories 186
    Making Connection: The Therapeutic Relationship 186
    The Viewing: Case Conceptualization and Assessment 187
    Targeting Change: Goal Setting 189
    The Doing: Interventions 189
Putting It All Together 192
    Case Conceptualization Template 192
    Treatment Plan Template for Individual 192
    Treatment Plan Template for Couple/Family 194
Research and the Evidence Base 195
Tapestry Weaving: Working with Diverse Populations 196
    Cultural, Ethnic, and Gender Diversity 196
    Sexual Identity Diversity 197
Online Resource 197
References 197
Symbolic Experiential Case Study: Cherokee Family Grieves Mother 198
    Symbolic-Experiential Case Conceptualization 198
    Symbolic-Experiential Treatment Plan 200

10 Intergenerational and Psychoanalytic Family Therapies 203

Lay of the Land 203
Bowen Intergenerational Therapy 203
    In a Nutshell: The Least You Need to Know 203
    The Juice: Significant Contributions to the Field 204
    Rumor Has It: The People and Their Stories 206
    The Big Picture: Overview of Treatment 207
    Making Connection: The Therapeutic Relationship 207
    The Viewing: Case Conceptualization and Assessment 207
    Targeting Change: Goal Setting 211

        The Doing: Interventions   211
        Interventions for Special Populations   213
    Putting It All Together   213
        Case Conceptualization Template   213
        Treatment Plan Template for Individual   214
        Treatment Plan Template for Couple/Family   215
    Psychoanalytic Family Therapies   217
        In a Nutshell: The Least You Need to Know   217
        The Juice: Significant Contributions to the Field   217
        Rumor Has It: The People and Their Stories   217
        The Big Picture: Overview of Treatment   218
        Making Connection: The Therapeutic Relationship   218
        The Viewing: Case Conceptualization and Assessment   219
        Targeting Change: Goal Setting   221
        The Doing: Interventions   221
    Putting It All Together   222
        Case Conceptualization Template   222
        Treatment Plan Template for Individual   222
        Treatment Plan Template for Couple/Family   224
    Research and the Evidence Base   225
    Tapestry Weaving: Working with Diverse Populations   226
        Gender Diversity: The Women's Project   226
        Ethnicity and Culture Diversity   226
        Sexual Identity Diversity   227
Online Resources   228
References   228
Intergenerational Case Study: African American Family Launching Children   229
    Intergenerational Case Conceptualization   229
    Intergenerational Treatment Plan   231

11   Cognitive-Behavioral and Mindfulness-Based Couple and Family Therapies   233

Lay of the Land   233
Behavioral and Cognitive-Behavioral Family Therapies   234
    In a Nutshell: The Least You Need to Know   234
    The Juice: Significant Contributions to the Field   234
    Rumor Has It: The People and Their Stories   235
    The Big Picture: Overview of Treatment   235
    Making Connection: The Therapeutic Relationship   236
    The Viewing: Case Conceptualization and Assessment   237
    Targeting Change: Goal Setting   240
    The Doing: Interventions   241
Putting It All Together   246
    Case Conceptualization Template   246
    Treatment Plan Template for Individual   247
    Treatment Plan Template for Couple/Family   249
Mindfulness-Based Therapies   250
    In a Nutshell: The Least You Need to Know   250
Gottman Method Couples Therapy   255
    In a Nutshell: The Least You Need to Know   255
    Debunking Marital Myths   256
    The Big Picture: Overview of Treatment   256
    Making Connection: The Therapeutic Relationship   257

The Viewing: Case Conceptualization and Assessment 257
   The Doing: Interventions 259
  Research and the Evidence Base 260
  Tapestry Weaving: Working with Diverse Populations 261
   Ethnic, Racial, and Cultural Diversity 261
   Sexual Identity Diversity 263
  Online Resources 264
  References 264
  Cognitive-Behavioral Case Study: Korean Family Conflict Over Grades 266
   Cognitive-Behavioral Case Conceptualization 267
   Cognitive-Behavioral Treatment Plan 269

12 Solution-Based Therapies 271

  Lay of the Land 271
  Solution-Based Therapies 272
   In a Nutshell: The Least You Need to Know 272
   Common Solution-Based Therapy Myths 272
   The Juice: Significant Contributions to the Field 273
   Rumor Has It: The People and Their Stories 273
   The Big Picture: Overview of Treatment 275
   Making Connection: The Therapeutic Relationship 276
   The Viewing: Case Conceptualization and Assessment 277
   Targeting Change: Goal Setting 279
   The Doing: Interventions 283
   Interventions for Specific Problems 286
  Putting It All Together 288
   Case Conceptualization Template 288
   Treatment Plan Template for Individual 288
   Treatment Plan Template for Couple/Family 290
  Solution-Oriented Ericksonian Hypnosis 291
   Difference From Traditional Hypnosis 291
   The Big Picture: Overview of Treatment 291
   The Doing: Interventions 292
  Research and the Evidence Base 292
  Tapestry Weaving: Working with Diverse Populations 293
   Ethnic, Racial, and Cultural Diversity 293
   Sexual Identity Diversity 294
  Online Resources 295
  References 295
  Solution-Based Therapy Case Study: Interfaith/Intercultural Couple Considers Marriage 297
   Solution-Based Case Conceptualization 297
   Solution-Based Treatment Plan 299

13 Narrative Therapy 301

  Lay of the Land 301
  Narrative Therapy 302
   In a Nutshell: The Least You Need to Know 302
   The Juice: Significant Contributions to the Field 302
   Rumor Has It: The People and Their Stories 303
   The Big Picture: Overview of Treatment 303
   Making Connection: The Therapeutic Relationship 304
   The Viewing: Case Conceptualization and Assessment 305
   Targeting Change: Goal Setting 306

The Doing: Interventions 307
Externalizing Questions 310
Interventions for Specific Problems 316
Putting It All Together 316
Case Conceptualization Template 316
Treatment Plan Template for Individual 317
Treatment Plan Template for Couple/Family 319
Research and the Evidence Base 320
Tapestry Weaving: Working with Diverse Populations 320
Ethnic, Racial, and Cultural Diversity 320
Sexual Identity Diversity 322
Online Resources 323
References 323
Narrative Case Study: Multiethnic Gay Couple 324
Narrative Case Conceptualization 325
Narrative Treatment Plan 327

14 Collaborative Therapy and Reflecting Teams 329

Collaborative Therapy and Reflecting Teams 329
In a Nutshell: The Least You Need to Know 329
The Juice: Significant Contributions to the Field 330
Rumor Has It: The People and Their Stories 331
The Big Picture: Overview of Treatment 332
Making Connection: The Therapeutic Relationship 333
The Viewing: Case Conceptualization and Assessment 335
Targeting Change: Goal Setting 336
The Doing: Interventions and Ways of Promoting Change 337
Reflecting Teams and the Reflecting Process 340
Putting It All Together 342
Case Conceptualization Template 342
Treatment Plan Template for Individual 343
Treatment Plan Template for Couple/Family 344
Clinical Spotlight: Open Dialogue, an Evidence-Based Approach to Psychosis 346
Outside the Therapy Room 346
Research and the Evidence Base 347
Tapestry Weaving: Working with Diverse Populations 348
Ethnic, Racial, and Cultural Diversity 348
Sexual Identity Diversity 349
Online Resources 350
References 350
Collaborative Case Study: Pakistani/Indian Couple with Sexual Abuse History 352
Postmodern Case Conceptualization 352
Postmodern Treatment Plan 354

15 Evidence-Based Treatments in Couple and Family Therapy: Emotionally Focused Therapy and Functional Family Therapy 357

Lay of the Land 357
Myths About Evidence-Based Treatments 357
Emotionally Focused Therapy 358
In a Nutshell: The Least You Need to Know 358
The Juice: Significant Contributions to the Field 359

Rumor Has It: The People and Their Stories   360
The Big Picture: Overview of Treatment   361
Making Connection: The Therapeutic Relationship   362
The Viewing: Case Conceptualization and Assessment   364
Targeting Change: Goal Setting   366
The Doing: Interventions   367

Putting It All Together   369
Case Conceptualization Template   369
Treatment Plan Template for Couple   370
Treatment Plan Template for Family   372

Tapestry Weaving: Diversity Considerations   373
Ethnic, Racial, and Cultural Diversity   373
Sexual Identity Diversity   373

Evidence Base   374

Functional Family Therapy   374
In a Nutshell: The Least You Need to Know   374
The Juice: Significant Contributions to the Field   375
Rumor Has It: People and Places   375
The Big Picture: Overview of Treatment   376
Making Connection: The Therapeutic Relationship   376
The Viewing: Case Conceptualization and Assessment   377
Targeting Change: Goal Setting   380
The Doing: Interventions   381

Putting It All Together   384
Case Conceptualization Template   384
Treatment Plan Template for Family   386

Tapestry Weaving: Diversity Considerations   387
Ethnic, Racial, and Cultural Diversity   387
Sexual Identity Diversity   388

Evidence Base   388

Online Resources   389

References   389

FFT Case Study: Mexican American Family with Substance-Abusing Teen   390
FFT Case Conceptualization   390
FFT Treatment Plan   392

16   Evidence-Based Group Treatments for Couples and Families   395

Lay of the Land   395

Brief Overview of Types of Groups   395
Psychoeducational Groups or "Classes"   395
Process Groups   396
Combined Groups   396

General Group Guidelines   396
Number of People   396
Open Versus Closed Groups   397
Group Selection: Who's in Your Group?   397
Group Rules   398

Evidence-Based Couple and Family Groups   398
Psychoeducational Multifamily Groups for Severe Mental Illness   398
Groups for Partner Abuse   400
Relationship Enhancement Programs   402
Parent Training   403

Summary 404
Online Resources 404
References 405

PART III Cross-Theoretical Case Conceptualization and Integration 407

17 Cross-Theoretical Case Conceptualization and Integration 409

Knowing Where to Look 409
    Case Conceptualization and the Art of Viewing 410
Theory-Specific vs. Cross-Theory Case Conceptualization 410
    Benefits of Cross-Theoretical Conceptualization 410
Cross-Theoretical Conceptualization vs. Integration 410
    Integration Options 411
Overview of Cross-Theoretical Case Conceptualization 413
Introduction to Client and Significant Others 413
Presenting Concern 414
Background Information 415
Strengths and Diversity Resources 416
    Personal or Individual Strengths 416
    Relational or Social Strengths and Resources 418
    Spiritual Resources 418
    Diversity Resources and Limitations 419
Family Structure 419
    Family Life Cycle Stage 421
    Boundaries: Regulating Closeness and Distance 421
    Triangles and Coalitions 423
    Hierarchy Between Child and Parents 423
    Complementary Patterns 424
    Satir's Communication Stances 424
    Gottman's Divorce Indicators 426
Problem Interaction Patterns 427
    Assessing Interaction Patterns 428
    Systemic Hypothesis 428
Intergenerational Patterns 429
    Attachment Patterns 431
Solution-Based Assessment 431
    Previous Solutions That *Did Not* Work 431
    Exceptions and Unique Outcomes: Previous Solutions That *Did* Work 432
    Miracle Question Answer 432
Postmodern and Cultural Discourse Conceptualization 433
    Dominant Discourses 433
Client Perspectives 434
Case Conceptualization, Diversity, and Sameness 435
Online Resources 436
References 436
Case Conceptualization Form 437

Name Index 445
Subject Index 447

Preface

The Purpose of This Book

Theory and Treatment Planning in Family Therapy is designed to introduce students to family therapy theories using a competency-based pedagogy. As an instructor in an accredited program and university that is required to measure student learning, I needed something that would enable me to effectively measure student learning. Although I created comprehensive assessment systems for measuring student mastery of competencies (Gehart, 2007, 2009), I realized that in order to enable students to actually learn the competencies I wanted to measure, they needed resources that meaningfully provided them with the detailed knowledge they need to actually develop real world skills. In short, I needed something more than a text that simply offered solid but old school "book knowledge"; I needed a resource that prepares my students for the realities of practicing therapy in the 21st century. This book provides the missing link between theory and practice that my students needed.

How This Book Relates to *Mastering Competencies in Family Therapy*

This text is an "essentials" version of *Mastering Competencies in Family Therapy* (Gehart, 2013). It includes the same theories, review of the evidence base, ethics discussion, and theoretical foundations section. The differences are as follows:

- *Theory and Treatment Planning in Family Therapy* is designed as an entry-level text for use in a theory or prepracticum skills class with students who are not currently seeing clients.
- *Theory and Treatment Planning in Family Therapy* does *not* include discussion and samples of clinical assessment, progress notes, or evaluation of progress.
- Unlike *Mastering Competencies, Theory and Treatment Planning in Family Therapy* teaches a form of theory-specific case conceptualization rather than emphasize cross-theoretical case conceptualization. A chapter on cross-theoretical case conceptualization is included as a final chapter, which educators can use as a culminating assignment for the class.
- The treatment plan in *Theory and Treatment Planning in Family Therapy* has fewer therapeutic tasks, making it more appropriate for students working with vignettes rather than actual clients.
- *Theory and Treatment Planning in Family Therapy* introduces theories at a slower pace for students learning them for the first time.
- The case studies in this book were created with a team of students and former students based on actual cases they saw in their first year of training to provide case studies that are truly relevant to students.

Text Overview

Using state-of-the-art pedagogical methods, this text is part of a new generation of textbooks, ones that are correlated with national standards for measuring student learning in mental health professions, including counseling, family therapy, psychology, and social work. Using a learning-centered, outcome-based pedagogy, the text engages students in an *active learning process* rather than deliver content in a traditional narrative style. More specifically, the text introduces family therapy theories using (a) theory-specific case conceptualization and (b) treatment planning. These assignments empower students to apply theoretical concepts and develop real world skills as early as possible in their training, resulting in greater mastery of the material. In addition, the text includes extensive discussions about how diversity issues and research inform contemporary practice of family therapy.

Furthermore, I use a down-to-earth style to explain concepts in clear and practical language that contemporary students generally appreciate. Instructors will enjoy the simplicity of having the text and assignments work seamlessly together, thus requiring less time in class preparation and grading. The extensive set of instructor materials—which include syllabi templates, detailed PowerPoints, test banks, online lectures, and scoring rubrics designed for accreditation assessment—further reduce educators' workloads. In summary, the book employs the most efficient and effective pedagogical methods available to family therapy theories, resulting in a win/win for instructors and students alike.

Appropriate Courses

A versatile book that serves as a reference across the curriculum, this text is specifically designed for use as a primary or secondary textbook in the following courses:

- Introductory or advanced family therapy theories courses as a primary or secondary text
- Prepracticum or fieldwork skills classes

Assessing Student Learning and Competence

The learning assignments in the text are designed to simplify the process of measuring student learning for regional and national accreditation. The case conceptualization and treatment plans in the book come with scoring rubrics, which are available on the student and instructor websites for the book at www.CengageBrain.com. Scoring rubrics are available for all major mental health disciplines using the following sets of competencies:

- *Counseling:* Council on the Accreditation of Counseling and Related Educational Programs (CACREP) standards
- *Marriage and Family Therapy:* MFT core competencies
- *Psychology:* Psychology competency benchmarks
- *Social work:* Council for Social Work Education accreditation standards

Rubrics are provided correlating competencies for each profession to the skills demonstrated on the learning assignments: case conceptualization and treatment plans.

Organization

This book is organized into three parts:

Part I Introduction to Competency, Theory, and Treatment Planning provides an introduction to competencies, research, ethics, philosophical foundations, and treatment planning.

Part II Family Therapy Theories covers the major schools of family therapy:

- Systemic Theories: MRI and Milan
- Strategic theory
- Structural family therapy
- Satir's human growth model
- Symbolic-experiential and internal family systems
- Intergenerational and psychodynamic theories
- Cognitive-behavioral and mindfulness-based family therapies
- Solution-based therapies
- Narrative therapy
- Collaborative therapy and reflecting teams
- Evidence-based couple and family therapies: Emotionally focused therapy and functional family therapy
- Evidence-based couple and family group therapies

Part III Cross-Theoretical Case Conceptualization and Integration:

- Cross-theory case conceptualization: This text introduces a comprehensive cross-theoretical approach to case conceptualization

The theory chapters in Part II are organized in a user-friendly way to maximize students' ability to use the book when developing case conceptualizations, writing treatment plans, and designing interventions with clients. The theory chapters follow this outline consistently throughout the book:

Each theory chapter includes the following:

- **In a Nutshell:** The Least You Need to Know
- **The Juice:** Significant Contributions to the Field: If there is one thing to remember from this chapter it should be …
- **Rumor Has It:** The People and Their Stories
- **The Big Picture:** Overview of the Therapy Process
- **Making Connection:** The Therapy Relationship
- **The Viewing:** Case Conceptualization
- **Targeting Change:** Goal Setting
- **The Doing:** Interventions
- **Putting It All Together**
  - Case Conceptualization Template
  - Treatment Plan Template for Individual
  - Treatment Plan Template for Couple/Family
- **Tapestry Weaving:** Working with Diverse Populations
  - Ethnic, Racial, Gender, and Cultural Diversity
  - Sexual Identity Diversity
- **Research and Evidence Base**
- **Online Resources**
- **Reference List**
- **Case Example:** Vignette with a theory-specific case conceptualization and treatment plan

Available with the Text

Instructors will find numerous resources for the book online at www.CengageBrain.com or at the author's sites (www.masteringcompetencies.com; www.dianegehart.com).

- **Online Instructor's Manual:** The Instructor's Manual (IM) contains a variety of resources to aid instructors in preparing and presenting text material in a manner that meets their personal preferences and course needs. It presents chapter-by-chapter suggestions and resources to enhance and facilitate learning, as well as

sample syllabi for how to use this book in a theory class, prepracticum skills class, or practicum class.
- **Cengage Learning Testing, Powered by Cognero:** Cognero is a flexible, online system that allows you to author, edit, and manage test bank content as well as create multiple test versions in an instant. You can deliver tests from your school's learning management system, your classroom, or wherever you want.
- **Online PowerPoint®:** These vibrant Microsoft® PowerPoint® lecture slides for each chapter assist you with your lecture by providing concept coverage using content directly from the textbook.
- **Digital Downloads:** The Digital Downloads include important forms and exercises to foster interactive learning exercises. These downloads include cross-theoretical case conceptualization and treatment plans.
- **CourseMate:** Available with the text, Cengage Learning's CourseMate brings course concepts to life with interactive learning, study, and exam preparation tools that support the printed textbook. CourseMate includes an integrated eBook, glossaries, flashcards, quizzes, videos, downloadable forms and more—as well as Engagement Tracker, a first-of-its-kind tool that monitors student engagement in the course.
- Online lectures by the author.
- Scoring rubrics precorrelated for national accreditation bodies:
  - *Counseling:* Council on the Accreditation of Counseling and Related Educational Programs (CACREP) standards
  - *Marriage and Family Therapy:* MFT core competencies
  - *Psychology:* Psychology competency benchmarks
  - *Social Work:* Council for Social Work Education accreditation standards

References

Gehart, D. (2007). *The complete marriage and family therapy core competency assessment system: Eight outcome-based instruments for measuring student learning.* Thousand Oaks, CA: Author. Available: www.mftcompetencies.com

Gehart, D. (2009). *The complete counseling assessment system: Eight outcome-based instruments for measuring student learning. Marriage, Couple, and Family Counseling Edition.* Thousand Oaks, CA: Author. Available: www.counselingcompetencies.com

Gehart, D. (2013). *Mastering competencies in family therapy* (2nd ed.). Pacific Grove, CA: Cengage/Brooks Cole.

Acknowledgments

*I would like to thank the following **content experts** who gave their time and energy to ensure that the information in this textbook was accurate and current:*

Wendel Ray and his doctoral students, **Todd Gunter** and **Allison Lux**: Philosophical Foundations and Systemic Therapies (Chapters 3 and 5)
Marion Lindblad-Goldberg: Structural Therapy/Ecosystemic Structural Family Therapy (Chapter 7)
Lynne M. Azpeitia: Satir Growth Model (Chapter 8)
Stephen Lankton: Satir Growth Model (Chapter 8)
Michael Chafin: Symbolic-Experiential Therapy (Chapter 9)
Richard Schwartz: Internal Family Systems (Chapter 9)
Michael Kerr and Cynthia Larkby: Bowen Intergenerational Therapy (Chapter 10)
Naomi Knoble: Gottman's Marriage Clinic Approach (Chapter 11)
Bill O'Hanlon: Solution-Based Therapies (Chapter 12)
Stephen Lankton: Ericksonian Hypnosis (Chapter 12)
Terry Trepper: Solution-Focused Therapy Research (Chapter 12)
Gerald Monk: Narrative Therapy (Chapter 13)
Harlene Anderson: Collaborative Therapy (Chapter 14)
Scott Woolley: Emotionally Focused Therapy (Chapter 15)
Thomas Sexton: Functional Family Therapy (Chapter 15)
Eric McCollum: Group Therapy (Chapter 16)
Ron Chenail: Competencies Assessment System
Thorana Nelson: Competencies Assessment System
William Northey: Competencies Assessment System
Julie Diaz: Clinical Forms

The following reviewers provided invaluable feedback on making this book work for faculty:

William F. Northey, Jr. (Bill Northey): Former AAMFT Staff
John K. Miller: University of Oregon
Joshua M. Gold: University of South Carolina
Brent Taylor: San Diego State
Randall Lyle: St. Mary's University
Cynthia T. Walley: Old Dominion University
Stephanie Hall: Monmouth University
Lisa Wilson: Hope International University
Tim VanderGast: William Paterson University
Forrest Kelly: West Texas A&M University

Teresa Wilcox: Hope International University
Michelle Hinkle: William Paterson University

The Case Study Team: Current and former students who helped draft the case studies in the book based on clients they saw in the first year.

Brooke Clark
Corie Lee Losiselle
Leyla Klumehr
Bryan Shyne
Manya Khoddami

Instructor and Student Materials:

Diana Pantaleo
Rena Jacobs
Kayla Caceres

The following students and colleagues assisted in researching and proofing the second edition:

Hiroko Okuishi
Alejandra Trujillo
Kayla Caceres

I would also like to thank the following people for their generous assistance:

Julie Martinez: My amazing and thoughtful editor and product manager.
Lori Bradshaw: For her diligent editorial assistance throughout.
Sean Cronin: For his guidance with the instructor and student materials.
Jitendra Kumar: For detailed and patient copy editing.
Joseph McNicholas: My husband, whose encouragement, support, and understanding made the process fun.
Michael and Alexander: My two sons who make it all worth while.

About the Author

Photo by Jones Photo Art

DR. DIANE R. GEHART is Professor in the Marriage and Family Therapy and Counseling Programs at California State University, Northridge. Having practiced, taught, and supervised for nearly 20 years, she has authored/edited:

- *Mindfulness and Acceptance in Couple and Family Therapy*
- *Collaborative Therapy: Relationships and Conversations That Make a Difference* (coedited)
- *Theory and Treatment Planning in Counseling and Psychotherapy*
- *Mastering Competencies in Family Therapy*
- *The Complete MFT Core Competency Assessment System*
- *The Complete Counseling Assessment System*
- *Theory-Based Treatment Planning for Marriage and Family Therapists* (coauthored)

She has also written extensively on postmodern therapies, mindfulness, mental health recovery, sexual abuse treatment, gender issues, children and adolescents, client advocacy, qualitative research, and counselor and MFT education. She speaks internationally, having given workshops to professional and general audiences in the United States, Canada, Europe, and Mexico. Her work has been featured in newspapers, radio shows, and television worldwide, including the BBC, National Public Radio, Oprah's *O Magazine*, and *Ladies Home Journal*. She is an associate faculty member at three international post-graduate training institutes: the Houston Galveston Institute, Taos Institute, and the Marburg Institute for Collaborative Studies in Germany. Additionally, she is an active leader in state and national professional organizations. She maintains a private practice in Thousand Oaks, California, specializing in couples, families, women's issues, trauma, life transitions, and difficult to treat cases. For fun, she enjoys spending time with her family, hiking, swimming, yoga, salsa dancing, meditating, and savoring all forms of dark chocolate. You can learn more about her work on www.dianegehart.com.

Author's Introduction

If you were hoping this book would provide a simple and straightforward set of guidelines for magically repairing relationships gone awry, let me start by saying you will be disappointed. Although magic wands are sometimes involved (see Chapter 12), the theories of family therapy are anything but linear, logical, or predictable. If you actually come to fully understand the concepts in this book, your way of looking at the world will forever be transformed. The assumptions and constructs you and most of Western society have used to view human problems will be systematically—or perhaps suddenly—dismantled. You will no longer be able to pretend that an individual's behavior can be understood or evaluated apart from his/her environment. Psychiatric symptoms such as depression, anxiety, or paranoia will not be seen as a sole individual's problem but rather a distinctive form of communication within a particular relational system. Furthermore, you will come to see how *you*, the therapist, significantly affect and cocreate what you observe in session. You will also learn how new frames and descriptions can suddenly liberate people from what seem intractable problems and inherent personality characteristics. As these ideas sink in, you will feel a bit like Alice tumbling down a rabbit hole into a strange new world—everything seems upside down and backwards. Reality will not be what it once was. You will witness seeming miracles. You will wonder and marvel at how humans construct and reconstruct their realities in and through relationships.

I should also warn you that family therapy theories will not put you in the position of an all-knowing guru who calmly sits outside the fray. No, you will join the family systems you intend to help. You will help cocreate new realities, not from the position of an expert but from the humble position as a collaborator in creating new meaning. Most of the time you will find therapy like Mr. Toad's wild ride, with many bumps, twists, turns, misfortunes, and unexpected good luck. You will always have theory to guide you but will not always know for certain where you will end up. But I promise you the journeys you will embark upon will leave you richer and wiser.

Enjoy the adventure!

Diane R. Gehart, Ph.D.
Westlake Village, California

Part I

Introduction to Competency, Theory, and Treatment Planning

CHAPTER 1

Competency and Theory in Family Therapy

Lights, Camera, Action!

Wilder and more unpredictable than anything Hollywood could conjure up, family therapy in the real world is enough to keep you on the edge of your seat. Some days it feels as though the room is lurched to a 45-degree angle and that the ship—once thought to be unsinkable—is going down. Other times your own head starts to spin in such a way that an exorcist might be warranted. By the time you begin to feel the Darth Vader death grip around your throat, you are ready to slip out to get some popcorn and catch your breath. But you can't quite do that when you are cast in the leading role in a therapy session! So, there you are, still sitting with them.

The entire Jackson family has their eyes fixed on the floor or a wall—anything but another person. The session is *not* going well. The otherwise jovial and free-spirited father has suddenly and harshly attacked his 8-year-old son for his lack of respect after the son used language and hand signals only allowed in R-rated movies. The heretofore demure and seemingly compliant mother then gives the father an equally punitive lambasting for his overreaction. The 6-year-old "perfect" sister is frozen: silent with tears welling. This all unraveled in less than 60 seconds, before which time you thought things were going pretty well. In fact, you were starting to fantasize about whom you might thank as you accept the award for Most Brilliant Family Therapist of the Year.

Now, *you* are the one expected to somehow rewrite the scene to ensure a reasonably happy—or at least not dramatic—ending. *How* are you going to do it? What are you going to *say*? *Which* issue are you going to address first? *Who* are you going to address first? What is your *plan*? You have some decisions to make, and less than 60 seconds to do so—no time for popcorn.

You might be surprised—and relieved—to learn there is more than one effective way to handle the situation, just as you can imagine there are many poor ways to respond. The purpose of this book is to help you develop a clear sense of the responses that are most likely to succeed. The options you will learn for handling such a situation and many others like it come from family therapy *theories*. These theories describe how generations of therapists have helped families like the Jacksons navigate the intimate, intense, and unpredictable adventure of helping couples and families learn to better express and experience their love for one another. The essence of becoming a competent therapist lies in the effective use of theory to inform what you do and say in session, and that is precisely the plot of this screenplay in which you will take center stage.

Competency and Theory: Why Theory Matters

Although much has changed in the past decade in mental health—better research to guide us, new knowledge about the brain, more details about mental health disorders, increased use of psychotropic medication—the primary tool that therapists use to help people has not changed: that tool is *theory*. Therapeutic theories provide a means for quickly sifting through the tremendous amount of information clients bring; then targeting specific thoughts, behaviors, or emotional processes for change; and finally helping clients to effectively make these changes to resolve their initial concerns. Even with fancy fMRIs, neurofeedback machines, and hundreds of available medications, no other technology has taken the place of theory. However, the transforming landscape in mental health care has changed how therapy theories are understood and used. Specifically, theory and how it is being used and understood has been recontextualized by two major movements in recent years: (a) the competency movement, which includes multicultural competency, and (b) research or evidence-based movement, which emphasizes the use of research to making in-session clinical decisions (discussed in more detail in the next chapter). These movements have not ended the need for theory but have instead changed how we conceptualize, adapt, and apply theory.

Why All the Talk About Competency?

All health professions, including mental health, have been abuzz in recent years with talk of *competencies,* detailed lists of the knowledge and skills professionals need to effectively do their job. The main source of this movement has been external to the field, from stakeholders who believe that professionals should not only be taught a consistent set of skills but their learning should be measured on real-world tasks (for a detailed discussion see Gehart, 2011). Thus, this movement is asking educators to shift their focus from conveying content to ensuring students know how to meaningfully apply the knowledge and skills of their given profession, such as effectively responding to the Jackson family outburst with grace and aplomb.

Each major mental health profession—including counseling, marriage and family therapy, psychology, psychiatry, psychiatric nursing, and chemical dependency counseling—has developed a unique set of competencies. Thankfully, there are many similarities across them. Two of the more commonly used sets of competencies specifically used in family therapy are the Marriage and Family Therapy Core Competencies developed by a task force commissioned by the American Association for Marriage and Family Therapy (Nelson, Chenail, Alexander, Crane, Johnson, & Schwallie, 2007) and the Marriage, Couple, Family, and Child Counseling standards developed by the commission that accredits counseling programs (CACREP, 2012). On nights when you have insomnia, you may find it helpful and interesting to read through what will be expected of you as a family therapist or counselor regardless of the title on your license.

These competencies are being used to more clearly define what family therapists must know and do in order to be competent. If you are new to the field, this actually will make the task of learning to be a family therapist far easier: the goals are now clearly defined. This book is designed to help you develop many of these essential competencies as quickly and directly as possible.

Competency and (Not) You

Although at first it may seem insensitive, the vernacular expression commonly used by my teen clients sums up the mindset of competency best: *"It's not about you."* It's not about *your* theoretical preference, what worked for *you* in your personal therapy, what *you* are good at, what *you* find interesting, or even what *you* believe will be most helpful. Competent therapy requires that *you* get outside of your comfort zone, stretch, and learn how to interact with clients in a way that works for *them*. In short, you need to be competent in a wide range of theories and techniques to be helpful to all of the clients with whom you work. If you allow me to go on, you might even begin to see how this makes some sense and might even be in your best interest.

Perhaps it is best to explain with an example: you will likely either have a natural propensity for generating a broad-view case conceptualization using therapy theories or have a disposition that favors a detail-focused mental health assessment and diagnosis; humans tend to be good either with the big picture or with details. However, to be competent, a therapist needs to get good at both even if one is easier, preferred, and philosophically favored. Similarly, you may prefer theories that promote insight and personal reflection; after all, that may be what works for *you* in *your* life. However, that may not work for your client and/or research may indicate that such an approach is not the most effective approach for your client's situation or cultural background. Thus, you will need to master theories of therapy that may not particularly interest you or even fit with your theory of therapy. Although at first you may not like this idea, I think that by the time you are done with the book, you might just warm up to it.

I first humbly learned this particular competency lesson when working with families in which the parents had difficulty managing the behaviors of their young children. I was never a huge fan of behaviorism, but it did not take too many hysterically screaming, clawing, and biting 2-year-olds before I was preaching the value of reinforcement schedules and consistency. Given my strong—admittedly zealous—attachment to my postmodern approach at the time, I have every faith that you will be driven either by principle (ideally) or by desperation (more likely) to move beyond your comfort zone to become a well rounded, competent therapist.

Common Threads of Competencies

Whether you are training to be a counselor, family therapist, psychologist, or social worker, you will notice that there are common themes across the various sets of competencies. You will want to take particular note of these:

- *Diversity and Multicultural Competence:* The use of therapeutic theory is always contextualized by diversity issues, which means that the application and applicability varies—sometimes dramatically—based on diversity issues, such as age, ethnicity, sexual orientation, ability, socioeconomic status, immigration status, etc.
- *Research and the Evidence Base:* To be competent, therapists must be aware of the research and evidence base related to their theory, client populations, and presenting problem.
- *Ethics:* Perhaps the most obvious commonality across sets of competencies is law and ethics. Without a firm grasp of the laws and ethical standards that relate to professional mental health practice, it is safe to say that you won't be practicing very long. A solid understanding of ethical principles such as confidentiality is a prerequisite for applying theory well.
- *Person-of-the-Therapist:* Finally, unlike most other professions, specific personal qualities are identified as competencies for mental health professionals, which will be discussed in more depth below.

Diversity and Competency

Over the past couple of decades, therapists have begun to take seriously the role of diversity in the therapy process, including factors such as age, gender, ethnicity, race, socioeconomic status, immigration, sexual orientation, ability, language, and religion. These factors inform the selection of theory, development of the therapy relationship, assessment and diagnosis process, and choice of interventions (Monk, Winslade, & Sinclair, 2008). In short, everything you think, do, or say as a professional is contextualized and should be informed by diversity issues. If you think effectively responding to diversity is easy or can be easily learned or that perhaps your instructors, supervisors, or some famous author has magic answers to make it easy, well, you are going to be in for an unpleasant surprise. Rather than a black-and-white still life, dealing with diversity issues is more like finger painting: there are few lines to follow, it is messy for everyone involved, and it requires enthusiasm and openheartedness to make it fun.

I have often heard new and experienced therapists alike claim that because they are from a diverse or marginalized group that they don't need to worry about learning more regarding diversity because they have lived it. Conversely, I have heard therapists from majority groups say things, such as "I don't have any culture." Both parties have much to learn on the diversity front. First of all, we are all part of numerous sociological groups that exert cultural norms on us, with the more common and powerful ones stemming from gender, ethnicity, socioeconomic class, religion, and age. Many, if not most, people belong to some groups that align more with dominant culture and some that are marginalized, resulting in a complex and often contradictory matrix of internalized beliefs, values, and aspirations. Additionally, some groups experience far more traumatic and painful forms of marginalization than others, with each individual within the group responding to these pressures differently.

To illustrate, some people experienced the process of coming out as gay as highly traumatic, especially if they were rejected by their families, and want therapists to address these issues gingerly while others find it insulting when therapists *assume* they feel oppressed due to sexual orientation because they have families, friends, and communities that are largely supportive. Furthermore, many Americans seem unaware that there is a very strong and distinct "American culture" of which they are a part; in fact, the various regions of America have very unique characteristics of which therapists need to be aware. As another example, midwestern men typically express their emotions far differently than men in California and therapists who expect the two types of men to handle emotions in a similar way are going to unfairly pathologize one or the other.

Suffice it to say, that competently handling diversity issues requires great attention to the unique needs of each person, and it is a career long struggle and journey that adds great depth and humanity to the person of the therapist. In this book, you will begin this journey by examining diversity issues related to each of the theories covered and start integrating these issues into case conceptualization and treatment planning. Diversity is so central to using theory, you will find discussions of diversity throughout each chapter in addition to an extended section at the end of each chapter covering various forms of ethnic, gender, and sexual identity diversity related to the implementation of the specific theory. In addition, you will be asked to concretely specify how diversity issues will inform treatment through case conceptualizations and treatment plans.

Research and Competency

Another common thread found in mental health competencies is understanding, and more importantly, *using* research to inform treatment and to measure one's effectiveness and client progress. In recent years, there has been a powerful movement within the field to become more evidence-based in mental health, which involves two key practices: (a) using existing research to inform clinical decisions and treatment planning, and (b) learning to use evidence-based treatments, which are specific and structured approaches for working with distinct populations and issues (Sprenkle, 2002). These movements are discussed in detail in Chapter 2, and issues related to the evidence base for each therapeutic theory are also discussed at the end of each theory chapter, with related evidence-based treatment highlighted. In addition, Chapters 15 and 16 cover leading evidence-based treatments in the field of couple and family therapy. If you were hoping to escape discussion of research in your theory text, you will be initially disappointed; but I hope by the end you find the integration an invigorating addition.

Law, Ethics, and Competency

I often quip with students entering the field that if they think therapists can cut corners with legal or ethical issues that they should transfer to a business program so that they can make some money without worrying about such details and avoid a felony prison sentence after working as an unpaid intern for four plus years. Okay, this is an exaggeration—there is such a thing as business ethics—but in the mental health fields the standards are much, much higher and strict. Therapists who fail to develop significant competence in legal and ethical issues will not last long. Although this book does not directly cover these issues, they are so central to the profession that even before you begin reading about theories and treatment planning, you need a brief introduction because you might just be tempted to run off and start applying the concepts and techniques in this book to your clients, friends, family, neighbors, pets, and self. All mental health professional organizations—such as the American Association for Marriage and Family Therapy, American Counseling Association, American Psychological Association, and the National Association of Social Workers—have codes of ethics which their members must follow. Thankfully, there is significant agreement between the various organizations resulting in general agreement on most key issues; federal and state laws also generally agree on the key principles. These issues are covered in more depth in Chapter 2.

Person-of-the-Therapist and Competency

Finally, being a competent therapist requires particular personal characteristics that are often difficult to define. Some qualities are basically assumed to be prerequisites for a professional—integrity, honesty, and diligence. These commonly take the form of following through on instructions the first time asked, raising concerns before they spiral into problems, staying true to one's word, etc. It is hard to establish competency in anything without these basic life skills.

The more subtle issues of the person-of-the-therapist come out when building relationships with clients. To begin with, the research is clear that clients need to feel heard, understood, and accepted by therapists, which often takes the form of offering empathy and avoiding advice giving (Miller, Duncan, & Hubble, 1997). Furthermore, therapists need to identify and work through their personal issues to avoid bias, inappropriately label a client, or what psychodynamic therapists call *countertransference* (see Chapter 10). Although more difficult to quantify, these issues often become quickly apparent by strong emotions or unusual interactions in relationships with clients, supervisors, instructors, and peers. Managing these well is part of being a competent therapist.

Finally, a more difficult aspect to define is *therapeutic presence*, a quality of self considered to have intrapersonal, interpersonal, and transpersonal elements, including elements of empathy, compassion, charisma, spirituality, transpersonal communication, patient responsiveness, optimism, and expectancies, making it elusive and difficult to operationalize (McDonough-Means, Kreitzer, & Bell, 2004). Clients—rather than a professional—are the best judges of this subtle quality, because in the end, it comes down to how the client experiences the therapist as a human being in the room. Although these competencies are more difficult to measure, they are nonetheless some of the more important to develop.

How This Book Is Different and What It Means to You

Theory and Treatment Planning in Family Therapy is a different kind of textbook. Based on new pedagogical model, learning-centered teaching (Killen, 2004; Weimer, 2002), this book is designed to help you *actively learn* the content and develop real-world competencies rather than simply deliver the content and hope that you will memorize it. Thus, learning activities are a central part of the text so that you have opportunities to apply and use the information in ways that facilitate learning. The specific learning activities in this book are (a) theory-specific case conceptualization, (b) cross-theory case conceptualization, and (c) treatment planning; these translate the theory learned in the chapter to practical client situations. This book teaches real world skills that you can immediately use to better serve your clients.

Also, this book is different in another way: it is organized by key concepts rather than general headings with long narrative sections. This organization—which evolved from my personal study notes for my doctoral and licensing exams back before I had email service (and, no, dinosaurs were not roaming the planet then)—facilitates the retention of vocabulary and terms because of the visual layout. Each year I receive numerous emails from enthusiastic newly licensed therapists thanking me for helping them to pass their licensing exams—they all say that the organization of the book made the difference. So, spending some time with this text should better prepare you for the big exams in your future (and if you have already passed these, you should be all the more impressed with yourself for doing it the hard way).

Lay of the Land

This book is organized into three parts:

Part I: Introduction to Family Therapy Theories provides an introduction to competencies, research, ethics, treatment planning, and philosophical foundations.

Part II: Family Therapy Theories covers the philosophical foundations and major schools of family therapy:

- MRI and Milan systemic theories
- Strategic theory
- Structural family therapy
- Satir's human growth model
- Symbolic-experiential and internal family systems
- Intergenerational and psychodynamic theories
- Cognitive-behavioral and mindfulness-based family therapies
- Solution-based therapies
- Narrative therapy
- Collaborative therapy and reflecting teams
- Evidence-based couple and family therapies: Emotionally focused therapy and functional family therapy
- Evidence-based couple and family group therapies

Part III: Cross-Theory Case Conceptualization:

- Cross-theory case conceptualization: This text introduces a comprehensive cross-theoretical approach to case conceptualization.

Anatomy of a Theory

The theory chapters in Part II are organized in a user-friendly way to maximize your ability to use the book to support you when developing case conceptualizations, writing treatment plans and progress notes, and designing interventions with clients. Theory chapters follow this outline:

Anatomy of a Theory

In a Nutshell: The Least You Need to Know

The Juice: Significant Contributions to the Field: If there is one thing to remember from this chapter it should be…

Rumor Has It: The People and Their Stories

The Big Picture: Overview of the Therapy Process

Making Connection: The Therapy Relationship

The Viewing: Case Conceptualization

Targeting Change: Goal Setting

The Doing: Interventions

Putting It All Together: Treatment Plan Template

- Treatment Plan Template for Individuals with Depression/Anxiety Symptoms
- Treatment Plan Template for Couples/Families with Conflict

(continued)

> *Tapestry Weaving:* Working with Diverse Populations
> - Ethnic, Racial, Gender, and Cultural Diversity
> - Sexual Identity Diversity
>
> *Research and Evidence Base*
>
> *Online Resources*
>
> *Reference List*
>
> *Case Example:* Vignette with a complete set of clinical paperwork described in Part III, including case conceptualization, clinical assessment, treatment plan, and a progress note.

In a Nutshell: The Least You Need to Know: The chapters begin with a brief summary of the key features of the theory. Although it may not be the absolute least you need to know to get an A in a theory class or help a client, it is the basic information you should have to memorize and be able to quickly articulate at any moment to help you keep your theories straight.

The Juice: Significant Contributions to the Field: In the next section, I use the principle of primacy (first information introduced) to help you remember one of the most significant contributions of the theory to the field of family therapy. In most cases, well-trained clinicians who generally use another approach to therapy are likely to be skilled and use this particular concept because it has shaped standard practice in the field. This section is your red flag to remember a seminal concept or practice for the theory. Feedback from students indicates this is often one of their favorite sections (I only hope that isn't because they skim the rest of the chapter; but, of course, *you* would never think of such a thing).

Rumor Has It: The People and Their Stories: In this section, you can read about the developers of the theory and how their personal stories shaped the evolution of the ideas. And, yes, some of the rumors are juicier than others. As the focus of this text is how therapy theories are actually used in contemporary settings, I have de-emphasized the history and development of the theory, but you will find brief summaries of such history here.

The Big Picture: Overview of the Therapy Process: The big picture provides an overview of the flow of the therapy process: what happens in the beginning, middle, and end, and how change is facilitated across these phases.

Making Connection: The Therapy Relationship: All approaches start by establishing a working relationship with clients, but each approach does it differently. In this section you will read about the unique ways that therapists of various schools build relationships that provide the foundation for change.

The Viewing: Case Conceptualization: The case conceptualization section will identify the signature theory concepts that therapists from each school use to identify and assess clients and their problems. This really is the heart of the theory and where the real differences emerge. *I encourage you to pay particularly close attention to these.* You can also read more about case conceptualization in Chapters 4 and 17.

Targeting Change: Goal Setting: Based on the areas assessed in the case conceptualization and the overall therapy process, each approach has a unique strategy for identifying client goals that become the foundation for the treatment plan.

The Doing: Interventions: Probably the most exciting part for most new therapists, the doing section outlines the common techniques and interventions for each theory. In some cases, a section for techniques used with special populations is included if these are notably different than those in standard practice.

Putting It All Together: Treatment Plan Templates: After graduation, you will probably thank me most for this section, which provides a template for a treatment plan that can be used for addressing depression, anxiety, or trauma with individual clients and conflict with couples and families. These plans tie everything in the chapter together (just imagine a little bow on top).

Tapestry Weaving: Working with Diverse Populations: This section reviews specific approaches for working with diverse populations using the theories covered in the chapter. Each chapter includes sections on ethnic and sexual identity diversity issues.

Research and Evidence Base: Finally, the chapters end with a brief review of the research and evidence base for each theory to offer a general sense of empirical foundations for the theory. In some cases, influential evidence-based treatments (see Chapter 2 for a definition) are highlighted.

Online Resources: A list of web pages and web documents are included for those who want to pursue specialized training or conduct further research on the theory.

Reference List: Many students pass right over reference lists and forget all about them. But if you have to do an academic paper or literature review on any of these theories, this should be your first stop. You might remember my historical difficulty with leaving the library. In this case, I had several hundred books and articles go through my 12′×12′ office while writing the editions of this book. Thus, you can certainly shorten the time it takes to locate key resources by pursuing these before you hit the library yourself (oh, I forgot, no one steps foot in these places anymore; I meant "surf" the library's web page while still in your bunny slippers).

Finally, each chapter ends with a case vignette, including a theory-specific case conceptualization and treatment plan.

Voice and Tone

Finally, I should mention that the voice and tone of this textbook is a bit different than your average college read. Hopefully, you have noticed by now that I am talking right at ya'. I also like to add some humor and have some fun while I write. Why? Well, I have more fun writing this way. But, more importantly, I want to engage you as if you were one of my students or supervisees learning about how to apply these ideas for the first time. Family therapies are relationship-based practices, one in which the parties co-construct knowledge together. Thus, it is hard for me to write about these ideas as a detached, faceless author, thereby perpetuating the myth of objectivity in knowledge construction (you'll better understand why I am worried about this after reading about the philosophical foundations of family therapy in Chapter 3). So, as I write, I am imagining you as full and real person eager to learn about how to use these ideas to help others. I am going to try to reach out to you, answer questions I imagine you have, and periodically tap you on the shoulder to make sure you are still awake☺.

Suggested Uses for This Text

Suggestions for Thinking About Family Therapy Theories

As you read the chapters in this book, you are going to be tempted to identify which ones you like the best and de-emphasize the ones to which you are less attracted. This may seem like a great idea at first, but here are some points to consider:

Favorite vs. Useful: The theories that the average therapist finds personally useful are probably not the same ones that the average client of a new therapist is likely to find useful. Many therapists are psychologically minded, meaning that they enjoy thinking about the inner world and how it works. However, most new therapists begin working in lower fee clients that serve diverse, multiproblem clients and families, many (but not all) of whom are not psychologically minded because they are often struggling with issues of survival and/or come from cultural traditions that place less value on analysis and understanding of the inner world. So, the theory you find most useful to you personally may not be a good fit for your first client.

Appreciative: The theories in this book are not casually chosen. They have become part of the standard cannon of theories because generations of therapists have found them helpful. Each has wisdom worthy of study. The one lesson I have learned over the years is that the more theories therapists understand, the better able they are to serve their clients because their understanding of the human condition and its concomitant problems is broader. Thus, I recommend approaching each theory with an attitude of searching for its most wide and useful parts. I facilitate this for you in the "Juice" section of each chapter that identifies the one concept I believe has near universal utility from the particular theory.

Common Threads: Family therapy theories are ironic: in one sense they are very different and inform distinct and mutually exclusive behaviors and attitudes. However, the better you understand one, the better you understand them all. In fact, some therapists, the common factors proponents, argue that theories are generally equally effective because they are simply different modes for delivering the same factors (Miller, Duncan, & Hubble, 1997; you will read more in Chapter 2). So, it is quite possible that commonalities across theories are *more important* than their differences.

Suggestions for Using This Book to Learn Theories

First, I recommend that you set aside an hour or two to read about a single theory from beginning to end (from In a Nutshell to Putting It All Together) to help get the full sense of the theory. Some chapters have a couple theories in one, so for these it is fine to read the chapter in chunks. Additionally, some learners may find it helpful to scan the treatment plan (either the template or example at the end of the chapter) or some other section first to provide a practical overview; that said, I have tried to organize the ideas in the way most people seem to prefer. But I encourage you to discover what works best for you as different learners have different strategies that work best for them. When you are done with a chapter, you might want to try completing a case conceptualization and treatment plan for yourself (you may have to make up a problem if you are nearly perfect) or someone else to get a sense of how this would work.

Finally, I strongly recommend that either after reading the chapter or after going to class you take good ol' fashioned notes. Yes, I mean it. I recommend that you type up (or if you prefer handwrite) a complete outline of the key concepts in your own words. Why do I advocate for such painful torture? All of us, myself included, when we read long, dense books such as this one fade in and out of alert attentiveness to what we are reading—often lapsing into more interesting fantasies or less interesting to do lists—and—gasp!—sometimes even skim large sections of the text (no, I am not surprised or offended). The only way to make sure that you really understand the concepts you read about is to put them in your own words and organize them in a way that make sense to you. If you need to take culminating exams or plan to pursue licensure, you will have to log the concepts in this book into your long-term memory, which requires more than cramming for a final exam. If you are new to graduate and professional school, I am sorry to be the bearer of the sad news that unlike undergraduate study where forgetting everything you learned the week after finals was generally not a problem. Being a mental health professional requires that you master and build upon what you learn and you will be expected to know what is in this book for the entire time you are active in the profession (seriously—and if you think that is bad, just wait until you get to a class on diagnosis—you'll have to memorize an even longer book). Thus, if your former study habits included all-night cramming, gallons of espresso (or other favorite caffeine delivery system), and little recall after the exam, you might want to try my note taking tip or some other strategy as you move forward.

Suggestions for Using This Book to Write Treatment Plans

I want to emphasize that the treatment plan format, templates, and examples in this book are just that: formats, templates, and examples. They do not represent the only approach or the only right approach, simply a solid approach based on the common standards and expectations. You most likely will work at a counseling agency or institution that uses another format, but the same general rules (the ones in Chapter 4) will still apply. That is why understanding the principles of how to write good goals and interventions is more important than memorizing the format.

Furthermore, don't use the templates and examples too rigidly. Feel free to modify the goal statements and techniques to fit the unique needs of your client. I have provided some relatively specific goals as an example of what might work and encourage you to radically tailor these for each client's unique needs. You will notice that treatment plans in the case study do not rigidly follow the templates; I encourage you to do the same.

Suggestions for Use in Internships and Clinical Practice

When working as an intern or licensed mental health professional, this book can be useful for teaching yourself theories and techniques in addition to learning how to complete clinical documentation. You will likely find that when you work with new populations and problems, you may be interested in considering how other therapy models might approach these situations. This book is designed to be a prime resource for quickly scanning to identify other possibilities. Alternatively, you might have a colleague or supervisor who uses a theory with which you are not familiar. You can use this book to quickly review that theory and avoid looking uneducated. In addition, this book is written to help you appreciate and find common ground across theories, which can be of particular benefit when working in a "mixed theory" context. However, to actually learn to practice any of these theories well, I strongly urge you to take advanced training from experts in that approach.

Suggestions for Studying for Licensing Exams

Licensing exams are not designed to be unnecessarily tricky or scary, simply to ensure that you have knowledge necessary to practice therapy without supervision *and not harm anybody*. And, it is a vocabulary test. If you have honestly engaged in your classes, done your homework, avoided cramming for tests and papers, and made it a priority to get decent supervision, you should have a strong foundation for taking your licensing exam. You should already have in your possession books (such as this) that cover all of the content to be studied for the exam. If your exam is to be taken upon finishing a lengthy post-Masters internship, you should use the entire two- to four-year period to read as many books as possible on the theories and materials covered by the exam (no novels for a few years).

I do not recommend that all of my students take long, expensive "review courses" because such courses are not necessary for those who are proactive in mastering the material on the exam long before they sign up to take it. If you start studying only after you are approved to take the test, you are starting about two to four years too late—and then, yes, you will need to take a crash course. My basic suggestion for studying for a mental health licensing exam is this: read an original text on each major theory during your post-degree internship, use the *Diagnostic and Statistic Manual* (DSM) to diagnose all clients whether you need to or not, and keep up with laws and ethics; then buy the practice exams (without the study guides) and take them until you consistently get 5% above the required passing score (e.g., 75% if the passing score is 70%). If you find that you are weak in a particular area, such as theory or DSM, use a text such as this, which is designed to the license review in mind. Once you consistently get 75% you are ready to take the test with the most learning and the least expense.

Suggestions for Faculty to Measure Competencies and Student Learning

This book is specifically designed to help faculty and supervisors simplify and streamline the onerous task of measuring student competencies as required by the various accreditation bodies. The forms and scoring rubrics for assessing student learning using counseling, psychology, social work, and family therapy competencies are available on the book's web page for instructors (see www.cengage.com). On this website, instructors will also find free online lectures, PowerPoints, sample syllabi, and a test bank (test banks are available only from your Cengage sales representative to maintain security of the questions). This text may be used as the primary or secondary text in a family therapy theories class or as a primary text in a prepracticum or practicum/fieldwork class. Because of its combination of solid theory and practical skills, it can easily be used across more than one class to develop student's abilities to conceptualize theory and write treatment plans, skills that are not likely to be mastered in a single class.

When designing a class to measure competencies and student learning using these treatment plans and case conceptualizations, I recommend initially going over the scoring rubrics with students so that they understand how these are used to clearly define what needs to be done and the expectations for the final product. I have found that is most helpful to provide two or three opportunities to practice case conceptualization and treatment planning over a semester to provide feedback and enable students to improve and build upon these skills in a systematic fashion. Specifically, I have a small group to present a case conceptualization and treatment plan with each theory studied based on a video the class watches on the theory; that way they have enough information to actually conceptualize the client dynamics and treatment. Then, the entire class can see an example and discuss the thought process of developing the plan. A later or final assignment for the class can be to independently develop a treatment plan for a case (either one assigned by the instructor, from a popular movie, personal life, or actual client). By the end of a semester with these activities, students will have developed not only competence but also confidence in their case conceptualization and treatment planning abilities.

Online Resources for Students

Students will find numerous useful resources for the text on the Cengage (www.cengage.com) and author websites (www.dianegehart.com; www.masteringcompetencies.com). These include:

- Online lectures: mp4 recordings of yours truly discussing content of select chapters
- Digital forms for the treatment plan and cross-theory case conceptualization
- Links to related websites and readings
- Glossary of key terms

Online Resources for Instructors

Instructors will find numerous resources for the book online on the Cengage site (www.cengage.com) or the author's sites (www.masteringcompetencies.com; www.dianegehart.com).

- Online lectures by the author
- Sample syllabi for how to use this book in a theory class, prepracticum skills class, or practicum class
- PowerPoints for all of the chapters
- Downloadable versions of the treatment plan and cross-theory case conceptualization
- Scoring rubrics for assignments correlated to each profession's competencies: counseling, family therapy, psychology, and social work
- Test bank
- Web quizzes

References

Council for Accreditation of Counseling and Related Educational Programs. (2009). *2009 standards.* Alexandria, VA: Author.

Gehart, D. (2011). The core competencies in marriage and family therapy education: Practical aspects of transitioning to a learning-centered, outcome-based pedagogy. *Journal of Marital and Family Therapy, 37,* 344–354. doi: 10.1111/j.1752-0606.2010.00205.x

Killen, R. (2004). *Teaching strategies for outcome-based education.* Cape Town, South Africa: Juta Academic.

McDonough-Means, S. I., Kreitzer, M. J., & Bell, I. R. (2004). Fostering a healing presence and investigating its mediators. *Journal of Alternative and Complementary Medicine, 10,* S25–S41.

Miller, S. D., Duncan, B. L., & Hubble, M. (1997). *Escape from Babel: Toward a unifying language for psychotherapy practice.* New York: Norton.

Monk, G., Winslade, J., & Sinclair, S. (2008). *New horizons in multicultural counseling.* Thousand Oaks, CA: Sage.

Nelson, T. S., Chenail, R. J., Alexander, J. F., Crane, R. Johnson, S. M., & Schwallie, L. (2007). The development of the core competencies for the practice of marriage and family therapy. *Journal of Marital and Family Therapy, 33,* 417–438.

Sprenkle, D. H. (2002). Editor's introduction. In D. H. Sprenkle (Ed.), *Effectiveness research in marriage and family therapy* (pp. 9–25). Alexandria, VA: American Association for Marriage and Family Therapy.

Weimer, M. (2002). *Learner-centered teaching: Five key changes to practice.* New York: Jossey-Bass.

CHAPTER 2

Research and Ethical Foundations of Family Therapy Theories

Lay of the Land

This chapter covers two key foundational elements of competent family therapy practice: the evidence base and ethics. The evidence-based section presents a multifaceted, pluristic approach to understanding what constitutes "evidence" and is divided into five major sections:

1. Research-informed clinician model
2. Evidence-based practice
3. Common factors research
4. Evidence-based treatments
5. Review of the MFT evidence

The ethics section covers legal and ethical issues that are particularly salient when working with couples and families. I have invited my colleague, Ben Caldwell, to write this section. He quotes legal and ethical codes by memory, with numbers, and in far more detail than most listeners would prefer. I should also warn you that he rarely offers the black-and-white answers we all hope for, but I am sure even seasoned therapists will learn a thing or two from this section.

Research and the Evidence Base

In the 21st century, all therapists are expected to be well versed in the evidence base for the treatments they use and the problems they treat, in much the same way that we expect our medical doctors to only use procedures and drugs that have been well researched for our particular medical condition. This may or may not come as a surprise to those new to the field, but in many ways mental health therapists have not been well versed in the evidence base for their field, especially when compared to other medical professions, even though the field has always had an evidence base. For many years, it has been as if researchers and clinicians spoke two different languages, with little communication between parties. In some cases, the research has been too specific or too vague to be useful to the average practitioner. In other cases, therapists work from philosophical positions that do not value research, at least in its more common forms. The good news is that translation efforts over the past two decades have made becoming an evidence-based practitioner easier than you might imagine. This chapter covers a general and realistic model for integrating research into clinical practice, the evidence-informed clinician model, as well as three strands of research that can inform daily practice—evidence-based practice, common factors, and evidence-based treatments. Finally, the end of this section introduces you to an excellent resource for quick and easy review of the MFT evidence base.

Research-Informed Clinician Model

For many years, the mental health practitioners, particularly psychologists, were encouraged to follow the scientist-practitioner model, which puts forth the ideal of integrating a research agenda with one's clinical work (Karam & Sprenkle, 2010). However, with the increasing sophistication of contemporary research, this ideal is rarely achieved by either researchers or clinicians. Nonetheless, research is increasingly expected to inform what therapists do in session. Thus, Karam and Sprenkle (2010) propose an alternative, the research-informed clinician model, which they argue is particularly well suited for Masters-level clinicians who generally have minimal training in research methods.

Rather than encouraging clinicians to conduct original research, the research-informed clinician model emphasizes integrating research into daily clinical practice and decision making. Such clinicians are knowledgeable about the practical implications of research, both in terms of how it informs therapy as well as concerns that clients may have. Karam and Sprenkle (2010) identify several ways that research-informed clinicians utilize the evidence base in their work:

- Locating and evaluating research findings to inform their clinical work, which should begin by becoming familiar with the evidence base for one's primary theory of choice. (*Note*: Each theory chapter in this book provides an overview of the related evidence base to get you started with this; the evidence-based practice section below further describes how to use the literature to inform clinical practice.)
- Using research to provide evidence-based psychoeducation to clients in session (e.g., research of John Gottman, see Chapter 11).
- Identifying ways research has either confirmed or disconfirmed commonly held assumptions by clinicians (e.g., treating couples with domestic violence issues, see Chapter 16).
- Using research to evaluate one's effectiveness as a clinician (see Chapter 16 in Gehart, 2014).
- Understanding the utility and limits of evidence-based treatments and practices (see evidence-based treatment section below).

In summary, the demands of third party payers and stakeholders have made the research-informed clinician model the preferred model for the first half of the 21st century, especially in government and insurance-based practice environments. In many ways, this is more a shift in awareness than practice. Virtually all clinicians have in one way or another employed the evidence base to some degree; however, many have been unable to articulate how. Moving forward, therapists will be increasingly asked to do so by all parties: clients, supervisors, employers, third-party payers, and the general public. In learning to do so, clinicians will become increasingly familiar and comfortable with drawing upon research to inform what they do—and thus, are more likely to use it. The next several sections in this chapter introduce the reader to several strands of research that can be used to inform clinical practice in various and distinct ways.

The Minimum Standard of Practice: Evidence-Based Practice (EBP)

More commonly used in the medical field, *evidence-based practice* (EBP) uses research findings to inform clinical decisions for the care of individual clients. In a Nutshell, evidence-based practice refers to knowing the evidence base related to a specific client's problem and contextual issues and using that information to make treatment decisions. For example, the research literature is very clear that systemic-structural family therapy approaches are the treatment of choice for adolescents with conduct and substance abuse disorders. Even if you are not formally trained in one of these approaches (see Chapters 5, 6, 7, and 15 to start your training), because the evidence base is so strong ethically you should use this knowledge to inform your treatment decisions. Think of it this way: If you were this child's parent, would you want to take your child to a therapist who has his/her own way of doing things based on theory and experience or to one who uses theory along with research on best practices to decide how best to help your child? You'd need a really strong referral to choose the first option.

Every therapist should strive to be an evidence-based practitioner: many would consider it an ethical obligation because the evidence base is quickly redefining standard practice (see ethics section below). Patterson, Miller, Carnes, and Wilson (2004) describe five steps of evidence-based practice for family therapists:

Step 1: Develop an answerable question to focus the search for information: e.g., which treatments are most effective for teens who cut to relieve emotional pain?

Step 2: Search the literature for the best empirical evidence to answer the question: e.g., search digital databases such as PsychInfo and scholar.google.com using the keywords *adolescents, self-harm,* and *treatment.*

Step 3: Evaluate the validity, impact, and applicability of the research to determine its usefulness in this case: e.g., is the study randomized? Were there comparison groups? What was the treatment effect size? Were the findings clinically relevant?

Step 4: Determine whether the research findings are applicable to the current client's situation: e.g., what are the potential benefits and risks of applying these findings with this client? Do I need to consider any diversity factors, such as age, ethnicity, class, or family system?

Step 5: After implementing the EBP, evaluate the effectiveness in this client's individual case: e.g., how did the client respond? Were there signs of improvement, no change, or worsening?

Becoming an evidence-based practitioner requires a willingness to continually learn and adapt one's practice to integrate the latest findings in the field. It also requires that therapists are responsive to their client's individual needs—even if researchers indicate a particular approach works for most clients with a specific condition, you need to evaluate whether it works for yours, and if not, you need to adjust your approach. Essentially, it comes down to making more informed decisions so that you are optimally efficient. As research becomes more clinician friendly and relevant, I anticipate that therapists will have a far closer relationship to their evidence base.

Heart of the Matter: Common Factors Research

Over the past decade, professional literature has been abuzz over the "common factors debate" (Blow, Sprenkle, & Davis, 2007; Sprenkle & Blow, 2004; Sprenkle, Davis, & Lebow, 2009). Common factors proponents contend that the effectiveness of therapy has more to do with the key elements found in all theories than with the unique components of a specific theory. To simplify the argument even further: the similarities matter more than the differences. This position is supported by meta-analyses (research on several research studies) of outcome studies in the field: when research studies control for confounding variables (such as therapist loyalty, comparison group, or measures of outcome), there is little evidence to support the superiority of one theory over another, both in psychotherapy (Lambert, 1992; Wampold, 2001) and in family therapy specifically (Shadish & Baldwin, 2002).

Within the common factors community, some (Miller, Duncan, & Hubble, 1997) emphasize the common factors while minimizing the role of theory, whereas others take a more moderate approach (Sprenkle & Blow, 2004), maintaining that theories are still important because they are the *vehicles through which* therapists deliver the common factors and because specific models may have an added benefit in certain contexts. Sprenkle and Blow (2004) point out that the common factors approach does not require therapists to relinquish therapeutic models but instead to understand their purpose differently. Rather than providing the "answer" to the client's problems, common factor proponents propose that using a structured treatment inspires confidence from clients in the therapeutic process, allowing therapists to coherently actualize common factors. From this perspective, a therapeutic model is better understood as a tool that increases therapist effectiveness rather than the "one and only true path" that resolves the client's problem.

Lambert's Common Factors Model

The most frequently cited common factors model is grounded in the work of Michael Lambert (1992). After reviewing outcome studies in psychotherapy, Lambert estimated that outcome variance (the degree to which change is attributed to a specific variable) could be attributed to four factors:

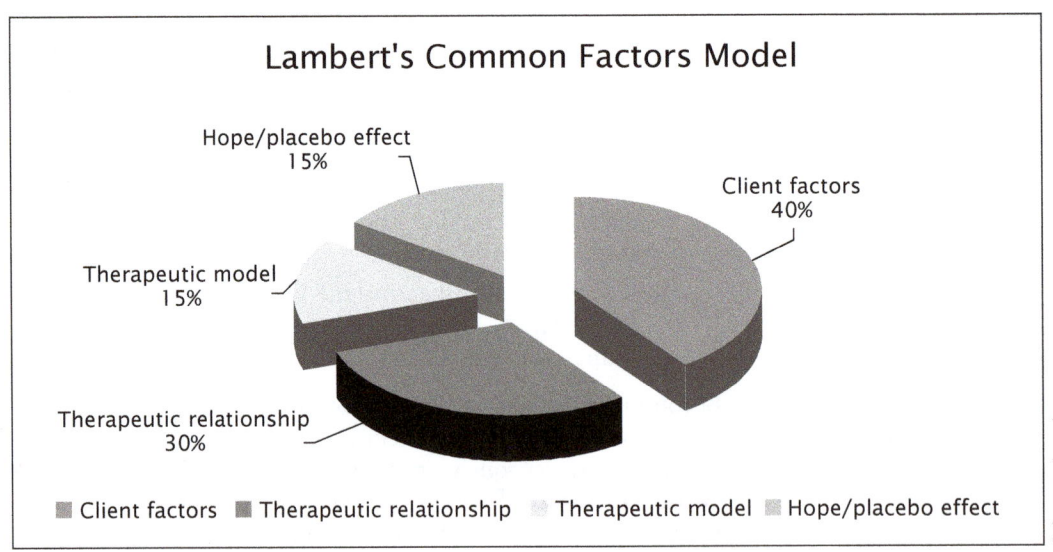

- *Client Factors:* An estimated 40%; includes client motivation and resources
- *Therapeutic Relationship:* An estimated 30%; the quality of the therapeutic relationship as the *client* evaluates it
- *Therapeutic Model:* An estimated 15%; the therapist's specific model for treatment and the techniques used
- *Hope and the Placebo Effect:* An estimated 15%; the client's level of hope and belief that therapy will help

Often these percentages are cited as facts, but although they are well-informed estimates based on a careful analysis of existing research, the numbers were not generated through an actual research study. They should be considered general trends in the research that inspires therapists to critically reconsider how they can help clients rather than exact percentages.

Wampold's Common Factors Model

Wampold (2001), who conducted a meta-analysis similar to Lambert's but compared only studies that included two or more actual therapy models (rather than comparing a model to the generic "treatment as usual" or a no-treatment control group), presents evidence for the following:

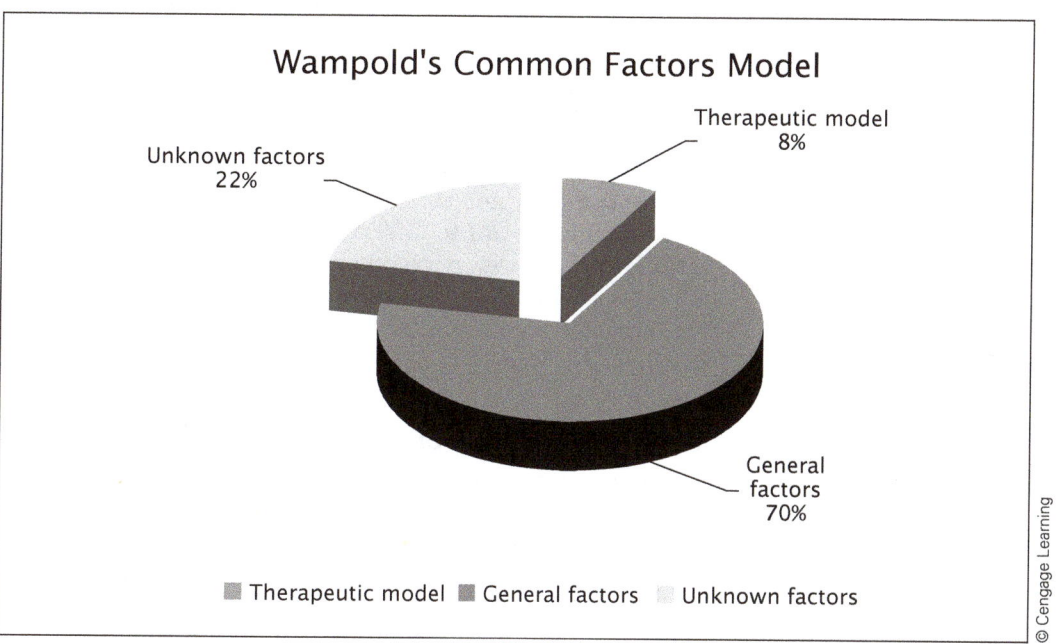

- *Therapeutic Model:* 8%; the unique contributions of a specific theory (compare with 15% in Lambert's model)
- *General Factors:* 70%; therapeutic alliance, expectancy, hope
- *Unknown Factors:* 22%; variance that is not related to known variables

Wampold's research further underscores that common elements across theories contribute more to positive therapeutic outcomes than the unique elements of a specific theory. Thus, research across theories continually indicates that general or common factors have the greatest impact on outcome; although this result may be due to the limits of research (Sprenkle & Blow, 2004) or other factors, it is the best information to date on the subject.

Client Factors

Lambert's (1992) research, which has been made most accessible to clinicians in the work of Miller, Duncan, and Hubble (1997), emphasizes the importance of activating client resources, such as by encouraging clients to create and use support networks and increasing client motivation and engagement in the therapeutic process. Tallman and Bohart (1999) propose that most theories work equally well because of the client's ability to adapt and utilize whatever techniques and insights the therapist may offer, the therapeutic process effectively becoming a Rorschach (ink blot test) that the client uses to create change.

Miller, Duncan, and Hubble (1997) describe two general categories of client factors:

1. *Client characteristics* include the client's motivation to change, attitude about therapy and change, commitment to change, personal strengths and resources (cognitive, emotional, social, financial, spiritual), and duration of complaints.
2. *Extra-therapeutic factors* include social support, community involvement, and fortuitous life events.

Therapeutic Relationship

In both Lambert's and Wampold's research, the quality of the therapeutic relationship appears to be more important than the specific model in predicting outcome, a finding that is consistent with much of the traditional wisdom in the field. In an effective relationship, the therapist accommodates to the client's level of motivation, works toward the client's goals, and demonstrates a genuine, nonjudgmental attitude. A particularly interesting—and humbling—finding is that the client's evaluation of the relationship is more strongly correlated with positive outcome than the therapist's evaluation (Miller, Duncan, & Hubble, 1997).

Despite the clear and consistent evidence for the importance of the therapeutic relationship, most outcome studies, especially those on evidence-based therapies, try to control for and factor out the impact of the therapist on treatment, thereby obscuring the role of the therapist in effective treatment (Blow, Sprenkle, & Davis, 2007). Perhaps this is done because it is hard to fully operationalize and measure the therapeutic relationship, or perhaps because researchers want a more scientific-sounding explanation (the treatment did it, not the relationship). Whatever the reason, the evidence-based treatment literature seems to undervalue and underestimate the importance of the therapeutic relationship. However, common factors research redirects therapists' attention to this important component. Therapists wanting to closely attend to relationship variables can use measures such as the Session Rating Scale (Duncan et al., 2003) to monitor the relationship on a weekly basis.

Therapeutic Model: Theory-Specific Factors

Theory-specific factors are what the therapist says and does to facilitate change while following his/her therapeutic model. These factors are what therapists and third-party payers consider important. However, as Lambert's research and estimations indicate, technique may not be as important as is typically assumed, actually being only half as important as the therapeutic relationship. However, it is still an influential factor over which therapists have significant control.

Hope and the Placebo Effect: Expectancy

Hope and expectancy, or the *placebo* effect, refer to clients' belief that therapy will help them resolve their problem. Lambert's (1992) emphasis on this factor heightens therapists' awareness of an often-neglected aspect of the therapeutic process, at least in the research literature (Blow, Sprenkle, & Davis, 2007). With this awareness, therapists can more consciously work to instill hope, which is particularly critical in the initial sessions.

Diversity and the Common Factors

Common factors can be particularly useful when working with diverse clients—whether culturally, sexually, linguistically, or in ability—because diversity always implies unique client resources and challenges, particularly for the therapeutic relationship, choice of approach, and strategies for instilling hope. For example, although gay, lesbian, bisexual, and transgendered clients are often ostracized in the general community, many have extensive informal and formal social support networks; the same is true of many ethnic groups and disabled or chronically ill people. Thus the societal challenge is partially offset by unique resources. Therapists can help clients leverage these resources to better manage the often daunting challenges of being different from the majority.

Similarly, with diverse clients the task of creating a therapeutic relationship in which the client feels accepted rather than judged requires more mindfulness and thoughtfulness because the therapist may not be aware of all the dynamics and traditions of these groups. Education on local diverse communities is of course necessary, but humility and admitting that you do not know the answer is often more important because it cultivates respect and openness (Anderson, 1997). When therapists proceed with curiosity and willingness to learn, they often discover distinct and effective means for instilling hope from within the client's culture and primary community, further strengthening the therapeutic relationship.

Do We Still Need Theory?

The natural question that follows from the common factors debate is: Do we still need theory? As noted by Sprenkle and Blow (2004), some therapists lean toward the "dodo bird verdict," suggesting that theory matters very little. The more moderate stance of Sprenkle and Blow (2004) emphasizes that "the models are important because they are the vehicles through which the common factors do their work" (p. 126).

Following this moderate position, theory still plays a critical role for new and seasoned clinicians, but not the role one might initially expect it to play. Rather than providing a system to help clients alleviate their symptoms and resolve their problems, a theory is a *tool* that helps the therapist to help the client. *Thus theory may be most relevant for the therapist—not the client.*

Theory gives therapists a system for interpreting the information they get about clients so that they can say and do things that will be useful. It also helps therapists know how best to relate and respond to clients. Without theory, it is easy to get lost in a sea of information, emotion, and challenging behaviors. Theory gives therapists a systematic way of dealing with the wide range of difficulties clients bring. Thus, choosing a theory involves identifying a theory that makes sense to the therapist and is useful to the therapist in navigating the "wild ride" that is psychotherapy. That said, future research may identify specific circumstances in which certain models work better for certain clients (Sprenkle & Blow, 2004).

Show Me Proof: Evidence-Based Treatments

Just in case you thought the debate was over, there is yet another thread in the theory debate that is pulling therapists in the apparently opposite direction from the common factors research: empirically supported treatments, often referred to as "evidence-based therapies." These therapeutic models, which were developed through research and randomized trials (Sprenkle, 2002), should not be confused with evidence-based *practice* (EBP; see above), although evidence-based treatments are sometimes referred to as evidence-based practices to keep us all thoroughly confused (this is one instance where the plural refers to something quite different than the singular).

When therapists, licensing boards, or funding institutions refer to therapy models as "evidence-based," they are generally referring to a set of standards that a 1993 task force of the American Psychological Association (APA) established for what was initially called *empirically validated treatments* (EVTs) and later called *empirically supported treatments* (ESTs), the change underscoring that a treatment is always in the process of being further studied and refined (American Psychological Association, 1993; Chambless et al., 1996). The APA established several categories for describing empirically supported therapies, and others have developed similar categories.

Empirically Supported Treatments and Their Kin
Empirically Supported Treatment Criteria

Empirically supported treatments (ESTs) meet the following criteria (Chambless & Hollon, 1998; Sprenkle, 2002):

- Subjects are randomly assigned to treatment groups.
- In addition to the group that receives the treatment being studied, there must also be *one* of the following:
  - A no-treatment control (usually subjects are on a waiting list)
  - An alternative treatment (for comparison; may be an unspecified approach: "treatment as usual")
  - A placebo treatment
- Treatment is significantly better than the no-treatment control and at least equally as effective as an established alternative.

- Treatment is based on a written treatment manual with specific criteria for including or excluding clients.
- A specific population with a specific problem is identified.
- Researchers use reliable and valid outcome measures with appropriate statistical measures.

Criteria for Additional Forms of Evidence-Based Treatments

In addition to the criteria for empirically supported treatments, criteria have been set for other evidence-based treatments:

- *Efficacious Treatments:* Meeting more stringent criteria, these treatments must meet the requirements for EBTs and in addition must undergo two independent investigations (studies conducted by someone who is not closely involved in the development of the treatment or invested in its outcome) (Chambless & Hollon, 1998; Sprenkle, 2002).
- *Efficacious and Specific Treatments:* Meeting the highest standards, these treatments must meet the criteria for efficacious treatments and in addition must be *superior* to alternative treatments in at least two independent studies (Chambless & Hollon, 1998; Sprenkle, 2002).

EST Pros and Cons

The advantages of ESTs are the following:

1. They have greater scientific support.
2. They have written manuals to guide treatment and are highly structured.
3. They target a specific population with a specific problem.

The disadvantages of ESTs are these:

1. They have limited applicability because they target a specific and therefore limited population.
2. They are expensive: therapists need highly specific training in the model and also need to be trained in a number of models to function effectively in most work environments.

Meta-Analytically Supported Treatments (MASTs)

A meta-analysis is a quantitative research method that combines results from multiple studies, generally by examining the effect size, or the outcome variance attributed to the treatment. Using meta-analytic studies, Shadish and Baldwin (2002) developed the following criteria for MASTs to broaden the type of research that can be used to establish efficacy while maintaining rigorous scientific standards:

- Effect sizes from more than one study of the treatment must be combined meta-analytically.
- All studies must be randomized comparisons of the treatment to a no-treatment control group.
- Meta-analysis must indicate a statistically significant effect size and a significant test.
- Meta-analysis must use sound methods (e.g., aggregating effect sizes).

Real-World Applications of ESTs and MASTs

In 2002, Shadish and Baldwin identified 24 family therapy theories that fit the criteria for a MAST, whereas only 5 met the criteria for an EST. This difference existed primarily because ESTs require (a) a written treatment manual and (b) a narrowly defined population with a specific problem, whereas MASTs allow for other forms of training and more general populations to demonstrate efficacy. The findings of the APA's 2005 follow-up report on ESTs highlight the necessity for more broadly defined standards for evidence-based treatments such as MASTs (Woody, Weisz, & McLean, 2005). This survey indicated that although a higher percentage of ESTs were taught in the classroom, clinical training in ESTs dropped significantly from 1993 to 2003. When asked to identify the reasons, supervisors cited "uncertainty about how to conceptualize training in

ESTs, lack of time, shortage of trained supervisors, inappropriateness of established ESTs for a given population, and philosophical opposition" (p. 9). Arguably, all but perhaps the last of these obstacles are clearly linked to the exact things that make ESTs unique: written treatment manuals and a narrowly defined population. Thus, although promising, ESTs have some practical limitations at this time, especially for the general practitioner.

Research in Perspective: Limitations of the Evidence-Based Movement

Therapists need to keep the evidence-based therapy movement in perspective. Almost all research indicates that any therapy is better than no treatment at all: that is one of the major ideas behind the common factors movement (Miller, Duncan, & Hubble, 1997; Sprenkle & Blow, 2004). The evidence-based therapy approach refines what we know and aims to develop better and more specific therapies; however, this does not mean that nothing in the field has ever been researched or even that what most therapists do on a daily basis isn't "evidence based." A more fair and realistic assessment is that family therapy and mental health therapies have an established history of meaningful research and our ability to do research more precisely has continually increased. More critically, what *has* changed in recent years is that the average practitioner is now expected to be able to articulate the evidence base for what he or she does in session. A research orientation is not new; however, our ability to conduct more meticulous and useful studies is improving as is the demand that therapists draw from it when working with clients.

Perhaps it is useful to reflect on the broader picture. More than in many other mental health disciplines, family therapy theories were developed through observational research (Moon, Dillon, & Sprenkle, 1990; Karam & Sprenkle, 2010). Teams of therapists observed sessions through one-way mirrors, developed hypotheses about what might work, tested these hypotheses, and then refined them as they went along. Rather than trying to prove a theory, these therapist-researchers were using outcomes to inform the development of a new frontier in mental health: working with couples and families. This type of research is rigorous in a different dimension than ESTs; namely, it can be usefully applied in everyday work settings by persons with standard training. Furthermore, all ESTs are based on "traditional theories" in the field; they are—in one sense—simply finely tuned versions of one or more traditional therapy models for a specific population or problem. Thus, the artificial dichotomy between theory and research or traditional theories and ESTs is nothing more than a misunderstanding of both. The next generation of clinicians—such as those using the research-informed model—will hopefully be more comfortable with both the evidence-based and theory worlds and better understand how the two support and inform each other.

Review of the MFT Evidence Base

Finally, I want to let you know about an academic goldmine that makes grad students and their faculty particularly giddy when they finally discover it: the January 2012 edition and the 38th volume of the *Journal of Marital and Family Therapy* (known to MFT nerds as *JMFT*; Sprenkle, 2012). You may not have had the best relationships with journal articles up to this point in your education—that is common. But I think this one is going to change all that. In a nutshell, this edition of *JMFT* includes 12 articles in which experts in the field summarize much of the current evidence base for you. It is an academic's fantasy come true. This is the third of such reviews of the MFT evidence base, making it the only mental health discipline I know of to have such a concise and easily accessible review of the associated research (and, trust me, I have looked). This single resource can make understanding evidence-based practice with couples, families, and children a breeze.

The areas of research reviewed include:

- Conduct disorder and delinquency with adolescents (Baldwin, Christian, Berkeljon, & Shadish, 2012; Henggeler & Sheidow, 2012)
- Child and adolescent disorders (Kaslow, Broth, Smith, & Collins, 2012)
- Affective disorders (e.g., mood disorders; Beach & Whisman, 2012)
- Treatment of couple distress (Lebow, Chambers, Christensen, & Johnson, 2012)
- Treatment of couples experiencing interpersonal violence (Stith, McCollum, Amanor-Boadu, & Smith, 2012)
- Relationship education for nondistressed couples (Markman & Rhoades, 2012)
- Family psychoeducation for severe mental illness (Lucksted, McFarlane, Downing, & Dixon, 2012)
- Family therapy for drug abuse (Rowe, 2012)
- Couple and family interventions for health problems (Shields, Finley, Chawla, & Meadors, 2012)
- Client perceptions of MFT (Chenail, St. George, Wulff, Duffy, Scott, & Tomm, 2012)

Legal and Ethical Issues in Couple and Family Therapy

Benjamin E. Caldwell, PsyD

I teach law and ethics for the couple and family therapy program at Alliant International University in Los Angeles. I know, for many of you, when you hear the phrase "I teach law and ethics," you brace yourself for cautionary tales of what you had better not do, or else the licensure police will come beating down your door.

This section is not like that. I promise. I will start with a brief overview of professional practice standards for psychotherapists (you may already be familiar with those; if not, there are a handful of excellent texts fully devoted to those standards). Then I will talk about some of the specific areas of difference for therapists working with couples and families. Finally, I will discuss technology and a therapist's refusal to treat certain clients—two controversial issues that are reshaping professional standards in family therapy today.

Lay of the Land: More than Just Rules

Here is my one (and only, I promise) paragraph of fear-mongering. People can sue you, or file a complaint against you to your licensing board, any time for any reason. They can sue you for looking at them funny. They can complain about you for breathing too loudly (or too quietly). They can sue you for not suing them first. If you base your entire practice on trying to never get sued or never get a complaint, you are trying to control something that is beyond your control. Now, whether someone will *win* that lawsuit, or whether a board will *do* anything about that complaint—that is another story. You can and should take reasonable steps to protect yourself, including knowing your state's laws, maintaining professional liability insurance, and consulting with colleagues, supervisors, and attorneys whenever necessary. But doing all of those things does not make you ethical. They are like a football player's shoulder pads: They are great for protection from injury, but they do not make you a great football player.

The fact is, being an ethical professional is about much more than simply knowing and following the rules. It is true that rules-based (and, as is often the case, fear-based) teaching about law and ethics does help you to know what the rules are—at least, until the rules change. But being an ethical professional is about knowing what to do when the legal and ethical rules governing our field *do not* tell you specifically what to do. There are times when those rules are not clear, or when they appear to contradict themselves. In those situations, simply knowing the rules will only leave a therapist confused and anxious, hoping that they do not do the wrong thing. It is in those instances when your character as a professional is most revealed. Being an ethical professional means engaging in a thoughtful and careful decision-making process that results in the best outcomes for your clients, within the accepted standards of the field.

Those accepted standards are largely consistent across mental health professions. This section is not meant to be a replacement for a full law and ethics course or textbook, and is not meant to cover the full scope of the legal and ethical issues in psychotherapy. (For a complete textbook focused on the legal and ethical issues in couple and family work, I would recommend Wilcoxon, Remley, and Gladding's *Ethical, Legal and Professional Issues in the Practice of Marriage and Family Therapy*.) Instead, this section is meant to offer particular guidance on those issues most relevant to—or most different in—couple and family therapy. Working with more than one person on a therapeutic issue raises unique concerns, and it is these concerns I will emphasize. A brief, broader overview is necessary to set the stage, but if you are already familiar with general professional ethics in psychotherapy, you can safely skip this part and pick up at "specific legal and ethical concerns in couple and family work."

The Big Picture: Standards of Professional Practice

Practicing any form of therapy in a professional manner means understanding three levels of rules that govern professional behavior:

- Laws
- Ethics
- Standard of care

Laws

First, there are laws. Laws are set either by the local, state, or federal government, in the form of legislation, or by judges, whose rulings in some cases establish specific responsibilities for professionals. *Tarasoff v. California Board of Regents*, which established a therapist's responsibility to intervene when a client poses an immediate danger to an identifiable victim, is an example of case law impacting psychotherapists. Laws are often about what you *must* do (like pass a licensing exam in order to practice independently) and *must not* do (like insurance fraud or sleeping with clients). They do not have exceptions, except those also spelled out in law. For example, you must maintain confidentiality *except* when a client is a threat to themselves or others, and other instances defined in state and federal law. *Regulations* are a subset of laws. Instead of being set by a state legislature, regulations are often set by licensing boards through state administrative processes. They do, however, hold the power of law.

Key point to remember: *Laws trump everything else*. If a law conflicts with an element of an ethical code, the therapist must abide by the law, but should do so in whatever manner allows for the most adherence to the ethical code.

Ethics

The second level of standards that govern professional behavior is *ethics*. Each of the major professional associations in mental health publishes a Code of Ethics that serves to guide responsible professional behavior. Counselors, social workers, and psychologists have ethics codes published by the American Counseling Association (www.counseling.org), National Association of Social Workers (www.nasw.org), and the American Psychological Association (www.apa.org), respectively. In family therapy, the American Association for Marriage and Family Therapy (www.aamft.org) and California Association of Marriage and Family Therapists (www.camft.org) each publishes their own ethics code. These codes go beyond the rules of the law, and define in more detail the expectations of family therapists. However, no ethical code can be expected to offer specific guidance on every possible scenario a therapist will encounter. Being an ethical practitioner is about much more than knowing the codes; it is about understanding *ethical reasoning* so that you can make the best choice of what to do when the legal and ethical rules are not clear.

There are a handful of *biomedical ethical principles* (Beauchamp & Childress, 2009) that may be used when weighing the ethics of decisions in mental health work:

- *Fidelity* refers to keeping promises and upholding loyalty. This can be challenging in couple and family work, where family members may attempt to get the therapist to take sides in their internal struggles.
- *Justice* refers to treating people fairly, bearing in mind that fairly does not always mean equally.
- *Autonomy* refers to the right of clients to make their own decisions and act independently. In family therapy, it is particularly important that a therapist respect a client's right to make their own decisions around romantic relationships, such as choosing to cohabit, separate, divorce, or marry; and around child care, including custody and visitation.
- *Beneficence* refers to the therapist's obligation to actively work to benefit clients. If a couple or family does not appear likely to benefit from ongoing treatment, the clients should be referred to another therapist.
- *Nonmaleficence* refers to avoiding harm to clients or others.

When ethical guidelines for handling a particular situation are unclear or appear to contradict, it is generally recommended that mental health professionals return to these general principles to assess the risks and benefits of their options.

Standard of Care

Finally, the third level of standards governing professional behavior is called the *standard of care*. California's definition of "reasonable suspicion" of child abuse, which triggers a therapist's mandated reporting responsibility, is one effort to define the standard of care in law:

> For purposes of this article, "reasonable suspicion" means that it is objectively reasonable for a person to entertain a suspicion, based upon facts that could cause a reasonable person in a like position, drawing, when appropriate, on his or her training and experience, to suspect child abuse or neglect. "Reasonable suspicion" does not require certainty that child abuse or neglect has occurred nor does it require a specific medical indication of child abuse or neglect; any "reasonable suspicion" is sufficient. (California Penal Code 11166(a)(1))

In essence, the standard says, "If another therapist would suspect child abuse with the information and training you have, you should suspect it too." This is what it means to have a standard of care: It is what most people at the same professional level are doing. Case documentation, such as treatment plans, works on a similar principle: While legal and ethical standards do not tend to specify what should be in progress notes, therapists tend to follow standard formatting and include similar content, because that is what their peers do. Even in elements of therapy for which there is no legal or ethical standard, failing to live up to the standard of care is seen as inadequate professional behavior.

There is no single place to look for a standard of care in writing. However, if you are unclear about the standard of care on any specific issue, the wise thing to do is consult with the colleagues and supervisors you most respect. They can best inform you what others in the profession are doing.

Specific Legal and Ethical Concerns in Couple and Family Work

There are several areas specific to couple and family work where the rules differ from individual therapy. These include identifying the patient, documentation, confidentiality, communicating with other systems, working with minors, child abuse reporting, and intimate partner violence.

Who Is Your Patient?

Perhaps the most difficult challenge in couple and family work is identifying whom exactly it is you are treating. In individual therapy, it is easy to identify your patient: Look at the person sitting in front of you. Couples and families, however, come to therapy with complex complaints. In some instances, they identify one member of the family as the problem who has brought them to you. In other instances, they say that the entire relationship or entire family is struggling. Who is the therapist treating?

The question is not just academic. How you answer it will have an impact on how you organize your treatment plan, how you document the case, how you handle conflict between family members, and even how payment takes place. Indeed, it is likely to reflect your underlying philosophy when it comes to couple and family work.

Therapists who specialize in treating couples and families often view themselves as treating the entire system as a unit. These therapists usually label themselves as "relational" or "systemic" in their work. Accordingly, they will keep a single file for the entire family, even if different combinations of family members are present in specific sessions. The treatment plan will be focused on systemic goals involving relationship or family functioning, goals that usually are agreed upon by all members of the couple or family attending therapy. Relational therapists do not ignore individual functioning, but rather they consider it in its interpersonal context.

Documentation

If the therapist is treating the family as a unit, they are likely to be keeping a single file on the entire family. That would mean one treatment plan for the entire family, and one progress note per session, no matter how many family members attended. That also means that under most state laws, the consent of every family member in treatment would be needed to release records to any individual family member requesting them.

If you are treating an individual, and other family members sometimes join in therapy to support the individual's progress toward their treatment goals, then it makes the more sense to document the session as part of the individual client's existing file, noting the involvement of the others. In this case, only the individual receiving treatment needs to consent to the release of their records.

Some work settings require therapists doing couple or family work to keep a separate and distinct file for each participant in the therapy. While this does allow for detailed recording of each person's behavior—something that can get lost when a single progress note is written for a session involving several family members—it also means the therapist will spend significant additional time documenting each session, and needs to be thoughtful about how they document family interaction in an individual's file. In this instance, each individual could consent to the release of his or her own records, but they could not request the release of other family members' files.

Confidentiality

Confidentiality is both a legal and ethical requirement for psychotherapists, and most of the time it is fairly straightforward: You cannot share what clients tell you. (My treatment contract for many years has included the line, "The therapy office is like Las Vegas. What you say here, stays here.") There are exceptions to this spelled out in the law: therapists are required to break confidentiality to report child abuse, and when necessary to stop clients who pose an imminent danger to themselves or others. There are many other exceptions to confidentiality spelled out in state and federal law, and it is important that your clients be made aware of both the general rule of confidentiality and its specific exceptions.

Confidentiality becomes more complex when working with couples and families. Consider a married couple where one partner is having an affair. If that partner tells you about the affair during a phone call, and then tells you something like "I really want to make my marriage work, but I am not ready to end the affair yet. Please don't tell my spouse," what should the therapist do? Revealing the secret could lead the couple to give up on treatment, and possibly their marriage. Keeping the secret, however, would mean colluding with the partner having the affair—and if the other partner found out later that you had known about the affair and not told them, they would likely feel betrayed.

If you are treating a couple or family, it is vital to have a policy around the holding of secrets. The Codes of Ethics for family therapists do not say what a therapist's policy around secrets should be, but rather they require that the therapist have a policy around secrets, inform clients of what the policy is, and stick to it. Two schools of thought have emerged around what the ideal secrets policy is:

1. *A "No Secrets" Policy*: This policy allows the therapist to communicate anything learned from any individual in the family to any of the other individuals in the family at any time. In short, it says that the therapist will not hold a secret for one family member. There are clear advantages to this policy: Holding secrets creates power imbalances in the therapy room, and could require that a therapist "play dumb" around such important issues as affairs or substance use. Refusing to hold secrets means that any such information, if it comes to the therapist, can be dealt with in the therapy setting.
2. *A Limited-Secrets Policy*: While family therapists typically do not advocate holding all secrets, there are many who believe it is advantageous to hold some information in individual confidence. Doing so allows for a more thorough and trustworthy assessment process. For example, a couple therapist may see each partner individually during the assessment stage of therapy with plans to work with the couple together afterward. By assessing individually, and being willing to hold secrets from that assessment, a therapist can encourage clients to be more honest about relational issues they may not feel comfortable discussing in front of their partner. Affairs, substance use, and domestic violence are all examples of subjects that clients may be more honest about individually than in front of their loved ones.

Whatever policy you choose, it is good to have it as part of your treatment contract when working with couples and families. It is good to discuss the policy with your clients to make sure they understand it. And, most importantly, once you have set your policy, you must follow it.

Communicating with Other Systems

Family therapists often handle issues around divorce, custody, parenting, and family functioning, and as a result often have connections to other systems, such as schools and court systems. In some cases, it may be that outside system that hires and pays the therapist, again raising the question of who, exactly, the "client" is. The therapist may be required to submit regular progress reports to the court; for example, on treatment goals that the clients had little to no say in.

All family members should be made aware as early as possible in therapy of the nature of the relationship the therapist has with third parties involved in the treatment. All family members should be informed about what information will be shared, with whom, and for what reason.

Working with Minors

Even if you are intending to do individual therapy, working with minors will make you a family therapist. Family members typically must give consent for the minor to receive treatment, they often want to be included in the therapy process, and they usually have a right to access treatment records. (Each of these varies a bit from state to state, so be sure you are familiar with your state's laws.)

State laws differ as to when a minor can consent on their own for therapy. Generally speaking, someone under the age of 18 cannot enter therapy without the consent of their parents. However, some states make exceptions to this. In California, for example, minors age 12 and older can independently consent for therapy if the therapist determines that the minor is mature enough to participate intelligently in treatment.

Whether it is part of an individual or family process, any work you do with a minor on their own should come with clear boundaries. Bear in mind here that family members may have conflicting motivations. Parents who are fearful for their child's well-being may want you to tell them everything that happens in an individual meeting with the child. The child, meanwhile, often will prefer a safe place to explore difficult emotional issues *without* the sense that their parents are always peering in. Many therapists resolve this conflict through a written agreement that defines what information will stay between the therapist and the minor, and what information will be shared with the family.

(As you may have guessed, this would be common among therapists using a "limited secrets" policy.) For example, if a minor is struggling to navigate peer relationships at school, the minor may not want the therapist to share that information with their parents—and it can be argued that the greater good is served by the therapist holding that information. (The child gets a safe place to be honest about their struggles, and the parents are not harmed by not having that information.) On the other hand, if a minor is engaging in drug or alcohol use, or otherwise endangering their physical health, that information would be shared with the parents. Again, whatever policy the therapist chooses to employ when it comes to secrets, the policy should be clear, in writing, and followed.

Child Abuse Reporting

In working with children and families, a therapist often will develop suspicion that child abuse has taken place. Sometimes this will arise directly from what clients tell the therapist; other times, the therapist makes this assessment based on physical or behavioral evidence of abuse. While state laws vary, psychotherapists typically must report physical abuse, sexual abuse, and neglect to local authorities so that the victim, as well as other potential victims, can be protected. It is vital to be aware of your state's rules for what must be reported, to whom, and how quickly once the therapist learns of the abuse.

Therapists who work with adolescents and their families should be particularly aware of their state's laws around the reporting of consensual sexual activity among minors. States may differentiate between consensual sexual activity that qualifies as criminal and activity that qualifies as abusive (for example, in some states there are some instances of statutory rape that would be considered criminal but not abusive). Since abuse *must* be reported, but criminal activity that is not abuse *cannot* be reported, every therapist must remain keenly aware of their state's current laws. The age of your client, the age of their partner, the nature of their relationship (for example, whether an older partner appears to be exploiting a younger one), and the specific activities they have engaged in all may be relevant to the question of whether their relationship can be considered abusive.

Intimate Partner Violence

Several researchers have suggested that couples therapists more thoroughly assess all couples entering therapy for current and past intimate partner violence (IPV). Without thorough assessment, such violence often goes unreported. Among couples seeking outpatient counseling, 36% to 58% have experienced male-to-female violence, and 37% to 57% have experienced female-to-male violence, in the past 12 months (Jose & O'Leary, 2009). Recent or ongoing violence can be a major impediment to successful couple therapy, and is considered a contraindication for emotionally focused therapy (Johnson, 2004), one of the best-validated approaches to couple work.

Generally speaking, therapists cannot break confidentiality to report IPV on their own. If children have witnessed violence between their parents, however, the situation becomes more complex, as does the therapist's responsibility to protect the children involved. In California, when children have witnessed domestic violence, a therapist can (but is not required to) report this to law enforcement as emotional abuse. Be sure you know your state's laws on child abuse reporting before filing a report based on witnessing violence in the home.

The treatment of IPV for those convicted of a first or second offense raises specific ethical questions, particularly around weighing potential risks against potential benefits of treating couples together. At present, treatment is often through court-mandated, gender-specific group therapy (Babcock, Green, & Robie, 2004). Such treatments appear to have small but meaningful impacts on recidivism (Stith, McCollum, Amanor-Boadu, & Smith, 2012). However, they also suffer from high dropout rates, and often draw recidivism data from arrest reports. This makes their measure of recidivism subject to underreporting.

For a therapist who treats couples, inevitably some of their clients will have a history of IPV. Other couples enter therapy with active, ongoing violence. There has been debate in the literature for years about how these couples should be treated. That debate follows many of the issues noted above, particularly around balancing the chance of benefit for couples with the risk of harm. Some consider the risk of harm too great to engage in any couple—or family-based treatment for at least several months after the last violent incident, noting concerns that some aggressors may become more violent when treated for IPV in a couple context. Others cite conflicting publications that support seeing couples together when their violence history has been low-level and mutual (e.g., Bograd & Mederos, 1999). Depending on where you practice, there may be state, county, or agency rules governing the treatment of domestic violence, particularly if one partner has been convicted of IPV.

Current Legal and Ethical Issues in Couple and Family Work

Professional standards change over time. Ideally, as a professional, you will take an active role in those changes. The AAMFT Code of Ethics (AAMFT, 2012) calls on family therapists to take an active role in developing or changing laws and regulations regarding family therapy to ensure such rules are in the public interest (principle 6.7). Often the changes in professional standards come about as a result of new forms of treatment or changes in the larger populations we serve. There are two specific issues that are currently reshaping the professional standards for couple and family therapy: technology and therapist values.

Technology

As videoconferencing and related technology have become more sophisticated, therapists have begun to utilize technology to provide services to clients who otherwise may not be able to come in to the office. Clients who live in rural areas without adequate health care services can be particularly helped by phone or videoconference therapy, as can clients with specific language needs that cannot be met by providers close to them.

Providing therapy services by phone or Internet is a challenging undertaking with individuals; with couples and families, it becomes even more challenging. Much of the information a therapist gathers in working with a couple and family has to do with the interaction between clients in session. When the couple or family can only be seen through a computer screen—or, if the therapy is taking place by phone, when the couple or family cannot be seen at all—it becomes much more difficult for a therapist to assess the dynamics being acted out in the room. It may be for this reason that research on therapy that takes place by phone or videoconference has so far focused almost exclusively on individual therapy (Barak, Hen, Boniel-Nissim, & Shapira, 2008).

For those therapists who do wish to attempt couple or family work by phone or Internet, there are several things to be aware of. First, a therapist's license only authorizes them to work in the state where they are licensed. A therapist licensed in Texas, for example, could *not* do telephone therapy with a couple in New York; the therapist would likely be seen as practicing in New York without a license.[1] Second, therapists providing phone- or Internet-based services must abide by additional ethical requirements, including (a) ensuring that electronic therapy is appropriate, considering the clients' needs and abilities; (b) informing clients about the potential risks and benefits of electronic therapy; (c) ensuring the security of the connection, to protect privacy and confidentiality; and (d) ensuring that the therapist has appropriate training and experience in using the technology. These requirements are present in both the AAMFT and ACA Codes of Ethics. The ACA code goes further, requiring that therapists also plan with clients for what to do if the connection is lost or a crisis occurs. Finally, there are no well-established protocols for couple and family therapy aided by technology.

---

[1] There have not yet been many test cases on this issue, though there is at least one instance of this rule being enforced. Christian Hageseth III, a psychiatrist in Colorado, prescribed antidepressant medication to a teenage patient in California through an Internet pharmacy (the psychiatrist and the patient never met). The antidepressant triggered thoughts of suicide in the patient, who ultimately killed himself. California pressed charges against the psychiatrist for practicing in California without a license, and Hageseth received jail time (Sorrel, 2009).

Technology is also impacting couple and family work in a very different way: by changing how couples and families relate to one another. While electronic communication allows families to maintain connection more easily over great distance, it also is frequently used to facilitate affairs and risky sexual behavior. The negative impacts of online activities are an increasingly common reason given by couples who seek therapy in my practice. However, the negative impacts of online relationships on couples and families are only beginning to be understood from a research perspective (Hertlein & Webster, 2008).

Therapist Values

At least three court cases have come about in the past several years around psychotherapists who said they were unwilling to treat gay and lesbian clients. These have led to a broader discussion in the field about where a therapist's obligations to serve clients end, and their right to practice based on their own values begins.

It is generally accepted that family therapy cannot be value-free. Any therapist brings their values into the therapy room. Indeed, a value for helping the larger community through healthy relationships is often part of why someone becomes a family therapist. In the interest of autonomy, therapists are usually taught to understand clients through the lens of the clients' values. The CAMFT Code of Ethics calls on therapists to be aware of their own personal values and not impose those values upon their clients. Similar clauses can be found in the ethics codes of APA, ACA, and NASW.

In the court cases, however, therapists have asserted that their values around sexuality preclude them from treating gay and lesbian clients. Julea Ward, a counseling student at Eastern Michigan University, refused to treat a lesbian client, and worked with her supervisor to ensure the client could be treated by another therapist at the same agency who did not share her conflict. EMU felt that her refusal to treat the client was discriminatory, and told Ward she must complete a remediation plan. She refused to do so, and was expelled from the program as a result. She then sued the university. Jennifer Keeton, a student at Augusta State University, filed a similar lawsuit after also being told she must complete a remediation plan to ensure she would not impose her values around sexuality on her clients. She had not refused to treat specific clients, but had made many statements in her classes that made it clear she would not work with gay or lesbian clients. Finally, Marcia Walden was suspended and soon fired after she refused to treat a CDC employee's same-sex relationship. Walden, who was working for a CDC contractor, had referred the case to a colleague who did not share Walden's opposition to same-sex relationships.

These cases each present strong examples of situations in which an ethical code can appear to be contradictory. On one hand, nondiscrimination clauses in all professional ethical codes in mental health suggest that clients cannot be turned away simply on the basis of their sexual orientation. On the other hand, competence clauses, also present in all professional ethical codes in mental health, require that therapists not work with clients that the therapist is not qualified to treat. If a therapist does not feel qualified to treat gay and lesbian clients, which clause becomes more important?

The court cases have yet to provide clear case law for therapists to follow. Keeton's case was dismissed by a federal district court in June 2012, at which time the other cases remained on appeal.

Conclusion

Even with the best preparation and knowledge, therapists working with couples and families will face situations that leave them unsure of how to proceed. Laws are typically made in reaction to events that have taken place, and so they are not going to cover every new situation that may emerge. Ethical codes are similarly imperfect, and a standard of care does not exist for every situation. Thankfully, as a professional therapist, you have a number of resources available that can help you with difficult decision making.

If you face legal or ethical questions in your work with couples and families, it is always better to ask questions before acting than after. Consulting with supervisors and colleagues can help give you a good idea of what other professionals would do in similar situations. Consulting with an attorney will help determine your legal responsibilities; your professional liability insurance carrier is likely to offer free legal consultation, and your professional association may provide the same. For questions about your ethical responsibilities, most professional associations have an ethics committee or ethics consultant who can provide advice to members.

Online Resources for Research

American Psychological Association Documents on ESTs:
www.apa.org/divisions/div12/journals.html

Common Factors Research:
http://scottdmiller.com/
http://heartandsoulofchange.com/

SAMSHA Registry of EvidenceBased Practices:
http://www.nrepp.samhsa.gov

Online Resources for Law and Ethics

American Association for Marriage and Family Therapy Code of Ethics:
http://www.aamft.org/imis15/content/legal_ethics/code_of_ethics.aspx

American Association for Marriage and Family Therapy Legal and Ethical Resources (membership required):
http://www.aamft.org/iMIS15/Professional/MFT_Resources/Legal_and_Ethics/Content/Legal_Ethics/Legal_Ethics.aspx

American Counseling Association Code of Ethics:
http://www.counseling.org/resources/codeofethics/TP/home/ct2.aspx

American Psychological Association Code of Ethics:
http://www.apa.org/ethics/code/index.aspx

Ben Caldwell's blog on research and policy issues in family therapy:
http://www.mftprogress.com

California Association of Marriage and Family Therapists Code of Ethics:
http://www.camft.org/AM/Template.cfm?Section=Code_of_Ethics&Template=/CM/HTMLDisplay.cfm&ContentID=11235

CPH and Associates (a professional liability insurance carrier) "Avoiding Liability Bulletin":
http://www.cphins.com/LegalResources/AvoidingLiabilityBulletin/tabid/106/Default.aspx

National Association of Social Workers Code of Ethics:
http://www.socialworkers.org/pubs/code/code.asp

Student Press Law Center information on the Jennifer Keeton case, including a link to the ruling:
http://www.splc.org/news/newsflash.asp?id=2403

References

American Association for Marriage and Family Therapy (2012). *AAMFT Code of Ethics*. Alexandria, VA: AAMFT.

American Psychological Association. (1993, October). *Task force on promotion and dissemination of psychological procedures: A report adopted by the Division 12 Board*. Retrieved August 24, 2008, from www.apa.org/divisions/div12/journals.html

Anderson, H. (1997). *Conversations, language, and possibilities: A postmodern approach to therapy*. New York: Basic Books.

Babcock, J. C., Green, C. E., & Robie, C. (2004). Does batterers' treatment work? A meta-analytic review of domestic violence treatment. *Clinical Psychology Review, 23*(8), 1023–1053.

Baldwin, S., Christian, S., Berkeljon, A., & Shadish, W. (2012). The effects of family therapies for adolescent delinquency and substance abuse: A meta-analysis. *Journal of Marital and Family Therapy, 38,* 281–304.

Barak, A., Hen, L., Boniel-Nissim, M., & Shapira, N. (2008). A comprehensive review and a meta-analysis of the effectiveness of internet-based psychotherapeutic interventions. *Journal of Technology in Human Services, 26*(2/4), 109–160.

Beach, S., & Whisman, M. (2012). Affective disorders. *Journal of Marital and Family Therapy, 38,* 201–219.

Beauchamp, T. L., & Childress, J. F. (2009). *Principles of biomedical ethics* (6th ed.). New York: Oxford University Press.

Blow, A. J., Sprenkle, D. H., & Davis, S. D. (2007). Is who delivers the treatment more important than the treatment itself? *Journal of Marital and Family Therapy, 33,* 298–317.

Chambless, D. L., & Hollon, S. D. (1998). Defining empirically supported therapies. *Journal of Consulting and Clinical Psychology, 66,* 7–18.

Chambless, D. L., Sanderson, W. C., Shoham, V., Johnson, S. B., Pope, K. S., Crits-Christoph, P., Baker, M., Johnson, B., Woody, S. R., Sue, S., Beutler, L., Williams, D. A., & McCurry, S. (1996). An update on empirically validated treatments. *Clinical Psychologist, 49*(2), 5–18. Available from www.apa.org/divisions/div12/journals.html

Chenail, R., St. George, S., Wulff, D., Duffy, M., Scott, K., & Tomm, K. (2012). Clients' relational conceptions of conjoint couple and family therapy quality: A grounded formal theory. *Journal of Marital and Family Therapy, 38,* 241–264.

Duncan, B. L., Miller, S. D., Sparks, J. A., Claud, D. A., Reynolds, L. R., Brown, J., & Johnson, L. D. (2003). Session Rating Scale: Preliminary psychometrics of a "working" alliance scale, *Journal of Brief Therapy, 3,* 3–12.

Gehart, D. R. (2014). *Mastering competencies in family therapy* (2nd ed.). Pacific Grove, CA: Cengage.

Henggeler, S., & Sheidow, A. (2012). Empirically supported family-based treatments for conduct disorder and delinquency in adolescents. *Journal of Marital and Family Therapy, 38,* 30–58.

Hertlein, K. M., & Webster, M. (2008). Technology, relationships, and problems: A research synthesis. *Journal of Marital and Family Therapy, 34,* 445–460.

Johnson, S. M. (2004). *The practice of emotionally focused couple therapy: Creating connection* (2nd ed.). New York: Guilford Press.

Jose, A., & O'Leary, K. D. (2009). Prevalence of partner aggression in representative and clinic samples. In K. D. O'Leary & E. M. Woodin (Eds.), *Psychological and physical aggression in couples: Causes and interventions* (pp. 15–35). Washington, DC: American Psychological Association.

Karam, E., & Sprenkle, D. (2010). The research-informed clinician: A guide to training the next-generation MFT. *Journal of Marital and Family Therapy, 36,* 307–319.

Kaslow, N., Broth, M., Smith, C., & Collins, M. (2012). Family-based interventions for child and adolescent disorders. *Journal of Marital and Family Therapy, 38,* 82–100.

Lambert, M. (1992). Psychotherapy outcome research: Implications for integrative and eclectic therapists. In J. C. Norcross & M. R. Goldfried (Eds.), *Handbook of psychotherapy integration* (pp. 94–129). New York: Wiley.

Lebow, J., Chambers, A., Christensen, A., & Johnson, S. (2012). Research on the treatment of couple distress. *Journal of Marital and Family Therapy, 38,* 145–168.

Lucksted, A., McFarlane, W., Downing, D., & Dixon, L. (2012). Recent developments in family psychoeducation as an evidence-based practice. *Journal of Marital and Family Therapy, 38,* 101–121.

Markman, H., & Rhoades, G. (2012). Relationship education research: Current status and future directions. *Journal of Marital and Family Therapy, 38,* 169–200.

Miller, S. D., Duncan, B. L., & Hubble, M. (1997). *Escape from Babel: Toward a unifying language for psychotherapy practice.* New York: Norton.

Moon, S. M., Dillon, D. R., & Sprenkle, D. H. (1990). Family therapy and qualitative research. *Journal of Marital and Family Therapy, 16,* 357–373.

Patterson, J. E., Miller, R. B., Carnes, S., & Wilson, S. (2004). Evidence-based practice for marriage and family therapies. *Journal of Marital and Family Therapy, 30,* 183–195.

Piercy, F. (2012). It is not enough to be busy. *Journal of Marital and Family Therapy, 38,* 1–2.

Rowe, C. (2012). Family therapy for drug abuse: Review and updates 2003–2010. *Journal of Marital & Family Therapy, 38,* 59–81.

Shields, C., Finley, M., Chawla, N., & Meadors, P. (2012). Couple and family interventions in health problems. *Journal of Marital and Family Therapy, 38,* 265–280.

Shadish, W. R., & Baldwin, S. A. (2002). Meta-analysis of MFT interventions. In D. H. Sprenkle (Ed.), *Effectiveness research in marriage and family therapy* (pp. 339–370). Alexandria, VA: American Association for Marriage and Family Therapy.

Sorrel, A. L. (2009). Doctor gets jail time for online, out-of-state prescribing. *American Medical News,* June 8. Retrieved from www.ama-assn.org/amednews/2009/06/01/prsd0601.htm

Sprenkle, D. H. (Ed.). (2002). Editor's introduction. In D. H. Sprenkle (Ed.), *Effectiveness research in marriage and family therapy* (pp. 9–25). Alexandria, VA: American Association for Marriage and Family Therapy.

Sprenkle, D. H., & Blow, A. J. (2004). Common factors and our sacred models. *Journal of Marital and Family Therapy, 30,* 113–129.

Sprenkle, D. H., Davis, S. D., & Lebow, J. (2009). *Beyond our sacred models: Common factors in couple, family, and relational psychotherapy.* New York: Guilford.

Sprenkle, D. (Ed.). (2012). Intervention research in couple and family therapy [Special edition]. *Journal of Marital and Family Therapy, 38*(1).

Stith, S., McCollum, E., Amanor-Boadu, Y., & Smith, D. (2012). Systemic perspectives on intimate partner violence treatment. *Journal of Marital and Family Therapy, 38,* 220–240.

Tallman, K., & Bohart, A. C. (1999). The client as a common factor: Clients as self-healers. In M. A. Hubble, B. L. Duncan, & S. D. Miller (Eds.), *The heart and soul of change: What works in therapy* (pp. 91–131). Washington, DC: American Psychological Association.

Wampold, B. E. (2001). *The great psychotherapy debate: Models, methods, and findings.* Mahwah, NJ: Erlbaum.

Woody, S. R., Weisz, J., & McLean, C. (2005). Empirically supported treatments: 10 years later. *Clinical Psychologist, 58,* 5–11.

CHAPTER 3

Theory-Specific Case Conceptualization and Treatment Planning

Treatment + Plan = ???

It was the first day at my new training site, one of the best in town. During the orientation, my supervisor went over the clinical paperwork that we would need to complete: intake forms, assessments, and treatment plans. Although I had learned about gathering intake and assessment information in my diagnosis class, I had never actually seen a treatment plan. Using sophisticated etymological skills that all graduate students rely on—treatment+plan—I deduced that this document would somehow describe my plans for how to treat the client. But how?

Like many interns, I was too embarrassed to ask and kept my ignorance quiet. I decided that if I could find a sample, I could probably fake it and not have to risk looking inexperienced (which I was) in the eyes of my well-respected supervisor. Thankfully, I was assigned cases handled by previous interns that included this mysterious document. So, I gathered up as many files as I could and found an uncomfortable yet well-worn seat in a poorly lit corner assigned to interns and tried to crack the code. *That* is how I learned to do treatment plans: secretly and shamefully. Thankfully, because you have this book in your hands, you have the opportunity to learn this once secret art with more dignity—and without a backache.

Just in case you missed it, the moral of the above story is to talk with your supervisors—no matter how brilliant they may appear—and do not be afraid to ask what seems like a silly question. Trust me, the chair was not the hardest part of learning treatment planning on my own.

Distinguishing a Therapist from Hairdresser: Theory-Specific Case Conceptualization

The first step in writing a good treatment plan is having a solid *case conceptualization*. When you really boil things down, one of the primary differences between a therapist, hairdresser, bartender, and good friend is how the person *views* the problem situation.

Hairdressers, bartenders, and good friends can be empathetic neutral third parties who listen well. But therapists do something different: they use theory to conceptualize the client's situation in ways that make it easier to improve the situation. That is the main purpose of theory. I promise you, once you have a clear case conceptualization, what to *do* in session is surprisingly clear.

Therapeutic case conceptualizations are developed from therapeutic theory. Each model has a different approach to case conceptualization; these differences are the primary distinctions between theory. The core of a theory-specific case conceptualization involves using the key concepts in a particular approach to view and understand a client's situation. For example, a structural family therapist (Chapter 7) assesses the family structure—boundaries, hierarchy, subsystems—to determine where and how to intervene. In contrast, a systemic family therapist (Chapter 5) assesses a family's interactional sequence—the rise of tension, the symptom, and return to "normal"—and uses this information to determine where in the sequence to intervene. Both forms of case conceptualization can be used to construct useful treatment plans for their respective theory. To use any theory effectively, it is essential to develop a thorough theory-specific case conceptualization.

Each theory chapter in this book includes a section called "The Viewing: Case Conceptualization." The concepts in this section can be used to create a narrative (written out) case conceptualization for that theory using the template and guides from the "Putting It All Together" sections. Finally, the end of each chapter includes a sample case conceptualization to get you started. In addition, Chapter 17 of this book includes a cross-theoretical case conceptualization that allows you to assess your case from multiple theoretical perspectives to create a deeper and richer understanding of the situation.

Writing a Theory-Specific Case Conceptualization

Taking the time to write a theory-specific case conceptualization is the most important and difficult step in the treatment planning process, making it the one most new therapists would prefer to skip. Why? It requires significant reflective and critical thinking; it is *not* formulaic. It will take some time to apply theoretical concepts to your client's actual situation. You will struggle. It most often will *not* be as clear as in the textbook (painful truth: nothing in life is). But in that struggle, you will actually learn how to use theory. You will become a competent therapist. Thankfully, to develop a theory-specific case conceptualization you only need to use the concepts described in "The Viewing: Case Conceptualization" sections of the theory in question. At the end of each theory chapter you will find a detailed example to give you a clue as to what the final product might look like. Nonetheless, applying the theory to your client will probably look quite different.

After completing the conceptualization, step back and ask yourself the following question:

- What two to three key patterns emerged in the case conceptualization?

These key themes will be the focus of the working phase and possibly closing-phase goals in the treatment plan below. Any potential crisis issues will be the focus of your initial phase goals. Closing-phase goals may also be informed by the specific approach's theory of health and normalcy if the theory has such elements. For example, some theories, such as structural therapy (Chapter 7), Satir's approach (Chapter 8), and the Bowen intergenerational therapy (Chapter 10), have theoretically defined long-term goals, such as clear interpersonal boundaries, differentiation, or self-actualization. When using such therapeutic models, therapists begin with these larger goals in mind and design the middle-phase goals to prepare clients for the closing-phase goals.

Identifying a Path: Treatment Planning

After completing your theory-specific case conceptualization, you are ready to develop a plan for addressing the problems you have identified in these other two documents. Treatment plans are fun: they are filled with hopes and dreams. In creating them, you have tremendous freedom but also the burden of responsibility. Because numerous good plans can be developed for any one client, you may choose which theory and techniques that are the best fit for a specific client, the specific problem, and the particular therapist-client relationship. As the therapist, you are responsible for shepherding an effective process and selecting a plan that is most likely to help the client; this plan should be based on clinical experience, current research, and standards of practice.

A Brief History of Mental Health Treatment Planning

The history of treatment planning in the field of marriage and family therapy is relatively short. The original theorists did not talk or write about treatment planning, and, in fact, if you search the literature you will find no form that would be accepted by a managed care company or county mental health agency for payment. If the approved approach to treatment planning did not come from the field of family therapy or even mental health more broadly, where did it come from? The short answer: the medical field.

Symptom-Based Treatment Plans

The type of treatment planning that most marriage and family therapists must complete to receive third-party payment and to maintain standard practice of care in the 21st century is derived from the medical model. Jongsma and his colleagues (Dattilio & Jongsma, 2000; Jongsma, Peterson, & Bruce, 2006; Jongsma, Peterson, McInnis, & Bruce, 2006; O'Leary, Heyman, & Jongsma, 1998) have developed the most extensive models. Referred to as *symptom-based treatment plans,* these documents focus solely on clients' medical symptoms. Most publications on treatment planning use a similar symptom-based model (Johnson, 2004; Wiger, 2005). But although these plans are relevant to those in the medical community, they do not help therapists conceptualize treatment in the most useful ways.

For example, if a parent brings a child to therapy who is having tantrums and the therapist develops a plan around the presenting problem (e.g., "reduce child's tantrums to less than one per week") and proceeds to deal directly with the tantrums without thoroughly conceptualizing the case, treatment is less likely to be successful. A systemic assessment will typically reveal that marital and/or parenting issues are contributing to the presenting problem, and couples therapy that targets tension in the marriage may actually be the best way to reduce the child's tantrums. The danger of symptom-based treatment planning is that the therapist will underutilize theory, focus on symptoms, and forget to assess the larger picture. Arguably, a good therapist would not do this; however, today's workplace realities include (a) heavy caseloads, (b) pressure to complete diagnosis and treatment plans by the end of the first session, and (c) highly structured paperwork and payment systems, all of which make it hard for a therapist to do a good job. Thus, symptom-based treatment planning, although convenient, may not be the best choice for today's practice environments.

Theory-Based Treatment Plans

Theory-based treatment planning, described by Gehart and Tuttle (2003), uses theory to create more clinically relevant treatment plans than the symptom model offers. Berman (1997) developed a similar approach for traditional psychotherapies. Both models include goals that are informed by clinical theories. However, I found that new trainees confuse theory-based goals and interventions because they use the same language. Furthermore, it is difficult for most students to address diagnostic issues and clinical symptoms in these theory-based plans because the language of these two systems is radically different. The solution was to develop a new, "both/and" model, called the "clinical treatment plan," that draws from the best of theory-based and symptom-based treatment plans and adds elements of measurability.

Clinical Treatment Plans

Clinical treatment plans provide a straightforward, comprehensive overview of treatment. In this text, I use a simplified model of the more extensive plans used in *Mastering Competencies in Family Therapy* (Gehart, 2010). The treatment plans in this book are intended for those first learning to do treatment planning, even before seeing clients in some cases (you are the lucky ones), whereas the treatment plans in *Mastering Competencies in Family Therapy* include details that are more relevant for those working with clients.

They include the following parts:

- *Therapeutic Tasks:* Describes treatment tasks that the therapist should perform at the initial, working, and closing phases of therapy. These tasks are informed by theory as well as ethical and legal requirements.
- *Client Goals:* Client goals are derived from the case conceptualization and are stated in theory-specific language. As goals, they should describe how specific relational dynamics will *change*. To help you keep this in the forefront of your mind, I start each with *decrease/increase*.
- *Interventions:* You should identify two to three interventions for achieving this goal using the therapist's chosen theory.

You will find completed examples of these at the end of each theory chapter. The basic template looks like this:

Treatment Plan

Therapist:_____   Case/Client:_____

Date:_____   Theory:_____

Initial Phase of Treatment
Initial Phase Counseling Tasks

1. Develop working counseling relationship. *Diversity note: [Describe how you will adjust to respect cultured, gendered, and other styles of relationship building and emotional expression.]*

   Relationship building approach/intervention:

   a. _____

2. Assess individual, systemic, and broader cultural dynamics. *Diversity note: [Describe how you will adjust assessment based on cultural, socioeconomic, sexual orientation, gender, and other relevant norms.]*

   Assessment strategies:

   a. _____

   b. _____

(continued)

Initial Phase Client Goals: Manage crisis, reduce distressing symptoms

1. ❑ Increase ❑ Decrease _____ (personal/relational dynamic using terms from theory) to reduce _____ (symptom).

   Interventions:

   a. _____

   b. _____

Working Phase of Treatment
Working-Phase Counseling Task

1. Monitor quality of the working alliance. *Diversity note: [Describe how you will attend to client response to interventions that indicate therapist using expressions of emotion that are not consistent with client's cultural background.]*

   a. *Assessment Intervention:* _____

Working-Phase Client Goals (Two to Three Goals). Target individual and relational dynamics using theoretical language (e.g., decrease avoidance of intimacy, increase awareness of emotion, increase agency).

1. ❑ Increase ❑ Decrease _____ (personal/relational dynamic using terms from theory) to reduce _____ (symptom).

   Interventions:

   a. _____

   b. _____

2. ❑ Increase ❑ Decrease _____ (personal/relational dynamic using terms from theory) to reduce _____ (symptom).

   Interventions:

   a. _____

   b. _____

3. ❑ Increase ❑ Decrease _____ (personal/relational dynamic using terms from theory) to reduce _____ (symptom).

   Interventions:

   a. _____

   b. _____

(continued)

Closing Phase of Treatment

Closing-Phase Counseling Task

1. Develop aftercare plan and maintain gains. *Diversity note: [Describe how you will access resources in the communities of which they are a part to support them after ending therapy.]*

    Intervention:

    a. _____

Closing-Phase Client Goals (*One to Two Goals*): Determined by theory's definition of health and normalcy

1. ❑ Increase ❑ Decrease _____(personal/relational dynamic using terms from theory) to reduce _____(symptom).

    Interventions:

    a. _____

    b. _____

2. ❑ Increase ❑ Decrease _____(personal/relational dynamic using terms from theory) to reduce _____(symptom).

    Interventions:

    a. _____

    b. _____

Writing Useful Therapeutic Tasks

Therapeutic tasks are included primarily to help therapists fully conceptualize these essential tasks early in their careers (although most older folks benefit from a reminder of basic competencies too). However, these will typically not be included in plans you send to insurance companies or third party payers. That said, they are an excellent thing to have documented if there are ever any legal or ethical questions about your performance.

As I believe in truth in advertising, therapeutic tasks are generally the easiest part of the treatment plan to develop because they are the most formulaic. Each theory has its own language and interventions for describing how to approach treatment tasks, such as creating a therapeutic relationship, and a good plan should reflect these differences. For example, a Bowen intergenerational therapist focuses on remaining nonreactive to clients, whereas a therapist using experiential family therapy has a more emotionally engaged approach to creating a therapeutic relationship. In addition, treatment tasks are one of the key places where therapists must adjust their approach to address diversity issues; as such, there is a place for you to indicate how you will do this.

Initial Phase

Perhaps not surprisingly, therapists have the most tasks in the initial phase of treatment because this is when they establish the foundation for therapy. Virtually all theories include at least two essential therapeutic tasks early in therapy, as shown in the form already presented:

- Establish a therapeutic relationship
- Assess individual, family, and social dynamics

Although each theoretical approach has different ways to do these two things, the cross-theory similarities make it easy for therapists to conceptualize this early phase of treatment. If problems arise in treatment, therapists can be sure that one of these initial tasks needs to be readdressed, most often the therapeutic relationship (thus, it is continued into the working phase; Batchelor & Horvath, 1999; Orlinsky, Rønnestad, & Willutzki, 2004).

Working Phase

In the working phase, the primary therapeutic task is to monitor the quality of therapeutic relationship as it is key to progress. If dissonance develops in the relationship or if the client is not making progress, the therapist needs to examine what might be the reason and how therapy can be adjusted to accomplish those goals. Family therapists rarely cite "resistance" as a valid reason for lack of progress. Steve de Shazer (1984) boldly declared the "death of resistance," arguing that lack of progress can never be blamed on the client. If initial interventions don't work, the therapist does not need to blame clients or himself/herself and instead should simply focus on figuring out what does work, and this most always requires attending to the quality and nuances of the therapeutic relationship. Similarly, systemic therapists have consistently admonished that, if clients are not progressing, the therapist has not yet developed a useful working hypothesis or found a way to usefully deliver it to clients (Selvini Palazzoli, Boscolo, Cecchin, & Prata, 1978; Watzlawick, Weakland, & Fisch, 1974). In the systemic approach, each "failed" intervention tells the therapist what doesn't work and, therefore, provides clues as to what might work and how to repair the therapeutic relationship.

Closing Phase

Put in bare-bones language, the primary task of the closing phase is for therapists to make themselves unnecessary in clients' lives (helps to be humble in this profession). During this phase of therapy, therapists work with clients to develop aftercare plans that include identifying (a) what they did to make the changes they have made, (b) how they will maintain their success, and (c) how they will handle the next set of challenges in their lives. Each therapeutic model has different ways of doing this, but there is a cross-theoretical consistency as well as a consistent logic to this task that is useful in most therapeutic situations. When this phase is done well, clients leave therapy feeling better able to handle the inevitable problems that will continue to arise in their lives.

Diversity and Treatment Tasks

For each treatment task, you should also note how you will address diversity issues, such as culture, ethnicity, race, sexual orientation, gender orientation, religion, language, ability, age, gender, etc. For example, with almost every client, you will need to slightly significantly adjust your relationship style for some form of diversity, such as age, ethnicity, or gender. Similarly, any assessment of functioning should always take into consideration diversity variables.

Here are examples of how diversity is addressed in treatment tasks:

- Use of humor with teens and men.
- More formal, respectful relational style with immigrants or clients from ethnic backgrounds that prefer such relations with professionals (*respecto* with Latino clients).

- Use of *personalismo* with Hispanic/Latino clients.
- Including spirituality and religious beliefs and resources, including connecting with priests, ministers, rabbis, or spiritual healers, or including specific religious practices that provide comfort or support.
- Including more extended family or tribe members with clients who come from backgrounds where the extended family/tribal system is the primary system.
- Use of present-focused, problem-focused approaches with clients who do not value exploring the past.
- Using interventions, assessments, and questions that enable Asian clients to avoid shameful discussions.
- Include culturally appropriate resources and persons in therapy.
- Assessing family-of-choice with gay, lesbian, bisexual, or transgendered clients.
- Accounting for the stress of marginalization and discrimination in assessment.

Writing Useful Client Goals

I want to be upfront from the start: writing useful client goals is a difficult task, *the most difficult part of writing a treatment plan and also the most meaningful*. The key is that they can be written only after one has done a thorough case conceptualization. To prepare for the challenges of writing client goals, you may want to try using the following Goal Writing Worksheet, which helps you link together the following:

- Client's reported problem
- Problematic relational dynamic identified in theoretical case conceptualization
- Psychiatric or relational symptoms

I promise you—in any given situation, you should be able to find two to four core dynamics that line up all three of these. Sometimes it takes awhile, but when you get it, all the seemingly random stories clients bring in from week to week all of a sudden seem clearly connected—and you will have a good idea of how to help.

Goal Writing Worksheet

Presenting Problem: These are used to help identify problem *dynamics*.

What does the client say is the problem(s)? Use the client's words and phrases as much as possible.

1. _____
2. _____
3. _____

Systemic Dynamic: These are used to write your *goals*.

Develop a case conceptualization based on "The Viewing: Case Conceptualization" section of each theory chapter. Identify two to four of the most salient problematic relational dynamics or discourse from the case conceptualization; these are the dynamics that are most likely to be contributing to the client's presenting problem. In some cases, you will see that certain dynamics overlap or are related; in these situations, try to summarize these overlapping dynamics into one point below.

1. _____
2. _____
3. _____

Symptoms

Identify two to four of the most salient psychological symptoms or issues (e.g., depression, anxiety, substance use, conflict with loved ones, isolation, loss of interest, hallucinations). List these below.

1. _____
2. _____
3. _____

Put It All Together

This keeps you honest: Do all the pieces fit together?

Presenting Problem	Dynamic	Symptom
1.		
2.		
3.		

Note: If for any reason you have symptoms that don't seem to be related to the dynamics you chose, review your case conceptualization again. The pieces should fit together.

Evidence-Base Practice (Optional): Helps you determine theory and technique

Use PsychInfo or a similar search engine to do a review of the research literature related to (a) the client's presenting problem, (b) diagnosis, (c) personal demographics/diversity factors, and/or (e) your intended therapy approach. Describe the key interventions, techniques, or guidelines below.

1. _____
2. _____
3. _____

Based on the most salient dynamics and evidence base as well as your client's needs, which theory and/or techniques do you plan to use with this case?
_____

The Goal Writing Process

The hardest part is always writing the goal. There are no clear rules for every situation because each client is unique. However, goal writing has three basic components:

Guidelines for Writing Useful Goals

1. *Start with a Key Concept or Assessment Area from the Theory of Choice:* Start with "increase" or "decrease" followed by a description using language from the chosen theory about what is going to change.
2. *Link to Symptoms:* Describe what symptoms will be addressed by changing the personal or relational dynamic.

(*continued*)

3. *Use the Client's Name:* Using a name (or equivalent confidential notation) ensures that it is a unique goal rather than a formulaic one.

Anatomy of a Client Goal

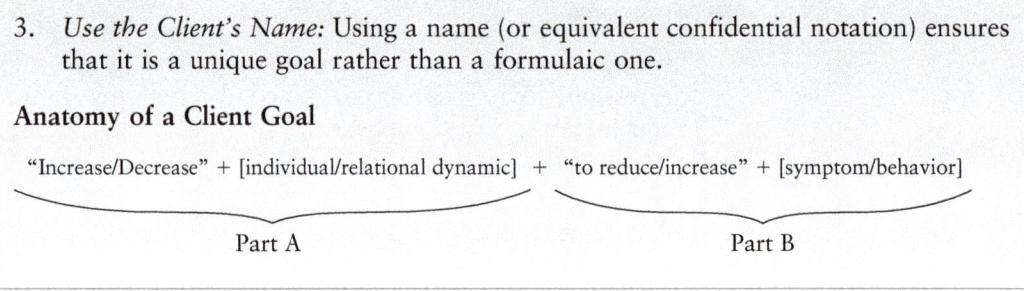

Part A gives the therapist a clear focus of treatment that fits with the theory of choice. Part B clearly links the changes in symptoms to the focus of treatment stated in A. Examples:

- *Increase* effectiveness of parental hierarchy between AF and CF *to reduce* frequency of CF's tantrums per week (structural therapy).
- *Reduce and interrupt* pursuer-distancer pattern between AF and AM *to reduce* AF's sense of hopelessness and AM's irritability (systemic therapy).
- *Increase* frequency of social interaction and reengagement in music and sports hobbies *to increase* periods of positive mood (solution-focused therapy).

Each part has a different function. Part A is most useful to therapists for conceptualizing treatment; Part B is most useful to third-party payers who require a medical model assessment. When therapists write goals that address both A and B, they allow themselves maximum freedom and flexibility to work in their preferred way while also answering the needs of third-party payers.

Initial Phase Client Goals

During the initial phase of therapy—in most cases the first one to three sessions—client goals generally involve stabilizing crisis symptoms, such as suicidal and homicidal thinking, severe depressive or panic episodes, and poor eating and sleeping patterns; managing child, dependent adult, and elder abuse issues; addressing substance and alcohol abuse issues; and stopping self-harming behaviors such as cutting.

In addition to stabilizing crisis issues, some theories have specific clinical goals that should be addressed in the initial phases. For example, solution-based therapists begin working on clinical symptoms in the first session by setting small, measurable goals toward desired behaviors as well as increasing clients' level of hope (O'Hanlon & Weiner-Davis, 1989).

Working-Phase Client Goals

Working-phase goals address the dynamics that create and/or sustain the symptoms and problems for which clients came to therapy. These are the goals that most interest third-party payers. The secret to writing good working-phase goals is framing them *in the theoretical language* used for conceptualization and then linking this language to the psychiatric symptoms. Using theoretical language enables therapists to document a coherent treatment using their preferred language of conceptualization rather than language that is geared for those prescribing medication.

For example, when a client is diagnosed with depression, many therapists include a goal such as "reduce depressed mood." Let's not kid ourselves: a person does not need a Master's degree and thousands of hours of training to come up with such a goal. This medical-model, symptom-based goal does not provide clues as to what the therapist will actually do. Furthermore, all documentation for the case will need to monitor the client's level of depression each week. In contrast, a clinical client goal should address the theoretical conceptualization that will guide the reduction of depression. For example:

- *Structural Therapy:* Reduce enmeshment with children and increase parental hierarchy to reduce episodes of depressed mood.

- *Satir's Communication Approach:* Reduce placating behaviors in marriage and at work to increase congruent communication and positive mood.
- *Narrative Therapy:* Reduce influence of family and societal evaluations of self-worth to increase sense of autonomy and reduce depressed mood.

Each of these goals addresses a client's depressed mood and provides a clear clinical conceptualization and sense of direction, being much more useful to therapists than the medical goal of "reduce depressed mood."

Closing-Phase Client Goals

Closing-phase client goals address (a) larger, more global issues that clients bring to therapy and/or (b) move the client toward greater "health" as defined by the therapist's theoretical perspective. As an example of the first type, clients may present with one issue, perhaps marital discord, and then later in therapy want to address their parenting issues. Similarly, often clients may present with depression or anxiety and in the later phase want to address relationship issues or an unresolved issue with their family of origin. A couple may present with several pressing issues, such as conflict and sexual concerns to be treated in the working phase, and in the later phase want to examine more global issues, such as redefining their identities and relational agreements.

The second type of client goal is driven by the therapist's agenda. Some approaches, such as Bowen intergenerational, humanistic, and structural therapies, have clearly defined theories of health that therapists work toward. Other approaches, such as systemic and solution-focused therapies, have less clearly defined long-term goals and theories. Closing-phase goals often include an agenda item that clients may not have verbalized. For example, differentiation, a long-term goal that is embedded in the theory of Bowenian intergenerational therapy, is too theoretical for clients to present.

Writing Useful Interventions

The final element of treatment plans is including interventions to support each therapeutic task or client goal. Once treatment has been conceptualized and therapeutic tasks and client goals have been identified, identifying useful interventions is generally quite easy. The interventions should come from the therapist's chosen theory and be specific to the client. The following points are guidelines for writing interventions:

> ### Guidelines for Writing Interventions
>
> - *Use Specific Interventions From Chosen Theory:* Interventions should be clearly derived from the theory used to conceptualize therapeutic tasks and client goals. If an intervention from another theory is integrated, the modifications should be clearly spelled out. For example, genograms can be adapted to solution-focused therapy (Kuehl, 1995).
> - *Make Interventions Specific to Client:* Use confidential notation (e.g., AF for adult female and AM for adult male) to make the goal as specific and clear as possible: for example, "Sculpt AF's pattern of pursuing and AM's tendency to withdraw."
> - *Include Exact Language When Possible:* Whenever possible, therapists should use the exact question or language they will use to deliver the intervention; for example, "On a scale of 1 to 10, how would you rate your current level of satisfaction in the marriage?"

Do Plans Make a Difference?

Of course, therapy rarely goes according to plan. Life happens; new problems arise; original problems lose their importance; new stressors change the playing field. But that does not make plans useless. Treatment plans help therapists in numerous ways:

- They help therapists think through which dynamics need to be changed and how.
- They provide therapists with a clear understanding of the client situation so that they can quickly and skillfully address new crisis issues or stressors.
- They give therapists a sense of confidence and clarity of thought that make it easier to respond to new issues.
- They ground therapists in their theory and in their understanding of how their theory relates to clinical symptoms.

All this is to say, do not be surprised when therapy does not go according to plan: instead, expect it. And know that the time you took to create a treatment plan makes you much better able to respond to the unplanned.

Online Resources

Symptom-Based Treatment Planners:
www.jongsma.com

Theory-Based Treatment Planning:
www.cengage.com/brookscole

References

Batchelor, A., & Horvath, A. (1999). The therapeutic relationship. In M. A. Hubble, B. L. Duncan, & S. D. Miller (Eds.), *The heart and soul of change* (pp. 133–178). Washington, DC: APA Press.

Berman, P. S. (1997). *Case conceptualization and treatment planning.* Thousand Oaks, CA: Sage.

Datilio, F. M., & Jongsma, A. E. (2000). *The family therapy treatment planner.* New York: Wiley.

de Shazer, S. (1984). The death of resistance. *Family Process, 23,* 11–17.

Gehart, D. (2010). *Mastering competencies in family therapy: A practical approach to theory and clinical case documentation.* Pacific Grove, CA: Brooks/Cole.

Gehart, D. R., & Tuttle, A. R. (2003). *Theory-based treatment planning for marriage and family therapists: Integrating theory and practice.* Pacific Grove, CA: Brooks/Cole.

Johnson, S. L. (2004). *Therapist's guide to clinical intervention: The 1-2-3's of treatment planning* (2nd ed.). San Diego, CA: Academic Press.

Jongsma, A. E., Peterson, L. M., & Bruce, T. J. (2006). *The complete adult psychotherapy treatment planner* (4th ed.). New York: Wiley.

Jongsma, A. E., Peterson, L. M., McInnis, W. P., & Bruce, T. J. (2006). *The child psychotherapy treatment planner* (4th ed.). New York: Wiley.

Kuehl, B. P. (1995). The solution-oriented genogram: A collaborative approach. *Journal of Marital and Family Therapy, 21,* 239–250.

O'Hanlon, W. H., & Weiner-Davis, M. (1989). *In search of solutions: A new direction in psychotherapy.* New York: Norton.

O'Leary, K. D., Heyman, R. E., & Jongsma, A. E. (1998). *The couples psychotherapy treatment planner.* New York: Wiley.

Orlinsky, D. E., Rønnestad, M. H., & Willutzki, U. (2004). Fifty years of process-outcome research: Continuity and change. In M. J. Lambert (Ed.), *Bergin and Garfield's handbook of psychotherapy and behavior change* (5th ed., pp. 307–393). New York: Wiley.

Selvini Palazzoli, M., Boscolo, L., Cecchin, G., & Prata, G. (1978). *Paradox and counterparadox.* New York: Jason Aronson.

Watzlawick, P., Weakland, J., & Fisch, R. (1974). *Change: Principles of problem formation and problem resolution.* New York: Norton.

Wiger, D. E. (2005). *The psychotherapy documentation primer* (2nd ed.). New York: Wiley.

CHAPTER 4

Philosophical Foundations of Family Therapy Theories

Lay of the Land

Before exploring the various models of family therapy, I want to briefly introduce you to their philosophical foundations. The two closely related philosophical traditions that inform family therapy approaches are *systems theory* and *social constructionism* (a particular strand of postmodernism). To some degree or another, all schools of family therapy have been influenced by these two theories, with traditional therapies drawing more heavily from systemic theory and more recent ones from social constructionist theory. The last portion of this chapter will address the question: who to marry and whom to date in the world of theory.

Systemically Influenced Family Therapies and Theories

- *Systemic and Strategic Theories:* Mental Research Institute (MRI), Milan, strategic therapies, and functional family therapy (Chapters 5, 6, and 15)
- *Structural Family Therapy* (Chapter 7)
- *Experiential Family Therapy Theories:* Satir's growth model, symbolic—experiential therapy, internal family systems, and emotionally focused couples therapy (Chapters 8, 9, and 15)
- *Intergenerational Theories:* Bowen's intergenerational and psychoanalytic therapies (Chapter 10)
- *Cognitive-Behavioral Family Therapies* (Chapter 11)
- *Early Solution-Based Therapies* (Chapter 12)

Social Constructionist Family Therapies

- *Later Solution-Based Therapies* (Chapter 12)
- *Narrative Therapy* (Chapter 13)
- *Collaborative Therapy and Reflecting Teams* (Chapter 14)

Systemic Foundations

Rumor Has It: The People and Their Stories

The Macy Conferences

Not so long ago (1940) in a place not so far away (New York), Josiah Macy (of Macy's department store fame) assembled an unexpected configuration of scholars and researchers to discuss how groups of things operate to form a system (Segal, 1991). This series of conferences in the early 1940s, the Macy Conferences, gave birth to *general systems theory* and *cybernetic systems theory*, which describe how biological, social, and mechanical systems operate. Rather than being developed by a single person, these ideas emerged from interactive dialogue and shared research that involved numerous cutting-edge experts and scholars. Their theories led to a new approach to psychotherapy—family therapy—that is not simply a modality (i.e., working with a family versus an individual) but a unique philosophical view of human behavior.

Systemic Theorists

> "The concepts that constitute Communication or Interactional [Systems] Theory emerged not from any one individual, but, rather were the product of the interaction between the members of what has become known as the Palo Alto Group [the Bateson Team]." —Weakland, 1988, p. 58

Gregory Bateson Gregory Bateson, who participated in the Macy Conferences with his then wife Margaret Mead, was a British anthropologist who explored cybernetic theory by studying intertribal interactions in New Guinea and Bali (Bateson, 1972, 1979, 1991; Mental Research Institute, 2002). Bateson's elegant and thoughtful articulations of cybernetic theory influenced numerous disciplines, including communications, anthropology, and family therapy. As part of his research on human communications, he assembled what was later known as the *Bateson Group*: Don Jackson, Jay Haley, William Fry, and John Weakland. For 10 years, he studied communication in families with members diagnosed with schizophrenia, provided consultation on cybernetic theory, and introduced team members to the trance work of Milton Erickson. The result was the *double-bind theory of schizophrenia* (Bateson, 1972), which reconceptualized psychotic behavior as an attempt to meaningfully respond in a family system characterized by double-bind communications. Bateson's prior anthropological research helped the team view problematic human behavior as a function of larger social systems rather than being purely intrapsychic.

Heinz von Foerster Another participant of the Macy Conferences, Heinz von Foerster was born in Austria and originally studied physics before developing his theories on cybernetic systems, second-order cybernetics, and radical constructivism, a postmodern theory that describes how an individual constructs his/her reality (Mental Research Institute, 2002). His work also contributed to the philosophical foundation for systemic therapies.

Milton Erickson Trained in medicine as a psychiatrist, Erickson was a master therapist, well known for his brief, rapid, and creative interventions and considered by many to be the father of modern hypnosis (Erickson & Keeney, 2006; Mental Research Institute, 2002). The early MRI team consulted with Erickson as they developed their brief approach to family therapy. Erickson's clinical innovations, brief approach, and emphasis on possibilities are reflected in their therapies. Erickson's work was also highly influential in the development of solution-based therapies (Chapter 12).

Bradford Keeney A family therapist, Keeney studied with Bateson in exploring the implications of cybernetics of cybernetics, or second-order cybernetics, which acknowledges the role of the observer (e.g., the therapist) on what is observed (Keeney, 1983, 1985). In his more recent anthropological work he has studied shamanism and the cybernetic (i.e., holistic) worldviews of native cultures, most notably the Kalahari Bushmen (Keeney, 1994, 1997, 1998, 2000a, 2000b, 2001a, 2001b, 2002a, 2002b, 2003). He has also used concepts from improvisational theater to develop *improvisational therapy* (Keeney, 1990) and, with his colleague Wendell Ray, has also developed *resource-focused therapy*, a strength-focused, systemic approach (Ray & Keeney, 1994).

The Spirit of Systemic Therapy

Systemic therapists are known for their ability to see the big picture at all times. Even when a client presents with individual issues, such as depression or anxiety, systemic therapists always view it within the larger relational contexts in which the symptom makes sense. In addition, they are known for their *irreverent* (see Chapter 5 for more details) attitude toward problems: whether the symptoms are conflict, feeling blue, drinking, psychosis, or an eating disorder, the systemic therapist is never flustered or views one as more "serious" than another or more an "individual" versus a "family" problem. Instead, all behavior simply is a means of communicating that makes sense within a particular relational system, with each person in the system doing the best they can. Thus, they always refrain from blaming one member of the family for a problem, perhaps the most difficult shift in perspective for new therapists.

Systemic therapists stay focused on the interconnection of behavior and its context—on the dance between people—rather than end their assessments with the labeling of individual pathology. They carefully attend to how behavior is always shaped by complex webs of relations, within the family, community, and larger society. They view much of what the average person might consider an "individual problem" as part of a much larger set of interactions, of which the problem and the individual is only a part. Their nonpathologizing, nonblaming view offers clients a refreshing and often liberating new way to think about their situation. Above all, they are pragmatic, always considering whether their interventions were useful or not, and if not, figuring out what might be.

Introduction to Key Concepts in Systemic Theory

Before I introduce you to specific therapeutic applications of systemic therapy in the following chapters, I believe you will find it helpful to become acquainted with the basic concepts of systemic theory. You will then notice that different systemic family therapy approaches emphasize particular elements and are less concerned with others. Additionally, although all of these ideas are related, there are two major threads of systemic theory that are emphasized in family therapy: (a) concepts related to systemic structure, and (b) concepts related to communication patterns. Tracking these two major themes may help you better see the patterns, interconnections, and potential applications of the concepts.

Key Concepts Related to the Structure of Systems
General Systems and Cybernetic Systems Theories

Systems theory has its origin in the cross-disciplinary study that began at the Macy Conferences, which were attended by rocket scientists (literally) building self-guided missiles, anthropologists studying intertribal interactions in Bali, and ecologists studying the interactions between species. These researchers discovered that, whether studying mechanical parts, social groups, or animals, they were noticing that systems operated using the same basic principles, which von Bertalanffy (1968) developed into *general systems theory*. Closely related but more focused on social systems was the *cybernetic systems theory* articulated by Gregory Bateson (1972), which has had the most influence on the field of family therapy.

Homeostasis and Self-Correction

The term *cybernetic* means "steersman" in Greek, which hints at the functional principles of cybernetic systems: they are *self-correcting*, and therefore able to "steer" their own course, in contrast to a computer, for example, which needs an outside entity to steer it (Bateson, 1972). What does a cybernetic system steer toward? *Homeostasis.*

Homeostasis, in the case of families, refers to the unique set of behavioral, emotional, and interactional norms that create stability for the family or other social group. Despite what the name might imply, homeostasis is not static but *dynamic*. Much like a gymnast constantly moving to maintain her balance on a beam, systems must be constantly in flux to maintain stability. In all living systems, it takes work to maintain balance, whether in mood, habits, weight, or overall health. The key to maintaining stability is the ability to self-correct, which requires feedback.

Negative and Positive Feedback

You can pretty much guarantee that *negative* and *positive feedback* will be on any multiple-choice test about family therapy. Why? Because the disciplinary use of the terms are the opposite of their colloquial use. So, remember, the negative and positive feedback questions are always *trick questions*—that is, if you haven't studied. Here's how to remember it:

Negative Versus Positive Feedback

- *Negative Feedback:* No new information to steersman = the waters are the same = homeostasis.
- *Positive Feedback:* Yes, new information is coming to the steersman = the waters are choppy, moving faster, colder = something is *changing*.

Negative feedback is "more of the same" feedback, meaning there is no new news or change (Bateson, 1972; Watzlawick, Bavelas, & Jackson, 1967). In contrast, like "positive test results" in medicine, positive feedback is news that things are not within expected parameters, which may be experienced as a problem or a crisis, depending on the situation. The problem or change could be due to what is generally considered bad news (death of a loved one, a fight with a spouse, a problem at work) or good news (graduation from college and needing to find a job, getting married and starting a new household, moving to a new city for a job). Both good and bad news can create positive feedback loops, which result in one of two options: (a) return to former homeostasis, or (b) create a new homeostasis.

In most cases, a system initially responds to positive feedback by trying to get back to its former homeostasis as quickly as possible. After a fight, most couples quickly want to make up and "get back to normal." After a death, most talk about how getting back to normal will take a while, depending on how significant the deceased person was. However, sometimes it is not possible to get back to the "old normal," and a new normal needs to be created. The new norm or homeostasis is also referred to as "second-order change."

First- and Second-Order Change

Second-order change describes when a system restructures its homeostasis in response to positive feedback and the rules that govern the system fundamentally shift (Watzlawick, Weakland, & Fisch, 1974). *First-order change* refers to when the system returns to its previous homeostasis after positive feedback. In first-order change, the roles can reverse (e.g., a former distancer could start pursuing), but the underlying family structure and rules for relating stay essentially the same: someone is pursuing someone. This type of first-order shift is frequent in the early stages of couples therapy when partners shift between being the pursuer and the distancer. For example, if in the beginning the woman was asking for more closeness from her husband and he was asking for more space, as therapy progresses they may shift roles. Although the problem may appear solved, functionally there has been no shift in the rules that regulate intimacy in the relationship; the partners have just changed roles, so it looks and feels different. A second-order shift with this couple would involve reducing the overall pursuer-distancer pattern and increasing each person's ability to tolerate more togetherness and more distance.

It is also important to remember that second-order change is not always necessary in therapy, depending on the client's situation. In therapy, first-order solutions make logical sense; second-order solutions seem odd and illogical because they are introducing new rules into the system. Here's a clinical confession: in actual practice the distinction between first- and second-order change is often difficult to discern. I sometimes like to say, "That was 1.5-order change." Perhaps that is because most of us change in small shifts. However, the concept of first- and second-order change enables the therapist to ask whether roles have merely shifted or whether there has been a fundamental change in the ability to negotiate more intense intimacy and tolerate greater independence.

Symmetrical and Complementary Relationships

Originating in Bateson's (1972) anthropological work, the distinction between symmetrical and complementary relationships is frequently used to understand family interactions. In *symmetrical* relationships, the parties have "symmetrical" or evenly distributed abilities and roles in the system: an equal relationship (Watzlawick et al., 1967). Conflict in symmetrical systems generally takes the form of two equals fighting until there is a winner: each is viewed and experienced as a relative equal and the outcome is not predictable. In family relationships, symmetrical dynamics are often seen in couples and similar-aged siblings.

In contrast, in *complementary* relationships each party has a distinct role that balances or complements the other, often resulting in a form of hierarchy. Conflict in these relationships is less frequent because there are clearly defined, separate roles. Complementary dynamics often become a problem with couples when their roles become exaggerated or rigid. Examples of common complementary dynamics include pursuer/distancer, emotional/logical, visionary/planner, and easygoing/organized. These dynamics can provide a counterbalance that is enjoyable and helpful, especially early in the relationship; however, often these roles become exaggerated and rigid, creating a feeling of "stuckness." By the time clients present for therapy, often appear to be deeply ingrained personality traits rather than roles that each person has taken on as part of the systemic dance. The family therapist's task is to see these rigid complementary roles as part of the larger system rather than as fixed personality structures. It is then much easier to have hope for change and to be creative in making change.

Key Concepts Related to Communication Patterns in Systems

"One Cannot Not Communicate"

The early work of the Bateson team resulted in Watzlawick et al.'s (1967) classic text, *Pragmatics of Human Communication,* in which they proposed the following axiom: "One cannot not communicate." In addition to blatantly ignoring high school English teachers' rules about double negatives, this axiom seems to contradict the most common problem presented by couples and families: "We can't (or don't) communicate." So, where did this claim come from? The Bateson team learned from their research with schizophrenic family members that even schizophrenic attempts to not communicate (e.g., nonsense, immobile states) still sent a message, often communicating a desire to not communicate. Since all behavior is a form of communication and it is impossible to not be engaged in some form of behavior (at least while we are alive), it follows that we are always communicating. As we all know, silence speaks volumes, as does nonsense, withdrawal, or a frozen pose; thus even the most creative attempts to not communicate send a message. More commonly, the claim "we just can't communicate" means that one person doesn't like what the other has to say and that the two are unable to reach agreement, at which point it is helpful to examine the anatomy of the communicated messages—namely, the report and command aspects of communication.

Communication: Report and Command (Metacommunication)

Each communication has two components—*report* (content) and *command* (relationship)—that help therapists conceptualize communication and, more importantly, miscommunication. The report is the content: the literal meaning of the statement. The command is the *metacommunication,* or the communication about how to interpret the communication (Watzlawick et al., 1967). The command aspect always *defines the relationship* between two people. For example, the same piece of advice (content), such as "You might want to wear sunscreen today," can be accompanied by a command that defines either a peer-peer relationship or a one-up/one-down relationship. This is where miscommunication, double-bind communication, and arguments come in.

> ### Elements of a Communicated Message
> Communicated message = Report: Data, information (primarily verbal)
>   + Command: Defining relationship (primarily nonverbal)

The concept of report and command helps explain why couples, families, friends, coworkers, and basically any two humans can have elaborate, drawn-out arguments over taking out trash, toilet seat lids, cat litter, toothpaste, and the recalled order of events at last night's party. These arguments, although appearing to be over "little things," are really about *how the relationship is being defined* in relation to the little things; thus they are about a big thing, namely, how to define each person's role in the relationship.

When arguing over trash and cat litter, couples are usually disagreeing with each other's message at the command (relationship) level, not the content. It often helps to move the discussion directly to the metacommunication level, communicating about the command aspect of the communication, which in the case of household chores may include power dynamics or perceived caring. By directly discussing the metacommunication aspects (e.g., when the wife tells her husband to take out the trash, he feels that she is treating him like a child), the couple can clarify these relational issues, at which point the content issues are usually quickly resolved. Since every communication, verbal or nonverbal, has both report and command functions, the process of "getting meta" is infinite, because the partners can then talk about the metacommunication (command) aspect of the first metacommunication.

Double Binds

Double-bind theory goes back to the Bateson Group's (Jackson, Haley, Fry, and Weakland) earliest research on families with a member diagnosed with schizophrenia (Bateson, 1972). Watzlawick et al. (1967) identify the following ingredients of a double-bind communication:

1. Two people are in an *intense relationship* that has high survival value, such as a familial relation, a friendship, a religious affiliation, a doctor-patient relationship, a therapist-client relationship, or a relationship between an individual and his/her social group.
2. Within this relationship a message is given that is structured with (a) a primary injunction (e.g., a request or order) and (b) a simultaneous secondary injunction that contradicts the first, usually at the metacommunication level.
3. The receiver of the contradictory injunctions has the sense that he/she *cannot escape* or step outside the cognitive frame of the contradictions, either by metacommunicating (e.g., commenting on the contradiction) or by withdrawing, without threatening the relationship. The receiver is made to feel "bad" or "mad" for even suggesting there is a discrepancy.

Common examples are the commands "love me" or "be genuine," in which one person orders another to have spontaneous and authentic feelings. In their research, the MRI team noticed that this type of communication characterized families who had a member diagnosed with schizophrenia. A common exchange in these families was a mother who gave her child a cold, distant hug (command aspect communicates distance) and then said, "Why are you never happy to see me?" (report aspect suggests closeness). No matter how the child responds, the mother can prove him/her wrong. Thus the "logical" response is a *nonresponse* or *nonsense response*, which characterizes schizophrenic behavior, such as word salad (spoken words that have no real meaning), loose associations (tangentially relating words or topics), or catatonic behavior (rigid, repetitive behavior that has no interactive meaning).

Although the double-bind theory does not account entirely for how schizophrenia develops or who develops it, it is still useful for clinicians working with families that get stuck—whether or not there is a member diagnosed with schizophrenia. Common examples of double binds in families that present for therapy are the following:

- Someone asks a partner or child to spontaneously "show love" in a specific manner (bring flowers, do chores), but when the person shows love in the way requested, the partner or parent says, "That does not count because I had to ask you to do it." This becomes a double-bind situation with no way for the person to show genuine feeling.
- A very strict parent makes all the child's decisions but says, "I trust you to make good decisions." When the child tries to comment on the incongruency, the parent reverts to "But I *do* trust you."

Identifying the double bind is the therapist's first step at intervening in these destructive patterns.

Epistemology

"The proposition 'I see you' or 'You see me' is a proposition which contains within it what I am calling 'epistemology.' It contains within it assumptions about how we get information, what sort of stuff information is, and so forth . . . certain propositions about the nature of knowing and the nature of the universe in which we live and how we know about it."
—Bateson, 1972, p. 478

Bateson's ideas about epistemology are foundational to all systemic family therapies. In the strict philosophical sense, *epistemology* is the study of knowledge and the process of knowing. From his cybernetic investigations, Bateson concluded that most of the propositions humans assume to be true are erroneous; they *appear* true because they capture one dimension of an interactional sequence, but they rarely include the broader awareness of how observer and observed reciprocally reinforce and impact each other. Thus a wife's complaint that her husband is cold and indifferent does not take into account how their ongoing series of interactions have impacted each of their behaviors and frames for interpretation. Family therapists pay careful attention to the family's epistemology, the operating premises that underlie their actions and cognitions (Keeney, 1983). They use the concept of epistemology to help untangle what seem to be contradictory and irreconcilable differences in experience between members of a system.

Second-Order Cybernetics

A later distinction in systemic literature, *second-order cybernetics* refers to applying systemic principles to the observing system, such as to the therapy system (therapist observing the family system). In the process of observing another system, a new observer-observed system is created: a second-order (or second-level) system. The therapist can no longer assume to be a neutral, unbiased observer, but is rather an active participant in creating what is observed.

The cocreation process happens in several different ways. First, a therapist's descriptions reveal *more about the therapist* than about the family in that any description reflects what information the therapist deems most valuable and useful. Second, *how* a therapist interacts with or treats a family significantly impacts the actions and attitudes of the family while in the therapist's presence. A therapist who engages a family in a detached, professional manner will elicit different behaviors than one who uses a playful, low-key style. Which is more "real"? Neither, or more accurately both. Each response is a "natural, honest" response for the family system *in the context* of a particular professional. Therapists who maintain an awareness of second-order cybernetic principles remain continually attuned to how their behavior is shaping that of the client and how their descriptions of clients reflect their own values. This attention to the cocreation of the therapist-client reality became the focus of social constructionist therapists.

Translating Systemic Theoretical Concepts Into Action
The Family as a System

The defining feature of systemic approaches is viewing the family as a system, an entity in itself, with the whole greater than the sum of its parts (Watzlawick et al., 1967). What does this really mean? Systemic therapists view the interactional patterns of the family as a sort of "mind" or organism that is not controlled by any single member or outside entity, such as a therapist. This view results in several startling propositions:

- *No Single Person Orchestrates the Interactional Patterns:* The rules that govern family interactions are not consciously constructed like the U.S. Constitution; instead they emerge through an organic process of interaction, feedback (reaction), and correction until a norm or homeostasis is formed. In fact, many of the arguments early in a relationship serve as feedback to shape the emerging relationship's homeostatic norms. In most cases, this whole process occurs with minimal metacommunication about how the relational rules are being formed.
- *All Behavior Makes Sense in Context:* Because all behavior is a form of communication, it makes sense in the context in which it is expressed, within the rules of that particular system. Thus, even the seemingly nonsensical communication of schizophrenia makes sense in the larger family system.
- *No Single Person Can Be Blamed for Family Distress:* Because no one consciously creates the rules but instead the patterns are mutually negotiated through ongoing interactions, it follows that no single person can be fully to blame for family problems. Although individuals do have moral and ethical obligations in cases of abuse, the interactions "make sense" within the broader relational context and rules.
- *Personal Characteristics Are Dependent on the System:* Although a member may display certain characteristics or tendencies, these are not inherent personality characteristics that exist independent of the system; rather, they emerge from the interactional patterns in the system. Thus, even when a family reports, "Suzie has always been like this" (e.g., angry, helpful, forgetful), the therapist takes this to be a statement more about the rules (possible rigidity) of the system than a truth about Suzie.

Social Constructionist Foundations of Family

Social constructionist philosophy is a particular strand of postmodern philosophy, which has influenced a wide range of disciplines, including art, theater, music, architecture, literature studies, cultural studies, and philosophy. Because their systemic foundations had already conceptualized reality from a relational perspective, family therapists were the first mental health professionals to embrace postmodern philosophy. Of the various postmodern schools—constructivist, social constructionist, structuralist, and poststructuralist—social constructionism has been the most influential in the development of new psychotherapy models, such as solution-focused, collaborative, and narrative therapies (see Chapters 13 and 14).

Systems Theory and Social Constructionism: Similarities and Differences

The move from a systemic view (particularly the second-order cybernetic perspective) to a social constructionist perspective can be seen as a natural evolution and continuation of systemic concepts, which describe how social interactions shape a person's experience of reality. Although the vocabulary and metaphors change, the *emphasis on relationships and the relational construction of reality does not*. The earliest writings in family therapy explored how people construct their lived reality through interpersonal relations (Bateson, 1972; Fisch, Weakland, & Segal, 1982; Jackson, 1952, 1955; Watzlawick, 1977, 1978, 1984; Weakland, 1951), laying the foundation for postmodern approaches. In fact, many of the original approaches that began systemically, such as Milan therapy and the MRI approach (see Chapter 5), over time evolved to a more constructionist form. Thus systemic and postmodern therapies have more shared views than differences, especially when compared with other nonrelational psychotherapies.

Systemic and social constructionist theories share the following assumptions:

- A person's lived reality is relationally constructed.
- Personal identity and an individual's symptoms are related to the social systems of which they are a part.
- Changing one's language and description of a problem alters how it is experienced.
- Truth can only be determined within relational contexts; an objective, outsider perspective is impossible.

Despite these and other similarities, there are notable differences that can be traced to a shift in metaphor. Systems theory uses a *systems* metaphor: a family is a system, a group of individuals who coordinate meaning and their understanding of the world. Social constructionist therapies use a *textual* metaphor: people narrate their lives to create meaning using the social discourses available to them. In addition, social constructionist therapies emphasize the role of the therapist in the coconstruction of the client's reality, much like systemic therapists' attention to second-order cybernetic dynamics, resulting in a different approach to relating to clients and their problems. Furthermore, constructionist therapists use clients' language and stories differently than systemic therapies to create interventions.

Rumor Has It: The People and Their Stories

Kenneth Gergen A social psychologist, Ken Gergen first introduced social constructionist ideas to the mental health professions in his 1985 article in *American Psychologist*. His work has laid the foundation for the development of social constructionist therapy approaches, most notably collaborative therapy (Anderson, 1997; Anderson & Gehart, 2007; Anderson & Goolishian, 1992) and to a lesser degree narrative therapy (Freedman & Combs, 1996; White & Epston, 1990). His work has included detailed applications of social constructionism to psychological and social issues (Gergen, 1999, 2001) and to postmodern ethics (McNamee & Gergen, 1999). His most recent work is on positive aging (Gergen & Gergen, 2007).

Sheila McNamee Working with Gergen, Sheila McNamee, a communications theorist, has been a leader in translating social constructionist ideas to therapy (McNamee & Gergen, 1992), including an in-depth exploration of ethical issues (McNamee & Gergen, 1999). She has also been at the forefront of developing social constructionist pedagogy (McNamee, 2007).

John Shotter John Shotter's social constructionist work focuses on how people coordinate *joint action* through shared meanings and understanding (Shotter, 1993). His work emphasizes the ethics of mutual accountability in social relationships (Shotter, 1984).

Michel Foucault Rejecting philosophical labels such as a *postmodernist, structuralist,* or *poststructuralist,* Michel Foucault (1972, 1979, 1980) was a prolific social critic and philosopher who described how power and knowledge shape individual realities in a given society. A significant influence on Michael White's narrative therapy (see Chapter 13), Foucault's work introduces the political and social justice ramifications of language and power in therapy.

Ludvig Wittgenstein An Austrian philosopher, Wittgenstein's (1973) philosophy of language is highly influential in postmodern therapies, notably solution-focused brief therapy and collaborative therapy (Chapters 12 and 14). He describes language as inextricably woven into the fabric of life and argues that language cannot be meaningfully removed from its everyday use, as it commonly is in philosophical and theoretical discussions.

Mikhail Bakhtin A Russian critic and philosopher, Bakhtin worked on dialogue and concepts of identity, emphasizing that the self is *unfinalizable* (can never be fully known) and that self and other are inextricably intertwined (Baxter & Montgomery, 1996).

The Postmodern Spirit

It is difficult to capture the spirit or general sense of postmodernists, but I am going to try. As you might imagine, they would say such an attempt to "accurately" portray anything is an impossible goal because words are always contexualized by their user's reality and can only convey that reality, not the essence of what is described. As such, the shift to a postmodern epistemology involves an ongoing awareness of how each moment of reality is constructed and how each person gives unique meaning to lived experience. When this realization infuses your view of the world—this may sound dramatic but I think it is fair to say—your life and relationships will never be the same again.

Simply keeping the fluid nature of meaning and reality foregrounded changes how you view and relate to others and any life experience. When your partner is upset about something that seems innocuous to you, you soften and become curious about how your partner is interpreting that event. When a client shares a fear or viewpoint that seems odd or surprising, you become intensely interested in how she came to that understanding of her life and listen to her story, fascinated with how meaning evolved. If ever a couple or family appears in your office with entirely different perspectives on the same situation, this seems only natural to you, and you help them to weave together these unique perspectives. Furthermore, you will no longer be able to maintain convictions and opinions like you used to—instead you will hold all of your views more tentatively, always open to evolving them further.

One of the greatest mistakes I see new therapists make is trying to use postmodern techniques without fully adopting the true spirit—or epistemological position—of postmodernism. As you might imagine, they are not very effective. For postmodern therapists are masters at seeing possibilities, hope, and strengths, where others do not. Their assumptions make them *optimists extraordinaire*.

Postmodern Assumptions

Skeptical of Objective Reality: "Whatever Exists Is Mute"

Postmodernists are skeptical about the possibility of identifying an *objective reality*, such as x is a healthy behavior and y is not (Gergen, 1985). They describe reality as "mute" (Gergen, 1998), meaning that events and things in life do not come with prepackaged meanings, such as marriage is good, fat is ugly, and cars are bad. Instead, meaning is constructed by communities of people.

Reality Is Constructed

Postmodernists view all "truths" and "realities" as *constructed* (you will notice and perhaps be irritated by the frequent use of quote marks to emphasize that a concept is a construction, not a truth). Language and consciousness are necessary to develop meanings and to determine the value of an object or thing (Gergen, 1985; Watzlawick, 1984). Different postmodern schools emphasize and analyze different levels of reality construction; however, they principally recognize that the construction of reality is a complex process that involves all of these levels:

- *Linguistic Level—Poststructuralism and Philosophy of Language:* Focuses on how words shape our reality rather than being a reflection of it, a premise shared by all forms of postmodern thinking.
- *Personal Level—Constructivism:* Focuses on how reality is constructed within an individual organism; most closely associated with later developments of MRI and Milan therapies (Watzlawick, 1984).

- *Relational Level—Social Constructionism:* Focuses on how reality is created in immediate relationships (Gergen, 1985, 1999, 2001); most closely associated with collaborative therapy.
- *Societal Level—Critical Theory:* Focuses on how reality is constructed at the larger, societal level; most closely associated with narrative therapy, which draws heavily on the work of Michel Foucault (1972, 1979, 1980).

Reality Is Constructed Through Language

Postmodernists generally agree that reality is constructed primarily through language. Language is not neutral: words have real effects in our lives (Gergen, 1985). Most importantly for therapeutic purposes, words are the primary medium for (a) fashioning our identities, and (b) identifying what is a problem and what is not (Anderson, 1997; Gergen, 2001). For example, a person can interpret the same set of events (e.g., losing one's cool) as "having a bad day" or "being a bad person," each having dramatically different implications for identity and the definition of the problem. This level of reality construction, used in all postmodern therapies, is emphasized by poststructuralists and constructivists.

Reality Is Negotiated Through Relationships

The meanings we attribute to life experiences are not developed alone but in relationships, with immediate friends and family and more broadly with society and the subcultures of which we are a part (Gergen, 1985). The meaning a person gives to a particular behavior, hair cut, job, family relation, sex act, or religious view is always embedded in a web of "local" (immediate) relationships as well as the larger societal dialogue about the particular issue. Thus, how a person views premarital sex, lying, or disciplining children develops through and within the multiple layers of outer dialogues. Postmodern therapists help clients untangle the dialogues around problems so that clients can determine those with which they choose to affiliate. This level of reality construction is emphasized in theories that emphasize social constructionism and critical theory.

Shared Meanings Coordinate Social Action

Shared meanings and values are needed to coordinate social action, or more simply, get along with others (Gergen, 2001; Shotter, 1993). Without agreed-upon meanings on what is polite and what is rude or what is good and what is bad, it would be impossible for humans to live together—there would be total chaos. Instead, groups of people coordinate meanings and values: we call this *culture*.

Tradition, Culture, and Oppression

We cannot make sense of our lives outside of tradition or culture, which refers to not only ethnicity and nationality but also any small or large group that has a set of norms. Cultural traditions create a framework for (a) making meaning of our individual lives and (b) successfully coordinating our actions with others. Culture provides a set of values that its members can use to interpret their lives, knowing whether they are living a "good" life. In addition, culture provides a framework for safely and effectively interacting with others, allowing for the shared meanings necessary for marriage, family life, commerce, recreation, and religion. However, selecting certain goods and values over others inevitably labels certain behaviors and qualities as bad and undesirable. If a culture values productivity, it views taking time to relax negatively; if a culture values family, it de-emphasizes individuality. Thus *all cultures are by their very nature oppressive* (Gergen, 1998), because—by definition—they must identify certain behaviors as acceptable and others as unacceptable. The degree to which a culture is oppressive is directly correlated with its ability to be reflexive.

Reflexivity and Humanity

Any given culture remains *humane* to the extent that it is *reflexive,* able to examine its effects on others and to question and doubt its values and meanings (Gergen, 1998).

Within any group of people, there are some people for whom the dominant cultural norms fit and others for whom they do not. The extent to which a culture listens and responds to the minority voices within it is the extent to which that culture maintains its humanity, growing and expanding to reduce the oppressive forces that are inescapable if humans are to live with one another.

Social Constructionism, Postmodernism, and Diversity

Postmodern philosophy, with its suspiciousness about singular "truths," has profoundly affected most current therapies because it heightens awareness of diversity issues. Postmodernists challenge the concept that norms cannot be fairly established because these norms are created by one group within the society and do not fairly capture the lived experience of others in that society and even less so the reality of other groups or cultures. This is readily seen with gender, socioeconomic status, age, culture, religion, and other factors. Postmodernism proposes that the behaviors, thoughts, and feelings of a white, middle-aged, Protestant male from the Northeast cannot be assumed to be the same as those of an adolescent son of Southeast Asian immigrants who are semi-migrant farmers in California's Central Valley. They both have their own reality and truth and their respective norms and definitions of the good life; therapists must meet each with this fact in mind.

Philosophical Overview

The philosophical foundations of family therapy are key to understanding and effectively implementing the specific approaches in this book. Each approach has found unique ways to use these philosophical concepts in the therapy session. Thus the same concepts have different practical expressions in the various theories. Nonetheless, they provide connecting threads that can be traced from one theory to the next, resulting in an undeniable kinship. In addition, these theoretical principles are valuable to you in another way: deciding which theory(ies) to use and how. The table below provides a quick comparison of the philosophical foundations for the theories covered in this book and identifies some of the key elements of the theory related to its philosophical foundations (please see individual chapters for definitions of the interventions).

Overview of Family Therapy Theories

	Theory	Philosophical Foundations	Focus of Intervention	Signature Intervention
Systemic Therapies	MRI Systemic	Cybernetics Systems	Interactional sequences	Reframe
	Milan Systemic	Second order cybernetic systems	Language	Circular questions
	Strategic	General systems	Power in interactional sequence	Directives
	Structural	General systems	Family structure	Enactments
	Satir Experiential	General systems Humanistic	Communication	Sculpting Communication
	Whitaker	General systems Humanistic	Family cohesion Personal growth	Therapy of absurd Affective confrontation
	Emotionally Focused Therapy	Attachment theory General systems Humanistic	Attachments Negative interaction cycle	Enactments

	Internal Family Systems	General systems	Internal family dynamics	Language of parts
	Bowen	General systems Evolutionary systems	Intergenerational dynamics	Genogram
	Psychodynamic	General systems Psychodynamic	Interlocking pathologies	Working through with insight
	Cognitive Behavioral Therapy	General systems Cognitive Behavioral	Function of symptom Family schemas	Psychoeducation
	Functional Family Therapy	General systems Cognitive Behavioral	Relational function of problem	Relational reframing Psychoeducation
Postmodern Therapies	Solution-Based Therapies	Cybernetic Postmodern	Solutions	Scaling questions
	Narrative	Social constructionism Critical theory	Dominant discourses	Externalizing
	Collaborative	Social constructionism	Local discourse	Not knowing Reflecting teams

Rock-Paper-Scissors and Other Strategies for Choosing a Theory

What theory should I use? Which is the best? Which is the best for me? Do I have to pick a theory? What if I like them all? Can't I just be eclectic? These are some of the first questions students ask as they begin to study family therapy theories. The answers are more complex than one would imagine, leaving honest instructors and supervisors no option but to respond with maddening "both/and" or "yes-and-no" answers (trust me, they do not do this just to torture students for recreation and sport—it really is an honest answer). But let me give you a hint: philosophical foundations are part of the answer.

How to Choose: Dating Versus Marrying

Much like parents' advice to their teenage children, in the first few years I recommend that you casually "date" a theory before you decide to settle down. I am always surprised by new therapists who feel this tremendous pressure to find the "perfect" or "right" theory for them immediately—much like teenagers who are convinced that their first love will be their lifelong partner: it's possible, just not the most common scenario. You might want to "play the field" for a while to learn what is out there and what works best for you.

Fortunately, theory dating generally ends better than romantic dating. After dating a theory, you are almost always forever enriched with new skills and knowledge, which is only sometimes true in the romantic realm. Additionally, the breakup part is almost always gentler. Thus, you can decide to try out a new theory every semester of your training or every year or so in your practice. After dating even two or three, your skill set and knowledge base will have significantly grown. You will have also learned more about who you are and your style as a therapist. At that point, you may find that it is time to settle down with one more than the others. When that happens, you are ready to define your philosophy.

Defining Your Philosophy

Once you have spent a few years dating, you may find that you are ready to settle down with one theory. Just as in love, there should be an engagement period during which you clearly define your commitment and get to know your new partner—and family of origin—more intimately. In the case of theory dating, this involves pursuing advanced training in your theory of choice, usually by going to intensive seminars or working with a supervisor who specializes in your theory. Just as in marriage, in which a commitment to one person entails a commitment to an entire family, once you decide to commit to a theory, you are also committing to the broader philosophy that is the theory's foundation. I believe therapists who are clear about their philosophy of what it means to be human (ontology) and how people learn and change (epistemology) are best positioned to handle the variety of problems with which skilled therapists must learn to work. If you only master the techniques (system of doing), then you are less well prepared for handling the variety of issues that highly competent therapists must master.

Although there are many ways to define philosophical foundations of family therapies, I find it simplest to begin by considering four major categories: modernist, humanistic, systemic, and postmodern, each having its own approach to defining truth, reality, the therapeutic relationship, and the therapist's role in the change process. The following table summarizes their differences.

Overview of Philosophical Schools

	Modernist	Humanistic	Systemic	Postmodern
Truth	Objective truth	Subjective truth	Contextual truth	Multiple, coexisting truths
Reality	Objective; observable	Subjective; individually accessible	Contextual; emerges through systemic interactions; no one person has unilateral control	Coconstructed through language and social interaction; occurs at individual, relational, and societal levels
Therapeutic Relationship	Therapist as expert; hierarchical	Therapist as empathetic other	Therapist as participant in therapeutic system	Therapist as nonexpert; coconstructor of meaning
Therapist's Role in Change Process	Teaching and guiding clients in better ways of being and interacting	Creating a context that supports natural self-actualization process	"Perturbing" system, allowing system to reorganize itself; no direct control of system	Facilitating a dialogue in which client constructs new meanings and interpretations

Modernism

Modernism is founded on logical-positivist assumptions of an external, knowable "Truth." In modernist approaches, the therapist assumes an unequivocal role as expert, as is common in individual and family forms of cognitive-behavioral and psychodynamic therapies (see Chapters 10 and 11; e.g., Dattilio & Padesky, 1990; Ellis, 1994; Scharff & Scharff, 1987).

> **Modernist Assumptions**
>
> - The therapist is an expert who assumes the primary responsibility for identifying pathology, problems, and goals, often assuming the role of teacher or mentor.
> - Theory and research are the primary sources of information for identifying problems and diagnosing.
> - The therapist uses theory and research to select treatment approaches; clients are expected to adapt to the selected treatment.

Two family therapy schools fit this category: psychodynamic and cognitive-behavioral therapies. Although broadly grounded in modernist assumptions about knowledge, each theory has its own unique position on the primary source of truth, the means through which it is best identified, and how best to define the therapeutic relationship.

Modernist Therapies

	Psychodynamic Therapies	Cognitive-Behavioral Therapies
Primary Source of Objective Truth	Therapist's analysis of client dynamics based on theory	Measurable, external variables
Means of Identifying Truth	"Reality check"; comparing client experience against external perceptions, events, etc.	Scientific experimentation; therapist's definition of "reality" and/or social norms (identified through research)
Therapeutic Relationship	Hierarchical; therapist indirectly leads client toward goals	Educational; therapist is straightforward in directing client toward goals

Humanism

Humanistic therapies (Chapters 8 and 9) are founded on a phenomenological philosophy that prioritizes the individual's subjective truth. They include Carl Rogers's (1951) client-centered therapy, Fritz Perl's gestalt therapy (Passons, 1975), Virginia Satir's (1972) communication approach, Carl Whitaker's symbolic-experiential therapy (Whitaker & Keith, 1981), and Sue Johnson's emotionally focused therapy (Johnson, 2004).

> **Humanistic Assumptions**
>
> - By nature, humans are essentially good.
> - All people naturally tend toward growth and strive for self-actualization, a process of becoming authentically human.
> - The primary focus of treatment is the subjective, internal world of clients.
> - Therapeutic interventions target emotions with the goal of promoting catharsis, the release of repressed emotions.
> - A supportive, nurturing environment promotes therapeutic change.

The work of Virginia Satir and Carl Whitaker most clearly illustrates this philosophical stance, which in family therapy is always combined with a systemic perspective that accounts for the effect of family dynamics on an individual's emotional inner life. Although Satir's and Whitaker's approaches are based on the same philosophical traditions, their therapeutic approaches have dramatically different styles and assumptions, including the best ways to address self-actualization, change, confrontation, and the therapist's use of self (referring to how therapists use their personhood in session).

Humanistic Therapies

	Satir's Communication Approach	Whitaker's Symbolic-Experiential Approach
Means of Promoting Self-Actualization	Emotionally safe and nurturing environment	Affective confrontation; "perturbing" the system
Change	Structured experiential exercises; role modeling	In vivo interactions with the therapist
Style of Confrontation	Gentle, educational	Direct, affective
"Authentic" Use of Self	Genuine caring for the client	Unedited and honest sharing of emotions and thoughts

Systemic Therapy

Rather than a formal philosophical school, systemic therapies are grounded in *general systems theory*, which stresses that living systems are open systems, connected with and embedded within other systems (von Bertalanffy, 1968), and *cybernetic systems theory*, which emphasizes a system's ability to self-correct to maintain homeostasis (Bateson, 1972). The latter is more influential in the development of specific therapeutic models, such as the Mental Research Institute's brief, problem-focused approach (Watzlawick, Weakland, & Fisch, 1974), strategic therapy (Haley, 1976; Madanes, 1981), and the Milan team's systemic approach (Boscolo, Cecchin, Hoffman, & Penn, 1987). Systems theories emphasize *contextual* truth, truth generated through repeated interpersonal interactions that set a "norm" and rules for behavior.

Systemic Assumptions

- One cannot *not* communicate; all behavior is a form of communication.
- An individual's behavior and symptoms always make sense in the person's broader relational contexts.
- All behaviors, including unwanted symptoms, serve a purpose within the system, allowing the system to maintain or regain its homeostasis or feeling of "normalcy."
- No one individual unilaterally controls behavior in a system; thus no one person can be blamed for problems in a couple or family relationship; instead, problematic behavior is viewed as emerging from the interactions between members of the system.
- Therapeutic change involves alternating the interaction patterns within the system.

Within the field of systemic family therapy, Bateson's (1972) distinction between first-order and second-order cybernetics had significant impact on how therapists worked with families. With *first-order cybernetics,* the therapist is an objective, neutral observer describing the family as an outsider. Such therapy relies on assessment instruments and the therapist's perception of the family system. *Second-order cybernetic* theory applies the rules of first-order cybernetics on itself, positing that the therapist cannot be an objective, outside observer but instead creates a new system with the family: the observer-observed or therapist-family system. This second-order system is subject to the same dynamics as the first, including the drive to maintain homeostasis and rules for relating that are mutually reinforced. Second-order cybernetic theory maintains that whatever the therapist observes in the family reveals more about the therapist's values and priorities than about the family's because any description exposes what the therapist pays attention to and what the therapist ignores or misses. Second-order cybernetics laid the foundation for the transition to postmodern therapy, specifically constructivism in the MRI and Milan schools (Watzlawick, 1984).

In general, all systems therapists are influenced by both first- and second-order cybernetic theory. In practice, therapists generally emphasize one level of systems analysis or another. Broadly speaking, strategic and structural therapies were based on first-order theory and the MRI and Milan approaches gravitated toward second-order and later constructivist approaches.

- *First-order cybernetic approaches* lean toward the modernist tendency to find a more objective form of truth. Therapists who practice systemic therapies with a first-order orientation use more assessment instruments of family functioning and rely heavily on the therapist's perception of the system to guide practice.
- *Second-order cybernetic approaches* lean more toward a postmodern approach to truth (see next section). Their focus is on how the therapist and client coconstruct a second-order system, which has its own unique set of rules for establishing truth.

Systemic Theories

	First-Order Cybernetics	Second-Order Cybernetics
Level(s) of Analysis	Family system	Family system (level 1) and therapist-family system (level 2)
Target of Interventions	Correcting interactional sequences	"Perturbing" or interrupting interactional sequences
Therapist's Role	Tends to appear as a knowledgeable expert	Cocreator of therapeutic system
Focus of Assessment	Behavioral sequences	Meaning-making systems (epistemology)

Postmodern Therapy

Postmodern therapies are based on the premise that objective truth can never be fully known because it must always pass through subjective and intersubjective filters.

Postmodern Assumptions

- The human mind does not have access to an outside reality independent of human interpretation; objectivity is not humanly possible.
- All knowledge and truth are culturally, historically, and relationally bound and therefore intersubjective: constructed within and between people.
- What a person experiences as "real" and believes to be "true" is shaped primarily through language and relationships.
- Language and the words used to describe one's experiences significantly affect how one's identity is shaped and experienced.
- The identification of a "problem" is a social process that occurs through language, both at the immediate local level and at the broader societal level.
- Therapy is a process of coconstructing new realities related to the client's personal identity and relationship with the problem.

Within family therapy, three schools of postmodernism are particularly influential (Anderson, 1997; Hoffman, 2002; Watzlawick, 1984):

- *Constructivism:* Constructivists focus on the construction of meaning within the individual organism, on how information is received and interpreted.
- *Social Constructionism:* Social constructionists focus on how people cocreate meaning in relationships. They emphasize how truth is generated at the local (immediate) relational level.
- *Structuralism and Poststructuralism:* Structuralists and poststructuralists focus on analyzing how meanings are produced and reproduced within a culture through various practices and discourses.

Postmodern Philosophical Foundations

	Constructivism	Social Constructionism	Structuralism and Poststructuralism
Level of Reality Construction	Individual organism	Local relationship	Societal, political
Associated Theories	Later MRI and Milan theories	Collaborative therapy; reflecting teams	Narrative therapy; feminist and culturally informed therapies
Focus of Interventions	Recasting interpretations with new language	Dialogues that highlight multiple meanings and interpretations	Deconstruction and questioning of dominant discourses (popular knowledge)
Therapist's Role	Facilitate alternative interpretations	Noninterventional; facilitate dialogical process	Help identify external and historical influences

Dancing with Others Once You Marry

Once you commit yourself to a theory and philosophical stance, it ironically becomes much easier to dance with others. As you master one theoretical approach and deepen your understanding of the philosophical assumptions underlying it, you are able to understand other theories at a greater depth. This is perhaps where the common factors come in. There are similar principles that seem to be at play in all theories, and the more intimate you are with one theory the better able you are to identify these factors in others. It is also the case that you can see more clearly the subtle differences in outcome from philosophical assumptions, word choices, and interventions that differ across theories.

As therapists become more aware of the set of philosophical assumptions underlying their theory, whichever school that might be, they learn to skillfully adapt and integrate ideas from other approaches in a way that is philosophically consistent with their own approach. When a therapist is "eclectic" or "integrative" in a way that is not grounded in a single philosophical set of assumptions, that therapist is going to confuse his/her clients. One week the therapist might use a modernist approach and is an expert who has answers and knows the best way to approach the problem. The next week the therapist might try to use a postmodern approach in which the client is expected to be the expert and participate more as an equal. The following week the therapist might then shift to systemic ideas that emphasize the importance of context in defining the problem. As you might well imagine, a client working with this therapist is going to be very confused because each week the *client is required to relate differently to the therapist and to assume a different level of participation.* The therapist is also sending contradictory messages as to what is the measuring stick for "truth," progress, and direction. However, if the therapist is able to keep the philosophical assumptions consistent throughout therapy—what is our measuring stick for truth? what are our roles?—then the therapist can adapt concepts and techniques from other approaches without sending conflicting messages to the client, thus effectively incorporating a wider range of practices within a coherent approach to therapy.

Online Resources

Brief Strategic and Systemic Therapy Network:
www.bsst.org

Ken Gergen's Web Page:
www.swarthmore.edu/SocSci/kgergen1/web/page.phtml?st=home&id=home

Mental Research Institute:
www.mri.org

John Shotter's Web Page:
www.pubpages.unh.edu/~jds/

The Taos Institute: Explores social constructionist practices in a wide range of disciplines:
www.taosinstitute.org

References

Anderson, H. (1997). *Conversations, language, and possibilities: A postmodern approach to therapy*. New York: Basic Books.

Anderson, H., & Gehart, D. (Eds.). (2007). *Collaborative therapy: Relationships and conversations that make a difference*. New York: Brunner/Routledge.

Anderson, H., & Goolishian, H. (1992). The client is the expert: A not-knowing approach to therapy. In S. McNamee & K. J. Gergen (Eds.), *Therapy as social construction* (pp. 25–39). Newbury Park, CA: Sage.

Bateson, G. (1972). *Steps to an ecology of mind*. San Francisco, CA: Chandler.

Bateson, G. (1979). *Mind and nature: A necessary unity*. New York: Dutton.

Bateson, G. (1991). *A sacred unity: Further steps to an ecology of mind*. New York: Harper/Collins.

Baxter, L. A., & Montgomery, B. M. (1996). *Relating: Dialogues and dialectics*. New York: Guilford.

Boscolo, L., Cecchin, G., Hoffman, L., & Penn, P. (1987). *Milan systemic family therapy*. New York: Basic Books.

Dattilio, F. M., & Padesky, C. A. (1990). *Cognitive therapy with couples*. Sarasota, FL: Professional Resources Exchange.

Ellis, A. (1994). *Reason and emotion in therapy (Rev.)*. New York: Kensington.

Erickson, B. A., & Keeney, B. (Eds). (2006). *Milton Erickson, M.D.: An American healer*. Sedona, AZ: Leete Island Books.

Fisch, R., Weakland, J., & Segal, L. (1982). *The tactics of change: Doing therapy briefly*. New York: Jossey-Bass.

Foucault, M. (1972). *The archeology of knowledge* (A. Sheridan-Smith, trans.). New York: Harper & Row.

Foucault, M. (1979). *Discipline and punish: The birth of the prison*. Middlesex: Peregrine Books.

Foucault, M. (1980). *Power/knowledge: Selected interviews and other writings*. New York: Pantheon Books.

Freedman, J., & Combs, G. (1996). *Narrative therapy: The social construction of preferred realities*. New York: Norton.

Gergen, K. J. (1985). The social constructionist movement in modern psychology. *American Psychologist, 40*, 266–275.

Gergen, K. J. (1998, January). *Introduction to social constructionism*. Workshop presented at the Texas Association for Marriage and Family Therapy Annual Conference, Dallas, TX.

Gergen, K. (1999). *An invitation to social construction*. Newbury Park, CA: Sage.

Gergen, K. (2001). *Social construction in context*. Newbury Park, CA: Sage.

Gergen, M., & Gergen, K. (2007). Collaboration without end: The case of the Positive Psychology Newsletter. In H. Anderson & D. Gehart (Eds.), *Collaborative therapy: Relationships and conversations that make a difference* (pp. 391–402). New York: Brunner/Routledge.

Haley, J. (1976). *Problem-solving therapy: New strategies for effective family therapy*. San Francisco, CA: Jossey-Bass.

Hoffman, L. (2002). *Family therapy: An intimate history*. New York: Norton.

Jackson, D. (1952, June). The relationship of the referring physician to the psychiatrist. *California Medicine, 76* (6), 391–394.

Jackson, D. (1955). Therapist personality in the therapy of ambulatory schizophrenics. *Archives of Neurology and Psychiatry, 74*, 292–299.

Johnson, S. M. (2004). *The practice of emotionally focused marital therapy: Creating connection* (2nd ed.). New York: Brunner-Routledge.

Keeney, B. (1983). *Aesthetics of change*. New York: Guilford.

Keeney, B. (1985). *Mind in therapy: Constructing systematic family therapies*. Basic Books.

Keeney, B. (1990). *Improvisational therapy: A practical guide for creative clinical strategies*. New York: Guilford.

Keeney, B. (1994). *Shaking out the spirits: A psychotherapist's entry into the healing mysteries of global shamanism.* Barrytown, NY: Station Hill.

Keeney, B. (1997). *Everyday soul: Awakening the spirit in daily life.* New York: Riverhead Books.

Keeney, B. (1998). *The energy break: Recharge your life with autokinetics.* New York: Golden Books.

Keeney, B. (2000a). *Gary Holy Bull: Lakota Yuwipi man.* Stony Creek, CT: Leete's Island Books.

Keeney, B. (2000b). *Kalahari Bushmen.* Stony Creek, CT: Leete's Island Books.

Keeney, B. (2001a). *Vusamazulu Credo Mutwa: Zulu High Sanusi.* Stony Creek, CT: Leete's Island Books.

Keeney, B. (2001b). *Walking Thunder: Diné medicine woman.* Stony Creek, CT: Leete's Island Books.

Keeney, B. (2002a). *Ikuko Osumi: Japanese master of Seiki Jutsu.* Stony Creek, CT: Leete's Island Books.

Keeney, B. (2002b). *Shakers of St. Vincent.* Stony Creek, CT: Leete's Island Books.

Keeney, B. (2003). *Ropes to god.* Stony Creek, CT: Leete's Island Books.

Madanes, C. (1981). *Strategic family therapy.* San Francisco, CA: Jossey-Bass.

McNamee, S. (2007). Relational practices in education: Teaching as conversation. In H. Anderson & D. Gehart (Eds.), *Collaborative therapy: Relationships and conversations that make a difference* (pp. 313–336). New York: Brunner/Routledge.

McNamee, S., & Gergen, K. J. (Eds.). (1992). *Therapy as social construction.* Newbury Park, CA: Sage.

McNamee, S., & Gergen, K. J. (1999). *Relational responsibility: Resources for sustainable dialogue.* Newbury Park, CA: Sage.

Mental Research Institute. (2002). *On the shoulder of giants.* Palo Alto, CA: Author.

Passons, W. R. (1975). *Gestalt therapies in counseling.* New York: Holt, Rinehart, & Winston.

Ray, W. A., & Keeney, B. (1994). *Resource focused therapy.* London: Karnac Books.

Rogers, C. (1951). *Client-centered therapy.* Boston: Houghton Mifflin.

Satir, V. (1972). *Peoplemaking.* Palo Alto, CA: Science and Behavior Books.

Scharff, D., & Scharff, J. S. (1987). *Object relations family therapy.* New York: Jason Aronson.

Segal, L. (1991). Brief therapy: The MRI approach. In A. S. Gurman & D. P. Knishern (Eds.), *Handbook of family therapy* (pp. 171–199). New York: Brunner/Mazel.

Shotter, J. (1984). *Social accountability and selfhood.* Oxford: Blackwell.

Shotter, J. (1993). *Conversational realities: Constructing life through language.* Thousand Oaks, CA: Sage.

von Bertalanffy, L. (1968). *General system theory: Foundations, development, applications.* New York: George Braziller.

Watzlawick, P. (1977). *How real is real?: Confusion, disinformation, communication.* New York: Random House.

Watzlawick, P. (1978/1993). *The language of change: Elements of therapeutic conversation.* New York: Norton.

Watzlawick, P. (Ed.). (1984). *The invented reality: How do we know what we believe we know?* New York: Norton.

Watzlawick, P., Bavelas, J. B., & Jackson, D. D. (1967). *Pragmatics of human communication: A study of interactional patterns, pathologies, and paradoxes.* New York: Norton.

Watzlawick, P., Weakland, J., & Fisch, R. (1974). *Change: Principles of problem formation and problem resolution.* New York: Norton.

Weakland, J. (1951). Method in cultural anthropology. *Philosophy of Science, 18,* 55.

Weakland, J. (1988, June 10). Personal interview with Wendel A. Ray. Mental Research Institute, Palo Alto, CA.

Whitaker, C. A., & Keith, D. V. (1981). Symbolic-experiential family therapy. In A. S. Gurman, & D. P. Kniskern (Eds.), *Handbook of family therapy* (pp. 187–224). New York: Brunner/Mazel.

White, M., & Epston, D. (1990). *Narrative means to therapeutic ends.* New York: Norton.

Wittgenstein, L. (1973). *Philosophical investigations* (3rd ed.; G. E. M. Anscombe, trans.). New York: Prentice Hall.

Part II

Family Therapy Theories

CHAPTER 5

Systemic Therapies: MRI and Milan

"Through time, you learn how to look at a system and appreciate it for what it is. Never expect the system to be different. It's important for the therapist and for the trainee to train themselves to see the system, to be interested in it, to appreciate this kind of a system without wanting to change it."

—Boscolo, Cecchin, Hoffman, & Penn, 1987, p. 152

Lay of the Land

Of all the theories in this book, "systemic theory" is the most challenging to present because of how it was initially practiced, how it has evolved over time, and how the term is used in various contexts. For starters, some people will refer to all family therapies that have systemic theory as a foundation as "systemic" (see Chapter 4), especially when these theories are discussed in comparison with more traditional counseling theories. Thus, some would say most of the chapters in this book are about systemic therapies. However, once you get into the field of family therapy or subspecialties of family psychology and counseling, the use of the term changes to reflect the more nuanced history of the ideas.

Most would agree that systemic therapy has its roots at the Mental Research Institute (MRI) in Palo Alto (also see Chapter 4 for more history). From this group, three other prominent forms of systemic therapy practice evolved: the Milan team, Jay Haley's strategic therapy (Chapter 6), and Virginia Satir's human growth model (Chapter 8). To further add to the lack of clear boundaries, Jay Haley and structural family therapy's founder, Salvador Minuchin (Chapter 7), also exchanged ideas in the development of their theories. Finally, over time, the practice of systemic therapy has been to draw relatively freely from these various forms of systemic therapy (with the exception of integrating Satir's approach), especially when developing systemic evidence-based therapies (see Chapters 2 and 15).

This evolving history allows for many options for how to teach and learn these ideas. Using a more historical approach, systemic theories can be taught by describing how each theory evolved to illuminate the more subtle distinctions between each. In contrast, a more pragmatic approach would be to use a more modern integrated systemic-structural approach. A middle ground might focus on an integrated systemic approach that draws from MRI, Milan, and strategic. In fact, I have tried all three variations (respectively, Gehart, 2010, 2013, 2014) and have found that no one approach is inherently superior to the other. As with beauty and many other things in life, it seems "the best approach" to learning systemic theories is in the eyes of the beholder, with most instructors preferring the approach that parallels how they themselves were introduced to the ideas (reminding me of the concepts of epistemology and second-order cybernetics that systemic theories use to understand communication, see Chapter 4). Because this book is designed to specifically teach family therapy theories with greater depth, I have opted for the more detailed historical approach to allow the reader to clearly trace the distinct styles in which systemic theories can be practiced. I believe this broader introduction enables the reader to identify with a style that might suit them best, even when integrating the various forms of systemic practice. However, other options are available for those who prefer a more contemporary, integrated approach (see the systemic therapy chapters in Gehart, 2013, 2010). I hope one of these approaches works for you.

So, let's begin our historic tour of systemic theories, which are generally attributed to three teams of therapists. Two are covered in this chapter (MRI and Milan), and the third (strategic) is covered in Chapter 6:

- *Mental Research Institute* (MRI; *a.k.a. The Palo Alto Group*): After the Bateson team concluded their groundbreaking research on family dynamics with schizophrenia, Richard Fisch and Don Jackson worked together to found the Mental Research Institute, which has since served as the most influential training center in family therapy, inspiring Jay Haley's strategic work, the Milan team's systemic approach, Virginia Satir's human growth model (see Chapter 8), and solution-focused brief therapy (see Chapter 12). The Brief Therapy Project at the MRI was designed to find the quickest possible resolution to client complaints, typically relying on action-based interventions (Watzlawick & Weakland, 1977; Watzlawick, Weakland, & Fisch, 1974; Weakland & Ray, 1995).
- *Milan Systemic Therapy:* After joining together to further Selvini Palazzoli's work with families who have anorectic or schizophrenic children, Mara Selvini Palazzoli, Gianfranco Cecchin, Giuliana Prata, and Luigi Boscolo formed the Milan team. Early in their work, they studied at the MRI and returned to Italy to design a therapeutic model that embodied the cybernetic systems theory of Gregory Bateson (1972, 1979; see Chapter 4). This model is called Milan systemic therapy, or long-term brief therapy. Milan therapists closely attend to how client language shapes family dynamics (Selvini Palazzoli, Cecchin, Prata, & Boscolo, 1978).
- *Strategic Therapy:* One of the original associates at the MRI, Jay Haley developed his own form of systemic therapy with his then wife, Cloe Madanes. Their approach focuses on the use of power and, in their later work, love in family systems (Haley, 1976).

MRI Systemic Therapy

In a Nutshell: The Least You Need to Know

MRI systemic therapy was developed with the intention of being a brief therapy, arguably the original brief therapy. MRI systemic therapists conceptualize the symptoms of individuals within the larger network of their family and social systems while maintaining a nonblaming, nonpathologizing stance toward all members of the family. The MRI approach is based on *general systems* and *cybernetic systems theories,* which propose that families are living systems characterized by certain principles, including *homeostasis,* the tendency to maintain a particular range of behaviors and norms, and *self-correction,* the ability to identify when the system has gone too far from its homeostatic norm and then to self-correct to maintain balance (see Chapter 4). MRI therapists rarely attempt linear, logical solutions to "educate" a family on better ways to communicate—this is almost never successful—but instead tap into the systemic dynamics to effect change. They introduce small, innocuous, yet highly meaningful alterations to the family's interactions, allowing the family to naturally reorganize in response to the new information.

The Juice: Significant Contributions to the Field

If you remember anything from this chapter, it should be:

Systemic Reframing

Reframing is a central technique that is found in most forms of systemic family therapy, such as structural therapy (see Chapter 7) and symbolic-experiential family therapy (see Chapter 9). The MRI team approached reframing from a constructivist position that is summarized in the following propositions (Watzlawick et al., 1974):

1. We experience the world through our categorizations of objects, people, and events ("If my husband does not bring flowers or offer other romantic gestures, he does not really love me").
2. Once an object, person, or event is categorized, it is very difficult to see it as part of another category ("Our relationship must be in trouble if he has stopped being romantic").
3. Reframing uses the same "fact" that supports one categorization to support another categorization; once a person sees things using this second perspective, it is difficult to see the original situation in the same way (e.g., "My husband's drop in romance may mean that he has become more comfortable and authentic in the relationship rather than playing courting games; it could be a sign of deepening commitment").

The basic component of a reframe is finding an alternative yet equally plausible explanation (categorization) for the same set of facts. Of course, the key is identifying the *client's current* worldview and finding an equally viable frame for the problem behavior from the *client's perspective*. Reframing in systemic family therapies typically involves considering the role of the symptom in the broader relational system, often highlighting how it helps maintain balance (homeostasis) in the relationship. Unlike in cognitive-behavioral therapies, clients are not expected to literally believe or adopt the proposed reframe; instead, it is hoped that the reframe will be *"news that makes a difference,"* allowing clients to generate useful understandings. For example, with certain couples it makes sense to reframe their arguments as a way to build passion and maintain connection in their relationship. Such a reframe is offered to a couple without expectation that it will have a specific effect, because the system is considered a unique entity that will make its own meaning. If the couple does not find the reframe helpful, perhaps not responding or disagreeing with it, the therapist uses this information to better understand their worldview and then identify other potentials for reframing the problem. In the case study at the end of this chapter, reframing is used to help an intensely divided blended family find a workable homeostasis.

Rumor Has It: The People and Their Stories

Don Jackson A brilliant clinician and founder of the MRI in 1958, Jackson was a principal figure in family therapy, especially in the development of concepts such as family homeostasis, family rules, relational quid pro quo, conjoint therapy, interactional theory, and, along with others at MRI, the double-bind theory (MRI, 2002; Watzlawick, Bavelas, & Jackson, 1967); he also was one of the first to question the myth of normality (Jackson, 1967).

John Weakland Originally trained as a chemical engineer, Weakland joined the Bateson Group and helped to articulate the application of communication theory, emphasizing the importance of basing theory on concrete and observable behaviors rather than inferences or constructs, which are not observable (MRI, 2002). Along with Haley, Weakland integrated Milton Erickson's (see Chapter 4) work at the MRI with the Brief Therapy Project.

Richard Fisch After initially proposing the creation of the MRI, Richard Fisch was appointed by Jackson to be the director of the new Brief Therapy Project, which was inspired by Erickson's brief hypnotic work; the goal was to develop a highly teachable form of brief psychotherapy (Watzlawick et al., 1974). Thus, Fisch spearheaded the development of the therapy model for which the MRI is most famous, studying how to influence others with words and indirect influence (Fisch & Schlanger, 1999; Fisch, Weakland, & Segal, 1982; MRI, 2002).

Paul Watzlawick Born in Austria, Paul Watzlawick was a communications theorist who cofounded the Brief Therapy Center at the MRI with Weakland and Fisch, with a goal of developing an ultra-brief approach to therapy (Watzlawick, 1977, 1978/1993, 1984, 1990). In his later writings, Watzlawick explored implications of radical constructivism for therapy and human communication in cleverly titled books such as *How Real Is Real?* (1977), *The Invented Reality* (1984), *Ultra Solutions: How to Fail Most Successfully* (1988), and *The Situation Is Hopeless but Not Serious: The Pursuit of Unhappiness* (1993).

Art Bodin One of the founding members of the Brief Therapy Project at the MRI, Art Bodin has continued his research on families, developing the Relationship Conflict Inventory (RCI) and the Teasing and Bullying Survey (TABS) as well as helping found the American Psychological Association's Division of Family Psychology.

William Fry A member of Gregory Bateson's original team studying paradoxical communication, Bill Fry continued research on the interactional elements of humor and nonverbal communication (MRI, 2002).

Jules Riskin The only surviving member of the original clinical staff at MRI, Jules Riskin developed one of the first methodologies for studying family interaction and conducted the first study of the normal family process, a decade before others in the field recognized the importance of such a project (MRI, 2002).

Wendel Ray Former director and senior research fellow at the MRI, Wendel Ray has been instrumental in ushering systemic therapies into the 21st century, working with Bradford Keeney (Chapter 4) to develop *resource-focused therapy,* a strength-based systems therapy (Ray & Keeney, 1994); with Milan team colleagues to further develop systemic concepts, such as *irreverence, cybernetics of prejudices,* and *eccentricity* (Cecchin, Lane, & Ray, 1992); and with colleagues at the MRI on theoretical and archival work (Weakland & Ray, 1995). He also serves as the Hammond Endowed Professor of Education and Professor of Family Systems Theory at the University of Louisiana at Monroe.

Barbara Anger-Diaz and Karin Schlanger Barbara Anger-Diaz and Karin Schlanger train therapists in brief therapy with Latino families; training and therapy are conducted in Spanish at the Latino Brief Therapy and Training Center at the MRI.

Giorgio Nardone Working closely with Watzlawick in his later years (Nardone & Watzlawick, 1993), Giorgio Nardone founded the Centro di Terapia Strategica in Arezzo, Italy, and the Brief Strategic and Systemic World Network. Founded in 2003, the latter brings together strategic and systemic therapy practitioners from around the world.

The Big Picture: Overview of Treatment
Interrupt (Don't Fix) Problem Interactional Sequences
MRI therapists focus solely on resolving the presenting problem, with the therapist imposing no other goals or agendas. Therapists see the presenting problem not as an individual problem but a relational one, specifically an *interactional* one (Ray & Nardone, 2009; Watzlawick & Weakland, 1977). Neither an individual nor a

relationship is considered "dysfunctional"; instead, the problem is viewed as part of the interactional sequence of behaviors that have emerged through repeated exchanges, with no one person to blame.

The first session involves getting a clear, behavioral description of the interaction sequence surrounding the problem, beginning with the initial exchange prior to the escalation of symptoms and ending with the interaction sequence that returns the system to homeostasis. Of particular interest to MRI therapists is the *attempted solution:* what has the client been doing to solve the problem that is not working (Watzlawick et al., 1974). *This* will be used to develop the intervention.

Once MRI therapists identify the interactional behavioral patterns and attempted solutions, they identify potential interventions to *interrupt* this sequence—not correct it. Much the way a school of fish cannot be shepherded but only interrupted and allowed to regroup, systemic and strategic therapists do not try to linearly instruct clients in preferred behaviors, because it rarely if ever works (Haley, 1987). Instead, they interrupt the problem sequence of behaviors, allowing the family to reconfigure itself around the new information that has been introduced to the system. For example, if a parent and child complain of frequent arguments as does the family in the case study at the end of the chapter, the therapist does not try to educate them on better communication techniques. Instead, the therapist interrupts the sequence by reframing the child's defiance as a veiled attempt to remain close to the parent because the child feels neglected and shocked by her mother's sudden remarriage.

Usually the therapist gives the family a task or reframe before they leave the session, with instructions to either complete the task or reflect on the new interpretation presented by the therapist. The next week, the therapist follows up on the prior week and then designs another task or reframe based on the response to the prior week's intervention. This process continues only as long as is necessary to resolve the presenting problem; then therapy is terminated. To recap, the general flow of systemic or strategic therapies is as follows:

The Process of Systemic Therapy

- *Assess the Interactional Sequence and Associated Meanings:* The therapist identifies the interactional behavior sequences that constitute the problem, including the actions and reactions of everyone in the system and the associated meanings.

- *Intervene by Interrupting the Interactional Sequence:* Using a reframing technique or a task, the therapist interrupts the sequence (avoids trying to fix or repair the sequence), allowing the family to reorganize itself in response to the perturbation. *Note: The differences between MRI, strategic, and Milan therapies are primarily seen in the preferred method of interrupting the interactional sequence.*

- *Evaluate Outcome and Client Response:* After the intervention, the therapist assesses the family's response and uses this information to design the next intervention.

- *Interrupt the New Pattern:* Then the therapist interrupts the new pattern with another intervention. This continues—interrupt behavioral sequence, allow family to reorganize and respond, and intervene again—until the problem is resolved.

Making Connection: The Therapeutic Relationship
Respecting and Trusting the System

MRI systemic therapists respect the family as a system, as an entity that has its unique epistemology, or way of knowing and understanding the world (Watzlawick et al., 1974). They have a deep, abiding trust that the system can reorganize itself without the therapist forcing change. Instead, the therapist provides opportunities for the family to reorganize itself. The symptoms are never seen as indicators of individual pathology but rather as the byproduct of family interactional sequences that have served a purpose. The importance of this trust is illustrated in the case study at the end of the chapter. In this case, the mother of teen suddenly remarries after over a decade of being widowed; the new husband is currently unemployed, and to compensate, the mother takes on a third job, making her daughter feel further neglected by the mother. In such a case, it is easy to judge the situation quickly, and therapists may be tempted to rush in to "fix" this "clearly ridiculous" situation; in contrast, a systemic therapist takes a more curious and open position, trusting the family will reorganize itself through a process that respects the system's integrity, allowing the family to reorganize their rules for relating in ways that organically emerge from the system.

Adapting to Client Language and Viewpoint

In the first meeting, the therapist tries to establish a positive, trusting relationship with the client (Nardone & Watzlawick, 1993; Watzlawick et al., 1974). One approach is to adapt to the clients' language, communicative style, and worldview, speaking logically with clients who focus on reason and more intensely with clients who have a more emotional manner of expression. First and foremost, the therapist respectfully engages with the client's representational framework or epistemology, including beliefs, values, and language. This is the inverse of traditional psychoanalysis, in which the patient must adapt to the language and viewpoint of the therapist.

Maneuverability

Maneuverability refers to the therapist's freedom to use personal judgment in defining the therapeutic relationship (Nardone & Watzlawick, 1993; Segal, 1991; Watzlawick et al., 1974). Therapists may choose to maintain an expert position or a one-down stance (see next section), depending on what would be most helpful to the family. Similarly, the therapist may be more distant or emotionally engaged, depending on the circumstances. Furthermore, the therapist may choose to be disliked by the client or be the "bad guy" in order to achieve the desired change in the family system, always attuned to whatever role might be most beneficial for the family. In the case study at the end of the chapter, maneuverability is particularly important as the family members have extreme positions, including the daughter's refusal to speak with her stepfather and the stepfather's lack of motivation to get a job or otherwise contribute to the family.

The Viewing: Case Conceptualization and Assessment
Interaction Patterns

When systemic therapists view a family, they focus on the interaction patterns *between* people (Boscolo, Cecchin, Hoffman, & Penn, 1987; Watzlawick et al., 1974). To illustrate this focus to new therapists, I have a "family of four" grab hold of a bright yellow rope to form a circle. Then I have them "dance" and move about in various patterns. Initially, the audience's eyes naturally focus on the movement of the dancers, our default, socialized habit of viewing interactions. I then have the family do the dance again, asking the audience to focus exclusively on the yellow rope and how it moves as it traces the interactional patterns of distance and closeness. This is a very gross and incomplete metaphor for how a systemic therapist looks at a family, but I think it helps get the point across: the focus is always on the yellow rope—the interaction—or what Milan therapists call the "game" (see below; Boscolo et al., 1987).

Family interactions can be hard to detect, especially early in training. The trick is tracing the homeostatic dance (how A responds to B and B in turn responds to A). When observing a family interact, therapists can assess the dance patterns by focusing less on the content of the conversation and more on the metacommunication (see Chapter 4) and interactions (movement of the rope). For example, if a couple is arguing about how to discipline a defiant child, the therapist will not focus on solving the problem with the child but rather on how the parents communicate: Does each share opinions, thoughts, and feelings? How does each respond to the other's divergent opinion? What strategies does each use to convince the other? Does one person get the last word?

By focusing on the interactional pattern for negotiating this type of problem, the therapist can identify where they are stuck.

Assessing interaction patterns involves tracing the behavioral sequences in the homeostatic dance in which the symptom is embedded. The interactional sequence is traced through four general phases: (a) things being normal (homeostasis), (b) tensions escalating (early positive feedback), (c) the symptomatic feelings and behaviors (positive feedback), and, finally, (d) self-correction with the ultimate return to normal (homeostasis). Depending on the situation, this sequence can take several minutes or several months. Clients typically only describe the symptom, but by tracing the symptom from homeostasis to homeostasis, systemic therapists have a much better sense of how to intervene.

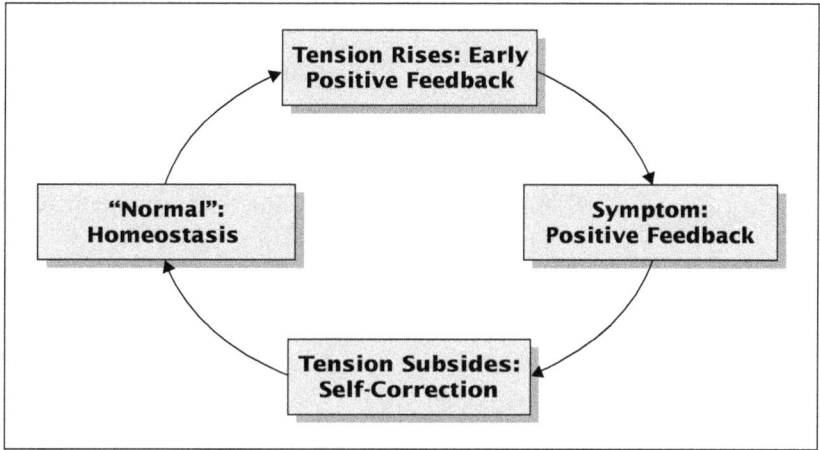

For example, if a client complains that she is anxious in public situations, a systemic therapist would explore the behaviors, interactions when the client feels "okay" or normal, then identify the behaviors, contexts, and relational and interactions when the anxiety starts to rise, what she does when she feels the anxiety at its height and how others respond, and then trace the behaviors and interactions until she feels "okay" or back to normal again. Similarly, when working with a couple who argues, the therapist first asks about what types of things they are doing when things are good between the two of them, what each starts to do as tensions rise, what each says and does in response to the other during the argument, and how they get back to a sense of "normal" again.

In addition to observing actual interactions, systemic therapists *ask* a series of questions to assess the interaction sequence. For example, with the couple who is having trouble with the defiant child, in addition to observing, the therapist would ask: What is happening just before the problem incident? What is each person saying and doing? How does the child respond? How does each parent and anyone else involved respond to the defiance? How does the child respond to this?

Identifying the Interactional Sequence

1. *Homeostasis (Normality)*: Can you describe what things are like when things feel "normal" and "okay"? What are things like before the tension starts?
2. *Start of Tension*: Can you tell me how things usually begin? Who does what and how do others respond? How does the first respond back?
3. *Escalation and Symptom*: Can you describe the escalation that leads to [the symptom]? Who does what and how does each person in the household respond?
4. *Return to Homeostasis*: Describe how things get back to normal?

Interactional/Systemic View of Problems

MRI therapists view all problems as fundamentally interactional (or systemic) rather than a form of individual pathology (Watzlawick et al., 1974). Otherwise stated, all "psychological" problems have a relational component, even if the interactional piece is not readily apparent. And if you didn't catch it, this *is* a radical perspective. However, they are *not saying* that there are never individual level biological or neurological issues or that an individual is not responsible for his/her actions. Instead, they draw our attention to the fact that even seemingly "individual pathology" always conforms to, is shaped by, and even shapes the social systems in which they are expressed. For example, seemingly "biologically based" symptoms, such as hallucinations, are significantly shaped by cultural context: not only the content but also the type of hallucination (Bauer et al., 2011). Systemic therapists point out that this is because the symptoms interact with and play a role in the larger family and social systems of which the identified patient is a part.

> ### Questions for Assessing Family Interactions
>
> To develop a concrete description of the interactional pattern, Nardone and Watzlawick (1993, p. 30) recommend that therapists answer the following questions:
>
> - What are the patient's usual, observable behavior patterns?
> - How does the patient define the problem?
> - How does the problem manifest itself?
> - In whose company does the problem appear, worsen, disguise itself, or not appear?
> - Where does it usually appear?
> - In what situations?
> - How often does the problem appear, and how serious is it?
> - What has been done and is currently being done (by the patient alone or by others) to resolve the problem?
> - Whom or what does the problem benefit?
> - Who could be hurt by the disappearance of the problem?

More-of-the-Same Solutions

When tracking interactional patterns, MRI therapists also focus on identifying *"more-of-the-same" solutions*, that is, solutions that perpetuate the problem (Watzlawick et al., 1974). For example, if the parents always respond to a child's defiance with some form of lecture and verbal punishment, this would be a more-of-the-same solution that is not working for the family. This "attempted solution" (i.e., verbal punishment) would be identified as the interactional behavior sequence that maintains the problem. After identifying the more-of-the-same behaviors and logic system (e.g., bad behavior is corrected with punishment), the therapist then identifies a behavior that represents a 180-degree shift in logic (e.g., cooperative behavior that is motivated by a strong emotional bond).

More-of-the-same solutions can further be described as mishandling the problem in one of three ways (Watzlawick et al., 1974):

- *Terrible Simplifications (Action Is Necessary but None Is Taken):* The client or family attempts to solve the problem by *denying* it: this solution is common with addictions, marital problems, and problematic family dynamics.
- *The Utopian Syndrome (Action Is Taken When It Is Not Necessary):* The client or family tries to change something that is either unchangeable or nonexistent: this solution is common with depression, anxiety, procrastination, perfectionism, and unrealistic demands on a relationship or a child.
- *Paradox (Action Is Taken at the Wrong Level):* Either a first-order solution is attempted for a problem that necessitates a second-order solution (e.g., parents are unable to adjust parenting techniques as child matures) or a second-order solution is attempted for a first-order problem (e.g., when people demand attitude or

personality changes and are not content with behavioral changes). This solution is common with schizophrenia, relational impasses, domestic violence, and abuse. This interactional sequence is characterized by double binds (see Chapter 4).

Metacommunication

As you may recall from Chapter 4 the MRI team described communication as having a report (the literal message) and command (nonverbal element that defines relationship) function. The command element is also referred to as *metacommunication*: communication about the communication (Watzlawick et al., 1967). When talking with families, systemic therapists listen for not only the content of the conversation but carefully attend to the command aspect, typically nonverbal clues as to how to interpret the verbal messages. Metacommunication often takes the form of voice tone, gestures, eye glances, or smirks and may reinforce the verbal message (congruent), satirize it (openly contradict), or create a double bind (by creating a context in which there is a contradiction that cannot be directly commented on). In the case study at the end of the chapter, the family's metacommunication provides a simple and clear illustration of the basic problem: the daughter refuses to acknowledge her stepfather as part of the family, and the stepfather's metacommunication says much the same; the mother, in contrast, has clearly placed the stepfather's need ahead of her daughter's and has "demoted" her daughter's position in the family, to which the daughter predictably rebels. This quick and straightforward tracking of metacommunication provides a rich conceptualization and a clear direction for how to intervene.

Complementary and Symmetrical Patterns

When assessing interaction patterns, systemic therapists also look for complementary and symmetrical patterns (Watzlawick et al., 1967). Complementary patterns become problems when they become rigid and exaggerated, such as the good parent/bad parent roles that often develop in families. Similarly in couple relationships, one partner might become the logical one and the other the emotional one. In other relationships, a symmetrical pattern develops that takes the form of competition. Some common complementary patterns include:

- Pursuer/distancer
- Logical one/emotional one
- Helpless one/rescuer
- Over-functioner/under-functioner
- Good parent (friend)/bad parent (disciplinarian)
- Social/recluse
- Oversexed/undersexed

These patterns typically are clearly evident in the interactional sequence. For example, a couple may report that most fights involve the wife pursuing the husband for attention in some form, and the husband responding with some form of withdrawal. Being aware of the basic complementary (or symmetrical) pattern helps the therapist trace the pattern across different issues the couples raises. In the case study at the end of the chapter, the mother and stepfather have both an over-functioner/under-functioner pattern as well as the good/bad parent patterns in the system. In addition, the mother is pursuing the daughter, who is withdrawing as part of her resistance to the new stepfather.

First and Second Order Change

When assessing clients, MRI therapists always view the situation through the lens of first and second order change (see Chapter 4 for basic definition). As you may remember, first order change typically involves a change in roles within the system (e.g., a pursuer becomes a withdrawer), but the basic interactional pattern stays the same (e.g., there is one pursuer and one withdrawer and the same level of intimacy is maintained). Second order change involves fundamentally changing the rules for relating in the system.

In most cases, this is the type of change that is necessary for problems to be resolved. So, therapists constantly track changes from week to week not just looking for some change but to assess whether it is first or second order change.

The Observation Team

The hallmark of the art of MRI systemic viewing has always been the *observation team* (Watzlawick et al., 1967; Watzlawick et al., 1974). Early in the development of these ideas, teams would sit behind a one-way mirror to observe the therapist working with the family. The team could see the systemic dance more rapidly and completely because the person in the room very quickly falls in sync with the family system and has a more difficult time seeing the entire dance. Anyone who has spent time behind a mirror and in front of it knows that you are always smarter behind it. The distance created by not being part of the interactional dance of the family increases one's ability to see the dance more clearly and quickly. Arguably the main reason family therapists still train with a mirror is to develop the ability to see the system's dynamics. This is not to say that a therapist in the room can't see the systemic dynamics; it is just harder and takes more practice.

Targeting Change: Goal Setting

Symptom-Free Interaction Patterns

MRI systemic therapists help the family to develop a new set of interaction patterns (i.e., homeostasis), a new game or dance that does not include symptoms (or any subsequent symptoms). Stated yet again, the goal is to create a family homeostasis that is problem-free—or at least does not involve facing the same problem over and over again (Watzlawick et al., 1974).

No Theory of Health

Unlike many earlier schools of therapy, such as psychodynamic or humanistic, systemic therapy does not have a predetermined definition of "healthy family functioning" that the therapist uses to define therapeutic goals. Cybernetic theory would only pose that *flexibility* characterizes healthy systems. As already stated, systemic therapists do not have a theory of health that defines how the family should look at the end of therapy. Instead, it is believed that the family system (not individual members) will reorganize itself from within to find a "functional," symptom-free interaction pattern in response to the *perturbations* or disruptions introduced by the therapist.

The Problem Is the Attempted Solution

Goals in the MRI brief therapy approach are developed using the following four-step procedure (Watzlawick et al., 1974):

1. *Define the Problem:* Use concrete, behavioral terms to describe the actions and reactions of all involved. For example, "My son is always defiant" is not a well-defined problem; a preferred problem description is "When I ask my son to do something, he says 'no' and when I push further he starts yelling and cursing; that's when I give in."
2. *Identify Attempted Solutions:* The therapist may ask, "How have you attempted to deal with this problem?" and listens for patterns in the various attempted solutions (e.g., threats that are not carried out, negotiating with child).
3. *Describe the Desired Behavioral Change:* MRI therapists take great care to develop concrete, behavioral goals that meaningfully focus clients' and therapists' attention in useful directions. The goals must be realistic, specific, and time-limited.
    - *Realistic:* Therapists need to be careful to avoid utopian goals, such as "Every time I ask my son to do something, he will obey." Instead, the therapist helps the client develop the more realistic goal of "increasing the frequency that my son politely carries out my first request."

- *Specific:* Goals describe clear behaviors and interactions (e.g., couple will greet each other with a hug and kiss and nonproblem talk for first 60 seconds) rather than vague intentions, such as "communicate better." As Watzlawick says, "It is the vagueness of goals that make their attainment impossible" (Watzlawick et al., 1974, p. 112).
- *Time Limited:* MRI brief therapists prefer to set a clear time limit for a specified goal, such as "within 4 weeks."
4. *Develop a Plan:* MRI therapists use two basic systemic principles:
    - The target of change is the *attempted solution.*
    - The tactic of change is to use the *client's own language* to speak directly to the client's view of reality.

Thus the plan is never linear or directive psychoeducation on how to "better communicate." Such plans and interventions are used in cognitive-behavioral therapy (see Chapter 8), not systemic therapy. In dramatic contrast, an MRI systemic approach targets the attempted solution rather than the presenting problem; again, this differs from solution-based therapies that target the preferred solution (see Chapter 12). Thus, rather than target a child's tantrums by teaching parents how to employ positive and negative reinforcement, systemic therapists identify the class of solutions used by the parents and develop an intervention that represents a 180-degree shift. If the parents respond by becoming embarrassed and giving in, the therapist designs an intervention in which the parents are emotionally unaffected and consistent in their consequences. On the other hand, if the parents typically respond with strict, harsh punishment, the therapist suggests an emotionally engaged and gentle approach.

The Doing: Interventions
Reframing
Described in detail in the Juice section above, reframing is a key technique in the MRI approach to facilitate second order change.

Less-of-the-Same Behavior Prescription
Once MRI therapists have identified the failed attempted solution(s), they identify what behavior would represent less of the same behavior or even a 180-degree shift: this alternative behavior will alter the interactional pattern in a meaningful way (Watzlawick et al., 1974). They would then "prescribe" this behavior; this is perhaps the most common technique in the approach, with clients leaving most sessions with some form of behavioral prescription. For example, if parents are responding to a child's tantrums with more and more and harsher and harsher punishment, a 180-degree shift might be a behavior that builds a stronger emotional bond with the child. Similarly, if a couple has a pursue/withdraw pattern, a 180-degree shift would be to have the pursuer curtail pursuing upon the first sign of rejection or not avoid pursuing regarding a certain topic all together. Note, the less-of-the-same behavior prescription is *not* a literal solution to the problem (as is typical of similar-looking interventions in cognitive-behavioral therapy), but rather is designed to interrupt the problem interaction sequence.

Therapeutic Double Bind: More-of-the-Same Behavior Prescription
The MRI *therapeutic double bind* is used to undo a double-bind message in a family or relationship (Watzlawick, Bavelas, & Jackson, 1967). The difference between a problem-generating double bind and a therapeutic double bind is simple: in the problem-generating form, no matter what you do, you are wrong; there is no escape. In a therapeutic double bind, no matter what you do, you do something different: something that moves you in a new direction.

A common double bind is when one party demands that the other show spontaneous displays of love and affection. For example, a wife may demand that her husband spontaneously express his love in more romantic ways: flowers, candlelight dinners, and gifts.

If he does what she asks, she can say that he only did it because she asked and that it wasn't spontaneous; if he does not do anything romantic, then she can say he obviously doesn't care because he didn't follow through. Either way he loses. A therapeutic double bind would be to have him show his romantic side in any way *other than* the specific ways his wife demands (more-of-the-same behavior—not doing what his wife asks—but in a new context). If he follows this directive literally, he will initiate new romantic behavior in the system; if he does not follow the instructions and instead chooses to use one of his wife's suggestions, then he is doing so *without the command* to do so. If you think it is tough to design such an intervention on your own, you are right—that is why the early family therapists preferred to work in teams. They could then more quickly assess the family's double binds and identify helpful therapeutic paradoxes.

Dangers of Improvement

A technique used by MRI and strategic therapists, the *dangers of improvement* involves asking clients to identify potential problems that might arise if the problem were resolved (Segal, 1991). For example, if a child becomes more independent in doing homework and getting chores done, what will the mother and father do to feel that they are parenting their son? If a couple suddenly stop arguing, how will they keep the flame of passion alive in their relationship? If a person stops being depressed and starts socializing again, how will he protect his solitude and quiet time? These questions undermine unrealistic and utopian worldviews that not only created the current problem but also are likely to be the source of new problems to come.

Restraining, Going Slow

Restraining or the directive to *go slow* is another common systemic and strategic intervention that carries a paradoxical flavor (Segal, 1991). When therapists "restrain" or instruct clients to "go slow," they warn clients to avoid changing too fast and encourage them to take change slowly. This has a paradoxical effect similar to a therapeutic double bind in that if the client complies, change will happen and the client will be better prepared for the setbacks that characterize most attempts at change. On the other hand, if the client has a rebellious response, he/she will attempt to rebel against slowness by working harder toward desired change. In either case, the change process is supported and "immunized" against setbacks.

Putting It All Together

Case Conceptualization Template
- *Interaction Sequence*: Describe each person's role (even seemingly minor players) in the interactional sequence, emphasizing the following:
  - Homeostasis
  - Rise of tension
  - Symptom/conflict
  - Return to homeostasis.
- *More of the Same Solutions:* What solutions are they trying that are not working? What might represent a 180-degree shift?
- *Metacommunication:* What is the metacommunication in problem interactions?
- *Complimentary Roles:* Pursuer/distance; under-/over-functioner; emotional/logical; good/bad parent, etc.
- *Hypothesized Homeostatic Function of Presenting Problem:* How might the symptom serve to maintain connection, create independence/distance, establish influence, reestablish connection, or otherwise help create a sense of balance in the family?
- *Describe Desired Behavioral Change:* In terms that are behavioral, realistic, specific, and time-limited.

Treatment Plan Template for Individual

MRI Initial Phase of Treatment with Individual

Initial Phase Counseling Tasks

1. Develop working counseling relationship. *Diversity Note: Adjust to respect cultured, gendered, and other styles of relationship building and emotional expression.* Relationship building approach/intervention:
   a. *Respect and trust the system* while *adapting client language* and maintaining therapeutic *maneuverability*.

2. Assess individual, systemic, and broader cultural dynamics. *Diversity Note: Adjust assessment based on cultural, socioeconomic, sexual orientation, gender, and other relevant norms.*
   Assessment strategies:
   a. Assess the *interactional sequence*(s) related to the problem, including rise of tension, symptom, return to homeostasis, *metacommunication*, and *complementary patterns*.
   b. Identify *more-of-the-same solutions*, including terrible simplifications, utopian syndrome, and paradox.

Initial Phase Client Goals

1. Increase possibilities for *viewing the problem* in such a way as to reduce depressed mood/anxiety.
   a. *Reframe* problem in relational context to shift meaning of symptoms.
   b. *Behavioral prescription* that represents a 180-degree shift to attempted solutions.

MRI Working Phase of Treatment with Individual

Working-Phase Counseling Task

1. Monitor quality of the working alliance. *Diversity Note: Attend to client response to interventions that indicate therapist's approach not connecting with client's culturally informed meaning systems.*
   a. *Assessment Intervention:* Monitor client's response to therapist *reframes* and other interventions to determine if therapist has been able to engage *client's meaning system*.

Working-Phase Client Goals

1. Increase new *interactional sequence patterns* to reduce depressive symptoms [specify]. Interventions:
   a. *Paradoxical behavioral prescriptions* that require client to enact particular depression symptoms at specific times to interrupt problem behavior sequence.
   b. *Behavioral prescriptions* that interrupt depressive symptoms by altering the where, when, how, and who of the symptoms.

2. Increase new *interactional sequence patterns* to reduce anxiety symptoms [specify].
   a. *Paradoxical behavioral prescription* that requires client to "worry" for set time period each day.
   b. *Behavioral prescriptions* that interrupt depressive symptoms by altering the where, when, how, and who of the symptoms.

(continued)

MRI Closing Phase of Treatment with Individual

Closing-Phase Counseling Task

1. Develop aftercare plan and maintain gains. *Diversity Note: Access resources in the communities of which they are a part to support them after ending therapy.*
   a. *Dangers of improvement* and *restraining* to prepare client to handle setbacks.

Closing-Phase Client Goals

1. Increase effectiveness of *relational interaction sequences* with partner/family to reduce depression and anxiety in specific relational contexts.
   a. *Dangers of improvement* intervention, noting how improvement will affect significant others.
   b. *Behavioral prescription* to have client slightly but meaningfully alter his/her half of problem interactions with others.

Treatment Plan Template for Couple/Family

MRI Initial Phase of Treatment with Couple/Family

Initial Phase Counseling Tasks

1. Develop working counseling relationship. *Diversity Note: Adjust to respect cultured, gendered, and other styles of relationship building and emotional expression.* Relationship building approach/intervention:
   a. *Respect and trust the couple/family system* while *adapting their language* and maintaining therapeutic *maneuverability*.

2. Assess individual, systemic, and broader cultural dynamics. *Diversity Note: Adjust assessment based on cultural, socioeconomic, sexual orientation, gender, and other relevant norms.*
   Assessment strategies:
   a. Assess the *problem interactional sequence*(s), including rise of tension, symptom, return to homeostasis, *metacommunication*, and *complementary* patterns; assess role of all member's in the household.
   b. Identify *more-of-the-same solutions*, including *terrible simplifications, utopian syndrome, and paradox*.

Initial Phase Client Goals

1. Increase the couple/family's possibilities for *viewing the problem* to reduce conflict.
   a. *Reframe* problem to show how each person's response is part of a larger systemic interaction pattern to shift meaning in system.
   b. *Behavioral prescription* that represents a 180-degree shift to attempted solutions (e.g., have pursuer withdraw and/or withdrawer pursue on alternating days or increase emotional connection between parents and child if using harsh punishment has failed).

MRI Working Phase of Treatment with Couple/Family

Working-Phase Counseling Task

1. Monitor quality of the working alliance. *Diversity Note: Attend to client's response to interventions that indicate therapist's approach not connecting with couple's culturally informed meaning systems.*

(continued)

a. *Assessment Intervention*: Monitor each client's response to therapist reframes and other interventions to determine if therapist has been able to engage family meaning system.

Working-Phase Client Goals

1. Increase new *interactional sequence patterns* related to [specific area of conflict] to reduce conflict.
   Interventions:
   a. *Paradoxical behavioral prescriptions* that require clients to enact conflict in new contexts.
   b. *Behavioral prescriptions* that interrupt the problem behavioral sequence by altering the where, when, how, and who of the conflict.
2. Increase new *interactional sequence patterns* related to [specific area of conflict] to reduce conflict and/or increase sense of emotional connection.
   a. *Relational reframing* of conflict that shifts each person's response in the interaction sequence.
   b. *Behavioral prescriptions* that interrupt conflict by altering the where, when, how, and who of the conflict.

Closing Phase of Treatment with Couple/Family

Closing-Phase Counseling Task

1. Develop aftercare plan and maintain gains. *Diversity Note: Access resources in the communities of which they are a part to support them after ending therapy.*
   a. *Dangers of improvement* and *restraining* to prepare couple/family to handle setbacks; include variations that will reduce blame of partner.

Closing-Phase Client Goals

1. Increase effectiveness of *relational interaction sequences* in [remaining areas of concerns] to increase sense of connection and relational satisfaction.
   a. *Dangers of improvement* intervention, noting how improvement may affect other aspects of couple/family's relationship and/or other relationships.
   b. *Behavioral prescription* to have clients slightly but meaningfully alter their behaviors in areas of concerns.

Milan Systemic Therapy

In a Nutshell: The Least You Need to Know

Milan therapy began as an attempt to put "pure" Bateson's cybernetic theory into systematic practice (Selvini Palazzoli et al., 1978). The team conducted much of its work with children diagnosed with severe disorders, such as schizophrenia or eating disorders. The approach is a "long-term brief therapy," meaning that the sessions were generally conducted once per month but generally therapists met with clients only 10 times. The Milan approach uses more language-based interventions, such as reframing and circular questions, in comparison to the MRI or strategic approaches, in which interventions tended to be more action-oriented. They referred to interaction patterns as "family games" (think playful games not manipulative games if you translate fairly from Italian) and used positive connotation to reframe symptoms. The approach aims to shift family interactions by shifting the family's epistemology and view of the problem. In later years, Cecchin and Boscolo worked closely with collaborative social constructionist therapists in evolving their ideas in a more postmodern direction (see Chapter 14).

The Juice: Significant Contributions to the Field
If you remember anything from this chapter, it should be:

Circular Questions
Regardless of which therapeutic model one chooses to practice, circular questions are perhaps one of the most useful techniques when working with more than one person in the room. Such questions help (a) assess and (b) make overt the overall dynamics and interactive patterns in the system, thereby reframing the problem for all participants without the therapist having to verbally provide a reframe, as already described in systemic reframing (Selvini Palazzoli et al., 1978; Selvini Palazzoli, Boscolo, Cecchin, & Prata, 1980). For example, I once had a family come to me complaining about a child's "anger problem." After they described the child's "problem" behavior, I inquired about how the rest of the family responded to the child's anger. The mother reported that she responded by getting frustrated and often yelled, and the father sharply corrected the child. As they listened to themselves describe their responses, both conceded that they responded to his anger with anger, and also noted that the younger sister disappeared into another room or somehow disengaged because of the intensity. As the parents began to look at their half of the interaction pattern, it became clear that anger was highly "contagious" for the three of them while it evoked anxiety in the daughter. These questions allowed the family to view their "son's anger problem" in the context of the broader interaction pattern, thus opening new options for relating.

Specific forms of circular questions (Cecchin, 1987) include the following:

- *Behavioral Sequence Questions:* Therapists use these questions to trace the entire sequence of behaviors that constitute the problem: "After John got mad, what did mom do? What did dad do? What did his sister do?" And after the response "What did John do next?," the therapist follows the sequence of interactions until homeostasis is restored (you may recognize this from assessing the "problem interaction pattern" in the case conceptualization described in Chapter 17).
- *Behavioral Difference Questions:* Therapists use behavioral difference questions when clients start labeling people and assuming that a particular behavior is part of a person's inherent personality. For example, if a child claims, "My mom's a nag," the therapist would ask, "What does she do that makes her a nag?" The description of problem behavior, "tells me to do things," is then compared with others: "How does your dad ask you to do things? Your teacher? How do *you* ask others to do things?"
- *Comparison and Ranking Questions:* Comparison and ranking questions are useful for reducing labeling and other rigid descriptions in the family: "Who is the most upset when Jackie has an episode? Who is the least affected? Who is the most helpful during these times? The least helpful?"
- *Before-and-After Questions:* When a specific event has occurred, before-and-after questions can be useful in assessing how the event affected the family dynamics: "Did you and your mother fight more or less after your dad got sick?"
- *Hypothetical Circular Questions:* Hypothetical questions are used to offer a scenario and have family members describe how each is likely to respond: "If mom were suddenly placed in the hospital, who would be the most likely person at her bedside? The least likely?"

Rumor Has It: The People and Their Stories
The Milan Team
The Milan team included *Mara Selvini Palazzoli, Gianfranco Cecchin, Luigi Boscolo,* and *Guiliana Prata* and was founded in 1967 by Selvini Palazzoli, who was studying anorexia in Italy at the time (Campbell, Draper, & Crutchley, 1991). They met weekly without pay, seeing families and studying the work of Bateson and the MRI group and frequently inviting Watzlawick for consultations. The team developed a unique approach that focused on meaning and ritual. In 1979, the team separated along interest and gender lines: Selvini Palazzoli and Prata, primarily researchers, wanted to continue their

investigation into treating families with psychotic members, while Cecchin and Boscolo worked together on clinical and training applications. The later work of Selvini Palazzoli and Prata centered on the *invariant prescription* (discussed later in this chapter), while Cecchin and Boscolo explored Keeney's *second order cybernetics* (1983) and related works, attending to language and the construction of meaning through multiple descriptors and eventually moving toward a more postmodern, social constructionist stance (see Chapters 13 and 14). Lynn Hoffman and Peggy Penn, featured in Chapter 14, worked closely with Boscolo and Cecchin.

The Big Picture: Overview of Treatment
Long-Term Brief Therapy

The Milan approach is highly structured, typically involving 10 sessions held once per month over ten months (Selvini Palazzoli et al., 1978; Selvini Palazzoli et al., 1980). Each session is divided into five segments:

- *Pre-session*: The team meets to discuss the family and develop hypotheses and interventions.
- *Session*: The conductor of the session meets with the family as the rest of the team observes from behind the mirror.
- *Intersession*: The conductor takes a break and develops an intervention with the help of the team, who typically sees the family dynamics more quickly than the person in the room.
- *Intervention:* The conductor of the session returns to the session and delivers the intervention.
- *Discussion:* After the family leaves, the team meets to discuss how the session went, refine their hypothesis, and decide how to proceed next time.

Session Overview

The following sequence of sessions was generally used when a family presented with a child as the identified patient.

Session 1
All family members are invited to allow the team to observe the interactional patterns and family game; from this they develop their initial hypothesis of the role of the symptom.

Session 2
Again, all family members are present, and the focus is on the children; however, after this only the parents are invited back.

Session 3
In this session with parents only, the team prescribes the invariant prescription.

Session 4
Again, only the parents return, and they become the focus of treatment. By changing the parental unit, symptoms in the children were expected to be addressed.

Session 5
The therapist addresses either the parents' success or failure to address the symptoms of the identified patient.

Sessions 6 to 10
The team continues to monitor the parents' success with the invariant prescription and the children's response; additional interventions are offered as needed to resolve the presenting complaint.

Making Connection: The Therapeutic Relationship
Neutrality and Multipartiality

In 1978 (Selvini Palazzoli et al., 1978; Selvini Palazzoli et al., 1980), the Milan team first described their therapeutic stance as one of *neutrality*, which has been one of the most misunderstood concepts related to their work (Boscolo et al., 1987). For the Milan team, neutrality connoted not only nonpartiality toward particular family members or problem descriptions but also multipartiality, the willingness to honor all perspectives.

In one sense, neutrality refers to the *pragmatic effect* the therapist has on the family, not the therapist's own feelings (Cecchin, 1987). Thus if, at the end of the session, the family cannot identify which "side" the therapist took, then the therapist has had the effect of being neutral. However, during the session, the therapist often appears to take a side by asking questions that align with a certain person's view of the problem; the therapist must counterbalance this by asking questions that align with each of the other perspectives so that at the end the therapist is viewed as neutral.

Neutrality also implies not becoming attached to particular meanings, descriptions, or outcomes (Boscolo et al., 1987). The Milan team carefully avoided buying into any one description of the problem, *including their own*. Neutrality extended to their own hypotheses and ideas about the family, and they avoided "falling in love" with their own ideas. Later, Cecchin, Lane, and Ray (1992) characterized this form of neutrality as a form of *irreverence* (discussed in next section) that allows for broad maneuverability. When therapists do not rigidly adhere to a particular problem description, they can see more possibilities for intervention rather than focus on the single solution that fits with their preferred hypothesis.

Curiosity and Esthetics

In 1987, Cecchin wrote an article to clarify common misconceptions about the Milan approach, particularly the concept of neutrality. He emphasizes that the therapeutic stance is primarily one of *curiosity*, a theme that is picked up in postmodern therapies, most notably with collaborative therapists (Chapter 14), with whom Cecchin has a close working relationship toward the end of his career. Therapeutic curiosity comes from a sincere interest in hearing the multiple, polyphonic voices within the system: "In this systemic orientation, we generate descriptions within a frame of curiosity rather than within a frame of true and false explanations" (Cecchin, 1987, p. 406). This curiosity is not a scientific curiosity about finding some singular truth but rather an *esthetic* curiosity about emerging patterns. To use a metaphor, systemic curiosity is not the curiosity of what a scientist learns from running a lab experiment but rather the curiosity you bring to experiencing complex art, dance, or poetry: a type of openness to the multiple possible layers of meaning.

In particular, Cecchin is interested in "the patterns that connect," a theme that draws upon Bateson's seminal cybernetic theories. So, when observing a family, the therapist's curiosity is directed to the patterns that connect, with an eye toward the multiplicity of possible patterns as well. More specifically, the therapist is curious about the multiple stories that families bring to therapy to describe a single interaction. In gathering multiple descriptions, a systemic therapist is *not* interested in the best or most accurate description but rather *how the multiple stories fit together—the pattern that connects!* Signs therapists are losing their sense of curiosity and neutrality include boredom with their work and psychosomatic symptoms related to work (Cecchin, 1987).

Irreverence

The concept of irreverence was not highlighted until later in the literature (Cecchin, Lane, & Ray, 1992), but—when properly understood—it clearly captures the therapist's relationship to the *problem* (not the client). The magic of this approach is derived from the therapist's irreverent relationship with problems.

What is there to be irreverent about? Like other systemic and strategic therapists, Milan systemic therapists are irreverent regarding the "catastrophic" appearance of problems. They do not give in to the appearance that a person has a "personality flaw," "illness," "unresolved childhood issue," or other deep, troubling problem, even though problems appear that way, especially to those who live them. But the systemic therapist knows that appearances are deceiving because problems are intimately connected with the relational and broader social context. In a different context, a person's problem behaviors would be different. The context and problem are always influencing each other. The art of irreverence is to not honor the problem as a mighty foe and assume it has more power than it does. Its power is entirely dependent on its context.

With experience, the therapist is able to see through the appearance of a "big, horrible" problem and instead see that the person, problem, and context form a fluid dance—and experience teaches that by changing only a few steps in the dance, the problem shifts, diminishes, and eventually disappears.

Irreverence is felt in the therapist's confidence and calm response to problem issues. It is driven not by disrespect but *fearlessness*. Whatever problem the client brings—whether the loss of a child or a teen's drama of the week—the therapist remains fearless, maintaining a deep sense of calm and faith in systemic processes, knowing that the problems are never as insurmountable as they appear and that systems are inherently self-correcting. Irreverence does not imply lack of empathy or sensitivity. Instead, it allows the therapist to maintain an openness, creativity, and flexibility to provide maximum benefit to the client.

Therapist's Unavoidable Influence

Grounded in a second order cybernetic perspective (see Chapter 4 for review), Cecchin, Lane, and Ray (1994) argue that the systemic therapist needs to actively monitor their influence on the client system. Because they are not neutral observers in the interactions they observe and in the responses they receive from clients, therapists are responsible for monitoring the effects of their beliefs (including their therapeutic theory), actions, and words on clients. What kind of system is the therapist cocreating with the client? Is it helping the client? Does the client feel respected and heard? Milan therapists have long been so true to their second order cybernetic perspective that they argue one should abandon their hypothesis about the family if it seems to be incorrect or unhelpful and to even abandon theoretical notions if they are not serving their purpose (using theory to abandon theory is a classic philosophical conundrum that you are free to ponder on sleepless nights: if a therapist uses theory to abandon theory, is she still practicing that theory?).

The Viewing: Case Conceptualization and Assessment
Family Games (Interactional Sequences)

Similar to MRI therapists, Milan therapists carefully assess the family's interactional sequences (see above), which they refer to as *family games*. Often, English speakers think that the Milan term *game* implies manipulation and ill intent, meanings the term may have in English. However, that is not the connotation of *game* in the systemic context, which focuses on the relational rules for how the family interacts, rules that are not consciously created but rather naturally emerge from the family interaction pattern.

In addition to observing actual interactions, Milan therapists ask circular questions to assess the interactional sequence; these questions can be used with individuals, couples, or families. For example, with the couple having trouble with the defiant child, the therapist would ask: "What is happening just before the problem incident? What is each person saying and doing? How does the child respond? How does each parent and anyone else involved respond to the defiance? How does the child respond to this? How do the parents respond to the child next?" and so on, until family homeostasis is restored. When working with only an individual, circular questions are often the only way to assess the interactional sequence.

The Tyranny of Linguistics

When assessing families, the Milan team paid particular attention to the family's word choices and expressions. They were mindful of how language, particularly descriptions of self and others, shape lived reality: they called this the "tyranny of linguistics" (Selvini Palazzoli et al., 1978). For example, the statements "I am depressed" or "he's an angry person" are global labels that create little space for noticing numerous moments throughout the day when other feelings and identities may be experienced. Instead, Milan therapists encourage descriptions of a person's *action*: times when I "do" depression or when he is "doing" anger. This subtle shift of words suddenly opens new opportunities for how people experience themselves while also allowing for descriptions of times when the depression is not as strong or a person is not angry.

For Milan therapists, positive labels, such as "intelligent" or "good," are as limiting and problematic as negative descriptors. Like negative descriptions, positive labels obscure the relational context that shapes them (Boscolo et al., 1987). Both positive and negative descriptors are avoided in favor of interactional behavioral descriptions.

Family Epistemology and Epistemological Errors

Drawing upon the work of Bateson, Milan therapists identify the family's epistemology, specifically the epistemological errors that appear to be maintaining the symptom (see Chapter 4 for review; Selvini Palazzoli et al., 1978). When examining epistemology, the therapist notes how the family *punctuates* events, meaning how events are described in terms of cause, effects, ordering, etc. One of the most common epistemological errors is to punctuate a systemic, mutually generated interaction as a one-directional causal event. For example, a husband may claim, "*Because she nags,* I withdraw." Most often, the wife will respond: "It's because you won't engage that I have to nag." Both descriptions fail to capture the systemic interaction that is going on because of how they are punctuating the sequence of events.

Targeting Change: Goal Setting
New Meaning

The overarching focus of Milan therapy is to generate new meanings and distinctions (punctuations) for the system that change the "games" (interactional behavior sequences) in the family. Firmly grounded in their cybernetic systemic perspective, Milan therapists do not believe they can "fix" or "correct" the family's interactions. Instead, they can only perturb or shake up the system so that it corrects itself (see Cybernetics in Chapter 4 if this does not sound familiar). Invariably, when families have their normal interaction patterns interrupted, they quite easily choose a better course of action. It seems that most of us consciously know what would be a more ideal response to a given situation but the pull of the systemic dynamics makes it hard to do so.

Goals

Like the other systemic therapies in this chapter, there are no predefined goals of health other than increased systemic flexibility; instead, the goals are simply symptom reduction by means of new systemic interaction patterns:

- The family integrates new information in such a way as to alter the "rules" of the family game so that no member has symptoms (Selvini Palazzoli et al., 1978).
- The family is able to maintain stability and cohesion as a system.

The Doing: Interventions
The Hypothesizing Process

> "*The construction of hypotheses is a continuous process, coevolving with the family's movement. The act of hypothesizing is best described using the concepts of cybernetic feedback loops, for as the family's response to the question modifies or alters one hypothesis, another is formed based on the specifics of that new feedback. This continuous process of hypothesis construction requires the therapist to reconceptualize constantly, both as an interviewer and team member.*" —Boscolo et al., 1987, p. 94

The process of generating a hypothesis, emphasized in the Milan approach, continues throughout therapy and generally has two phases:

1. *Hypothesizing for Conceptualization:* This is the behind-the-mirror process of developing and revising hypotheses that the therapist and team use to guide their "viewing" and provide an overall focus and direction for therapy.
2. *Hypothesizing as Intervention:* When the therapist and team believe it will benefit the family, the hypothesis they have been developing behind the mirror is shared verbally with the family as an in-session intervention.

As already discussed, a hypothesis usually defines the role of the symptom in maintaining the family's homeostasis (Boscolo et al., 1987). For example, a teen's delinquent behavior may serve to bring the parents closer together; without the child's "problem" behavior, the couple might experience other difficulties that would be more difficult for the family to manage. Hypotheses can also highlight how the family wants to simultaneously keep things the same and change, two counterbalancing forces. When therapists deliver hypotheses to the family, the purpose is to "create news of a difference" and shift how they think about the problem, thus creating new possibilities for change. Hypotheses need to be worded so that they are different from how the family is currently viewing the situation yet not too different so that they react against it. Instead, hypotheses need to be *plausible* in the family's current worldview (Watzlawick et al., 1974).

The Milan team identified three common types of hypotheses (Boscolo et al., 1987):

- *Hypotheses About Alliances:* These describe alliances and coalitions: who is on whose team.
- *Hypotheses About Myths and Premises:* These identify unrealistic or problematic myths and premises that are contributing to the problem (myths of the perfect marriage, ideal child, etc.).
- *Hypotheses That Analyze Communication:* These track problematic communication patterns, such as double binds.

Positive Connotation (Milan-Style Reframe)

Milan therapists originally developed positive connotations to not contradict themselves when prescribing paradoxical interventions (Selvini Palazzoli et al., 1978). Therapists use positive connotation to respect both the family's fear of change and their request for change. They *interpret the behavior* of each member of the family positively, as having an underlying benevolent motivation (Boscolo et al., 1987). A common positive connotation is reframing a child's problematic behavior as a way to keep the parents together and reframing the parents' arguments about how to handle this behavior as a way to show their dedication to the family and each other. Perhaps the most important outcome of positive connotation is the effect it has on the therapist, enabling the therapist to view members of the system less judgmentally and with greater hope.

Circular Questions

Described in Juice above, circular questions are one of the primary interventions used in Milan therapy. These questions often were so effective in revealing the structure of the system to the family in a language they understood that it often makes other reframes, such as positive connotation, unnecessary (Boscolo et al., 1987).

Counterparadox

Similar to the MRI therapeutic double bind (see above), Milan counterparadox is used to therapeutically respond to the paradoxes, or double binds, that families and couples create for themselves (Selvini Palazzoli et al., 1978; Watzlawick, Bavelas, & Jackson, 1967). Counterparadox messages request that the family not change even though they have come in for change; often this has the effect of amplifying the problem behavior to such a degree that the family spontaneously gives it up because its absurdity becomes undeniable (Boscolo et al., 1987). Counterparadox is particularly appropriate with families who are not cooperative with other forms of intervention.

Rituals

Milan therapists refer to their behavioral prescriptions as *rituals*. Rituals were behavioral assignments given to the family that address the double-binding communication and designed to help the family shift meanings associated with the problem. In the early years, rituals were often used to reinforce positive connotations or counterparadox. For example, to reinforce a positive connotation of a child's symptomatic behavior, the therapist may ask that each night at dinner each family member thank the symptomatic member for his/her sacrifice for the family and cite the specific benefit for the speaker (Boscolo et al., 1987). However, this type of ritual was often experienced as negative or blaming. In later years, Cecchin and Boscolo began to prefer a more subtle form of ritual that addressed double-binding communication by putting the conflicting directives of the double bind in a sequence. They might do this by having the family act as if one set of messages or opinions were true on odd days, and then on even days acting as though another set of messages or opinions were true. For example, a woman caught in a triangle with her husband and child might be asked to act like a wife to her husband on odd days, a mother to her child on even days, and then on the seventh day to "be spontaneous." How a family responds to the ritual was taken as new information that is used to refine the working hypothesis of systemic interactions and to design a more meaningful intervention next session.

Invariant Prescriptions

One of the Selvini Palazzoli's earliest therapeutic innovations and the focus of her long-term research, the *invariant prescription* is just what it sounds like: an intervention that is not varied across families (Selvini Palazzoli, 1988). Used primarily with families whose children are labeled anorexic or schizophrenic, the intervention severs covert coalitions between a parent and a child. Parents are instructed to arrange to go on a date (or other outing) and to not tell the children where they are going or why. The desired effect is to create a secret between the parents, ending inappropriate coalitions by creating a clear boundary between the parents as a unified team and the symptomatic child. The child loses the status of being a special confidant of the one parent, lessening his/her emotional burden and resulting in fewer problem symptoms. Although originally designed for severe pathology such as anorexia and psychosis, this intervention is also effective with modern parents who overemphasize transparent and open communication with their children to the extent that the children use this information to "manipulate" them.

Putting It All Together

Case Conceptualization Template

- *Family Games/Interaction Sequence:* Describe each person's role (even seemingly minor players) in the interactional sequence, emphasizing the following:
  - Homeostasis
  - Rise of tension
  - Symptom/conflict
  - Return to homeostasis
- *Tyranny of Linguistics:* What descriptions are being used to describe people as having inherent characteristics (nouns) rather than as behaviors (verbs)?
- *Epistemology:* What are the epistemological errors that characterize the interactions? Is there one-directional, blaming descriptions of the problem that do not recognize the systemic dynamics? What problematic meanings are making the client stuck?
- *Hypothesis/Positive Connotation:* What is the role of the symptom in maintaining family homeostasis? How might the symptom serve to maintain connection, create independence/distance, establish influence, reestablish connection, or otherwise help create a sense of balance in the family?

Treatment Plan Template for Individual

Milan Initial Phase of Treatment with Individual

Initial Phase Counseling Tasks

1. Develop working counseling relationship. *Diversity Note: Adjust to respect cultured, gendered, and other styles of relationship building and emotional expression.* Relationship building approach/intervention:
   a. Use *neutrality* and *curiosity* to help client feel understood and *irreverence* toward the problem labels.

(continued)

2. Assess individual, systemic, and broader cultural dynamics. *Diversity Note: Adjust assessment based on cultural, socioeconomic, sexual orientation, gender, and other relevant norms.*
   Assessment strategies:
   a. Use *circular questions* to develop a *hypothesis* about how *family games* and interaction sequences are related to the client's "doing" of depression and anxiety.
   b. Identify how *epistemological errors* and related *linguistic descriptions* help to maintain the depressive and anxious symptoms.

Initial Phase Client Goals

1. Increase client's ability to *reframe* [depressive/anxiety symptom] as systemic thereby changing its *meaning* and their *punctuation* of the problem to reduce depression and anxiety.
   a. *Positive connotation* of each person's role in the conflict (client's and those outside of session).
   b. *Circular questions* to enable client to see systemic interaction cycle and client's role in it.
   c. Deliver *hypothesis* to client that reframes symptoms from a systemic perspective, highlighting how each person's behavior is contextualized by others.

Milan Working Phase of Treatment with Individual

Working-Phase Counseling Task

1. Monitor quality of the working alliance. *Diversity Note: Attend to client response to interventions that indicate therapist using expressions of emotion that are not consistent with client's cultural background.*
   a. *Assessment Intervention:* Observe client responses to ensure that client sees the therapist as essentially *neutral,* not taking client's side or side of someone else in the system; monitor *therapist influence* to ensure having a positive effect on client.

Working-Phase Client Goals

1. Increase *new meanings* and develop alternative *linguistic punctuations* of problem behavior to reduce depression and anxiety.
   a. *Circular questions* to identify differences, see behavioral sequences, and consider hypothetical situations to shift meaning associated with problem behaviors and how they fit into relational patterns.
   b. *Rituals* designed to slightly alter client's behavior sequences related to symptoms.
2. Increase satisfying *relational interactions* related to [specific symptom] to reduce depression and anxiety.
   a. *Rituals* prescribed to alter client's half of interaction in problem relational cycle.
   b. *Circular questions* to shift meanings associated with particular area of conflict, emphasizing the other person's possible view of situation.

Milan Closing Phase of Treatment with Individual

Closing-Phase Counseling Task

1. Develop aftercare plan and maintain gains. *Diversity Note: Access resources in the communities of which they are a part to support them after ending therapy.*
   a. Begin therapy with clearly defined end point, and develop *rituals* to sustain new behaviors.

(continued)

Closing-Phase Client Goals

1. Increase satisfying *relational interactions* [in a different area of life] to reduce depression and increase sense of wellness.
   a. *Circular questions* to develop new meanings related to area of life.
   b. *Rituals* designed help client to shift meaning associated with the problem.

Treatment Plan Template for Couple/Family

Milan Initial Phase of Treatment with Couple/Family Conflict

Initial Phase Counseling Tasks

1. Develop working counseling relationship. *Diversity Note: Adjust to respect cultured, gendered, and other styles of relationship building and emotional expression.* Relationship building approach/intervention:
   a. Use *neutrality* and *curiosity* to help all parties feel understood and *irreverence* toward the problem labels.

2. Assess individual, systemic, and broader cultural dynamics. *Diversity Note: Adjust assessment based on cultural, socioeconomic, sexual orientation, gender, and other relevant norms.*
   Assessment strategies:
   a. Use *circular questions* to develop a *hypothesis* about how *family games* and interaction sequences maintain the couple/family's conflictual interactions.
   b. Identify how *epistemological errors* and related *linguistic descriptions* help to maintain the problem interactions.

Initial Phase Client Goals

1. Increase couple/family's ability to *reframe* conflict as systemic thereby changing its *meaning* and their *punctuation* of the problem to reduce family conflict.
   a. *Positive connotation* of each person's role in the conflict.
   b. *Circular questions* to enable family to see systemic interaction cycle.
   c. Deliver *hypothesis* to family that reframes their conflict from a systemic perspective, highlighting how each person's behavior is contextualized by others.
   d. Only parents invited to third session.

Milan Working Phase of Treatment with Family

Working-Phase Counseling Task

1. Monitor quality of the working alliance. *Diversity Note: Attend to client response to interventions that indicate therapist using expressions of emotion that are not consistent with client's cultural background.*
   a. *Assessment Intervention:* Note family/couple responses to ensure that they see the therapist as essentially *neutral*; monitor *therapist influence* to ensure having a positive effect on client.

Working-Phase Client Goals

1. Decrease influence of *outside coalitions* to reduce family conflict.
   a. *Invariant prescription* to severe coalitions with children and/or outside others.
   b. *Rituals* to interrupt couple/family interactions that trigger coalitions.

2. Increase *new meanings* and develop alternative *linguistic punctuations* of problem behavior to reduce conflict.

(continued)

a. *Circular questions* to identify differences, see behavioral sequences, and consider hypothetical situations to shift meaning associated with problem behaviors in couple/family.
   b. *Counterparadox* to address paradoxes inherent in labeling of family members.
3. Increase satisfying *family interactions* [*games* related to specific topic] to reduce family conflict and increase sense of connection.
   a. *Rituals* prescribed to alter conflict cycle.
   b. *Circular questions* to shift meanings associated with particular area of conflict.

Milan Closing Phase of Treatment with Family

Closing-Phase Counseling Task

1. Develop aftercare plan and maintain gains. *Diversity Note: Access resources in the communities of which they are a part to support them after ending therapy.*
   a. Begin therapy with clearly defined end point, and develop *rituals* to sustain new behaviors.

Closing-Phase Client Goals

1. Increase satisfying *family interactions* [*games* related to a different area of relationship] to reduce family conflict and increase sense of connection.
   a. *Circular questions* to develop new meanings related to area of conflict.
   b. *Rituals* designed help couple/family to shift meaning associated with the problem, often by acting from different epistemological positions on alternating days of the week.

Research and the Evidence Base

Systemic therapy began as a research project: the Bateson team began by studying communication in families with members diagnosed with schizophrenia. The tradition of observational research has been integrated into required standard training in systemic and strategic models in the form of observation teams. Although there has been less systematic research on outcomes of specific systemic models, such as strategic, Milan, or MRI, there is consistent and growing research on the effectiveness of evidence-based systemic approaches for specific conditions, such as adolescent substance abuse, adolescent conduct issues, depression related to relationship distress, severe mental illness, and couple distress (Sprenkle, 2012). These evidence-based treatments use key theoretical concepts and techniques from systemic family therapies and apply them to a specific population.

Systemic therapies are quickly emerging as a leading force in the realm of evidence-based treatments. Numerous empirically supported treatments incorporate key elements of systemic and strategic therapies, including:

- *Multisystemic Family Therapy* (Henggeler, Schoenwald, Borduin, Rowland, & Cunningham, 1998; see section later in this chapter)
- *Brief Strategic Family Therapy* (structural ecosystemic therapy, structural eco-developmental preventive interventions; see Chapter 6; Szapocznik & Williams, 2000)
- *Ecosystemic Structural Therapy* (see Chapter 7; Lindblad-Goldberg, Dore, & Stern, 1998)
- *Multidimensional Family Therapy* (Liddle, Dakof, & Diamond, 1991)
- *Functional Family Therapy* (see Chapter 15)
- *Emotionally Focused Couples Therapy* (see Chapter 15)

Although each of these approaches is unique, they all incorporate identifying the problem interaction sequence and in some way interrupting or altering it using one or more systemic or strategic techniques to address the needs of a specific population.

Clinical Spotlight: Multisystemic Therapy

Multisystemic therapy (MST) was developed in the 1970s to treat serious juvenile offenders (Multisystemic Therapy Services, 1998). A family-based treatment model, MST draws from strategic, structural, socioecological, and cognitive-behavioral therapy models. In addition, MST also places significant emphasis on the adolescent's and family's broader social network, removing offenders from problematic social networks, improving school and/or vocational performance, and developing a strong support network for the child and the family (Henggeler, 1998; Henggeler & Borduin, 1990; Multisystemic Therapy Services, 1998).

Goals

The overarching goals of MST include the following:

- Decrease antisocial behavior and other clinical problems.
- Improve functioning in family relations.
- Improve functioning in school and/or work contexts.
- Minimize out-of-home placements, including incarceration, residential treatment, and hospitalization.

Case Conceptualization

When conceptualizing treatment, MST therapists consider risks and protective factors in the following domains: individual, family, peer, school, and neighborhood and community (Multisystemic Therapy Services, 1998).

	Risks	Protective Factors
Individual	Positive attitude toward antisocial behavior, psychological symptoms, hostility, low intellectual functioning and verbal skills	Intelligence, eldest child, easy-going personality, pro-social values, problem-solving skills
Family	Lack of parental supervision, ineffective or inconsistent discipline, lack of emotional expression, conflict, parental difficulties	Connection to parents and family, supportive family, strong parental relationship
Peer	Association with socially deviant peers, poor relational skills, few pro-social friends	Connections with pro-social friends
School	Poor performance, little interest, little support	Committed to education, goals, acceptable performance
Neighborhood and Community	High family mobility, criminal subculture, disorganized	Involvement in religious or social organizations, strong support network

Principles of Intervention

1. *Finding the Fit:* The therapist assesses how the adolescent's problems systemically fit within the broader family, peer, school, and community culture.
2. *Focus on Positives and Strengths:* Therapeutic interactions emphasize strengths and potential strengths, in both the individual adolescent and the family.
3. *Increasing Responsibility:* Interventions aim to increase responsible behavior with all family members, promoting parental involvement and helping teens accept responsibility for their choices.
4. *Focus on the Present, on Action, and on Clarity:* Interventions are action oriented and present focused and target specific, easily defined problems that are easily tracked and measured, such as achieving a specific grade point average or adhering to curfew.

5. *Targeting Sequences:* Consistent with its systemic foundation, sequences of behaviors are the target of change; these sequences can be between family members or with peers and the larger social systems.
6. *Developmentally Appropriate:* Interventions are developmentally appropriate for youth, promoting step-by-step development of the competencies and skills needed for success as an adult.
7. *Continuous Effort:* Interventions, by design, require daily or weekly effort from the family.
8. *Evaluation and Accountability:* Rather than blame the family if an intervention does not work, MST continually assesses the effectiveness of interventions and adjusts as necessary to ensure success.
9. *Generalization:* Treatment is designed to help the adolescent and family generalize their skills and abilities to solve problems in other life area.

Tapestry Weaving: Diversity Considerations

Ethnic, Racial, and Cultural Diversity

Because systemic and strategic (Chapter 6) family therapies do not rely on a theory-based definition of health and normalcy, they adapt relatively easily to different cultural groups and subpopulations. Furthermore, drawing on systemic and constructivist foundations, these therapies aim to work from *within* the client's worldview; when this is successfully achieved, the therapist adapts the language and interventions to the client's values and beliefs.

Of particular interest, two groups of systemic therapists have systematically adapted and used the approach with Latino/Hispanic clients. The Latino Brief Therapy Center at the MRI uses the non-normative and nonpathologizing principles of the model to guide their work, creating unique meaning and interventions with each client, rather than rely on cultural stereotypes of the extremely diverse Latino community (Anger-Díaz, Schlanger, Rincon, & Mendoza, 2004). Similarly, brief strategic family therapy, an evidence-based approach that uses elements of systemic and structural family therapies (see Chapter 6 for detailed description), was developed specifically to treat Hispanic youth with high risk behaviors and substance issues; the approach has also been found to be effective with African American youth (Robbins, Horigian, Szapocznik, & Ucha, 2010; Santisteban, Coatsworth, Perez-Vidal, Mitrani, Jean-Gilles, & Szapocznik, 1997).

In addition to these specific treatment approaches, McGoldrick, Giordano, and Garcia-Preto (2005) provide descriptions of how systemic ideas and approaches can be modified to meet the needs of families from 46 different ethnicities, an invaluable resource when working with cultural groups other than one's own (and perhaps your own; read and see). Although these descriptions must be general and do not capture the unique lived experience of a particular client that may be in front of you at a given moment, they can give you, the therapist, a clue about how best to understand and engage the client (and most clients appreciate you having at least a clue about their background as well as giving them room to share their own unique experience and beliefs).

Sexual Identity Diversity

Because of its nonpathologizing and non-normative assumptions and its attention to larger systemic dynamics, systemic therapies have been widely used with clients who identify as gay, lesbian, bisexual, transgendered, or questioning (GLBTQ). Butler (2009) identifies five principles for adapting systemic therapy approaches for these clients.

1. *Understanding Heterosexism:* Therapists need to be aware that gender and sexual minorities have *daily* experiences of being marginalized, judged, and ignored, and rarely see their reality reflected in advertisements, movies, and other social forums.
2. *Therapist Self-Reflectivity:* Heterosexual therapists need to engage in personal self-reflection of their experience of being privileged and having their sexual and gender orientations generally approved by society.

3. *Locating Your Position, Transparency, and Self-Disclosure:* Based on the work of feminist therapists, Butler recommends disclosing one's gender and sexual orientation when working with GLBTQ clients. Such openness allows for frank discussions about areas of similarity and differentness, helping to address issue of power and hierarchy. However, too much self-disclosure can be nonproductive.
4. *Client as Expert, and Therapist as Curious:* Drawing upon collaborative therapy principles (see Chapter 14), therapists should assume a curious stance, viewing the client as the expert in their lives and lifestyle.
5. *Connecting to Wider Systems:* Finally, therapists should consider the larger social systems of which their GLBTQ clients are a part and encourage them to seek positive, supportive communities.

Butler (2009) also discusses how therapists need to adjust their approach to GLBTQ couples. First, therapists should be aware that sex, gender roles, monogamy, family of origin, family of choice, and ex-partners generally have very different meanings with these couples. Furthermore, gender roles are thought to have more impact than sexual orientation per se in these relationships, with internalized gender roles often greatly restricting same-sex relationships and being a particular issue in couples with a transgendered partner. Therapists can help couples to deconstruct their gender role expectations and assumptions in ways that strengthen the partnership. Finally, therapists should keep in mind that the family life cycle looks different for most GLBTQ individuals, couples, and families (Goldenberg, 2009).

Adolescents Coming Out to Family

Another common issue that family therapists address is adolescents coming out to their families (Butler, 2009). First and foremost, therapists should be aware that gay, lesbian, bisexual and transgendered youth have higher rates of suicidality, substance abuse, self-harm, depression, anxiety, and school issues; so it is important to monitor the adolescent's safety and address these issues. Before coming out, most adolescents are generally able to accurately predict their parents' responses based on casual comments about sexuality over the years. Therapists can use DeVine's (1984) stage theory for the family's process of coming out to help families negotiate this transition.

- *Subliminal Awareness:* The child's sexuality is suspected and sometimes provoked by inquiries about dating and same-sex friendships.
- *Impact:* The child announces his/her sexuality and suspicions are confirmed.
- *Adjustment:* The family struggles to maintain its old homeostasis while the child is expected to hide or deny his/her sexuality.
- *Resolution:* The family comes to accept the child's sexuality identity.
- *Integration:* The family shifts their values related to gender and sexual norms.

Many parents whose child comes out report a "loss" of their dreams of having a wedding or grandchildren, and much of their initial response may actually reflect the personal losses they are projecting. Therapists can help redirect parents to focus on their child's real needs at this time while also developing a more hopeful view of their future family life.

Online Resources

Brief Strategic and Systemic World Network:
www.bsst.org

Mental Research Institute:
www.mri.org

Multisystemic Therapy:
www.mstservices.com

References

*Asterisk indicates recommended introductory readings.

Anger-Díaz, B., Schlanger, K., Rincon, C., & Mendoza, A. (2004). Problem-solving across cultures: Our Latino experience. *Journal of Systemic Therapies, 23*(4), 11–27. doi:10.1521/jsyt.23.4.11.57837

*Bateson, G. (1972). *Steps to an ecology of mind*. San Francisco, CA: Chandler.

Bateson, G. (1979). *Mind and nature: A necessary unity*. New York: Dutton.

Bauer, S. M., Schanda, H., Karakula, H., Olajossy-Hilkesberger, L., Rudaleviciene, P., Okribelashvili, N., & Stompe, T. (2011). Culture and the prevalence of hallucinations in schizophrenia. *Comprehensive Psychiatry, 52*(3), 319–325. doi:10.1016/j.comppsych.2010.06.008

*Boscolo, L., Cecchin, G., Hoffman, L., & Penn, P. (1987). *Milan systemic family therapy*. New York: Basic Books.

Butler, C. (2009). Sexual and gender minority therapy and systemic practice. *Journal of Family Therapy, 31*(4), 338–358. doi:10.1111/j.1467-6427.2009.00472.x

Campbell, D., Draper, R., & Crutchley, E. (1991). The Milan systemic approach to family therapy. In A. S. Gurman & D. P. Knishern (Eds.), *Handbook of family therapy* (pp. 325–362). New York: Brunner/Mazel.

*Cecchin, G. (1987). Hypothesizing, circularity, and neutrality revisited: An invitation to curiosity. *Family Process, 26*(4), 405–413.

Cecchin, G., Lane, G., & Ray, W. (1992). *Irreverence: A strategy for therapist survival*. London: Karnac.

Cecchin, G., Lane, G., & Ray, W. A. (1994). Influence, effect, and emerging systems. *Journal of Systemic Therapies, 13*(4), 13–21.

DeVine, J. L. (1984). A systemic inspection of affectional preference orientation and the family of origin. *Journal of Social Work and Human Sexuality, 2*, 9–17.

Fisch, R., & Schlanger, K. (1999). *Brief therapy with intimidating clients*. New York: Jossey-Bass.

*Fisch, R., Weakland, J., & Segal, L. (1982). *The tactics of change: Doing therapy briefly*. New York: Jossey-Bass.

Gehart, D. (2010). *Mastering competencies in family therapy: A practical approach to theory and clinical case documentation*. Pacific Grove, CA: Brooks/Cole.

Gehart, D. (2013). *Theory and treatment planning in counseling and psychotherapy: A competency-based approach for applying theory in clinical practice*. Pacific Grove, CA: Brooks/Cole.

Gehart, D. (2014). *Mastering competencies in family therapy: A practical approach to theory and clinical case documentation* (2nd ed.). Pacific Grove, CA: Brooks/Cole.

Goldberg, A. E. (2009). Lesbian, gay, and bisexual family psychology: A systemic, life-cycle perspective. In J. H. Bray & M. Stanton (Eds.), *The Wiley-Blackwell handbook of family psychology* (pp. 576–587). Hoboken, NJ: Wiley-Blackwell. doi:10.1002/9781444310238.ch40

Haley, J. (1976). *Problem-solving therapy: New strategies for effective family therapy*. San Francisco, CA: Jossey-Bass.

*Haley, J. (1987). *Problem-solving therapy* (2nd ed.). San Francisco, CA: Jossey-Bass.

Henggeler, S. W. (1998). *Multisystemic therapy*. Charleston, NC: Targeted Publications Group. Downloaded May 2, 2008, from www.addictionrecov.org/paradigm/P_PR_W99/mutisys_therapy.html

Henggeler, S. W., & Borduin, C. M. (1990). *Family therapy and beyond: A multisystemic approach to treating the behavior problems of children and adolescents*. Pacific Grove, CA: Brooks/Cole.

Henggeler, S. W., Schoenwald, S. K., Borduin, C. M., Rowland, M. D., & Cunningham, P. B. (1998). *Multisystemic treatment of antisocial behavior in children and adolescents*. New York: Guilford.

Robbins, M. S., Horigian, V., Szapocznik, J., & Ucha, J. (2010). Treating Hispanic youths using brief strategic family therapy. In J. R. Weisz & A. E. Kazdin (Eds.), *Evidence-based psychotherapies for children and adolescents* (2nd ed.) (pp. 375–390). New York: Guilford Press.

Jackson, D. D. (1967). The myth of normality. *Medical Opinion and Review, 3*, 28–33.

*Keeney, B. (1983). *Aesthetics of change*. New York: Guilford.

Liddle, H. A., Dakof, G. A., & Diamond, G. (1991). Adolescent substance abuse: Multidimensional family therapy in action. In E. Daufman & P. Kaufman (Eds.), *Family therapy of drug and alcohol abuse* (pp. 120–171). Boston, MA: Allyn & Bacon.

Lindblad-Goldberg, M., Dore, M., & Stern, L. (1998). *Creating competence from chaos*. New York: Norton.

McGoldrick, M., Giordano, J., & Garcia-Preto, N. (Eds.). (2005). *Ethnicity and family therapy* (3rd ed.). New York: Guilford.

Mental Research Institute. (2002). *On the shoulder of giants*. Palo Alto, CA: Author.

Multisystemic Therapy Services. (1998). *Multisystemic therapy*. Downloaded May 2, 2008, from www.mstservices.com/text/treatment.html

*Nardone, G., & Watzlawick, P. (1993). *The art of change: Strategic therapy and hypnotherapy without trance*. New York: Jossey-Bass.

Ray, W. A., & Nardone, G. (Eds.). (2009). *Paul Watzlawick: Insight might cause blindness and other essays*. Phoenix, AZ: Zieg, Tucker, & Theisen.

Ray, W. A., & Keeney, B. (1994). *Resource focused therapy*. London: Karnac Books.

Santisteban, D. A., Coatsworth, J., Perez-Vidal, A., Mitrani, V., Jean-Gilles, M., & Szapocznik, J. (1997). Brief structural/strategic family therapy with African American and Hispanic high-risk youth. *Journal of Community Psychology, 25*(5), 453–471. doi:10.1002/(SICI)1520-6629(199709)25:5<453::AID-JCOP6>3.0.CO;2-T

Segal, L. (1991). Brief therapy: The MRI approach. In A. S. Gurman & D. P. Knishern (Eds.), *Handbook of family therapy* (pp. 171–199). New York: Brunner/Mazel.

Selvini Palazzoli, M. (Ed.). (1988). *The work of Mara Selvini Palazzoli*. New York: Jason Aronson.

*Selvini Palazzoli, M., Boscolo, L., Cecchin, G., & Prata, G. (1980). Hypothesizing-circularity-neutrality: Three guidelines for the conductor of the session. *Family Process, 19*(1), 3–12.

Selvini Palazzoli, M., Cecchin, G., Prata, G., & Boscolo, L. (1978). *Paradox and counterparadox: A new model in the therapy of the family in schizophrenic transaction*. New York: Jason Aronson.

Sprenkle, D. (Ed.). (2012). Intervention research in couple and family therapy [Special edition]. *Journal of Marital and Family Therapy, 38*(1).

Szapocznik, J., & Williams, R. A. (2000). Brief strategic family therapy: Twenty-five years of interplay among theory, research and practice in adolescent behavior problems and drug abuse. *Clinical Child and Family Psychology Review, 3*(2), 117–135.

Watzlawick, P. (1977). *How real is real? Confusion, disinformation, communication.* New York: Random House.

Watzlawick, P. (1978/1993). *The language of change: Elements of therapeutic conversation.* New York: Norton.

Watzlawick, P. (Ed.). (1984). *The invented reality: How do we know what we believe we know?* New York: Norton.

Watzlawick, P. (1988). *Ultra solutions: How to fail most successfully.* New York: Norton.

Watzlawick, P. (1990). *Munchhausen's pigtail or psychotherapy and "reality" essays and lectures.* New York: Norton.

Watzlawick, P. (1993). *The situation is hopeless but not serious: The pursuit of unhappiness.* New York: Norton.

Watzlawick, P., Bavelas, J. B., & Jackson, D. D. (1967). *Pragmatics of human communication: A study of interactional patterns, pathologies, and paradoxes.* New York: Norton.

Watzlawick, P., & Weakland, J. H. (1977). *The interactional view: Studies at the Mental Research Institute, Palo Alto, 1965–1974.* New York: Norton.

*Watzlawick, P., Weakland, J., & Fisch, R. (1974). *Change: Principles of problem formation and problem resolution.* New York: Norton.

Weakland, J., & Ray, W. (Eds.). (1995). *Propagations: Thirty years of influence from the Mental Research Institute.* Binghamton, NY: Haworth Press.

Systemic Case Study: Blended Family Issues

Olga (AF42), of German and Swedish decent, and Olga's mother, Gertrude (AF62), are seeking help for Elizabeth (CF16), Olga's 16-year-old daughter. Olga and Gertrude report that for the past six months Elizabeth has been very defiant, especially toward her, and has been experiencing symptoms of depression such as insomnia, lack of appetite, loss of interest in activities she used to enjoy, and no longer hangs out with her friends. Her exceptional grades have slowly started to slip because Elizabeth has stopped doing her homework, claiming she no longer has an interest in schoolwork and wants to drop out. Elizabeth's father died from cancer while her mother was pregnant with her and her mother did not start dating until she was 15 years old.

Her mother met an Italian American man (AM) online six months prior to this meeting and within two months they were married. Immediately after the wedding he moved in to the house with Elizabeth, Olga, and Gertrude. Elizabeth has never spoken to her stepfather. She says she is angry with him because he "stole" her mother away from her and her mother no longer has any time for her. Gertrude is happy that her daughter has found love again and has developed a close relationship to Olga's new husband. Previous to Olga's marriage Elizabeth never had many rules to follow, but since her stepdad moved in she has enforced very strict rules such as curfew and chores around the house. Her mother also took on a third job when she got married to make extra money to support them. Elizabeth's stepdad is currently unemployed and not actively seeking employment, which bothers her.

During the session Elizabeth's mother constantly interrupted her daughter and yelled back at her as if she is her sister or a friend and not her mother. Elizabeth and her mother both say that they have not had a conversation with each other since her mother married her stepdad without it ending up in an argument. Elizabeth often complains about her stepfather, which sparks an argument between her mother and herself. Gertrude has mostly tried to stay out of the issues between Elizabeth, Olga, and her husband because she doesn't know what else to do. Elizabeth's mother and grandmother report that they often hear Elizabeth crying in her room and when Olga tries to comfort her Elizabeth is extremely defensive.

Systemic Case Conceptualization

Interactional Sequence
Homeostasis: CF is alone in her room avoiding interaction with her mother and stepfather.
Start of Tension: Mother knocks on CF's door and demands that she come out and spend time with the family for dinner.

Escalation and Symptom: CF yells at her mother and says she doesn't want to come out because she cannot stand being around her stepfather. CF's mother yells back and says awful comments such as "I wish I never had you" and calls her daughter names. Daughter yells insults back about stepfather. Mother storms away.

Return to Homeostasis: Mother goes back to CF's room and apologizes to her and allows her to eat her dinner alone in her room.

More of the Same Solutions

CF chooses to isolate herself and avoid her mother and refuses to speak with her stepfather; this just makes her mother angrier and her isolation often escalates into an argument.

AF42 tries to solve the problem by at first pursuing her for connection and then after an argument apologizing to her daughter for yelling at her and calling her names; she then allows her daughter to continue to isolate herself in her room.

AM and AF62 (mother) ignore the entire exchange and also does not take action to engage CF.

Metacommunication

CF's metacommunication clearly indicates that she does not accept the new mother-daughter relationship as the mother has tried to define it since getting married.

CF's metacommunication with her stepfather indicates that she does not in any form accept him as her mother's husband or her stepfather.

AF42 metacommunication is trying to redefine the relationship in more hierarchical ways than they had in the past; it also defines AM as the head of the household.

AM's metacommunication suggests he does not see himself having a provider or parenting role, although he does expect CF to do her chores and be responsible.

AF62 metacommunication is supportive of her daughter's choices.

Complementary and Symmetrical Patterns

Between AF42 and CF: The complementary pattern that CF and her mother take on is that of the pursuer/distancer. Most fights involve the mother pursuing her daughter to interact with herself and her stepfather, and CF responds with some form of withdrawal such as staying in her room and choosing to never communicate with her stepfather.

Between AF42 and AM: AF42 is over-functioning and AM is under-functioning. AF42 is the good parent; AM is the bad parent.

Hypothesis: Role of Symptom

CF's behavior is an attempt to maintain the same sense of closeness as well as the same level of freedom/hierarchy she shared with her mother for years and suddenly lost after AF shifted all of her energies to her new husband.

Describe the Desired Behavioral Change

Within four weeks CF and AF42 can have conversations without arguing, slamming doors, etc.

AF42 and CF will spend at least one hour per week doing a shared favorite activity without AM.

CF and AM will speak amicably.

AM will contribute more actively to the household; AF will work less.

Systemic Treatment Plan

MRI Initial Phase of Treatment with Family

Initial Phase Counseling Tasks

1. Develop working counseling relationship. *Diversity Note: Use humor to connect with teen; consider couple's cultural differences; respect blended family dynamics as well as memory of deceased father.*
    a. Respect and *trust the family system* while maintaining a *neutral position*, especially in relation to stepfather.

(continued)

2. Assess individual, systemic, and broader cultural dynamics. *Diversity Note: Consider differences in couple's cultural backgrounds, specifically differences in valuing independence vs. interdependence; transition to single-parent household to blended family.*
   a. Assess the *problem interactional sequence*(s) and related *metacommunication*, considering the role of all members in the family related to AF/CF arguments and larger family dynamics. Encourage family to use nonblaming, behavioral descriptions of each person's actions.
   b. Identify the *complementary pattern* of pursuer/distancer between CF and her mother and over-/under-functioning of AF/AM.
   c. Identify *more-of-the-same solutions,* solutions that perpetuate the problem.
   d. Obtain *behavioral description* of desired change.

Initial Phase Client Goals

1. Increase family's possibilities for viewing the problem in such a way as to reduce depressed mood/anxiety and conflict.
   a. *Reframe* the problem in relational context to shift meaning of symptoms, perhaps focusing on how both CF and AM appear depressed, with CF's desiring for meaningful connection with AF and AM's needing to have a meaningful role in the family.
   b. *Behavioral prescription* that represents a *180-degree shift* to attempted solutions, such as stopping AF42 from pursuing connection with CF and instead make low-pressure offers that the daughter is free to decline or accept. AF62 should be encouraged to pursue interaction with her granddaughter as a 180-degree shift from ignoring the situation. AM can be encouraged to take a more active role in some way.

MRI Working Phase of Treatment with Family

Working-Phase Counseling Task

1. Monitor quality of working alliance. *Diversity Note: Attend to connection with teen and parents; ensure AM feels respected and AF has sense of dignity and privacy.*
   a. Monitor family's response to reframes, behavioral prescriptions, and other interventions to determine if you have been able to engage the family meaning system.

Working-Phase Client Goals

1. Increase positive, nonconflictual *interactional sequences* between CF and AF to reduce conflict and reduce CF's depression.
   a. Use *behavioral prescriptions* such as encouraging CF's mother to ignore CF's rebellious behavior or leaving small symbolic gifts with notes for CF.
   b. Utilize the *paradoxical behavioral prescription* that exaggerates each person's pursuer/distance position.

2. Reduce *over-/under-functioner pattern* between AF and AM to reduce conflict with CF.
   a. *Paradoxical behavioral prescriptions* that exaggerate AM's lack of contribution and AF's over-functioning in the family.
   b. *Relational reframing* that link how AM and CF share depressive symptoms while AF over-functions, and possibly is also depressed.

3. Increase *interactional sequences* that enhance emotional connection between AF and CF to reduce CF's depression and parent/child conflict.
   a. *Relational reframing* of conflict by suggesting to mother that her daughter may be acting out because she wants to spend more time with her mother and not because she wants to make her mother angry.

(continued)

b. *Behavioral prescriptions* such as suggesting that CF become the pursuer and politely request that she spend time with her mother instead of withdrawing from her.

Closing Phase of Treatment with Couple/Family

Closing-Phase Counseling Task

1. Develop aftercare plan and maintain gains. *Diversity Note: Identify support from families of origin and church.*
   a. *Dangers of improvement* and *restraining* to prepare family to handle setbacks; include variations that will reduce blame.

Closing-Phase Client Goals

1. Increase sense of connection between all members of the family to increase relational satisfaction and reduce depression.
   a. *Behavioral prescriptions* that allow family to safely explore simple ways of entire family to interaction that are minimally threatening and define CF and AM relationship in ways both accept.
   b. *Paradoxical behavioral prescriptions* that allow CF, AF42, and AM to experience the absurdity of their positions in nonthreatening ways.

CHAPTER 6

Strategic Therapy

"If successful therapy is defined as solving the problems of the client, the therapist must know how to formulate a problem and how to solve it. And if he or she is to solve a variety of problems, the therapist must not take a rigid and stereotyped approach to therapy. Any standardized method of therapy, no matter how effective with certain problems, cannot deal successfully with the wide range that is typically offered to a therapist. Flexibility and spontaneity are necessary."

—Haley, 1987, p. 8

Strategic Therapy

In a Nutshell: The Least You Need to Know

An original member of the MRI team, Jay Haley moved on to develop his own approach with his wife Cloe Madanes. Strategic therapy shares many similarities with the MRI approach: both are grounded in general systems and cybernetic theories, both are brief therapy approaches, and both use uniquely crafted behavioral prescriptions to effect change. The primary differences lie in case conceptualization and his particular approach to behavioral prescription, *directives.* Being heavily influenced by hypnotherapist Milton Erickson (see Chapter 4, Haley's approach uses more metaphors than the MRI approach, including the metaphor of power, over which Haley and Bateson had significant disagreements (for one of the most interesting debates in the field read Bateson, 1972; Dell, 1989; Haley, 1987). In addition, Bateson's friendship with Minuchin (Chapter 7) also influenced his conceptualization approach, which includes the family life cycle, structure, and hierarchy (power by another name). A strategic session is characterized by the use of enigmatic, creative, and sometimes dramatic directives: behavioral prescriptions that aim to interrupt the problem interaction sequence.

The Juice: Significant Contributions to the Field

If you remember anything from this chapter, it should be:

Directives

Directives are the most basic of strategic techniques (Haley, 1987; Madanes, 1991); however, they are also the most frequently misunderstood by those new to the practice. In essence, directives are directions for the family to complete a specific task, usually between sessions but sometimes within the session. The tasks are rarely "logical" or linear solutions to the problem; instead they somehow "perturb" the system's interaction patterns to create new interactions (Haley, 1987). Thus, if a couple are arguing, the therapist will *not* ask them to set the timer so that each person speaks for 5 minutes followed by the other summarizing or responding for the next 5 minutes. That would be a logical or linear directive such as those found in cognitive-behavioral family therapy (Chapter 11). Nor will the therapist ask the couple to simply stop or learn a communication technique; the assumption is that if they could do this, they would have already done so. Instead, the therapist asks them to argue but to change one or two key elements, such as the place, timing, or turn-taking style. A directive may be to have their "normal" argument while fully clothed in the bathtub or to have it after rearranging furniture to simulate a courtroom.

Directives get people out of their ruts with the *smallest change possible:* this is key because most of us resist change even if we want it (paradoxical isn't). When these directives work, clients generally experience a simultaneous shift in emotions, insight, and behavior. I like to compare it to a chocolate-vanilla swirl soft serve ice cream: the changes are perfectly synchronized and come fully integrated together. From the clients' perspective, directives jolt them awake from their usual life patterns, and generally the clients know exactly how to shift themselves to make the necessary changes. Unlike insight in traditional psychodynamic therapy, directives create visceral "ah-ha" moments because clients are in the midst of the action that needs to change. It is a bit like sudden enlightenment in the Zen tradition—hence, I like to think of strategic therapists as the Zen Masters and Mistresses of the field. In the case study at the end of the chapter, the therapist uses directives to help a Chinese American couple navigate differences in opinion about the role of the wife's parents helping to raise their two young children.

Rumor Has It: The People and Their Stories
Strategic Therapists
Jay Haley and Cloe Madanes Jay Haley, one of the original members of the Bateson team, student of Milton Erickson, and cofounder of the Brief Therapy Project at MRI, developed his own systemic approach, strategic therapy (Haley, 1963, 1973, 1976, 1980, 1981, 1984, 1987, 1996; Haley & Richeport-Haley, 2007; MRI, 2002). He and his wife, Cloe Madanes (1981, 1990, 1991, 1993), founded the Family Therapy Institute in Washington, D.C. Their approach is based on the concepts of hierarchy, power, and love, and uses *directives* for interventions.

Eileen Bobrow Eileen Bobrow founded and directs the Strategic Family Therapy and Training Center at the Mental Research Institute, which prior to her arrival focused primarily on the brief therapy model developed at the MRI.

Jim Keim A research fellow at the MRI, Jim Keim (1998) uses strategic therapy to work with families with oppositional-defiant and conduct-disordered children.

Jose Szapocznik Jose Szapocznik and his colleagues at the Center for Family Studies in Florida developed the empirically supported brief strategic family therapy (BSFT) to address drug abuse problems with Cuban youth in Miami (Szapocznik & Williams, 2000).

The Big Picture: Overview of Treatment
Initial Interview
The initial interview in strategic therapy is highly structured, having five formal stages: (a) the social stage, (b) problem stage, and (c) interaction staged, (d) goal-setting stage, and (e) task-setting stage (Haley, 1987). In addition, Haley describes how the initial phone contact is also a key element of setting up the therapy process.

Phone Contact In strategic therapy, the initial interview is considered to begin with the very first contact, typically by phone. Haley (1987) recommends gathering the following information:

- Names, addresses, and phone numbers of the relevant persons
- Listing of who lives in the household and their ages
- Employment status
- Prior therapy experiences
- Referring party
- One to two sentence statement about the presenting problem

Haley recommends gathering the information in a matter-of-fact way and asking for all members of the household come to the initial appointment.

Social Stage Strategic therapists begin the initial interview with a social stage: everyone is to be personally greeted and made comfortable before discussing the problems. Lasting only a few minutes, this stage involves the therapist helping the family to feel comfortable, keeping in mind that many families enter therapy feeling embarrassed for not having resolved the problem on their own. During this phase, the therapist carefully observes for the following:

- Who sits near whom? This *might* indicate possible alliances.
- What is the mood of the family? Are they appearing cheerful, unhappy, angry, reluctant, or desperate?
- What are the parent-child relations like? Do the kids seem closer to one parent than another?
- What is the relationship like between the adults? Do there seem to be differences in how they view the problem child?
- How does the family respond to the therapist? Are they cautious, curious, or eager to gain favor?

During this first social stage, the therapist is assessing interactions and mood (Haley & Richeport-Haley, 2007). The therapist does not share these observations and holds them *tentatively,* because these observations may easily be wrong. Haley is well aware that the therapy context is special, and families do not act the same in it as they do at home or in other social contexts. So, any inferences must be made tentatively.

Problem Stage After each person has been engaged the therapist in a courteous social exchange, the therapist shifts to a more businesslike mode, and inquires about what brings them to therapy. He recommends therapists beginning by openly sharing what they have been told and explain that they want everyone to come in to get their opinions about the situation. Often, by clarifying the therapist's position, the family is more willing to talk freely. Haley does recommend adjusting the introduction to the educational level of the family. Some options for asking about the problem include (Haley, 1987):

- *What Is Your Problem?:* This question defines therapy as a place where problems are addressed.
- *What Is It You Want From Me?:* This question reduces the possibility of the family's report and makes the therapy context more personal.
- *What Changes Do You Want?:* This question frames therapy as a place of change.
- *Why Are You Here?:* This question allows the family to discuss either the problem or change.

The more general and ambiguous the question, the more opportunities the family has to reveal their point of view; the more specific the question, the more the therapist can focus the discussions. Therapists choose the type of question that they think best fits a particular family and situation.

Who to Ask First Strategic to the core, Haley considers numerous issues—hierarchy in the family, investment in change, influence in returning to therapy—before deciding whom to ask the question he has strategically chosen to start the interview. In general, he recommends first engaging the adult (to respect the hierarchy) who is *less involved* in the problem; he recommends that the person who has the *most influence* in bringing the family back be treated with the most respect. Typically, the father is the less involved adult, and the mother has the most influence in bringing the family back. Haley cautions that therapists should be particularly careful about avoiding gender-based coalitions, by inadvertently showing greater agreement or understanding with the parent of the same gender as the therapist. He also cautions against starting with a child, because the child may feel on the spot or blamed. In the case study at the end of the chapter, the therapist is faced with a cultural dilemma if the grandparents and couple show up: asking the couple first favors an "American" family structure, thus siding with the father; if the therapist asks questions of the grandparents first, the therapist would be reinforcing traditional Chinese family norms, thus siding with the wife and her parents. In such a case, the therapist will need to take opportunities to allow both sensibilities to be voiced and respected.

How to Listen The therapist should listen from a position of helpful interest and avoid offering advice or asking about how someone feels about the situation: instead the focus is on facts, behaviors, and opinions. The therapist also avoids interpretation or attempts to create insight. If someone interrupts another, the therapist briefly allows the interruptions enough to observe the interaction but then intervenes to allow the initial person speaking to finish. After the first person speaks, the therapist then continues until all members of the family have a chance to share their thoughts on the issue.

Interaction Stage After each family member has had a chance to respond to the therapist's question, the therapist then directs the family to discuss it among themselves, removing him/herself from the discussion long enough to watch the family interact. Ideally, the therapist can even invite the family to *engage* in the problem in the therapy room. The goal of this stage is to *see* the family structure and interaction patterns through the interaction. Who sides with whom? Who has the most influence? The least? Who is connected and who is disconnected?

Goal Setting Stage Next, the therapist reengages in the conversation by getting a clear statement of the change that all family members agree to, including the identified patient, often the problem child. This becomes the therapeutic contract. To identify a meaningful change goal, the therapist must work with the family to define the problem in such a way that it is solvable. Saying I "don't want to be depressed or anxious anymore" is not considered a solvable problem. Even before insurance companies started mandating it, Haley advised that problems and goals be defined in terms that are observable, measurable, or countable so that it is very clear when change has occurred. Unlike some other systemic therapists, Haley recommends keeping the problem focused on the identified patient and the presenting problem rather than shift the definition of the problem to a more systemic view of the situation; he believes the family will be more willing to stay engaged in therapy by focusing therapy as they expect.

Task Setting Finally, a skilled strategic therapist often has developed a directive by the end of the first session, in which case, the therapist may send the family home with a simple task to complete over the week to begin the process of intervention.

Subsequent Sessions Subsequent sessions involve following up on the directive given for homework and then developing another directive based on their response (see Directives in The Doing section below). The therapist continues this until the symptoms are resolved. That is why they call it a brief therapy.

Making Connection: The Therapeutic Relationship
Strategic Positioning
Similar to maneuverability in the MRI approach, in strategic therapy, the therapist's role and demeanor shifts depending on the needs of the particular client or family. The position is always *strategic:* what response from the therapist is most likely to promote change? (Haley, 1987) In fact, if it seems that the family may benefit from joining together in not liking the therapist, a strategic therapist is generally willing to serve in that role too (that is dedication). Two of the most commonly used positions are social courtesy and taking the one-down position.

Social Courtesy
Haley (1987) describes the initial stage of therapy as the social stage, a time during which the therapist engages in casual social conversation, about the weather or traffic, to make clients comfortable and reduce their sense of shame: "the model for this stage is the courtesy behavior one would use with guests in the home" (p. 15). Before discussing the problem, the therapist ensures that all members have been properly greeted as people, and thus engages them from a more casual and down-to-earth position. When working with diverse clients, such as the Chinese family in the case study at the end of the chapter, the therapist should adapt this stage to be culturally appropriate, which in this situation would entail a more formal greeting and chit-chat that was careful to avoid any potential topics that would cause embarrassment and focus on highlighting the family's competencies.

The One-Down Stance or Helplessness
Used in all forms of systemic and strategic therapies, the *one-down stance* is used to increase clients' motivation, often paradoxically by claiming, "I'm not sure if I am able to handle such a problem" (Segal, 1991). This move is often helpful with clients who act as if their situation is hopeless; when the therapist instead takes the hopeless stance, the client is then motivated to find hope. Systemically this works because in most systems there is a counterbalance: if one person is hopeless, the other feels compelled to be hopeful to maintain a balance. This same dynamic is also observed between couples in crisis: generally one person will manage the crisis, allowing the other one to more fully feel the panic and trauma.

Beyond serving this paradoxical purpose, the one-down stance also expresses a certain attitude toward the family system. Systemic therapists view the system as an entity with its own rules and integrity that must be respected, much like a mountain climber must respect the awesome forces of nature or the sailor an ocean. Like a family system, nature and the ocean are not things the therapist can control; instead, there is a deep respect for their power and ways. Thus the one-down stance is a sincere and genuine position for a systemically trained therapist.

The Viewing: Case Conceptualization and Assessment
Strategic Conceptualization
Madanes (1991) identifies six ways to think about a problem in strategic therapy; these dimensions can be used to conceptualize the role of a symptom in the family system or in an individual's broader social world. One or more of these is used, depending on the case.

- *Involuntary Versus Voluntary:* Clients generally present by viewing the problem as involuntary; strategic therapists instead view the symptoms as voluntary, with the exception of organic illness. For example, they see arguing, worrying, or being depressed as behaviors or solutions clients choose.
- *Helplessness Versus Power:* Although symptomatic people appear helpless, their symptoms also generate significant power, allowing them to make otherwise unreasonable demands, receive more care and attention, or excuse otherwise unacceptable behavior. For example, agoraphobia can be seen as an attempt to keep the family close.

- *Metaphorical Versus Literal:* Symptoms may be viewed as a metaphor for another problem sequence of behaviors in the system. For example, a child's defiance can be seen as a metaphor for the same defiance the husband feels but does not express toward the mother; or binging can be seen as a metaphor for rejecting the mother's attempt to over nurture the child.
- *Hierarchy Versus Equality:* Therapists may conceptualize the presenting problem as a demand for more or less hierarchy, depending on the family's set of circumstances. For example, caving in to a child's tantrums allows the child to be at the top of the power hierarchy.
- *Hostility Versus Love:* Many family interactions—rejecting a lover because one feels unworthy, disciplining a child, pursuing a partner for sex or communication—can be viewed as motivated by either hostility or love. Strategic therapists may use either interpretation, depending on the therapeutic situation. For example, a husband's constant pursuit for sex may be a sign of love and caring *or* a power move because he feels he has none, depending on the situation. Similarly, a wife's refusal of sex could be motivated by love for husband (e.g., only wanting sex when she really has feelings) or family (e.g., gives too much and is too tired) *or* a power move to gain control because she feels she has none.

Stages of Family Life

When conceptualizing a family's situation, strategic therapists attend the family's stage of life and where they may be stuck (Haley & Richeport-Haley, 2007). Symptoms often indicate a stage-of-life problem, which the therapist can use identify where and how to intervene. Life stages include:

- Birth and infancy
- Early childhood
- School age
- Adolescence
- Leaving home
- Becoming a parent
- Becoming a grandparent
- Old age

At any one of these transitions, a person may get stuck and have difficulty transitioning to the next phase because each requires the family to renegotiate how it balances independence and interdependence. In some cases, one member begins the transition sooner than another or there is not agreement on how to transition.

Hierarchy and Power

Strategic therapists pay attention to and respect *family hierarchy*: who has the influence and in what arenas of family life? Depending on the family and culture this could be the parents or the grandparents (Haley, 1987). Often each parent has different forms of influence, but sometimes one parent clearly has more influence in the family. In such cases, the therapist tries to engage the less involved parents. More so than relying on verbal report of who has "power," the therapist closely observes the family's interaction to see how requests/demands are responded to and who tends to get their way in an argument. For example, a father may be described as "the head of the household" but often the mother holds more influence on actual decisions made in the family. Hierarchy cannot be accurately identified from a single interaction but must be repeatedly observed across interactions. Most families have fairly complex patterns of hierarchy and power that the therapist traces and uses to develop interventions.

When observing hierarchy between parents and children, the therapist focuses on whether the child actually acts on parent requests or whether the parents cave in to the child's demands. An effective parent-child hierarchy involves a parent making a request and the child willingly complying with minimal fuss on most issues. An *ineffective parent hierarchy* describes a parent who is unable to engage their child's cooperation on most requests. An *excessive hierarchy* describes authoritarian parenting style in which the rules are very rigid, often age-inappropriate, and although the child is compliant, there is significant emotional distance and/or tension in the relationship. The case study at the end of the chapter presents a particularly complex hierarchical family struggle. The couple is at odds over whether to follow a more traditional Chinese family hierarchy in which the parents maintain significant influence over adult children or whether to follow a more American structure, in which the grown children are fully autonomous. This situation is further complicated by the couple living with the wife's parents, not the husband's, as is traditional.

Strategic Humanism

The latest developments in strategic therapy emphasize increasing the family's ability to love and nurture rather than dominate and control, which is consonant more generally with the evolution of the field of family therapy (Madanes, 1993). Thus case conceptualizations focus more on the family's unsuccessful attempts to show love rather than on their attempts to control one another. For example, Jim Keim's (1998) work with oppositional children aims to increase parental nurturing behaviors to help them reestablish a more effective hierarchy, highlighting that a nurturing parent can be as much of an authority as a disciplinarian; once some authority has been reestablished in the nurturing realm, Keim instructs parents to begin enforcing behavioral rules.

Targeting Change: Goal Setting
Strategic Goals

> "The main goal of therapy is to get people to behave differently and so to have different subjective experiences."—Haley, 1987, p. 56

Strategic therapy does not have a predefined set of long-term goals for individual or family functioning, other than to promote a change that alters people's subjective experiences (mood, thoughts, and behaviors). Madanes (1990, 1991) conceptualizes all problems brought to therapy as stemming from an existential dilemma between love and violence, because these two experiences tend to be closely correlated in human affairs. Thus the ultimate strategic goal is to help clients find ways to love without dominating, intruding upon, or harming the other. This goal may be achieved by doing the following (Madanes, 1991):

- Correcting the couple or family hierarchy (either increasing or decreasing it)
- Reducing intrusion or increasing engagement by changing a parent or partner's level of involvement
- Reuniting family members
- Changing who is helpful and how, including empowering children to be appropriately helpful
- Repenting for an injustice and forgiving
- Increasing the expression of compassion and unity

The Doing: Interventions
Directives

A defining strategic intervention, directives are behavioral tasks that the therapist gives to clients to alter their interaction patterns. Haley (1987) identifies two general types of directives: straightforward and indirect. *Straightforward directives* are used when the therapist has the power and influence to get people to do what is asked; *indirect directives* are used when the therapist has less authority in the eyes of the client. Indirect directives generally take the form of paradoxical or metaphorical tasks.

Technically, straightforward directives mean giving good advice, but Haley is quick to admonish against them: "Giving good advice means the therapist assumes that people have rational control of what they are doing. To be successful in the therapy business, it may be better to drop that idea" (Haley, 1987, p. 61). Therefore, most straightforward directives aim to change the way the family interacts by introducing *new action*. Often therapists are tempted to ask clients to stop doing something; Haley warns against this also: "If the therapist tells someone to stop usual behavior, he must usually go to an extreme or get other family members to cooperate and change their behavior" (p. 60). Thus most strategic directives re-sequence interaction patterns by requesting *small behavioral or contextual changes,* such as having the *other* parent discipline a child who has broken house rules or asking a couple to add a 10-second pause between exchanges in a fight. These smaller changes are much easier to accomplish. The differences between straightforward and indirect directives are outlined in the following table.

Therapeutic Directives

	Straightforward Directives	Indirect Directives
Type of Task	• Do something different: alter behavioral sequence • Stop a behavior (rarely used) • Good advice; psycho-education (rarely used)	• Paradoxical tasks • Metaphorical tasks
Type of Therapeutic Relationship	• Therapist has influence; well accepted as expert	• Therapist less well accepted as expert
Type of Problem	• Client able to feel control over small behaviors requested as part of task	• Client feels has very little control

Straightforward Directives Haley maintained that "the best task is one that uses the presenting problem to make a structural change in the family" (Haley, 1987, p. 85). Designing straightforward directives involves several steps:

1. *Assess the Situation:* The therapist identifies the interaction sequence of behaviors that constitute the presenting problem; for example, the child comes home after curfew; the father angrily enforces a harsh punishment; the mother defends the child to the husband and does not enforce the punishment over the week.
2. *Target a Small Sequence Change:* The therapist finds one small behavioral change in the sequence that the family can reasonably make that would alter the problem sequence; for example, have the mother enforce the curfew; have the daughter propose a "fair" punishment for breaking curfew.
3. *Motivate the Family:* Haley (1987) emphasizes the importance of motivating the family *before* the task is issued. For straightforward tasks are usually done by appealing to the shared goal. Begin by first talking about how everyone wants the problem behavior to stop: "Everyone agrees that they want the arguing at home to stop."
4. *Give Precise, Doable Instructions for the Directive:* When describing the task, the therapist needs to be extremely precise—when, where, how, who, on which day of the week—while accounting for weekend changes in schedules, weekday homemaking tasks, and special events during the week that the task is assigned. Tasks should be clearly *given* rather than suggested: "Next week, when Susan comes in late from curfew or breaks any other house rule, mom, I want you to enforce the consequences, and dad, you are not to be part of the discussion with Susan and your wife. If you need to discuss what happened, you will discuss it with your wife alone."
5. *Review the Task:* The therapist asks the family members to repeat their parts: "Just to review and make sure we all understand, can each of you tell me what you are going to do this week?"
6. *Request a Task Report:* The following week, the therapist asks the family how it went. Generally, one of three things happens: (a) they do it, (b) they don't do it, or (c) they partially do it (Haley, 1987). If clients complete the task, they are congratulated. If they have partially done the task, the therapist does not excuse them too quickly because that would send a message that undermines the therapist's authority. If they fail to attempt the task, Haley uses two responses, one nice and

one not so nice. In the nice response, the therapist says, "I must have misunderstood you or your situation to ask that of you—otherwise you would have done it" (Haley, 1987, p. 71). In the not-so-nice approach, the therapist emphasizes that the family has missed an opportunity to make changes for their own good and that this is a failure and a loss for them (not that they disappointed the therapist); the therapist does not ask them to redo the task even if they say they want to. Instead, the therapist uses this experience to increase motivation for the next task.

Indirect Directives: Paradoxical Interventions Perhaps the most misunderstood of strategic and systemic interventions, *paradoxical intervention,* also called *symptom prescription,* involves instructing clients to engage in the problem behavior in some fashion, such as by assigning a couple to argue from 7:00 to 7:15 P.M. on Tuesday and Thursday. Symptom prescription and other paradoxical interventions are "paradoxical" because they do not follow linear logic—or at least at first blush. You will know when it is time to use one of these because when it is appropriate to use paradox it will not seem paradoxical—at least in the mind of the therapist—it will be the only logical, obvious action to take. Paradox makes sense with two general types of problems: (a) when making any other therapeutic changes disrupts the family's current level of stability or (b) when the problem seems "uncontrollable" from the client's perspective: I can't keep myself from worrying, nagging, eating, fighting, etc.

- *Paradox with Families That Avoid Change:* When family members stabilize around one member being the problem, they are often resistant to the therapist's attempts to change that perception because it may create more discomfort than they are currently experiencing; in these situations paradox can be useful (Haley, 1987). The therapist may restrain or caution against certain changes: "Perhaps you need to argue to keep your passion strong; if you stopped, things might get worse." Or the therapist might choose to paradoxically encourage relapse to prevent it (Haley, 1976). Paradoxical tasks are difficult to deliver because the therapist must communicate several messages at once:
    - I want to help you resolve your problem.
    - I am sincerely concerned about you.
    - I think you can be normal, but perhaps you cannot be.
  When paradox is successful, change is usually spontaneous.
- *Paradox with Uncontrollable Symptoms:* If clients have symptoms that they claim they absolutely cannot control, it makes sense to use *symptom prescription* as an intervention. Symptom prescription changes the *context* of the problem behavior. If the context changes, the *meaning* of the behavior must change. When the meaning changes, thoughts, feelings, and subsequent behaviors automatically change also. A common example is using paradox for worrying, which is particularly difficult to treat because it is a cognitive and emotional process that often has few associated behaviors. When clients are given the directive to set an egg timer and worry for 10 minutes at a particular time in the day, worrying radically changes from a vague, free-floating experience to a consciously chosen activity that they can voluntarily start and, most often quickly discover, can also stop at will. When the context is changed, the meaning and experience of worrying change, often creating significant movement on a symptom that seemed totally out of the client's control.

Indirect Directives: Metaphorical Tasks Inspired by Milton Erickson's trance work and therapeutic style, the metaphoric task is used when it is not appropriate to explicitly address a problem (Haley, 1987). Haley believes that metaphors need to be acted on, not just talked about, to create change. Metaphoric tasks involve four stages:

1. The therapist identifies an area of the client's life with similar dynamics to an area the therapist wants to change (e.g., discussing adopting a pet with a child who has been adopted).
2. The therapist uses a story or conversation to discuss how adoption works (e.g., what happens when the dog gets sick).

3. The therapist takes a position about how things should change in the metaphoric area (e.g., child is ready to adopt first pet).
4. A task is usually assigned in the metaphoric area (e.g., child's parents are encouraged to help him adopt a pet).

Pretend Techniques

"Fake it 'til you make it" sums up the spirit of pretend techniques used by strategic therapists. In these techniques, clients are asked to "pretend" they have achieved their goal for a designated period of time (from minutes to days) to help make their desired changes (Madanes, 1991). When they fake a behavior, even for a short period of time, often there is a genuine change in perspective, feeling, or behaviors. For example, when a couple are asked to fake being in love for an evening, they often feel those genuine feelings return. Systemically, this technique works by triggering old, preferred interaction patterns that are already there or, if this is an entirely new behavior, by creating new interaction sequences that are now available within the system. In either case, brief periods of pretending introduce new interactional sequences, creating new options for behavioral, emotional, and cognitive change.

Ordeal Therapy

Inspired by the work of Milton Erickson, *ordeals* are often used in strategic therapy when the client feels helpless in controlling a symptom such as arguing, worrying, overeating, smoking, nail biting, and drinking. They are based on a simple premise: "If one makes it more difficult for a person to have a symptom than to give it up, the person will give up the symptom" (Haley, 1984, p. 5). Rather than try to develop linear, logical means for stopping the behavior (that would be a cognitive-behavioral approach), the strategic therapist allows the symptom with a twist: the client must complete another task, an "ordeal," first or afterwards.

The ordeal need not be directly related to the undesired activity but often carries a metaphoric relation to it. For example, if a person is trying to reduce his/her "emotional eating" to soothe difficult emotions, the ordeal would target the behavior and internal tension in either a linear or logical way (e.g., engage in a favorite hobby, write in a journal, before eating) or in an indirect or nonlogical way (e.g., perform a random act of kindness for a stranger or loved one, clean the house before engaging in eating). In most cases, ordeal therapy is less about creating a horrific, unappealing ordeal to stop undesired behavior and more about shaking up or perturbing the systemic pattern so that new behavioral sequences can evolve. To return to the family dance metaphor, the ordeal isn't a behavioral deterrent or punishment as much as a new piece of furniture in the middle of the dance floor that forces the system to make changes to navigate around it. As the system adjusts to this rather small and innocuous twist, it must change to create new steps and is able to do so with less fear and resistance than if it were told to stop dancing its favorite (or at least best-rehearsed) dance.

Putting It All Together

Case Conceptualization Template

- *Interaction Sequence:* Describe each person's role (even seemingly minor players) the interactional sequence, emphasizing the following:
  - Homeostasis
  - Rise of tension
  - Symptom/conflict
  - Return to homeostasis
- *Strategic Conceptualization/Hypothesis:* Develop a systemic hypothesis using one of the following metaphors:
  - Involuntary versus voluntary
  - Helplessness versus power
  - Metaphorical versus literal

- Hierarchy versus equality
- Hostility versus love
• *Family Life Stage of Development:* Identify family life cycle stage and how the person/family may be struggling to balance independence and interdependence related to developmental issues:
  - Birth and infancy
  - Early childhood
  - School age
  - Adolescence
  - Leaving home
  - Becoming a parent
  - Becoming a grandparent
  - Old age
• *Hierarchy and Power:* How does the symptom express problems related to hierarchy and power in relationships.

Treatment Plan Template for Individual

Strategic Initial Phase of Treatment for Individual

Initial Phase Counseling Tasks

1. Develop working counseling relationship. *Diversity Note: Adjust to respect cultured, gendered, and other styles of relationship building and emotional expression.* Relationship building approach/intervention:
   a. *Social courtesy* to create initial context of safety and *strategic positioning* to facilitate change over course of therapy.

2. Assess individual, systemic, and broader cultural dynamics. *Diversity Note: Adjust assessment based on cultural, socioeconomic, sexual orientation, gender, and other relevant norms.*
   Assessment strategies:
   a. Assess *problem interaction sequence*, noting the affect of *stage of family life, power, and alliances.*
   b. Conceptualize problem using one or more strategic frames, such as voluntariness, helplessness, hierarchy, love, etc.

Initial Phase Client Goals

1. Increase ability to see *voluntary* nature of depressive and anxiety symptoms to reduce depressed mood and anxiety.
   a. *Paradoxical directives* that require client to perform depression/anxiety symptoms in new context.
   b. *Directives* that instruct client to make a small but meaningful alternation to symptom pattern.

Strategic Working Phase of Treatment with Individual

Working-Phase Counseling Task

1. Monitor quality of the working alliance. *Diversity Note: Attend to client response to interventions that indicate therapist using expressions of emotion that are not consistent with client's cultural background.*
   a. *Assessment Intervention:* Monitor level of *therapist influence* and adjust interventions accordingly.

(continued)

Working-Phase Client Goals

1. Decrease client's sense of seeming *helplessness* related to [presenting symptoms] to reduce depression and anxiety.
   a. *Paradoxical tasks* that exaggerate helplessness and reveal hidden benefits.
   b. *Metaphorical tasks* that make overt the covert power associated with being symptomatic.

2. Increase client's *sense of voluntary control* over "uncontrollable" symptom [specify] to reduce depression and anxiety.
   a. *Ordeal therapy* that alters sequence of behaviors related to depression and anxiety.
   b. *Pretend techniques* to increase desired behaviors.

Strategic Closing Phase of Treatment with Individual

Closing-Phase Counseling Task

1. Develop aftercare plan and maintain gains. *Diversity Note: Access resources in the communities of which they are a part to support them after ending therapy.*
   a. *Restraining* and *dangers of improvement* to help client prepare for setbacks.

Closing-Phase Client Goals

1. Increase ability to see how *love* motivates own and other's behaviors to increase sense of well-being and connection to others.
   a. *Reframe problem interactions* with others focusing on how love motivates interaction.
   b. *Directives/metaphorical tasks* designed to highlight the hidden motivation of love.

2. Increase ability to [complete developmental tasks] based on client's place in the *family life cycle* to reduce chance of relapse.
   a. *Directives* designed to rebalance independence and interdependence based on stage of family life.
   b. *Pretend techniques* to help client master new developmental challenges.

Treatment Plan Template for Couple/Family

Strategic Initial Phase of Treatment with Couple/Family

Initial Phase Counseling Tasks

1. Develop working counseling relationship. *Diversity Note: Adjust to respect cultured, gendered, and other styles of relationship building and emotional expression.*
   Relationship building approach/intervention:
   a. Social courtesy with all clients to create initial context of safety and strategic positioning to facilitate change over course of therapy.

2. Assess individual, systemic, and broader cultural dynamics. *Diversity Note: Adjust assessment based on cultural, socioeconomic, sexual orientation, gender, and other relevant norms.*
   Assessment strategies:
   a. Assess *problem interaction sequence*, noting the affect of *stage of family life, power, and alliances* (in the family and with outside others).
   b. Conceptualize problem using one or more *strategic frames*, such as voluntariness, helplessness, hierarchy, love, etc.

(continued)

Initial Phase Client Goals

1. Increase ability to see *loving motivations* that underlie problem interaction cycle to reduce conflict.
   a. *Strategic reframing* that recharacterizes each person's role in the interaction as having an ultimate, underlying motivation of love.
   b. *Directives* that instruct couple/family to make a small but meaningful alternation to symptom pattern.

Strategic Working Phase of Treatment with Couple/Family

Working-Phase Counseling Task

1. Monitor quality of the working alliance. *Diversity Note: Attend to client response to interventions that indicate therapist using expressions of emotion that are not consistent with client's cultural background.*
   a. *Assessment Intervention*: Monitor level of *therapist influence* and adjust interventions accordingly.

Working-Phase Client Goals

1. Decrease couple/family's *problem interactions* that are attempts to define the balance of *power* in the relationship to reduce conflict.
   a. *Metaphorical tasks* that make overt the covert power dynamics in the relationship.
   b. *Directives* that make small changes to the behavioral sequences related to power in the relationship.
2. Increase client's *sense of voluntary control* over "uncontrollable" elements of conflictual interactions to reduce conflict.
   a. *Ordeal therapy* that alters sequence of behaviors related to couple/family conflict.
   b. *Pretend techniques* to increase desired behaviors.

Strategic Closing Phase of Treatment with Couple/Family

Closing-Phase Counseling Task

1. Develop aftercare plan and maintain gains. *Diversity Note: Access resources in the communities of which they are a part to support them after ending therapy.*
   a. *Restraining* and *dangers of improvement* to help prepare for setbacks, including variations that will reduce the tendency to blame others for setbacks.

Closing-Phase Client Goals

1. Increase ability to see how *love* motivates own and other's behaviors in the relationship to increase sense of well-being and connection.
   a. *Reframe problem interactions* focusing on how love motivates interactions, even problem interactions.
   b. *Directives/metaphorical* tasks designed to highlight the hidden motivation of love.
2. Increase ability to [complete developmental tasks] for *family life stage* to reduce chance of relapse.
   a. *Directives* designed to rebalance independence and interdependence based on stage of family life.
   b. *Pretend techniques* to help master new developmental challenges.

Research and the Evidence Base

Similar to other systemic approaches, strategic therapy has been used in the development of several evidence-based treatments, including:

- *Multisystemic Family Therapy* (see Chapter 5)
- *Brief Strategic Family Therapy* (structural ecosystemic therapy, structural eco-developmental preventive interventions; see Clinical Spotlight section in this chapter; Szapocznik & Williams, 2000)
- *Multidimensional Family Therapy* (Liddle, Dakof, & Diamond, 1991)
- *Functional Family Therapy* (see Chapter 15)

Although there has been less systematic research on outcomes of specific systemic models, such as strategic, there is consistent and growing research on the effectiveness of evidence-based systemic approaches for specific conditions, such as adolescent substance abuse, adolescent conduct issues, depression related to relationship distress, severe mental illness, and couple distress (Sprenkle, 2012).

Clinical Spotlight: Brief Strategic Family Therapy

Drawing on structural and strategic therapies, Jose Szapocznik and his colleagues at the Center for Family Studies developed brief strategic family therapy (BSFT) to address drug abuse problems with Cuban youth in Miami (Szapocznik & Williams, 2000). Expanded and adopted to treat African Americans and other Hispanic populations, this therapy is recognized as an evidence-based approach; a complete manual is available through the National Institute for Drug Abuse website (www.nida.nih.gov/TXManuals/bsft; Szapocznik, Hervis, & Schwartz, 2003). BSFT is based on three central concepts: *systems*, *structure* (patterns of interaction), and *strategy*

Goals

Brief strategic family therapy has two goals:

- Reduce or eliminate child drug use
- Change family interactions that are supporting the problem behaviors (youth drug use)

Case Conceptualization

In comparison with other evidence-based approaches, BSFT focuses primarily *within* family dynamics using structural and strategic family therapy concepts (Santisteban, Suarz-Morales, Robbins, & Szapocznik, 2006; Szapocznik et al., 2003):

- *Structure and Organization:* The therapist uses traditional structural concepts such as subsystems, hierarchy, leadership, and coalitions to assess the structure, organization, and flow of information in the family.
- *Resonance:* Using the structural therapy concepts of boundaries, the therapist assesses emotional resonance within the broader context of cultural norms: enmeshed (high resonance) and disengaged (low resonance).
- *Developmental Stage:* The therapist uses the family's ability to adapt its structure to support members in their current life stage of development (e.g., increasing autonomy as children grow).
- *Life Context:* The therapist assesses the effects of the family's broader social life, such as extended family, community, school, peers, and courts.
- *Identified Patienthood:* The more the family believes the identified patient is to blame for all its problems, the more difficult it is to treat the family.

- *Conflict Resolution:* The therapist assesses the family's style of conflict resolution:
  - *Denial:* Conflict is not allowed to emerge: "We have no problems."
  - *Avoidance:* When conflict arises, it is quickly stopped or covered up, such as by procrastinating, minimizing, or postponing difficult conversations.
  - *Diffusion:* When the problem is brought up, the subject is switched to another problem topic, often as a personal attack against the person who raised the issue.
  - *Conflict Emergence Without Resolution:* Conflict occurs but no resolution is reached.
  - *Conflict Emergence With Resolution:* The family is able to resolve the conflict.

Principles of Intervention

Interventions are deliberately chosen to target the aspects of family interactions that are most likely to achieve the desired outcome.

- *Joining:* The therapist uses structural family therapy, joining, to connect with the family system.
- *Enactments:* Structural enactments are used to assess family functioning and restructure family interactions.
- *Working in the Present:* Interventions target current interactions with minimal focus on the past.
- *Reframing Negativity:* Therapists reframe negative interpretations to promote caring and concern within the family.
- *Reversals:* Therapists may coach one or more family members to do or say the *opposite* of what is typically done or said.
- *Working With Boundaries and Alliances:* Standard structural techniques are used to either loosen or strengthen boundaries to better meet developmental needs.
- *Detriangulation:* The therapist may remove a third, less powerful person from a conflict between two others.
- *Opening Closed Systems:* Systems in which open conflict is not allowed must be "opened" to allow effective expression and resolution of differences.

Tapestry Weaving: Diversity Considerations

Ethnic, Racial, and Cultural Diversity

As strategic therapy is closely related to the MRI and Milan systemic approaches, these approaches share similar principles when used with diverse clients. Strategic therapists are careful to avoid prescribing what "normal" families should be like (Walsh, 1993). Haley went as far as to claim:

> To say that a person with a symptom is behaving in a way that is out of the ordinary implies that there is an ordinary way to behave. To discuss such a proposition thoroughly would require an analysis of a particular culture and the range of individual deviation permissible within that culture before a person is considered out of the ordinary. (Haley, 1990, p. 6)

Haley (1990) was careful to clarify that his work was primarily done with persons in Western cultures, and his writing and examples should be interpreted with this cultural context in mind. Instead of using theories of normalcy, the focus is on developing a customized approach to effecting systemic change: "Since in strategic family therapy a specific therapeutic plan is designed for each problem, there are no contraindications in terms of patient selection and suitability" (Madanes, 1991, p. 396). The emphasis on a clear systemic hierarchy is a value embraced by many cultures, especially traditional and immigrant cultures; however, this emphasis on hierarchy may be an issue of difference with clients from particular backgrounds.

Hispanic Adolescents and Their Families

Brief strategic family therapy (BSFT), an evidence-based approach highlighted in the Clinical Spotlight section previously, is an approach that was designed for Hispanic (predominantly Cuban and Central and South American) and later African American youth with conduct and substance abuse issues. This approach integrates many elements of traditional strategic therapy—such as reframing and paradoxical reversals—to effectively treat diverse families with troubled teens. Over the years, the developers and practitioners of BSFT have noted the following as unique needs for this population:

- *Family as Treatment Unit:* Because Hispanic cultures place a high value on family integrity, including the family when treating youth is particularly important. However, when this is not possible, BSFT will work with an individual, an adaptation called one-person BSFT (Robbins, Horigian, Szapocznik, & Ucha, 2010).
- *Valuing Family Hierarchy and Traditions:* BSFT therapists work with and respect the hierarchy that characterizes Hispanic families and couples (Hervis, Shea, & Kaminsky, 2009; Robbins et al., 2010). The therapist empathizes with and supports the presenting framework.
- *Active Therapist Role:* These families respond particularly well to active therapists who were directive and present-oriented, providing a sense of leadership and expertise for the family (Robbins et al., 2010).
- *Americanism vs. Hispanicism:* The therapists attend to the varying levels of American vs. Hispanic values that the parents and children embraced in a given system as well as potential differences between the spouses. Depending on their level of acculturation, Hispanic parents were more likely to value family cohesion and parental authority whereas adolescents were more likely to espouse American values of individualism, which is magnified by the normal adolescent process of seeking independence (Robbins et al., 2010).
- *Highlighting Problem Interactions:* Rather than dilute interventions, therapists highlight problem interactions, requiring clients to stop their autopilot mode and look and listen to what is going on in the room (Hervis et al., 2009).
- *Reframing and biculturalism:* Many of the reframes offered by BSFT provide a bicultural view—a both/and position—that makes room for both traditional Hispanic as well as American values (Hervis et al., 2009).

Sexual Identity Diversity

Because of its nonpathologizing and non-normative assumptions and its attention to larger systemic dynamics, strategic therapy, like other systemic therapies, has been widely used with clients who identify as gay, lesbian, bisexual, transgendered, or questioning (GLBTQ). The principles described in Chapter 5 for working with GLBTQ and adolescents coming out apply to strategic therapy. In addition, specific applications for working with transgendered youth and their families from a systemic approach have also been identified.

Transgendered Youth

There is little research or literature to guide therapists when working with transgendered persons, especially youth. However, unlike gay, lesbian, and bisexual youth, transgendered youth often cannot hide their difference, and thus it is more likely to be a point of family discussion whether the child wants it public or not (Coolhart, Baker, Farmer, Malaney, & Shipman, 2012). The majority of transgendered youth (59%) report negative parental reactions to their gender identity, at least initially, with those more gender nonconforming reporting greater conflict with parents (Grossman, D'Augelli, Howell, & Hubbard, 2005). However, parental acceptance is important for transgendered individuals, correlated with adult life satisfaction and self-esteem (Erich, Tittsworth, Dykes, & Cabuses, 2008).

Coolhart and colleagues (2012) outline an assessment tool for therapists working with transgendered youth and their family. They recommend working with the family of transgendered youth whenever possible, attending to systemic dynamics. The assessment tool comprises questions that cover nine domains:

- *Early Awareness and Family Context:* These questions address the general family structure and history as well as both parents' and youth's early gender experiences. For example: "Please provide additional information about how your family's context in terms of racial, ethnic, and national communities interfaces with beliefs about variant expressions of gender and sexuality" (Coolhart et al., 2012, p. 16).

- *Parents' Attunement with Youth's Affirmed Gender:* These questions address how the parents and family have responded to the youth's transgender identity. For example: "When your child disclosed (or you discovered) their transgender identity, what was this experience like for you individually, as a couple, and as a family as a whole?" (Coolhart et al., 2012, p. 17).
- *Current Gender Expression:* These questions address the youth's current preference for expressing gender. For example: "Have you dressed privately in clothing typically associated with your affirmed gender? When did you first begin to dress in private? How did it feel when you first began to dress in private? Did anyone else know? If so, what was their reaction?" (Coolhart et al., 2012, p. 17).
- *School Context:* These questions explore the youth's experience and possible harassment at school related to gender identity. For example: "What is your current gender expression at school? Are you 'out' to any of your friends?" (Coolhart et al., 2012, p. 18).
- *Sexual Relationships/Development:* These questions are typically asked to the youth in private and address sexual orientation, abuse, and activity: "How do you identify your sexual orientation? Have you ever experienced or witnessed sexual or physical abuse?" (Coolhart et al., 2012, p. 19).
- *Current Intimate Relationship(s):* If the youth is involved in a current relationship, the therapist inquires about the relationship and the partner's knowledge of the youth's gender identity and possible interest in transitioning.
- *Physical and Mental Health:* These questions assess for physical and mental health issues that may need addressing, including the potential for self-harm and substance abuse.
- *Support:* The questions aim to identify sources of support for the family and youth, including family, friends, church, neighbors, support groups, Internet connections, etc.
- *Future Plans/Expectations:* This last set of questions explores whether the youth has plans for gender transition, hormone therapy, surgery, children, etc.

Online Resources

Brief Strategic and Systemic World Network:
www.bsst.org

Brief Strategic Family Therapy Manual:
www.nida.nih.gov/TXManuals/bsft

Brief Strategic Family Therapy Training:
www.brief-strategic-family-therapy.com

Mental Research Institute:
www.mri.org

Multisystemic Therapy:
www.mstservices.com

Strategic Therapy (Jay Haley):
www.jay-haley-on-therapy.com/html/strategic_therapy.html

Strategic Therapy (Cloe Madanes):
www.cloemadanes.com

Strategic Therapy (Eileen Bobrow):
www.briefstrategicfamilytherapy.com

References

*Asterisk indicates recommended introductory readings.

Bateson, G. (1972). *Steps to an ecology of mind*. San Francisco, CA: Chandler.

Coolhart, D., Baker, A., Farmer, S., Malaney, M., & Shipman, D. (2012). Therapy with transsexual youth and their families: A clinical tool for assessing youth's readiness for gender transition. *Journal of Marital and Family Therapy* [Early View Version]. doi: 10.1111/j.1752-0606.2011.00283.x

Dell, P. F. (1989). Violence and the systemic view: The problem of power. *Family Process*, 28(1), 1–14. doi: 10.1111/j.1545-5300.1989.00001.x

Erich, S., Tittsworth, J., Dykes, J., & Cabuses, C. (2008). Family relationships and their correlations with transsexual well-being. *Journal of GLBT Family Studies*, 4(4), 419–432.

Grossman, A. H., D'Augelli, A. R., Howell, T. J., & Hubbard, S. (2005). Parents' reactions to transgender youths' gender nonconforming expression and identity. *Journal of Gay and Lesbian Social Services*, 18(1), 3–16.

Haley, J. (1963). *Strategies of psychotherapy*. New York: Grune & Stratton.

Haley, J. (1973). *Uncommon therapy: The psychiatric techniques of Milton H. Erickson, M.D.* New York: Norton.

Haley, J. (1976). *Problem-solving therapy: New strategies for effective family therapy*. San Francisco, CA: Jossey-Bass.

Haley, J. (1980). *Leaving home: The therapy of disturbed young people*. New York: McGraw-Hill.

Haley, J. (1981). *Reflections on therapy*. Chevy Chase, MD: The Family Therapy Institute of Washington, DC.

Haley, J. (1984). *Ordeal therapy*. San Francisco, CA: Jossey-Bass.

*Haley, J. (1987). *Problem-solving therapy* (2nd ed.). San Francisco, CA: Jossey-Bass.

Haley, J. (1990). *Strategies of psychotherapy*. Rockville, MD: Triangle Press.

Haley, J. (1996). *Learning and teaching therapy*. New York: Guilford.

Haley, J., & Richeport-Haley, M. (2007). *Directive family therapy*. New York: Hawthorne.

Hervis, O. E., Shea, K. A., & Kaminsky, S. M. (2009). Brief strategic family therapy: Treating the Hispanic couple subsystem in the context of family, ecology, and acculturative stress. In M. Rastogi & V. Thomas (Eds.), *Multicultural couple therapy* (pp. 167–186). Thousand Oaks, CA: Sage.

Keim, J. (1998). Strategic family therapy. In E. Dattilio (Ed.), *Case studies in couple and family therapy* (pp. 132–157). New York: Guilford.

Liddle, H. A., Dakof, G. A., & Diamond, G. (1991). Adolescent substance abuse: Multidimensional family therapy in action. In E. Daufman & P. Kaufman (Eds.), *Family therapy of drug and alcohol abuse* (pp. 120–171). Boston, MA: Allyn & Bacon.

Madanes, C. (1981). *Strategic family therapy*. San Francisco, CA: Jossey-Bass.

Madanes, C. (1990). *Sex, love, and violence: Strategies for transformation*. New York: Norton.

Madanes, C. (1991). Strategic family therapy. In A. S. Gurman & D. P. Knishern (Eds.), *Handbook of family therapy* (pp. 396–416). New York: Brunner/Mazel.

Madanes, C. (1993). Strategic humanism. *Journal of Systemic Therapies*, 12(4), 69–75.

Mental Research Institute. (2002). *On the shoulder of giants*. Palo Alto, CA: Author.

Robbins, M. S., Horigian, V., Szapocznik, J., & Ucha, J. (2010). Treating Hispanic youths using brief strategic family therapy. In J. R. Weisz & A. E. Kazdin (Eds.), *Evidence-based psychotherapies for children and adolescents* (2nd ed.) (pp. 375–390). New York: Guilford Press.

Santisteban, D. A., Coatsworth, J., Perez-Vidal, A., Mitrani, V., Jean-Gilles, M., & Szapocznik, J. (1997). Brief structural/strategic family therapy with African American and Hispanic high-risk youth. *Journal of Community Psychology*, 25(5), 453–471. doi:10.1002/(SICI)1520-6629(199709)25:5<453::AID-JCOP6>3.0.CO;2-T

Santisteban, D. A., Suarz-Morales, L., Robbins, M. S., & Szapocznik, J. (2006). Brief Strategic Family Therapy: Lessons learned in efficacy research and challenges to blending research and practice. *Family Process*, 45, 259–271.

Segal, L. (1991). Brief therapy: The MRI approach. In A. S. Gurman & D. P. Knishern (Eds.), *Handbook of family therapy* (pp. 171–199). New York: Brunner/Mazel.

Szapocznik, J., Hervis, O. E., & Schwartz, S. (2003). *Brief strategic family therapy for adolescent drug abuse* (NIH Publication No. 03-4751). NIDA Therapy Manuals for Drug Addiction. Rockville, MD: National Institute for Drug Abuse.

Sprenkle, D. (Ed.). (2012). Intervention research in couple and family therapy [Special edition]. *Journal of Marital and Family Therapy*, 38(1).

Szapocznik, J., & Williams, R. A. (2000). Brief strategic family therapy: Twenty-five years of interplay among theory, research and practice in adolescent behavior problems and drug abuse. *Clinical Child and Family Psychology Review*, 3(2), 117–135.

Walsh, F. (1993). Conceptualization of normal family process. In F. Walsh (Ed.), *Normal family process* (pp. 3–72). New York: Guilford.

Strategic Case Study: Intergenerational Conflict in Chinese American Family

Tim Wong (AM34), a 34-year-old third generation Chinese American and his wife Fang Wong (AF31), a 31-year-old first generation Chinese American came to counseling because they are having frequent heated arguments about her parents living in the house with them. Tim and Fang have been married for five years and have a 3-year-old daughter named Nancy (CF3) and a 6-month-old daughter named Betty (CF1). Although Tim goes to work five days a week, he has a very caring and loving relationship with both of his daughters. Fang describes that she considers her husband a "family man" and would do anything for his daughters and his wife. Fang also has a loving and affectionate relationship with her daughters. Before Tim and Fang had children, they decided that they wanted to have an authoritative parenting style with their children. Fang hated that she was raised in an authoritarian home and ensured that it would be different for her daughters.

Soon after Fang had her second child her parents, Li (AF55) and Chen (AM55), moved in with her family to help take care of her 3-year-old daughter and newborn. Tim was skeptical on having Fang's parents move in to the house but Fang persuaded him to try it out for a while because she is really stressed about going back to work and needs the extra help at home. Fang's mother has always been an authoritarian parent who is very strict. She is often demanding and non-responsive. Ever since Fang's parents moved in her mother has been extremely overbearing, frequently correcting Tim and Fang on their parenting skills because their techniques are very different. Fang often feels irritated by her mother but just brushes it off because her mother has acted this way since she was born. Tim on the other hand is constantly at battle with Li and argues frequently with Fang about asking her parents to find another home to live in. Tim insists that they hire a nanny but Fang and her mother strongly believe it is better to have a family member help raise a child than a stranger. They have been arguing about this for over four and a half months and are worried that Li living in the house is negatively affecting their marriage. Fang's father Chen has agreed to attend counseling sessions, and his wife Li has reluctantly agreed to attend.

Strategic Case Conceptualization

Interaction Sequence

- *Rise of Tension:* When taking care of the children AM does something wrong according to AF55 such as "make the diaper too tight" or "not discipline his daughter firmly enough." AM gets angry at AF55 but instead of confronting her he later takes his anger out on AF31 and tells her she needs to ask her mother to stop criticizing him or to move out.
- *Conflict:* AF31 becomes angry with AM34 and yells at him for taking his anger out on her. AF31 tells AM34 that if he has an issue with her mother he needs to deal with it on his own. AM yells back at AF and leaves the room without resolution.
- *Return to Homeostasis:* After AM cools off he comes back in to the room and acts as if everything has been resolved.

Strategic Conceptualization/Hypothesis

- *Hierarchy Versus Equality:* Since AF55 and AM55 have moved in to the household there has been a battle over who is at the top of their hierarchy. Based on traditional cultural values that value elders, AF55, AM55, and to a large extent AF31 view AF31's parents as the appropriate heads of the household. AM34 is unaccepting of this, adopting a more American position. In addition, his refusal to accept her parent's authority may also be related to the fact that traditionally in Chinese culture, the female moves into the husband's house, so there is less cultural precedent for this arrangement.

Family Life Stage of Development

- *Parents With Young Children:* AF31 and AM34 are struggling with cultural variations with the transition to parenthood. AF31 is leaning toward a more traditional family form—multigenerational household—and AM34 is leaning toward a more American form of the nuclear family with parents, not the elders, having the position of greatest authority.

Hierarchy and Power

- Most of the struggles in the family surround issues of power. AM is not accepting of AF55 being the person with greatest influence regarding child rearing in the household. AF31 generally supports her mother, but also wants to eventually have a less hierarchical relationship with her child than her mother had with her. AM34 is not accepting of AF55's current level of power and actively tries to pursue AF31 to support him in changing this, but without success, thus further exaggerating his sense of powerlessness.

Strategic Treatment Plan

Strategic Initial Phase of Treatment with Couple/Family

Initial Phase Counseling Tasks

1. Develop working counseling relationship. *Diversity Note: Avoid language that might create sense of shame; respect cultural variations of family hierarchy; allow for more subtle forms of emotional expression.*
   Relationship building approach/intervention:
   a. Personally greet AF31 and AM34 and engage them in social conversation. Make sure each feels comfortable and reduce his or her sense of shame.
   b. Determine if they want to invite AF55 and AM55 to first or later session.

2. Assess individual, systemic, and broader cultural dynamics. *Diversity Note: Consider Chinese family structure and norms for intergenerational relationships and parenting.*
   a. Assess *problem interaction sequence*, noting the affect of *stage of family life, power, and alliances* within the family, especially across generations.
   b. Identify theme of *power* and *hierarchy* within the multigenerational family, exploring cultural meanings and norms.

Initial Phase Client Goals

1. Increase ability to see loving motivations that underlie problem interaction cycle to reduce conflict.
   a. *Reframe* AF31, AM34, and AF55's roles in the interaction by highlighting how each one is motivated by love for the others and what each believes to be best for the family.
   b. **Directives** that interrupt the problem sequence such as a symbolic gesture each time AF55 offers "sage" advice.

Strategic Working Phase of Treatment with Family

Working-Phase Counseling Task

1. Monitor quality of the working alliance. *Diversity note: Attend to client response to interventions that indicate clients feel safe and not shamed.*
   a. Monitor therapist influence and adjust their interventions accordingly. For example: If AM34 is feeling like therapist is siding with AF31 and AF55, therapist needs to takes measures to become more neutral or take measures to align with AM equally.

Working-Phase Goals

1. Decrease AM34, AF31, and AF55's *problem interactions* that are attempts to define the balance of *power* in the relationship to reduce conflict.
   a. *Paradoxical interventions* that require AM34 and AF31 to exaggerate being helpless children.

(continued)

b. *Metaphorical tasks* that highlight AM34's sense of powerlessness, AF55's sense of being the only competent parent, and AF31's sense of being torn between AM34 and AF55.
   c. *Directives* that allow for alternation of which person is in charge of parenting on a regular schedule (e.g., Mondays AF55, Tuesdays, AF31, Wednesdays, AM34, Thursdays AM55).
2. Increase interactions that highlight each member's sense of love and commitment to each other to reduce conflict.
   a. Use *pretend techniques* to increase desired behaviors, such as showing appreciation and acceptance of each style of parenting.
   b. *Directives* that include simple acts of appreciation and sharing of cultural traditions.

Strategic Closing Phase of Treatment with Couple/Family

Closing-Phase Counseling Task

1. Develop aftercare plan and maintain gains. *Diversity Note: Identify resources in Buddhist temple to which they belong and the local Chinese community.*
   a. Use *restraining* by encouraging family to take things slowly and avoid changing too fast.
   b. Discuss the *dangers of improvement* such as if AF55 being so involved with parenting AF31 and AM34 will have to take more responsibility, etc.

Closing-Phase Client Goals

1. Increase shared understanding and agreement of how power and parenting roles will be distributed within the family to reduce conflict.
   a. *Reframing* that allows family to appreciate both Chinese and American versions of intergenerational relations and parenting.
   b. *Directives* that allow the family to explore various forms of power and hierarchy within the family.
2. Increase AF34 and AF31 ability to adapt level of interdependence and independence within couple and with extended family as new parents to increase sense of well-being and reduce conflict.
   a. *Directives* designed to rebalance independence and interdependence based on stage of family life as new parents and cultural.
   b. *Pretend techniques* to help each explore behaviors that enable them to increase their adaptation to being parents.

CHAPTER 7

Structural Family Therapy

"Training in family therapy should therefore be a way of teaching techniques whose essence is to be mastered, then forgotten. After this book is read, it should be given away, or put in a forgotten corner. The therapist should be a healer: a human being concerned with engaging other human beings, therapeutically, around areas and issues that cause them pain, while always retaining great respect for their values, areas of strength, and esthetic preferences. The goal in other words is to transcend technique."

—Minuchin & Fishman, 1981, p. 1

Lay of the Land

Structural therapy is primarily associated with the work of Salvador Minuchin, and the majority of this chapter focuses on his work. In addition, the chapter highlights evidence-based treatments that draw heavily from traditional structural family therapy: ecosystemic structural family therapy.

In a Nutshell: The Least You Need to Know

As the name implies, structural therapists map family structure—boundaries, hierarchies, and subsystems—to help clients resolve individual mental health symptoms and relational problems (Minuchin & Fishman, 1981). After assessing family functioning, therapists aim to restructure the family, realigning boundaries and hierarchies to promote growth and resolve problems. They are active in sessions, staging enactments, realigning chairs, and questioning family assumptions. Structural family therapy focuses on strengths, never seeing families as dysfunctional but rather as people who need assistance in expanding their repertoire of interaction patterns to adjust to their ever-changing developmental and contextual demands.

The Juice: Significant Contributions to the Field

If you remember a thing or two from this chapter, they should be these:

Juice 1: Boundaries, or Rules for Relating

Boundaries are one of the few family therapy terms that have trickled into the vernacular, so often that your clients come in talking about them. At first glance, the term seems two-dimensional, and much like Goldilocks, you are tempted to sum things up by saying they are too rigid, too weak, or just right. However, as you begin to work with the idea of boundaries, you quickly learn that boundaries are far more complex than they initially appear. But let's start with the simple definition.

Boundaries are rules for managing physical and psychological distance between family members, for defining the regulation of closeness, distance, hierarchy, and family roles (Minuchin & Fishman, 1981). Although they may sound static, they are organic, living processes. Structural therapists identify three basic types of boundaries:

- *Clear Boundaries:* Clear boundaries are "normal" boundaries that allow for close emotional contact with others while simultaneously allowing each person to maintain a sense of identity and differentiation (Colapinto, 1991). Each culture has a unique style of balancing closeness and distance, with different appropriate outward expressions of this balance. For example, some cultures require more physical space for clear boundaries than do others.
- *Enmeshment and Diffuse Boundaries:* Diffuse or weak boundaries lead to relational enmeshment (for those who like to be technically correct, boundaries are diffuse and relationships enmeshed). Families with overly diffuse boundaries do not make a clear distinction between members, creating a strong sense of mutuality and connection at the expense of individual autonomy (Colapinto, 1991). When talking with an enmeshed family, therapists typically see family members doing the following:
    - Interrupting one another or speaking for one another
    - Mind reading and making assumptions
    - Insisting on high levels of protectiveness and overconcern
    - Demanding loyalty at the expense of individual needs
    - Feeling threatened when there is disagreement or difference

    How can you tell the difference between clear and close versus diffuse boundaries? Simple. If boundaries are diffuse, the family will report symptoms and problems in one or more individuals and/or complaints about family interactions. Moreover, behaviors that constitute problematic boundaries in one cultural context may be clear in another cultural context (Minuchin & Fishman, 1981). Immigrant and other bicultural families present special problems because there is more than one cultural context at play. Thus, although identifying problem boundaries seems straightforward at first, it quickly becomes murky in actual practice, requiring therapists to proceed mindfully and respectfully and to attend to each family's unique situation. In the case study at the end of the chapter, the family is experiencing an increase in enmeshed boundaries after a divorce, resulting in the mother's overinvolvement with the son as well as inappropriate coalitions that each parent forms with the children.

- *Disengagement and Rigid Boundaries:* Rigid boundaries lead to relational disengagement. Autonomy and independence are emphasized at the expense of emotional connection, creating isolation that may be more emotional than physical (Colapinto, 1991). These families have excessive tolerance for deviation, often failing to mobilize support and protection for one another. Therapists working with disengaged families notice the following:
    - Lack of reaction and few repercussions, even to problems
    - Significant freedom for most members to do as they please
    - Few demands for or expressions of loyalty and commitment
    - Consistently using parallel interactions (e.g., doing different activities in the same room) as substitutes for reciprocal interactions and engagement

Again, rigid boundaries cannot be accurately assessed without taking cultural and developmental variables into consideration; unless members are experiencing symptoms or problems, there is little ground for identifying boundaries as overly rigid.

Juice 2: Enactments

Perhaps the most distinctive of structural interventions, enactments are techniques in which the therapist prompts the family to reenact a conflict or other interaction (Colapinto, 1991; Minuchin, 1974; Minuchin & Fishman, 1981). Regardless of the therapy model you choose to use in the end, enactments are one of the most important techniques for therapists to master. Why? Because most couples and families are going to start arguing in your office whether you ask them to or not, so you better be prepared! Enactments are one of the best ways to handle this.

Minuchin preferred enactments to talking about interactions because often people describe themselves as one way but behave quite differently, not because they are malicious or hypocritical but because it is often difficult to see clearly how our behavior looks from the outside (Minuchin & Fishman, 1981). Enactments are used to both assess and alter the problematic interactional sequences, allowing the therapist to *map, track,* and *modify* the family structure. As therapists become more experienced, they require only a few minutes of watching a family interact to know where and how to *restructure* the family. Restructuring may take the form of creating a clearer boundary in enmeshed relationships (e.g., stopping people from interrupting and speaking for one another), increasing engagement by encouraging the expression of empathy or direct eye contact, or improving parental effectiveness by helping the parent successfully manage a child's in-session behavior.

An enactment occurs in three phases, as a "dance in three movements" (Minuchin & Fishman, 1981, p. 81):

1. *Observation of Spontaneous Interactions: Tracking and Mapping:* When talking with the family, the therapist closely follows both content and process, listening for the rules and assumptions that coordinate the family's interactions, such as demands for overconnectedness, extreme disconnection, or hierarchical confusion, as well as strengths and resources. The therapist tracks *actual transactions* more closely than verbal accounts (Colapinto, 1991), while developing a hypothesis that *maps* the family's boundaries and hierarchy (Minuchin & Fishman, 1981). Once therapists identify an area for change, they are ready to invite the family into the active phase of enactment.

2. *The Invitation: Eliciting Transactions:* The invitation for an enactment is issued in two ways: either the therapist directly asks the family to engage in an enactment, or the family spontaneously starts an enactment of at-home behavior, usually in the form of an argument (Colapinto, 1991). Obviously, the therapist does not need to do much when the family spontaneously begins; if the family does not, the therapist must issue an explicit invitation to "show" the problem:

    "Can you reenact what happened last night?" or

    "Please show me what happens at home when he is 'defiant'; can you act out an incident of defiance that happened last week so I have a good idea of what the problem really is?"

3. *Redirecting Alternative Transactions:* This is the most important part. It really is not therapeutic to ask a family to start enacting problem behaviors if the therapist does not jump in and help redirect the behavior to clarify boundaries and hierarchies. How therapists redirect the interaction depends on the particular interaction that needs to be changed. Redirection often involves the following:
    - Stopping family members from interrupting or speaking for one another
    - Directing two people to directly engage each other while asking a third member to allow the other two to communicate
    - Encouraging emotional understanding and connection between disengaged parties
    - Rearranging chairs physically to increase or decrease emotional closeness
    - Requesting parents to actively establish an effective hierarchical position with a child

Enactments are beneficial because they provide live practice with new interactions and family patterns, increasing the likelihood of transitioning in-session gains and insights into everyday family life. They also reduce the illusion that the problem belongs to a single person; when a family demonstrate the problem in front of the therapist, it becomes clear that the reported problem does not belong to a single person but to the larger family unit. Finally, enactments increase the family's sense of competence and strength by helping them to successfully engage in new preferred behaviors (Minuchin & Fishman, 1981). In the case study at the end of the chapter, the therapist uses enactments to help the divorced family clarify boundaries, severe coalitions, and help the parents learn to coparent as a divorced couple.

Rumor Has It: The People and Their Stories
Salvador Minuchin
Trained as a pediatrician and child psychiatrist, Salvador Minuchin is considered the progenitor of structural family therapy (Colapinto, 1991; Minuchin, 1974). Minuchin lived and worked on three continents; he was born and raised in Argentina and then lived in Israel during two periods of his life before settling in the United States. In 1954, after returning from a two-year period of working with displaced children in Israel, he began his psychiatry training with Harry Stack Sullivan, whose psychoanalytic work emphasized interpersonal relationships. After his training, Minuchin accepted a position at the Wiltwyck School for delinquent boys and suggested to his colleagues—Dick Auerswald, Charlie King, Braulio Montalvo, and Clara Rabinowitz—that they see the entire family. With no formal models to follow, they used a one-way mirror to observe each other and developed a working model as they went along. In 1962, Minuchin visited the Mental Research Institute, where Haley, Watzlawick, Fisch, and others were at the forefront of developing family therapy approaches. There he befriended Jay Haley, who developed the strategic approach to family therapy (see Chapter 6); the mutual influence of this friendship is evident in the work of both men.

From 1965 to 1976, Minuchin served as the director of the Philadelphia Child Guidance Clinic, and in 1975 he founded the Family Therapy Training Center (later renamed the Philadelphia Child and Family Therapy Training Center). In 1967, Minuchin, Montalvo, Guerney, Rosman, and Schumer published *Families of the Slums,* considered the first book to describe structural therapy and a book that discussed diversity issues before the term *multiculturalism* was coined. Over the years, Minuchin and his colleagues have written numerous books that detail how they have developed and refined this model to address changing cultural contexts and specific diagnoses (Minuchin, Rosman, & Baker, 1978). Minuchin is still an active leader in the field, continuing to teach new generations of therapists (Minuchin, Nichols, & Lee, 2007). His influential students and colleagues include Harry Aponte, Jorge Colapinto, Charles Fishman, Jay Lappin, and Michael Nichols.

Harry Aponte
Harry Aponte (1994, 1996) attends to issues of spirituality, poverty, and race in the practice of structural family therapy.

Marion Lindblad-Goldberg
Marion Lindblad-Goldberg succeeded Minuchin as the director of the Philadelphia Child and Family Therapy Training Center in 1986 and still serves as its director. She and her colleagues (Lindbald-Goldberg, Dore, & Stern, 1998) developed the empirically supported treatment called *ecosystemic structural family therapy* (ESFT), which they emphasize in current training programs at the center.

The Big Picture: Overview of Treatment
Minuchin (1974) identifies three main *phases* of structural therapy:

1. Join the family and accommodate to their style (build an alliance).
2. Map the family structure, boundaries, and hierarchy (evaluate and assess).
3. Intervene to transform the structure to diminish symptoms (address the problems they identified in the assessment).

Generally, therapists alternate between phases two and three many times, revising and refining the map and hypotheses about family functioning until the problems are addressed and resolved. I like to think of it as similar to the golden rule with shampoo: "lather, rinse, repeat" until you achieve the desired effect.

Who Attends Therapy?
To be able to assess the system, structural therapists prefer to begin therapy with the entire family, but they do not insist on it (Colapinto, 1991). However, once the family

system has been assessed, the therapist often meets with specific subsystems and individuals to achieve structural goals. For example, often sessions with the couple alone are necessary to strengthen the boundaries between the couple and parental subsystems and to sever cross-generational coalitions.

Making Connection: The Therapeutic Relationship
Joining and Accommodating

Structural family therapists have a unique term for the therapeutic relationship: *joining* (Minuchin, 1974; Minuchin & Fishman, 1981; Minuchin & Nichols, 1993). They "join" the system in the sense that they *accommodate* to its style: how people talk, what words they use, how they walk, and so forth. *Mimesis*, a Greek term that means "copy" (as in *mimeograph;* if you're too young to remember that, consider yourself lucky), has also been used to refer to the process of accommodating the family's way of being. In the historical context of psychotherapy, this is a radical concept, because unlike in psychodynamic, cognitive-behavioral, and even experiential therapies, the therapist does not take a superior role.

Minuchin (1974) compared the process of joining a family to an anthropologist studying a new culture, which always begins by sitting back and observing patterns, habits, and behaviors before beginning to address one's own agenda, in this case to alleviate the family's distress. The process of joining can also be likened to falling in rhythm with the family. Do they talk fast or slow? Do they talk over one another or wait for clear pauses to speak? Do they use teasing and humor, or are their words gentle and soft? A successful structural therapist needs to have a wide repertoire of social skills to successfully join with families, especially when working with diverse populations.

Joining as an Attitude

Colapinto (1991) emphasizes that joining is more of an attitude than a technique; it is the glue that holds the therapeutic system together through the often turbulent and challenging journey called therapy (Minuchin & Fishman, 1981). The attitude of joining requires (a) a strong, clear sense of connection and affiliation (e.g., curiosity, openness, sensitivity, acceptance) and (b) an equally clear sense of distance and differentiation (e.g., questioning, dissenting, promoting change).

Therapeutic Spontaneity

Structural therapists strive to cultivate therapeutic spontaneity, which does not refer to a do-as-you-please attitude but rather a relationally and contextually responsive expression of self: "the therapist's spontaneity is constrained by the context of therapy" (Minuchin & Fishman, 1981, p. 3). Therapeutic spontaneity refers to the ability to flow naturally and authentically in a variety of contexts and situations. Much in the way riding a bike becomes natural after a painful period of training wheels and falls, therapeutic spontaneity is cultivated and shaped through the training process, which increases therapists' repertoire for "being natural" in a wide range of clinical situations.

Therapist's Use of Self

According to Minuchin, therapists must use themselves to relate to the family, varying from being highly involved to professionally detached (Minuchin & Fishman, 1981). They may be clearly detached from family interactions so that they can clarify boundaries or prescribe a specific intervention, maintain a moderate level of connection to coach the family in new interactions, or assume a fully engaged position by taking sides with one family member to "unbalance" the system (an intervention discussed later in the chapter). The therapist is highly flexible, adapting to each family's needs and cultural norms.

"Making It Happen"

> "The primary injunction from the model to the therapist can be summarized in three words: 'Make it happen.'" —Colapinto, 1991, p. 435

The therapist's job is to find a way to help the family achieve desired change, and he/she must do whatever it takes to make this happen. Therefore, therapists' roles can vary widely: they can be the "producer" who ensures conditions that make therapy possible, the "stage director" who pushes the family toward more functional patterns, the "protagonist" who directly uses himself/herself to alter stuck family interactions, or the "narrator" or "coauthor" who collaboratively helps the family revise their script. Thus therapists need to be open to playing whichever role will be most beneficial for a particular family in a given session, rather than being wedded to their own favorite roles.

Recent Adaptation: A Softer Style

In his later work, Minuchin has described a change in approach: "I have moved from being an active challenger—confronting, directing, and controlling—to a softer style, in which I use humor, acceptance, support, suggestion, and seduction on behalf of the same goals" (Minuchin et al., 2007, p. 6). Despite this change, Minuchin has not abandoned the expert role or the goal of achieving change in the present. A recent study that analyzed structural family therapy sessions found that therapist empathy was not only readily evident but appears to be a key ingredient in facilitating change (Hammond & Nichols, 2008).

The Viewing: Case Conceptualization and Assessment

Structural family therapists conceptualize and assess the following factors:

Structural Assessment

- Role of symptom in the family
- Subsystems
- Cross-generational coalitions
- Boundaries
- Hierarchy
- Complementarity
- Family development
- Strengths

Role of the Symptom

Structural therapists identify three possible relationships between the symptom and the family system (Colapinto, 1991):

1. *Family as Ineffectual Challenger of Symptom:* The family is *passive*. In order to maintain a highly enmeshed or disengaged family structure, it fails to challenge the symptomatic member.
2. *Family as "Shaper" of Individual's Symptoms:* The family structure shapes the individual's experience and behaviors.
3. *Family as "Beneficiary" of the Symptom:* The symptom performs a regulatory function in maintaining the family structure.

As in virtually all forms of family therapy, the *symptom bearer* or *identified patient* (IP) is never seen as the sole source of the problem, and instead the family interaction patterns are targeted for intervention. In the case study at the end of the chapter, the divorce is viewed as shaping the individual's behaviors, including the son's substance use and dropping grades.

Subsystems

Minuchin (1974) conceptualized a family as a single system that also had multiple *subsystems*. Some subsystems can be found in almost every family: couple, parental, sibling, and each individual as a separate subsystem. In addition, in some families other influential subsystems develop along gender lines, hobbies, interests (sports, music), and even personalities (serious versus fun-loving). When assessing a family, generally the most important subsystem issues to consider are (a) whether there is a clear distinction between the parental and couple subsystems and (b) whether there is a clear boundary between the parental and child/sibling subsystems. Alternatively stated, is there an effective parental hierarchy?

Cross-Generational Coalitions: Problematic Subsystems

One type of subsystem is particularly damaging: *cross-generational coalitions* (Minuchin & Fishman, 1981; Minuchin & Nichols, 1993). A cross-generational coalition is a subsystem that forms between a parent and child *against* the other parent or other key caretaker. This is a common family dynamic; often a mother has grown closer to her children and has unresolved marital or parenting conflicts with her spouse. The inverse, a father in coalition with the children against the mother, is also frequent, as is a family that divides into "teams," with the father and mother heading their team in overt or covert opposition to the other. These coalitions are especially common in divorces, and in fact, both parents typically try to create these coalitions simultaneously against the other: thus the children become the rope in a tug of war. These coalitions are often *covert*, meaning they are not directly addressed or spoken about in the family but are evident by secrets between the parent and child ("don't tell your mom/dad about this") or comments that compliment the child and disparage the other spouse ("I am so glad you didn't inherit your father's/mother's gene for X"). These coalitions can also involve other caretakers, such as grandparents or parentified children. In the case study at the end of the chapter, the family has developed extremely problematic cross-generational coalitions, one between father and son against the mother and the other between the mother and daughter against the father. Although this is common after divorce, it is highly disruptive to everyone's functioning, as evidenced in this family by the son's substance use and dropping grades.

Boundary Assessment

As described in the Juice section above, structural therapists assess interpersonal boundaries between all subsystems and individuals in the family and between the family and external systems, such as families of origins, friends, etc.

Hierarchy

When working with reported problems in child behavior, therapists must first assess the parental hierarchy so that they know how to intervene (Colapinto, 1991; Minuchin, 1974; Minuchin & Fishman, 1981). There are three basic forms of parental hierarchy:

- *Effective*: When the parental hierarchy is appropriate and effective, parents can set boundaries and limits while still maintaining emotional connection with their children.
- *Insufficient*: When the parental hierarchy is insufficient, parents are not able to effectively manage the child's behavior and often adapt a permissive parenting style. This style is easy to identify in the therapy office: the parents are not able to keep younger children from tearing up the waiting room and office, or their teens act as though they have the right to set their own curfews and rules. Often the parents hope that the therapist will "teach" their children to listen, but this often requires intervening more with the parents than the children. Generally, these parents have enmeshed boundaries with their children, but not always. As with boundaries, the outward expression of effective versus insufficient can only be determined by examining the cultural context, family life stage of development, and symptomatic behavior.
- *Excessive*: When there is excessive hierarchy, the rules are developmentally too strict and unrealistic and consequences are too severe to be effective. In this situation, there is almost always a rigid boundary between children and parents. These parents need assistance in developing age-appropriate rules and expectations and in developing a stronger emotional bond with their children.

Complementarity

Much like systemic therapists (Chapter 5), structural therapists assess for rigid complementary patterns between family members (Colapinto, 1991). Like a jigsaw puzzle, family members develop complementary roles: the over-/under-functioner, good/bad child, understanding/strict parent, logical/emotional partner, and so forth. Over time, these systemically generated roles become viewed as inherent personality characteristics that seem unchangeable. The more exaggerated and rigid these roles become, the less adaptable the individuals and family become. Structural therapists recognize the mutually reinforced patterns and target the ones that need to change for members to grow.

Family Development

Rather than a static entity, the family is viewed as continually growing and changing in response to predictable stages of development as well as unexpected life events such as death, a move, or divorce (Minuchin & Fishman, 1981). Minuchin and Fishman (1981) identify four major stages in family development:

1. Couple formation
2. Families with young children
3. Families with school-age or adolescent children
4. Families with grown children

At each stage the members need to renegotiate boundaries to define the levels of closeness and differentiation that will support individual members' growth needs. Families often get stuck transitioning from one stage to another if they fail to renegotiate boundaries and hierarchy as the family develops.

Strengths

Minuchin strongly feels that therapists should avoid labeling families as dysfunctional and instead recognize their strengths, particularly their cultural and idiosyncratic strengths (Minuchin & Fishman, 1981; Minuchin & Nichols, 1993). He powerfully argues against seeing the family as an enemy of its individual members, as is frequent in psychological literature, encouraging therapists to recognize how the family provides support, protection, and a foundation for its members. Family strengths, such as a strong connection to extended family or community, are identified and used to promote the goals of individual and family growth as well as to reduce symptoms.

Targeting Change: Goal Setting

> *"A well-functioning family is not defined by the absence of stress, conflict, and problems, but by how effectively it handles them in the course of fulfilling its function. This, in turn, depends on the structure and adaptability of the family."* —Colapinto, 1991, p. 422

Structural therapists target similar goals for all families (Colapinto, 1991; Minuchin, 1974):

- *Clear boundaries* between all subsystems that allow for connectedness and differentiation congruent with the family's cultural contexts
- Clear distinction between the *marital/couple subsystem* and the *parental subsystem*
- *Effective parental hierarchy* and the severing of cross-generational coalitions
- A family structure that promotes the *development and growth of individuals and the family*

The Doing: Interventions

Enactments and Modifying Interactions

As described in the Juice section above, enactments are one of the most frequently used and impactful interventions that structural therapists use.

Systemic Reframing

As the family begins to describe their problems, therapists reflect on their understanding using *systemic reframing* (Colapinto, 1991; Minuchin, 1974; Minuchin & Fishman, 1981).

A systemic reframe takes into account that all behavior has reciprocal antecedents: person A affects person B's response, which then affects person A's response, ad infinitum (A · B). Reframes often highlight complementary relationships in the family, such as the pursuer/distancer pattern. Reframing usually involves removing the blame from one person (the identified patient) and "spreading" blame equally by describing how each person's response contributes to the problem dynamic. Once this is done, blame becomes a moot point.

Systemic reframing involves piecing together each member's description of the problem and reframing it to reveal the broader systemic dynamic. Thus, if the wife complains that her husband never listens to her, and he complains that she is always nagging him about something, the therapist can systemically reframe their descriptions to highlight how the more she pushes him to listen and interact, the more he withdraws; and the more he withdraws, the more she feels compelled to pursue him for interaction.

How to Generate Systemic Reframes

- Assess broader interactional patterns (complementary relationships, hierarchy, boundaries, etc.)
- Redescribe the problem (use interactional patterns to describe the problem in a larger context)

Boundary Making

Boundary making is a special form of enactment that targets over- or underinvolvement to help families soften rigid boundaries or strengthen diffuse boundaries (Colapinto, 1991; Minuchin, 1974). Structural therapists use this technique to direct who participates and how. By actively setting boundaries, therapists interrupt the habitual interaction patterns, allowing members to experience underutilized skills and abilities. Boundary making may involve several different directives:

- Asking family members to change seats
- Asking family members to move seats further or closer together or turn toward one another
- Having separate sessions with individuals or subsystems to strengthen subsystem boundaries
- Asking one or more members to remain silent during an interaction
- Asking questions that highlight a problem boundary area (e.g., "Do you always answer for your son when he is asked a question?")
- Blocking interruptions or encouraging pauses for less dominant persons to speak

Challenging the Family's Worldview

Challenging the family's worldview and unproductive assumptions typically involves verbally questioning operational assumptions in the family system, whether overtly spoken or covertly acted upon (Colapinto, 1991; Minuchin, 1974; Minuchin & Fishman, 1981). Common assumptions that create problems for individuals, couples, and families include the following:

- "Kids' needs come first."
- "It's better to keep the peace than start conflict."
- "It is easier to sacrifice my needs than ask for what I want."
- "If I give here, you should give there."
- "It's better for the kids for us to stay in this unhappy marriage."

Structural therapists often challenge these assumptions by overtly questioning whether they are actually having the effect family members anticipate. The challenge can be delivered softly or strongly, depending on what will be most effective in the particular family structure. In the case study at the end of the chapter, the therapist challenges the parents' worldview that they no longer have to work together to parent their teenage children; the therapist unites the parents by explaining how their ongoing conflict is significantly contributing to their son's drug use and school issues and helping them develop a vision of what successful parenting in a divorced family looks like.

Intensity and Crisis Inductions

Intensity and crisis inductions are interventions that use affect to create structural shifts in hierarchy and boundaries, especially when the family is having trouble "hearing" the therapist with other interventions (Minuchin, 1974; Minuchin & Fishman, 1981; Minuchin & Nichols, 1993). Because families differ in the degree of loyalty they demand to their reality, they need different levels and styles of intensity, depending on the issue being discussed. Intensity involves turning up the emotional heat by using tone of voice, pacing, and word choice to break through rigid and stuck interactional patterns. For example, a therapist may say to a couple who claims to have no time for a weekly date because of their children's numerous after-school activities, "Do you think your children would prefer to be in soccer and have divorced parents or to have fewer activities and an intact family?"

Closely related to intensity, *crisis induction* in structural therapy is used with families who chronically avoid a conflict or problem (Colapinto, 1991). For example, in families with anorectic children, the therapist may bring the symptom into the room by staging a meal and having the family deal with it. Similarly, with alcohol or substance abuse issues, the therapist often induces a crisis so that the family will acknowledge and finally address the problem. The therapist can then help the family develop new interactions and patterns.

Unbalancing

Unbalancing is used for more extreme difficulties in hierarchy or when the identified patient is being scapegoated. This intervention is used to realign boundaries between subsystems (Minuchin, 1974; Minuchin & Fishman, 1981). Therapists use their expert position to temporarily "join sides" with individuals who are being scapegoated or with subsystems that need to develop stronger boundaries by arguing their cause or helping to explain their perspective to others. At first glance, this may seem to go against the general rule of neutrality that characterizes structural therapy specifically and psychotherapy more generally. However, unbalancing is done only briefly and with specific realignment goals in mind, generally only after more direct interventions, such as enactments and challenging assumptions, have failed.

Expanding Family Truths and Realities

Each family develops a unique worldview that defines its realities and truths. When working with a highly rigid family structure, structural therapists directly challenge these beliefs and realities (already discussed under Challenging the Family's Worldview; Minuchin & Fishman, 1981). However, whenever possible, structural therapists cite these beliefs to *expand* the family's functioning in new directions. For example, they might say, "Because you obviously have such deep concern for your child, you are parents who are likely to understand that the child needs space to grow in order to really flourish" or "Since you are willing to go to such lengths to be helpful, it seems you are probably able to be helpful in an even more challenging way: allowing him to make his own mistakes." Rather than introduce an entirely foreign concept, the structural therapist takes the family's fundamental premise that has been supporting the problem and redirects its logic to support an alternative set of behaviors and interactions, allowing the family to maintain its core beliefs but use them in new ways.

Making Compliments and Shaping Competence

Minuchin and Fishman (1981) strongly caution therapists that professional training creates a "search and destroy" (diagnose and treat) approach to psychopathology that often blinds therapists to family strengths and positive interaction patterns. Instead, therapists should augment and reinforce the family's natural positive patterns and strengths. *Compliments* are used to bolster behaviors that support families in moving toward their goals, and *shaping competence* involves noticing small successes along the way to reaching goals. For example, families usually improve after enactments and refrain from interrupting or speaking for each other in later sessions. Therapists can shape competence by noticing these changes in session or as they are reported from week to week.

Shaping competence also involves refusing to function for the family in session. For example, rather than taking responsibility for having children focus and behave during a session, the therapist asks the parents to do so. If a child is kicking the furniture or gets up to play with a toy during a family session, rather than correct the child, the therapist asks the parents to have the child stop. Similarly, if the therapist is trying to strengthen the parental hierarchy, the therapist asks the parents to answer questions first and recognizes their authority by directing children to ask parents for permission to do such things as go to the bathroom or get some water.

Putting It All Together

Case Conceptualization Template
- *Role of Symptom in the Family:* Describe relation of family and symptom of symptom:
  - Family as ineffectual challenger of symptom
  - Family as "shaper" of individual's symptoms
  - Family as "beneficiary" of the symptom
- *Subsystems:* Describe:
  - Parental, including grandparents, stepparents, parentified children
  - Couple: Is this system distinct from the parental subsystem?
  - Sibling, including stepsiblings
  - Other significant subsystems based on gender, interests, etc.
- *Cross-Generational Coalitions:* Describe any cross-generational coalitions
- *Boundaries:* Describe boundaries in family system between individuals and subsystems:
  - Enmeshed
  - Clear
  - Disengaged
- *Hierarchy:* Describe parent-child hierarchy and other salient hierarchies in the system.
  - Effective (authoritative)
  - Insufficient (permissive)
  - Excesive (authoritarian)
- *Complementarity:* Describe complimentary roles, such as pursuer/distancer, over-/under-functioner, logical/emotional, good/bad parent, good/bad child, etc.
- *Family development:* Identify family life cycle stage and any issues in meeting developmental needs:
  - Couple formation
  - Families with young children
  - Families with school-age or adolescent children
  - Families with grown children
- *Strengths:* Identify family strengths that may be unique to family or related to diversity factors.

Treatment Plan Template for Individual
The following treatment plan template can be used with individuals with depressive or anxiety symptoms.

Structural Initial Phase of Treatment with Individual

Initial Phase Counseling Tasks

1. Develop working counseling relationship. *Diversity Note: [Describe how you will adjust to respect cultured, gendered, and other styles of relationship building and emotional expression.]*
   a. *Join* with client, adapting to gender/culture/class norms for relating, expressing emotion, conversational tempo, etc.
2. Assess individual, systemic, and broader cultural dynamics. *Diversity Note: [Describe how you will adjust assessment based on cultural, socioeconomic, sexual orientation, gender, and other relevant norms.]*
   a. Assess structure of family of origin and family of procreation (or current partnership), including *subsystems, coalitions, boundary patterns, hierarchy, complementary relationships, family stage of development, and strengths.*
   b. Identify the *role of the symptom* (e.g., depression/anxiety) in the client's family system.

Initial Phase Client Goals

1. Increase *strength of diffuse* [or flexibility if rigid] *boundaries* with/in [specify person/context] to reduce depressed mood and anxiety.
   a. *Challenge client's world view* as it relates to maintaining unclear boundaries.
   b. *Shape competence* by helping client transfer strategies for clear boundaries in one area of life to the area where boundaries are a problem.

Structural Working Phase of Treatment with Individual

Working-Phase Counseling Task

1. Monitor quality of the working alliance. *Diversity Note: [Describe how you will attend to client response to interventions that indicate therapist using expressions of emotion that are not consistent with client's cultural background.]*
   a. *Assessment Intervention*: Adjust *joining* style and *accommodate* to system when client verbally or nonverbally demonstrates problems in connection to therapist.

Working-Phase Client Goals

1. Decrease [specify extreme *complementary roles/behaviors*] to reduce depressed mood and anxiety.
   a. *Intensity and crisis induction* to enable client to see the negative effects of complementary/extreme roles/behaviors.
   b. *Expanding client truths* to use elements of complementary role to reduce its extremity.
2. Increase *strength of diffuse* [or flexibility if rigid] *boundaries* with/in [specify another person/context] to reduce depressed mood and anxiety.
   a. *Enactments* to have client practice setting clearer boundaries.
   b. *Compliments* to shape new behaviors that establish clear boundaries.
3. Increase *clarity and appropriateness of role* in family of origin, partnership, and/or family of procreation to reduce anxiety.
   a. Assign *boundary making* moves for client to use to clarify role out of session.
   b. *Challenge client's worldview* about what "must be done" in specific relationships.

(continued)

Structural Closing Phase of Treatment with Individual

Closing-Phase Counseling Task

1. Develop aftercare plan and maintain gains. *Diversity Note: [Describe how you will access resources in the communities of which they are a part to support them after ending therapy.]*
   a. *Shape competence* to help client identify how to handle setbacks as they arise.

Closing-Phase Client Goals

1. Increase ability to maintain levels of *independence appropriate to current family life stage* of development to reduce depressed mood and increase sense of well-being.
   a. *Challenge client's worldview* to adapt to tasks associated with new stage of development.
   b. *Compliment* behaviors that support stage-appropriate independence.

2. Increase ability to maintain levels of *interdependence and relational connection appropriate to current family life stage* to reduce anxiety and increase sense of well-being.
   a. *Enactments* to explore new, stage-appropriate ways of relating.
   b. *Shaping competence* to enable client to build on current skills to develop new ones.

Treatment Plan Template for Couple/Family

The following treatment plan template can be used with couples or families experiencing conflict.

Structural Initial Phase of Treatment with Couple/Family

Initial Phase Counseling Tasks

1. Develop working counseling relationship. *Diversity Note: [Describe how you will adjust to respect cultured, gendered, and other styles of relationship building and emotional expression.]*
   a. *Join* with system, adapting to gender/culture/class norms for relating, expressing emotion, conversational tempo, etc.

2. Assess individual, systemic, and broader cultural dynamics. *Diversity Note: [Describe how you will adjust assessment based on cultural, socioeconomic, sexual orientation, gender, and other relevant norms.]*
   a. Assess structure of system, including *subsystems, coalitions, boundary patterns, hierarchy, complementary relationships, family stage of development, and strengths*.
   b. Identify the *role of the symptom* (e.g., conflict or identified patient's symptoms) in the system.

Initial Phase Client Goals

1. Sever *coalitions* (cross-generational or with external third parties) and to reduce conflict.
   a. *Intensity and crisis induction* to enable clients to see the negative effects of coalitions.
   b. *Boundary making* in session to expose and realign coalitions.

(continued)

Structural Working Phase of Treatment with Couple/Family

Working-Phase Counseling Task

1. Monitor quality of the working alliance. *Diversity Note: [Describe how you will attend to client response to interventions that indicate therapist using expressions of emotion that are not consistent with client's cultural background.]*
   a. Assessment Intervention: Adjust joining style and accommodate to system when any member of the system verbally or nonverbally demonstrates problems in connection to therapist.

Working-Phase Client Goals

1. Increase *strength of diffuse* [or *flexibility if rigid*] *boundaries* between [person/subsystem A] and [person/subsystem B; add more if necessary] to reduce conflict.
   a. *Challenge family's worldview* as it relates to maintaining unclear boundaries.
   b. *Enactments* to enable couple/family experience relating with clear boundaries.

2. Increase [or decrease] *parental hierarchy* and *separate the spousal from parental subsystems* to reduce conflict.
   a. *Enactments* to have couple/family learn how to interact with clear hierarchy and subsystem boundaries.
   b. *Boundary making* to realign hierarchy and subsystems.

3. Decrease *complementary roles* in system and increase *clarity and appropriateness of roles* to reduce conflict.
   a. *Boundary making* to decrease complementary roles.
   b. *Challenge couple/family's worldview* about who is capable of what in system.

Structural Closing Phase of Treatment with Couple/Family

Closing-Phase Counseling Task

1. Develop aftercare plan and maintain gains. *Diversity Note: [Describe how you will access resources in the communities of which they are a part to support them after ending therapy.]*
   a. *Shape competence* to help couple/family identify how to handle setbacks as they arise.

Closing-Phase Client Goals

1. Increase ability to maintain levels of *independence appropriate to current family life stage* of development to reduce conflict and increase sense of well-being.
   a. *Challenge couple's/family's worldview* to adapt to tasks associated with new stage of development to reduce conflict and increase sense of intimacy.
   b. *Compliment* behaviors that support stage-appropriate independence.

2. Increase ability to maintain levels of *interdependence and relational connection appropriate to current family life stage* to reduce conflict and increase intimacy.
   a. *Enactments* to explore new, stage-appropriate ways of relating.
   b. *Shaping competence* to enable couple/family to build on current skills to develop new ones.

Research and the Evidence Base

Likely due to their elegant simplicity and clarity, the core components of structural therapy have been used to develop several empirically supported treatments, especially those targeting youth:

- *Brief Strategic Family Therapy* (and two related models: structural ecosystemic herapy and structural ecodevelopmental preventive interventions; Szapocznik & Williams, 2000)

- *Ecosystemic Structural Family Therapy* (Lindblad-Goldberg, Dore, & Stern, 1998)
- *Multisystemic Family Therapy* (Henggeler, Schoenwald, Borduin, Rowland, & Cunningham, 1998)
- *Multidimensional Family Therapy* (Liddle, 2002)
- *Functional Family Therapy* (Sexton, 2011; see Chapter 15)
- *Emotionally Focused Therapy* (Johnson, 2004; see Chapter 15)

These empirically supported treatments generally target adolescents from diverse families and integrate structural therapy components to assess and restructure the family. The most commonly used elements include the concept of interpersonal boundaries, appropriate family hierarchy, and enactments to facilitate relational change.

Clinical Spotlight: Ecosystemic Structural Family Therapy

Ecosystemic structural family therapy (ESFT), an empirically supported adaptation of structural family therapy (Minuchin, 1974), was developed by Marion Lindblad-Goldberg and her colleagues at the Philadelphia Child and Family Training Center (formerly the Philadelphia Child Guidance Clinic) to treat children and adolescents with severe emotional or behavioral problems and their families within the context of their communities (Lindblad-Goldberg, Dore, & Stern, 1998). ESFT has addressed a wide range of child and adolescent clinical problems across all levels of severity and in diverse treatment settings. In the in-home or community setting, ESFT targets youth who are either at risk of out-of-home placement or who have already spent time in inpatient or residential settings. The families of these youth tend to be compromised by trauma-induced parental substance abuse, conflictual relationships, emotional disturbance, and the absence of emotional or concrete support.

Case Conceptualization

ESFT is a bio/developmental/systemic trauma-informed clinical model that examines the biological and developmental influences of family members as well as current and historical familial, cultural, and ecological influences. It is based on the fundamental assumption that both child and parental functioning are inextricably linked to their relational environment.

ESFT therapists are guided by five interrelated constructs:

- Family structure
- Family emotion regulation
- Individual differences (historical, biological, cultural, developmental)
- Affective proximity (emotional attachment between parent and child and between parents)
- Family development

Goals

The primary targets of therapeutic change are the following:

- Parental executive functioning
- Child coping skills
- Coparent alliances
- Nonadaptive emotional attachment patterns
- Emotional regulation
- Extrafamilial supports to family members

Interventions

The necessary agents of change in ESFT are (a) the family members in partnership with the therapist and (b) the family-therapist entity in partnership with extrafamilial helpers. ESFT incorporates techniques from many different models of psychotherapy to create relational change. It is the relational objective that determines an intervention's appropriateness. Structural techniques that are used to reorganize or restructure the way family members relate to one another include boundary making and rebalancing power or clarifying hierarchy. Like the structural family therapy model, ESFT emphasizes building strong therapeutic alliances with all family members. The most common structural interventions used in ESFT are behavioral enactments and validation of family members' strengths. The most common ESFT intervention is *enactment* to help family members practice new ways of relating: adjust the necessary emotional proximity between family members, learn to regulate emotions, and learn to tolerate distress. Other commonly used techniques address thinking, beliefs, or knowledge in the family; these techniques include reframing, constructing adaptive narratives, psychoeducation, and the use of rituals.

Tapestry Weaving: Working with Diverse Populations
Cultural, Ethnic, and Socioeconomic Diversity

> "Every family has elements in their own culture, which if understood and utilized, can become levers to actualize and expand the family members' behavioral repertoire. Unfortunately, we therapists have not assimilated this axiom. Though we pay lip service to the strengths of the family, and talk about the matrix of development and healing, we are trained as psychological sleuths. Our instincts are to 'search and destroy': pinpoint the psychological disorder, label it, and eradicate it." —Minuchin & Fishman, 1981, pp. 262–263

Minuchin and his colleagues developed the structural family therapy model to work with poor, ethnically diverse, urban families because they did not find that traditional insight-oriented approaches were effective with this population (Colapinto, 1991; Minuchin et al., 1967). From its inception, structural family therapy has attended to the dynamics and needs of diverse families, especially those with children who are having difficulties. In addition to Minuchin, structural therapy proponents, such as Harry Aponte (1994, 1996), have been leading voices in the field of family therapy on issues of spirituality, race, and poverty. Because Minuchin and many of the proponents of structural family therapy were themselves from diverse and immigrant backgrounds, they were aware of the strengths of diverse families. Structural family therapy employs an active, engaged approach in which the therapist often takes an expert stance in relation to the family, an approach that often fits with the values of traditional cultures.

Structural therapy has been widely used and studied with Hispanic, African American, and Asian Americans. Most notably, brief strategic family therapy (Chapter 6) and structural ecosystemic therapy (see research and the evidence base below) have been studied and adapted for working with both Hispanic and African American families. In fact, brief strategic family therapy was initially developed for use with Cuban youth with conduct issues; the family focus fits well with Hispanic valuing of family relationships. Current versions of the model, culturally informed and flexible family-based treatment, specifically add content and themes for Hispanic families, specifically acculturation/immigration issues, intrafamilial conflict related to acculturation, and the experience of discrimination in the larger culture (Santisteban & Mena, 2009). In addition, Santisteban and Mena (2009) also identify open and honest communication about risky sexual behavior as particularly difficult in Hispanic families due to religious and cultural norms; thus, this is a topic that needs cultural sensitive attention and intervention when working with female and male Hispanic youth and their families.

The careful consideration of culture in brief strategic and structural ecosystemic therapies have also identified different needs of ethnic groups. In one study, researchers compared outcomes of structural ecosystemic therapy with Hispanic vs. African American drug-using youth, with Hispanics benefiting differently from the treatment (Robbins,

Szapocznik, Dillon, Turner, Mitrani, & Feaster, 2008). Whereas Hispanic youth experienced a decrease in substance use, African youth did not but they did report improved racial socialization and family functioning; the researchers propose that these differences may be related to the relative power of each ethnic group in the Miami community, with African Americans being more disenfranchised. Additionally, in another study of HIV-positive African American women structural ecosystemic therapy was found to be better than the person-centered therapy and community controls for reducing psychological distress and family-related hassles but did not significantly improve family support. These sometimes puzzling culture-specific findings remind clinicians to be cautious and attentive when working with diverse families.

Structural family therapy has also been used with Asian families. Kim (2003) describes structural family therapy as ideal for Asian Americans, especially first generation families: "Its useful concepts, such as hierarchy and its advocacy for a parental executive system, boundaries, and subsystems make it ideal for and compatible with Asian American cultural and family values" (p. 391). Specifically, Kim suggests using the structural concept of parental hierarchy to understand the common Asian value of *filial piety,* the value of high respect for and obedience to parents and elders. In most Asian families, children are expected to sacrifice their own desires for the family, a value that is at odds with many social norms in American society, thus a frequent source of conflict in Asian immigrant families. When working with these families, Kim recommends reframing children's disobedience in terms of children seeking peer approval in the dominant culture rather than a betrayal and devaluing of the family, which is how it appears to many Asian parents. The goal with such families is to help them develop more flexible boundaries to allow members to successfully flow between both cultural worlds. Yang and Pearson (2002) describe successfully using similar structural practices with families in China to reduce relapse in families with a schizophrenic member.

Sexual Identity Diversity

Little specific information has been written on using structural therapy with gay, lesbian, bisexual, and transgendered couples or families. However, general research on gay and lesbian families indicates that their basic structure and dynamics are similar to (not statistically different from) those of heterosexual couples (Gottman, 2008), and therefore the same general family structural considerations are generally the same: having clear boundaries, an effective parental hierarchy, and a separation of the spousal/parental subsystems. Research has also found no difference between children raised by same-sex vs. heterosexual parents on measures of well-being, self-esteem, peer relations, and social adjustment (Biblarz & Savci, 2010). Lesbian families are generally found to have highly egalitarian parenting practices, and they tend to equal or surpass heterosexual couples in time spent with children, parenting skill, warmth, and affection (Biblarz & Savci, 2010). In contrast, gay male couples often must redefine their definitions of masculinity and fatherhood and that the development of their family structures are more complex and unique to each couple; however, like lesbian parents, they are more likely to equally share parenting duties and styles than heterosexual couples (Biblarz & Savci, 2010).

When working with gay/lesbian families, therapists also need to consider the unique experience and pressures on these couples and families. Fitzgerald (2010) identifies issues common when working with gay/lesbian families that therapists must consider: children defending their parents, children needing to determine to whom it is safe to share openly about their family, the process of parents coming out to their children, and parents feeling the need to be "perfect" in society's eyes.

Online Resource

Philadelphia Child and Family Therapy Training Center:
www.philafamily.com

References

*Asterisk indicates recommended introductory readings.

Aponte, H. J. (1994). *Bread and spirit: Therapy with the new poor: Diversity of race, culture, and values.* New York: Norton.

Aponte, H. J. (1996). Political bias, moral values, and spirituality in the training of psychotherapists. *Bulletin of the Menninger Clinic, 60*(4), 488–502.

Biblarz, T. J., & Savci, E. (2010). Lesbian, gay, bisexual, and transgender families. *Journal of Marriage and Family, 72*(3), 480–497. doi:10.1111/j.1741-3737.2010.00714.x

*Colapinto, J. (1991). Structural family therapy. In A. S. Gurman & D. P. Kniskern (Eds.), *Handbook of family therapy* (Vol. 2, pp. 417–443). New York: Brunner/Mazel.

Gottman, J. M. (2008, April). *Marriage counseling: Keynote address.* Annual Conference of the American Counseling Association, Honolulu, HI.

Fitzgerald, T. (2010). Queerspawn and their families: Psychotherapy with LGBTQ families. *Journal of Gay & Lesbian Mental Health, 14*(2), 155–162. doi:10.1080/19359700903433276

Hammond, R. T., & Nichols, M. P. (2008). How collaborative is structural family therapy? *The Family Journal, 16*(2), 118–124. doi:10.1177/1066480707313773

Henggeler, S. W., Schoenwald, S. K., Borduin, C. M., Rowland, M. D., & Cunningham, P. B. (1998). *Multisystemic treatment of antisocial behavior in children and adolescents.* New York: Guilford.

Johnson, S. M. (2004). *The practice of emotionally focused marital therapy: Creating connection* (2nd ed.). New York: Brunner/Routledge.

Kim, J. M. (2003). Structural family therapy and its implications for the Asian American family. *The Family Journal, 11*(4), 388–392. doi:10.1177/1066480703255387

Liddle, H. A. (2002). *Multidimensional family therapy treatment for adolescent cannibis users.* Rockville, MD: Substance Abuse and Mental Health Services Administration.

Lindblad-Goldberg, M., Dore, M., & Stern, L. (1998). *Creating competence from chaos.* New York: Norton.

Minuchin, S. (1974). *Families and family therapy.* Cambridge, MA: Harvard University Press.

*Minuchin, S., & Fishman, H. C. (1981). *Family therapy techniques.* Cambridge, MA: Harvard University Press.

Minuchin, S., Montalvo, B., Guerney, B. G., Rosman, B., & Schumer, F. (1967). *Families of the slums.* New York: Basic Books.

Minuchin, S., & Nichols, M. P. (1993). *Family healing: Tales of hope and renewal from family therapy.* New York: Free Press.

Minuchin, S., Nichols, M. P., & Lee, W. Y. (2007). *Assessing families and couples: From symptom to system.* New York: Allyn & Bacon.

Minuchin, S., Rosman, B., & Baker, L. (1978). *Psychosomatic families: Anorexia in context.* Cambridge, MA: Harvard University Press.

Robbins, M. S., Szapocznik, J., Dillon, F. R., Turner, C. W., Mitrani, V. B., & Feaster, D. J. (2008). The efficacy of structural ecosystems therapy with drug-abusing/dependent African American and Hispanic American adolescents. *Journal of Family Psychology, 22*(1), 51–61. doi:10.1037/0893-3200.22.1.51

Santisteban, D. A., & Mena, M. P. (2009). Culturally informed and flexible family-based treatment for adolescents: A tailored and integrative treatment for Hispanic youth. *Family Process, 48*(2), 253–268. doi:10.1111/j.1545-5300.2009.01280.x

Sexton, T. L. (2011). *Functional family therapy in clinical practice: An evidence-based treatment model for working with troubled adolescents.* New York: Routledge.

Szapocznik, J., & Williams, R. A. (2000). Brief Strategic Family Therapy: Twenty-five years of interplay among theory, research and practice in adolescent behavior problems and drug abuse. *Clinical Child and Family Psychology Review, 3*(2), 117–134.

Yang, L., & Pearson, V. J. (2002). Understanding families in their own context: Schizophrenia and structural family therapy in Beijing. *Journal of Family Therapy, 24*(3), 233–257. doi:10.1111/1467-6427.00214

Structural Case Study: African American Family Adjusts to Divorce

Taylor, a 16-year-old African American (CM16), has come to counseling because of frequent pot use, dropping grades, and constant fighting at home with his mom after the divorce of his parents about a year ago. Taylor claims that the pot use began four years ago and that he would smoke on occasion up until a year ago where it became a daily habit. Taylor explains that he uses pot to help cope with stress from school and mostly stress from home. Taylor is a gifted athlete and has a rigorous schedule due to hours of practice and taking honors classes. He at times feels so overwhelmed that he claims that he "just wants to give up" and drop out of school to relive the constant stress he feels to "be the best" at everything. He claims that he just wants to work on cars and eventually own his own mechanic shop. He explains that his mom, Tenisha (AF45, African American), is overinvolved and has high expectations that are impossible to achieve. He describes his mother as uptight and temperamental; on the other hand he describes his father,

Michael (AM49, African American) as laid back, fun, and easy to talk to. Taylor explains that his father understands him best and feels like he can talk to him about his struggles; he also explains that his father knows what his mother can be like and tries to stand up for him when arguments occur. Michael also confided in Taylor before the divorce and has continued since the divorce; Michael also tries to get Taylor to side with him when arguments occur between him and Tenisha. Taylor has mentioned that he would rather live with his father instead of his "crazy" mother. Taylor has an older sister, Tamara (CF19), that he lives with as well; he feels that they get along well but that he is unable to talk to her because she often sides with their mom.

Structural Case Conceptualization

- *Role of Symptom in the Family:* Describe relation of family and symptom of symptom:
  Family as "Shaper" of Individual's Symptoms: The family has divided mother and daughter against father and son. This has left CM16 feeling as if he needs to pick sides. He feels overwhelmed by the stress that he has resorted to smoking pot and has lost focus in sports and academics. Both parents are worried about their son's lack of motivation and frequent pot use, an issue that brings them together as a family yet also causes constant arguments because the parents blame one another for their son's recent issues. CM16's father blames the mom for being too aggressive and overinvolved, and CM16's mom blames the father for being too passive and easy on CM16, allowing him to get away with everything. When this interaction occurs CM16 disengages and uses negative coping mechanisms.
- *Subsystems:*
  - *Couple*: Although AM49 and AF45 are legally divorced, emotionally their conflict is unresolved and continues to play out in their relationshps with the children.
  - *Parental*: When they first had their children, there was an effective parental hierarchy, both parents were in control but AF45 maintained much of the control and AM49 supported that. After the divorce, the parental coalition deteroated. AM49 no longer supports AF45 with her role as a mother and a disciplinarian and belittles her in front of CM16. This gives CM16 a false sense of power and he feels like he has the authority to talk back to his mom.
  - *Sibling*: CM16 and CF19 also have a sibling subsystem, this taught CM16 how to share, cooperate, and support within a peer group.
- *Cross-Generational Coalitions:*
  - *AM/CM:* Since the divorce, CM16 and AM49 have formed a cross-generational coalition against AF45. This began during a period of high tension between AM49 and AF45 when AM49 would confide in CM16 about AF45 and the fighting that occurred. The coalition grew stronger when AM49 and AF45 divorced; AM49 continues to confide in CM16 and enourages him to side with him when an arguments erupt. AM49 also sides with CM16 when he argues with AF45 and belittles AF45's role as a mother.
  - *AF/CF:* Similarly, AF45 and CF19 have formed a covert cross-generational coalition against AM49. AF45 has very little family and only a few friends. AF45 lacks a sufficient support system; therefore when the divorce was overwhelming she sought out support from her daughter. She would often cry to CF19 about her struggles and about how awful AM49 was being during the separation. This caused CF19 to have empathy for AF45 and the felt-need to be her mother's main source of support.
- *Boundaries:*
  Especially since the divorce, the family system has enmeshed boundaries that are related to their presenting concerns. Especially since the divorce, AF45 has enmeshed boundaries with both CM16 and CF19. She is overprotective and overconcerned in CM16's life and hasn't given him much space to individuate, which is important at his stage of development. She often tries to make descisions for him, because she explains that she knows what CM16's needs. AF45 also has enmeshed boundaries

with CF19, relying too heavily on her for support. AM49 also has enmeshed boundaries with CM16, inappropriately forming a coalition and using him as an ally.

- *Parental Hierarchy:*
Early on in AF45s and AM49's marriage they maintained an effective hierarchy, although AF45 was always more of the disciplinarian and AM49 supported that role. The hierarchy started to diminish about four years ago when AM49 and AF45 were constantly fighting. AM49 belittled AF45's role as a mother and disciplinarian; he would often side with the children and let them talk down to AF45. The current hierarchy is insufficient between AM and children is insufficient, whereas the hierarchy AF is trying to establish with CM16 is too rigid for his stage of development. In addition, CM16 now displays a false sense of power within the family due to his coalition with AM.

- *Complementarity:*
AF45 and AM49 take the role as the good parent/bad parent. AF45 was the disciplinarian and is looked at as the "bad" parent. AM49 is the "easy going" parent that lets the kids get their way and avoids disciplinary actions. In addition, CM16 and CF19 currently have the good/bad child roles.

- *Family Development:*
Family with School-Aged or Adolescent Children: AF45 needs to adjust her parental boundaries with both her children. She is still overinvolved and allows little room for her children to individuate. She needs to give more responsibility and freedom to her adolescent children. CM16 describes that AF45 allows little time for him to hang out with friends because she feels it is more important that he focus on his schoolwork and athletics.

- *Strengths:*
The family has a strong religious tradition and have a history of being a relatively peaceful and supportive family. Both of the children have excelled in school and extracurriculum activities, with the parents highly involved in these activities. AM49 and AF45 have a history of cooperatively parenting together. Divorced couple is willing to come to family therapy together to help son.

Structural Treatment Plan

Structural Initial Phase of Treatment with Family

Initial Phase Counseling Tasks

1. Develop working counseling relationship. *Diversity Note: Acknowledge the dedication of parents to come to family session for CM16 even though they are divorced. Be respectful of cultural norms and pressures, including experiences of discrimination. Look for opportunities to use religious faith to support family. Use humor to connect with teens; respect each parent's authority and autonomy.*
    a. *Join* with system, adapting to gender/culture/class norms for relating, expressing emotion, conversational tempo, etc.
2. Assess individual, systemic, and broader cultural dynamics. *Diversity Note: Adjust expectations for recency of divorce, cultural/religious norms, and extended family/support system.*
    a. Assess structure of system, including *subsystems, coalitions, boundary patterns, hierarchy, complementary relationships, family stage of development, and strengths*.
    b. Identify the *role of the symptom* (e.g., conflict or identified patient's symptoms) in the system.

(continued)

Initial Phase Client Goals

1. Sever *coalitions* between AM49 and CM16 and AF45 and CF19 and to reduce conflict and reduce pot use.
   a. *Intensity and crisis induction* to enable clients to see the negative effects of coalitions. Help AM49 realize how his coalition with CM16 is affecting the family and how it is negatively affecting CM16 (frequent pot use and anger toward AF45). Help AF45 understand the effects of coalition with CF19 and how the coalitions are separating the family.
   b. *Boundary making* in session to expose and realign coalitions.

Structural Working Phase of Treatment with Family

Working-Phase Counseling Task

1. Monitor quality of the working alliance. *Diversity Note: Monitor verbal and nonverbal cues that may indicate one or more family members are not feeling respected by therapist or that therapist is taking side.*
   a. Adjust *joining style* and *accommodate* to system when any member of the system verbally or nonverbally demonstrates problems in connection to therapist.

Working-Phase Client Goals

1. Increase *strength of diffuse boundaries* between AF45 and CM16, AF45 and CF19, AM49 and CM16, AM49 and CF19 to reduce conflict.
   a. *Challenge family's worldview* regarding boundaries and explore alternative ways of being emotionally close while respecting independence of others.
   b. *Enactments* to enable family experience relating with clear boundaries, especially between AF45 and CM16.

2. Increase effectiveness of divorced couple's *coparenting coalition and balance hierarchy* to reduce conflict and substance use.
   a. *Challenge couple's worldview* to increase their awareness of detrimental their conflict is to their children and solidify their commitment to coparenting cooperatively for the well-being of their children (couple only sessions).
   b. *Enactments* to have parents learn how to work together to continue to support each other as parents (couple only sessions).

3. Decrease *complementary roles* of good parent/bad parent in system and increase *clarity and appropriateness of roles* to reduce conflict.
   a. *Boundary making* with family to decrease complementary roles of good/bad parent.
   b. *Challenge family's worldview* about who is capable of what in system. Encourage both AF45 and AM49 to take complementary roles as loving, caring parents that discipline their children appropriately.

Structural Closing Phase of Treatment with Family

Closing-Phase Counseling Task

1. Develop aftercare plan and maintain gains. *Diversity Note: Explore possibilities for family to receive support from the church they are affiliated with.*
   a. *Shape competence* to help family identify how to handle setbacks as they arise.

(continued)

Closing-Phase Client Goals

1. Increase ability to maintain levels of *independence appropriate to current family life stage of development* and *adjust to divorce* to reduce conflict and increase sense of well-being.
   a. *Challenge family's worldview* to adapt divorce while preparing to launch children. Help family develop vision of a sense of family that continues for children even after marriage ends.
   b. *Compliment* behaviors that support stage-appropriate independence and functional adjustment to divorce.

2. Increase ability to maintain levels of *interdependence and relational connection appropriate to current family life stage* to reduce conflict and increase sense of well-being.
   a. *Enactments* to explore new, stage-appropriate ways of relating and solving problems as a family with divorced parents.
   b. *Shaping competence* to enable family to build on current skills to develop new ones.

CHAPTER 8

Satir's Human Growth Model

"Life is not the way it's supposed to be. It's the way it is. The way you cope with it is what makes the difference."

—Virginia Satir

Lay of the Land

Experiential family therapies include three approaches, the Satir growth model, symbolic-experiential therapy, emotionally focused couples therapy, and internal family systems.

- *The Satir Growth Model:* Focuses on family communication and family structure using warmth and support (covered in this chapter)
- *Symbolic-Experiential Therapy:* Focuses on symbolic meanings and emotional exchanges within the family using a balance of warmth and confrontation to promote change (Chapter 9)
- *Emotionally Focused Therapy (EFT):* A leading evidence-based approach to couples therapy that uses experiential, systemic, and attachment theories (EFT will be covered in detail in Chapter 15 (evidence-based therapies), but I mention it here because it has a strong experiential foundation).
- *Internal Family Systems:* Originally developed with trauma and abuse survivors, this integrative approach works with a person's internal "parts" using systemic principles.

Common Assumptions and Practices
Targeting Emotional Transactions

Whereas systemic, strategic, structural, and cognitive-behavioral family therapists primarily track *behavioral* interaction sequences, experiential family therapists focus more on the *affective* or *emotional* layer of those same interactions—while still attending to behavior and cognition (Johnson, 2004; Satir, Banmen, Gerber, & Gomori, 1991; Whitaker & Bumberry, 1988). Assessment and intervention target the emotional exchanges between family members and significant others in relation to the presenting problem.

Warmth, Empathy, and the Therapist's Use of Self

More so than strategic, structural, and intergenerational family therapists, experiential family therapists use warmth and empathy in building relationships with clients (Johnson, 2004; Satir, 1988; Satir et al., 1991; Schwartz, 1995; Whitaker & Bumberry, 1988). Therapists use themselves—their personhood—to make this strong affective connection with clients. This approach creates a sense of safety that allows clients to explore areas of emotional vulnerability.

Individual and Family Focus

Experiential family therapies address individual and family concerns as distinct sets of problems (Johnson, 2004; Satir et al., 1991; Schwartz, 1995; Whitaker & Bumberry, 1988). In contrast, systemic, structural, and intergenerational therapies conceptualize individual systems as part of the family system, assuming that if the family system is treated the individual symptoms will be resolved. Experiential therapists may not entirely disagree with this perspective but are much more deliberate in treating problems at the individual level.

The Satir Growth Model

In a Nutshell: The Least You Need to Know

One of the first prominent women in the field, Virginia Satir began her career in family therapy at the Mental Research Institute (MRI; see Chapter 5) working alongside Jay Haley, Paul Watzlawick, Richard Fisch, and the other leading family therapists in Palo Alto (Satir, 1983, 1972). She eventually left the MRI to develop her own ideas, which can broadly be described as infusing humanistic values into a systemic approach. She brought a warmth and enthusiasm for human potential that is unparalleled in the field of family therapy. Her therapy focused on fostering individual growth as well as improving family interactions. She used experiential exercises (e.g., family sculpting; see later section on Sculpting), metaphors, coaching, and the self of the therapist to facilitate change (Satir et al., 1991; Satir & Baldwin, 1983). Her work is practiced extensively internationally, with Satir practitioners connecting through the Satir Global Network.

The Juice: Significant Contributions to the Field

If you remember one thing from this chapter, it should be:

Communication Stances

The communication stances in the Satir growth model offer a clinician of any theoretical orientation an efficient and effective means of conceptualizing how best to communicate and interact with a client (Satir, 1983, 1988; Satir et al., 1991). Satir described five communication stances: congruent, placator, blamer, superreasonable, and irrelevant. Each stance either acknowledges or minimizes three realities: self, other, and context. Four of the stances—placator, blamer, superreasonable, and irrelevant—are *survival stances* that were used to "survive" as a child during difficult times. Everyone uses one of these to some extent because all children are put in situations they are not ready to handle as they move from one life stage to another. Survival stances often fit together like puzzle pieces within a family, with people assuming complementary stances to create balance. In all cases, the goal is to move people toward more congruent communication, communication in which they respectfully balance the needs of *self* and *others* while responding appropriately within and acknowledging the *context*.

At first glance, these stances appear too simplistic to be of much clinical relevance. I believed this myself until I started teaching case conceptualization (see Chapter 17). Over time, however, I began to appreciate their remarkable sophistication and the insights they offer. Identifying a client's communication stance can help therapists design interventions more effectively and use almost every utterance to move clients toward their goals, whether working solely from a Satir approach or from an entirely different approach. Therapists can also communicate with a wide range of clients, adjusting their comments and interventions to accommodate the client and using consistent and focused language to reinforce therapeutic movement with every communication, from scheduling an appointment to wording interventions from any model. In the case study at the end of the chapter, the therapist works with a Persian family in which the mother and son take on placating roles whereas the father typically uses a blaming stance and sometimes even a super reasonable stance, often justifying his position by saying, "This is the way it is done [in our culture]."

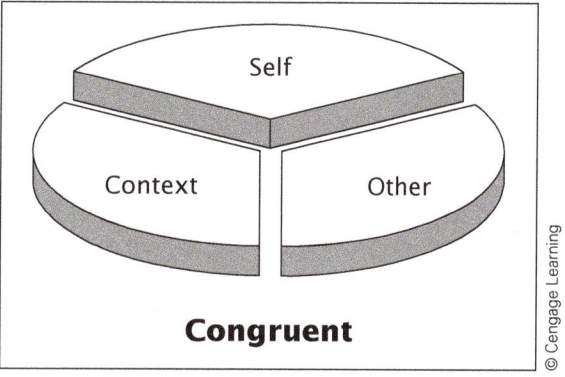

Communication Strategies for Survival Stances

Unlike the congruent stance, each survival stance minimizes one or more essential parts of the total picture, as illustrated by the dark shading in the following diagrams.

Placating Stance People who tend to use the placating stance manage relational distress by focusing on the needs of others at the expense of their own self-worth (Satir et al., 1991). As the name suggests, placators are pleasers: they actively do everything they can to make others happy—or at least not angry. They tend to avoid conflict at all costs—and those costs generally include their own happiness and health.

In her earlier work, Satir described placators as not acknowledging context in addition to self (Bandler, Grinder, & Satir, 1976), but in later work indicates that they may acknowledge context as well. After consulting with colleagues who worked with Satir (Stephen Lankton and Lynn Azpieta, personal communication), it seems there is some variability but general agreement that placators can vary in terms of how much they attend to context, with those less responsive to it having more difficulties. For example, a student who dutifully tries to give the professor everything she wants without question or making waves is a placator who acknowledges the context, which is the formal academic environment. In contrast, a placator who tries to get an A grade by baking cookies, offering to run errands, or otherwise disregarding academic norms is not as sensitive to context. So, if you work with these ideas long enough, you can see how subtle they can be.

Because all forms of placators have people-pleasing tendencies, therapists use less directive therapy methods, such as multiple-choice questions and open-ended reflections, to require them to voice their opinion and take a stand. Often this is quite painful and scary for placators. With clients who tend toward placating, therapists should carefully avoid giving opinions, making it seem that they have an opinion, or offering too much personal information. Clients will use this type of information to know what parts of themselves to hide and what parts to foreground to gain therapist approval. Never underestimate placators; they are skilled in the art of people pleasing. Research indicates that some clients make up things to give the impression that therapy is progressing (Gehart & Lyle, 2001). Not until the placator regularly and openly *disagrees* with the therapist has rapport been established.

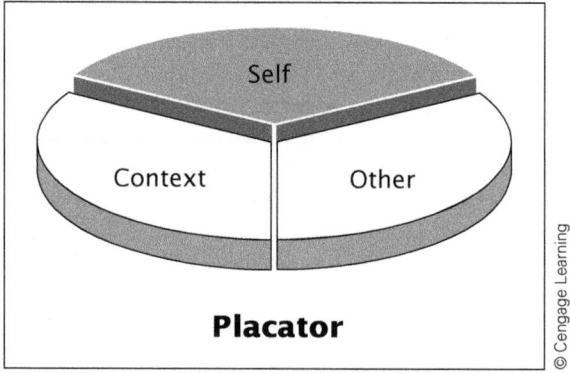

Placator

Blaming Stance The polar opposite of placating, the blaming stance involves asserting one's position and power at the expense of others' needs and wants (Satir et al., 1991). Blamers are quick to see other people and external circumstances as the source of their woes: "If she didn't do X, then I wouldn't feel Y or do Z" or "If things I didn't have to do X, then I wouldn't feel Y or do Z." When a person takes on a blaming position, there is an excuse for everything. Although it may seem as though these people have self-esteem, they do not; the blaming stance is taken out of fear not confidence. Those who use it often tend to feel alone and unsuccessful. In contrast, those using a congruent stance to communicate their needs are able to do so from a place of calm and maintain a sense of self-esteem because their behavior is not harming others.

Similar to placating, this stance was described as ignoring context in Satir's earlier work (Bandler, Grinder, & Satir, 1976) and as acknowledging it in her later work (Satir et al., 1991). In the case of blamers, those that are less sensitive to context are readily identifiable when they have inappropriate outbursts in public or otherwise judge or dictate in ways clearly incongruous for the circumstances (look for live examples at your local grocery store or coffee house). Whereas those who are more sensitive to context are able to modulate their fault finding to be within acceptable norms—but generally annoying behavior nonetheless—for a given setting (look for the eye rolling and insincerely polite requests to correct the slightest imperfections at the same locations).

Therapists seek to increase blamers' awareness of others' thoughts and feelings and help them learn how to communicate their personal perspectives in ways that are respectful of others. With these clients, direct confrontation often strengthens the therapeutic relationship (counter to what one might expect). Most blamers lose respect for "wimpy" (think placating) therapists who do not speak their minds honestly and directly, a skill that blamers have mastered. Blamers generally prefer more upfront and direct communication than is generally tolerated in polite society.

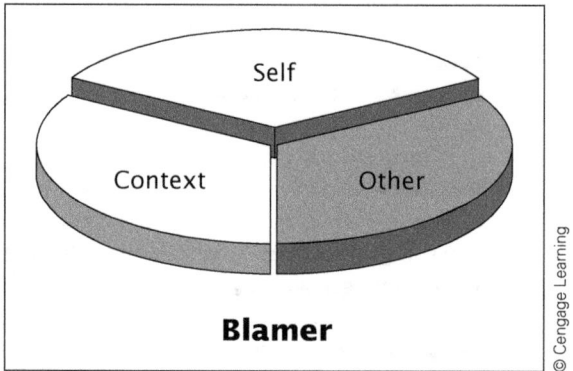

Blamer

Superreasonable Stance Less frequent than the first two, the superreasonable stance involves acknowledging context at the expense of both self and other (Satir et al., 1991).

This generally takes the form of being nonemotional and fixated on a particular system of logic, which can be simply practicalities or at times systems of religious rules. When working with superreasonable clients, logic and rules reign supreme. Therapists must refer to context to gain validity in their world. The goal with this stance is to help clients value the internal, subjective realities of themselves and others.

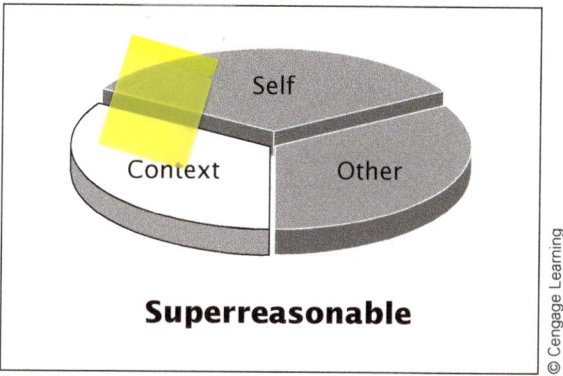

Irrelevant Stance Finally, the irrelevant stance involves avoiding acknowledging self, other, or context when relational distress arises, generally through humor or tangential conversation (Satir et al., 1991). In many cases, those who take on the irrelevant stance are viewed as amusing, spontaneous, entertaining, or cheerful—because they distract from any form of tension or negativity. They may even seem to have no problems because nothing upsets them. However, those who use this stance frequently do not have a high sense of self-worth or the capacity for intimacy because they are never really present, and often feel confused and out of balance.

The irrelevant stance creates a unique challenge for therapists because there is no consistent grounding in self, other, or context for the therapist to use in understanding and communicating with the client. Instead, the therapist must spend time "floating" along with the client's distractions to identify the unique "anchors" of the client's reality that the therapist can tap into. Often the first step is to make the therapeutic relationship a place of utmost safety so that there is less need for distracting communication. As treatment progresses, the therapist works with irrelevant clients to increase their ability to recognize the thoughts and feelings of self and others and to acknowledge the demands of context. Progress is typically slower with those who use this stance frequently.

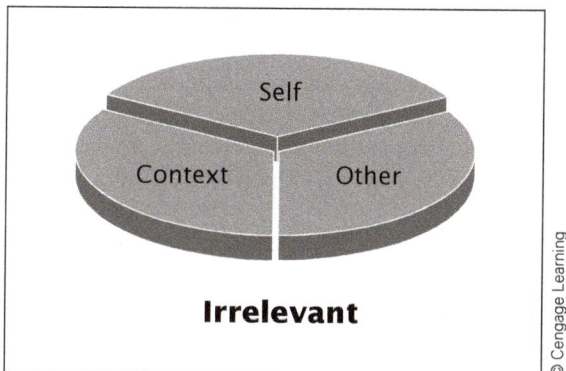

Rumor Has It: The People and Their Stories

Virginia Satir A true pioneer, Virginia Satir was one of the first therapists to work with entire families. She began her private practice in 1951, and by 1955 she was training therapists at the Illinois Psychiatric Institute to work with families. She then joined the newly established MRI in Palo Alto, California, to continue her research. A grant from the National Institute of Mental Health in 1962 provided funding to establish the first family therapy training program. In 1964, she published her first book, *Conjoint Family Therapy*, which outlines the key aspects of her model. She left the MRI to become the director of Esalen Institute in Big Sur, California, offering workshops facilitating personal growth. She also founded the AVANTA network (also called the Virginia Satir Global Network) to connect practitioners of the model.

John Banmen Having trained and worked closely with Virginia Satir (et al., 1991), Banmen now teaches the Satir growth model internationally, particularly in China, Taiwan, and Hong Kong where Satir's work is highly influential, and continues to publish on current applications of her work (Banmen, 2002, 2003).

Maria Gomori Having trained and studied with Satir (Satir et al., 1991), Maria Gomori (2002) now runs the Satir Professional Development Institute of Manitoba, which offers comprehensive training in the model.

Lynne Azpeitia Having trained and worked closely with Satir, Lynne continues training the next generation of Satir practitioners and developing strength-oriented applications for working with nonclinical populations (Azpeitia, 1991, 1995).

The Big Picture: Overview of Treatment

Satir et al. (1991) use a six-stage model of change that is based on Satir's research at the MRI on cybernetic systems and that draws from humanistic principles, including the assumption that people naturally strive toward growth. Her six-stage model describes how the therapy process helps families move toward a *second order change* in the family structure (see Chapter 3). The model also emphasizes that the therapist perturbs the system, shakes it up, and respects its ability to naturally reorganize itself in a more useful way, rather than attempting to direct and control the system from the outside. Thus, even though Satir uses education to coach clients on how to better communicate, the aim is not for clients to literally follow her instructions but rather to adapt and respond to the instructions in a way that works for their system. The six stages are as follows:

1. *Status Quo:* This is a state of homeostasis that includes at least one symptomatic member.
2. *Introduction of Foreign Element:* A foreign element, which may be a life crisis, tragedy, or therapeutic intervention, gets the system off balance.
3. *Chaos:* The new perspective creates a *positive feedback* loop that throws the system into a state of chaos; at this point the "natural" response is to feel uncomfortable, and in almost all cases the family tries to regain the status quo (stage 1), which may or may not be possible.
4. *Integration of New Possibilities:* Eventually the family system interprets the new information in a meaningful way; the therapist needs to be respectful of how the system uses the information and responds to therapist-client interactions, honoring and trusting the system's autonomy.
5. *Practice:* The system develops a new set of interaction patterns based on the new information. This may or may not look like what the therapist expects, but the therapist asks two key evaluative questions: (a) Are the symptoms improving? (b) Is each person able to self-actualize and grow?
6. *New Status Quo:* This is a state of new homeostasis that does not include a symptomatic member and that allows all members to grow and flourish.

In most cases, therapy involves going through these six stages several times, with the discomfort and relative sense of chaos diminishing each time the client cycles through, becoming increasingly comfortable with change.

Making Connection: The Therapeutic Relationship
Humanistic and Systemic Foundations
Satir et al. (1991, pp. 14–15) state four primary assumptions about people and therapy; the first two reflect *humanistic* assumptions and the latter two a *systemic* view.

> **Assumptions of Satir's Growth Model**
> 1. People naturally tend toward positive growth (humanistic principle).
> 2. All people possess the resources for positive growth (humanistic principle).
> 3. Every person and every thing or situation impacts and is impacted by everyone and everything else (systemic principle).
> 4. Therapy is a process that involves interaction between therapist and client; in this relationship, each person is responsible for him/herself (systemic and humanistic principle).

These assumptions clarify the therapist's role in the process (the therapist is "responsible for him/herself"), but also how the therapist believes the therapeutic process works: clients already possess the natural inclination toward growth and the resources for it, and it is the therapist's job to activate these tendencies. These assumptions inform the role of the therapist as a *guide* to the process of becoming more fully human.

Therapeutic Presence: Warmth and Humanity
Carl Rogers (1961, 1981), a leader in humanistic, experiential therapies, based his client-centered approach on three therapist qualities: (a) congruence or genuineness, (b) accurate empathy, and (c) unconditional positive regard. These conditions are the theoretical foundation of the warmth and humanity for which Satir is legendary (Satir et al., 1991). Her way of being in the world radiated an unshakable hope and deep-felt respect for her clients. Her presence in the room put people at ease and allowed them to feel secure enough to nondefensively relate to one another. She created a safe haven that made it easy for her clients to address the issues in their lives. How she did this is a much more difficult question to answer. This quality of being, or *therapeutic presence,* is difficult to define, and there are few methods for systematically developing it (Gehart & McCollum, 2008). The more *congruent* therapists are—that is, able to communicate authentically while responding to the needs of both self and others—the better they can create the therapeutic warmth and humanity that characterized the work of Satir. When working with diverse clients such as the Persian family in the case study at the end of the chapter, the therapist must create safety using the client's cultural values and norms. In the case study, the immigrant father would need to be engaged using a more formal and emotionally restrained approach, whereas the son and mother are more receptive to more American forms of warmth and emotional expression.

Making Contact
Satir et al. (1991) describe establishing a therapeutic relation as "making contact," which refers to a series of connections both within the therapist and between the therapist and the other. Making contact begins with the therapist being in contact with himself/herself, including all resources of the self as defined in the *self-mandala:* physical, intellectual, emotional, sensual, interactional, nutritional, contextual, and spiritual. In addition, the therapist prepares to meet another "miracle," a fellow human being. Then, the therapist works to make contact with each client in the room, engaging them on "all channels": mind, body, and spirit. Thus open-body positioning and congruent communication are critical. Once the therapist has made contact with each person, the therapist works to help family members make contact with each other and ultimately with others in their broader social system. The therapy process cannot move forward until the therapist has made contact with the client. When contact is made, clients feel valued for who

they are, regardless of their problem, and they feel free to make mistakes in the therapist's presence. Making contact involves the following:

- Making direct eye contact with clients
- Touching clients (e.g., shaking hands)
- Sitting or standing at the same physical level so that eye contact is easy (i.e., leaning down to talk with children)
- Asking each person's name and how he or she prefers to be called (Satir et al., 1991)

Empathy

Satir practitioners convey their understanding of clients' subjective, inner realities by expressing *empathy*, an accurate understanding of another's emotional reality. Empathy does not mean that a therapist takes the client's side, avoids confronting inconsistencies, or ignores the client's responsibility. Don't get me wrong; clients *like* therapists who take their side, don't confront them, and ignore their responsibility in the problem situation. It feels good to be "validated" in this way, but this type of "validation" is detrimental to therapeutic progress. Similarly, comments that imply a client has a "right" to feel a certain way can also shut down the process of reflecting on the client's responsibility for his/her half of an interaction. On the other hand, expressing empathy emphasizes that it is not wrong or right, normal or abnormal, to feel a certain way—that this is simply how the client feels and this is his/her "truth," at least for the moment (e.g., "It sounds like you felt betrayed"). Honoring the client's unique experience without validating it as right or normal makes it much easier to move the client toward seeing that the other person has his/her unique experience that is also "true" for that person, creating a context from which to understand how these two (or more) realities may collide and create problematic interactions. The therapist working with the family in the case study at the end of the chapter has a very challenging task of expressing empathy for the father's traditional values that lead him to reject his son who has come out as gay and at the same time have empathy for the son's sense of loss and rejection due to his father's response. Thus, if the therapist in any way indicates one person has a more true reality than the other, she is likely to lose connection with the other.

Conveying Hope

In joining with clients, therapists instill hope that they can change and that things can be better, even if they seem hopeless at the moment. Satir et al. (1991) emphasize that clients must have faith that change is possible before they can move forward with treatment.

Establishing Credibility

Clients also need to believe that therapy and their particular therapist can help them resolve their problems. Therapists establish credibility by making a personal connection (making contact) and by being confident and competent (modeling high self-esteem; Satir et al., 1991).

The Viewing: Case Conceptualization and Assessment

In the Satir growth model, both family functioning and individual functioning are assessed.

Assessment of Family Functioning

- Role of the symptom in the system
- Family dynamics
- Family roles
- Family life chronology
- Survival triad

> **Assessment of Individual Functioning**
> - Survival stances
> - Six levels of experience: the iceberg
> - Self-worth and self-esteem
> - Mind-body connection

Role of the Symptom in the System

Much like other systemic therapists, Satir (1972) viewed the symptom as having a role in the family system. For example, a child's exaggerated acting out, such as with drug use or sexual activity, may serve to reduce tension in the marriage by getting the parents on the same page about the child's issue. Similarly, a person's depression may be a means of avoiding an unpleasant confrontation with a spouse or boss. Basically, symptoms always have an emotional function in the family system, even if they are consciously and logically unwanted. The question is, why this particular symptom in this particular family (or relationship)? If the therapist can understand how the symptom makes emotional sense in the family dance, it will be easier to help the family find ways to interact successfully without the need for the symptom.

Family Dynamics

Along with assessing family roles, Satir et al. (1991) identify problematic family dynamics:

- *Power Struggles:* These can be within the family and couple or with extended family members.
- *Parental Conflicts:* These can involve parents disagreeing about how to parent and care for children.
- *Lack of Validation:* The family openly expresses little emotional support or validation.
- *Lack of Intimacy:* There is minimal sharing of significant personal information and one's emotional life.

Family Roles

Satir (1972, 1988) assessed each person's role in the family system to understand the function of the problem. Possible roles include the following:

- The martyr
- The victim or helpless one
- The rescuer
- The good child or parent
- The bad child or parent

Family Life Chronology

The family life chronology (Satir, 1983; Satir et al., 1991) is a timeline that includes the major events in an individual's or family's life:

- Births and deaths
- Important family events: marriages, moves, tragedies, major illnesses, job loss
- Important historical events: wars, natural disasters, economic downturns

This chronology gives both the therapist and client a "big picture" view of the context for the potential problem and provides clues as to what old wounds may be fueling the current problems, as well as what potential strengths and resources might exist.

Survival Triad

Another area for assessment is the survival triad (Satir, 1988)—the child, mother, and father—and the quality of the relationship between all three. Satir asserted that it was in this primary triad that a child learns how to be human (Azpeitia, 1991). Is there an emotional bond between each parent and the child, or do the parents have significantly different levels of connection with the child? Because the survival triad should serve as a nurturing system for the child, when the child is experiencing difficulty, the therapist considers how the nurturing function of these relationships can be improved.

Survival Stances

Useful in understanding individuals and relational dynamics, Satir practitioners assess each member of the family's survival stance (see Juice).

Six Levels of Experience: The Iceberg

Satir practitioners use the *six levels of experience* to help clients *transform* their feelings about feelings to make lasting change (Satir et al., 1991). These levels are likened to an *iceberg*, with the behavior the only visible layer and the other five layers unseen beneath the surface. The six layers of the iceberg are as follows:

- *Behavior:* The behavior on the surface; the external manifestation of the person's inner world.
- *Coping:* Defenses and survival stances: placating, blaming, superreasonable, and irrelevant; these come out in times of stress, and a person may use different stances in different relational contexts.
- *Feelings:* Present feelings that are strongly past-based, using past events to interpret the present.
- *Perceptions:* Beliefs, attitudes, and values that inform one's sense of self; most perceptions are formed when very young and are based on a limited view of reality.
- *Expectations:* A strong belief about how life should go, how people should behave, and how one should perform; most expectations are formed while young and are often unrealistic and/or may not apply to a particular situation.
- *Yearnings:* Universal longings to be loved, accepted, validated, and confirmed.

In therapy Satir practitioners use the six levels of experience to assess the motivation behind problematic behaviors and interactions. By understanding the *yearnings* as well as *perceptions* and *expectations* that fuel problematic behavior, coping, and feelings, therapists are able to help clients transform their feelings about feelings to more effectively meet the underlying longing for love and acceptance. In the case study at the end of the chapter, the identification of yearnings enables the therapist and the family to have more empathy for and understanding of each person's vastly different position.

Self-Worth and Self-Esteem

Satir was one of the first to recognize the importance of self-worth and self-esteem, and she always assessed a person's level of self-esteem (Satir, 1972). Clinically, it is generally unhelpful to assess self-esteem in an all-or-none fashion (i.e., high vs. low), as is frequently done by teachers, parents, and concerned others who believe they have identified the secret cause behind a child's poor behavior. Instead, it is more useful to consider the specific *aspects of the self* that a client *values* and the aspects of which he/she is ashamed. For example, a child may value and have confidence in her ability to make friends but have less confidence in her scholastic abilities.

More recent research on self-esteem indicates that *self-compassion,* or acceptance of one's strengths and weaknesses, is a better indicator of happiness than self-esteem, which can be artificially high (Neff, 2003). People who overestimate their abilities and worth often have high self-esteem but significant problems in their interpersonal and work or school relationships because they have unrealistic expectations of "what is due them." The best indicator of health, self-compassion is one's ability to accept strengths and weaknesses in oneself and others. People who are judgmental, impatient, or intolerant of others' weaknesses are almost always equally harsh on themselves; conversely, those who are hard on themselves are almost always equally harsh on others, even if it is not verbalized. As self-compassion and self-worth rise, people become more realistic and tolerant of their own and others' weaknesses while realistically assigning and assuming responsibility for their actions. In the case study at the end of the chapter, the therapist pays particular attention to the son's sense of self-worth as he comes out to his family.

Mind-Body Connection

Satir et al. (1991) also consider the mind-body connection: how emotional issues may be manifesting in the body, either *symbolically* or *functionally*. For example, if a person is feeling burdened, this emotional feeling may manifest symbolically as stooped shoulders ("should-ers") or as a "burdened" posture. Similarly, if a person is feeling overwhelmed, this may manifest functionally in often being ill or exhausted. In addition, the role of nutrition and exercise is assessed. Finally, Satir maintained that the way the body is used indicates the person's communication stance:

- *Congruent Communication and Self-Esteem*: Open and relaxed body postures
- *Placating*: Timid and reserved postures
- *Blaming*: Pointing, angry, and stiff postures
- *Superreasonable*: Cold and distant postures
- *Irrelevant*: Hyper and distracted

Targeting Change: Goal Setting

At the most general level, the goal of Satir's growth model is transformation: to achieve optimal realization of a person's full potential (Azpeitia, 1991). This goal translates into two broad sets of practical goals for treatment planning:

1. Relational, family, or systemic goals
2. Individual goals

Satir therapists' attention to individual goals reflects their humanistic foundations and is unique among systemically based therapists; symbolic-experiential therapists, who share this humanistic foundation, also have systemic and individual goals (see discussion of symbolic-experiential therapy later in chapter).

Relationally Focused Goals: Congruent Communication

The heart of relational goals in Satir's approach is congruent communication: the ability to communicate authentically while responding to the needs of both self and others. Specifically, the goal is to help the family develop ways for all members to communicate so that the system's homeostasis no longer needs the initial symptoms (problems) to maintain balance. The following are examples of specific goals (Satir, 1983; Satir et al., 1991).

Examples of Relationally Focused Goals
- Increase congruent communication in relationships with spouse, parent, child, etc.
- Change family rules and "shoulds" to general guidelines

Individually Focused Goals: Self-Actualization

The overarching individual goal is consistent with other humanistic approaches: to promote the self-actualization of all members of the system. Self-actualization means fulfilling one's potential and living an authentic and meaningful life. The more a person self-actualizes, the greater his/her sense of self-worth and self-esteem. Examples of specific goals (Satir et al., 1991) are as follows.

Examples of Individually Focused Goals
- Increase a sense of self-worth and self-compassion
- Reduce defensiveness and the use of survival stances

The Doing: Interventions

Therapist's Use of Self

Therapists' use of self—being authentically who they are—is one of the most essential interventions in Satir's approach (1988). By being authentic, therapists provide a role model for how to communicate congruently and also show the effects of increased self-actualization. Furthermore, the use of the therapist's self in therapy, which may include self-disclosure, creates a safe relationship in which clients can practice communicating congruently without negative consequences and learn to tap into more authentic modes of expression.

Ingredients of an Interaction

The foundation for all other interventions (Azpeitia, 1991), the *ingredients of an interaction* details the internal communication process and can be used to teach clients about internal and relational processes (Satir et al., 1991). The ingredient questions are used to help people better understand their interactions with others and can be used with individuals, couples, or families. When a client shares a troubling interaction with another, the therapist uses the following seven questions to walk the client through the "ingredients of an interaction" (Azpeitia, 1991; Satir et al., 1991):

1. *What do I HEAR AND SEE?* The therapist prompts clients to describe behaviorally what happened without adding interpretations; this is similar to "videotalk" in solution-based therapies (Chapter 12) (e.g., "My child did not take out the trash immediately upon my asking her to do so").

2. *What MEANINGS do I make of what I hear and see?* After obtaining a clear, behavioral description, the therapist facilitates a discussion of how the clients interpreted these behaviors; meanings often connect back to past experiences (e.g., "My child does not respect me; I would never have spoken to my parents that way").

3. *What FEELINGS do I have about the meanings I make?* Next, the therapist helps clients identify specific feelings about the meanings and interpretations (e.g., "This makes me angry and hurt"). Therapists use "facilitation of emotional expression" techniques, described in detail in next section.

4. *What FEELINGS do I have ABOUT THESE FEELINGS?* Then the therapist asks clients about whether they can accept and tolerate the feelings; the more congruent a person is, the more accepting he/she is of various feelings. If a person is not accepting, survival stances will be triggered (e.g., *congruent*: "I don't like these feelings, but I know that they are natural"; *incongruent*: "I don't like these feelings and need my child to do what I say so that I don't have such feelings anymore"). Subsequent interventions target this level of feeling, not the emotions attached to the interpretation.

5. *What DEFENSES do I use?* When people use incongruent communication, they respond in the interaction using defense mechanisms such as projection, denial, ignoring, or one of the communication stances (e.g., "I get angry and yell at my child for not following my request"; "I blame my spouse for spoiling the child").

6. *What RULES FOR COMMENTING do I use?* Each person's family of origin and early significant others impart rules for commenting on interpersonal interactions that may limit self-worth, restrict choices, and determine what actions are "allowed" and "appropriate"; most of these rules are unspoken and brought into future relationships without reflection on whether they are appropriate or useful (e.g., "It is not okay for a parent to be vulnerable with a child"; "Parents should always be in charge"). These rules are transformed using "softening" techniques (described in Softening Family Rules below).

7. *What is my RESPONSE in the situation?* How does the client respond behaviorally and verbally (e.g., "I tell my child 'You don't respect me' in an angry, hostile tone and stay angry even after the child does the chore"; "I also get angry at my spouse for not being stricter with the children")? Incongruent and problematic responses are targeted for change.

Facilitating Emotional Expression

Satir therapists work with clients to help them express difficult-to-articulate emotions related to the presenting problem (Satir, 1991). If clients complain about family members or a difficult life situation, experiential therapists listen closely for the emotions, expressed or unexpressed, that are related to the problem circumstance. They listen not just for the surface emotions (e.g., "I am angry that my partner was late") but also for deeper levels of emotion (e.g., "My partner does not care about me") and use questions and empathetic reflections to focus clients on those deeper emotions.

Examples:

- If a client tells a story without identifying their emotions directly:
  - "As I listen to you explain what happened to you during that argument, I am getting a sense that you were feeling [name the emotion]. Does that sound right?"
  - "That sounds like a very difficult situation. Can you share some of the emotions you have been experiencing as you go through this?"
- If a client does identify emotions, the therapy can help explore them with questions such as:
  - "You say you were feeling [emotion]. I am wondering if part of you is also feeling [name another emotion that client seems to be experiencing]."
  - "Can you say more about what feeling [name emotion] feels like for you."

Softening Family Rules

Satir et al. (1991) coached families in softening rigid family rules by changing them to *guidelines*. For example, rather than saying "I should not get angry," a client or family was encouraged to revise the limiting rule to "When I am angry, I will express this anger in a way that is respectful to the others and myself." In addition, Satir encouraged families to have as few rules as possible and to be flexible, adapting them for each context and as children's developmental needs change.

Communication Enhancement: Coaching, Role Play, and Enactment

A hallmark of Satir's approach, coaching clients in how to have authentic, congruent communication in session, involves the "ingredients of an interaction" (previously discussed) combined with specific communication coaching strategies (Satir, 1988; Satir et al., 1991). When coaching communication, Satir had clients turn their chairs toward one another and gave the clients prompts, such as "Tell your partner how you feel about what happened on Saturday night." If the client was able to congruently address his/her partner about this, she then prompted the partner to respond in kind, continuing this process until the problem was resolved. When a person had trouble communicating congruently and reverted to a survival stance, Satir interrupted the conversation and suggested how to rephrase the statement or how to make the client's nonverbal communications more congruent (e.g., "Now, can you say that starting the statement with 'I' instead of 'you'?" or "Can you show the emotion you say you are feeling when you speak to your wife?").

Common Areas of Communication Coaching

- Asking clients to start statements using "I" rather than "You"
- Taking full responsibility for one's feelings rather than blaming others
  (e.g., "Instead of 'You made me feel …,' try 'When X happened, I felt …'")
- Encouraging clients to be direct and honest rather than expecting the other to read between the lines
- Identifying double binds (e.g., "You asked your husband to show more affection, but when he does, you are upset and say he only did so because you asked")

In addition, the "ingredients of an interaction" technique (see previous discussion) was used to help coach clients through what they heard and saw, how they interpreted and felt about it, and the feelings about feelings and family-of-origin rules for communicating about what happened.

Sculpting or Spatial Metaphor

Satir's most distinctive intervention is family sculpting, which is done either with the family or in a group setting (Satir, 1988; Satir et al., 1991). Sculpting involves putting family members in physical positions that represent how the "sculptor" sees each person's role in the family. For example, if a child sees a parent as blaming or harshly punitive, he may sculpt that parent with a harsh look and angry pointing finger and perhaps himself as the child cowering or hiding. Typically, either the therapist or the client may direct the sculpting. If family members are directing the sculpting, each person in the family is given an opportunity to sculpt the family as he or she sees it. Sometimes the sculptor assigns each family member a line to say that represents how he or she might be feeling, thinking, or viewing the situation. Usually, however, the essence of this intervention is to give a nonverbal, symbolic depiction of the family process from each person's perspective.

In most cases sculpting is a highly effective nonverbal confrontation that bypasses cognitive defenses. Through the sculpting process, a person is able to literally *see* how he or she is contributing to the problematic family process far more quickly than when the same information is provided with words. For example, if a person places herself far away from the rest of the family because she feels scapegoated or ostracized, this often communicates the emotional reality of her situation far more effectively than if she were to use words, which would generally be responded to with a verbal rationalization. When all family members are sculpting, it is generally best to let each person sculpt how he/she sees the family before allowing the family members to discuss one another's sculpting. Therapists encourage family members to respect each person's subjective experience as illustrated in the sculpting and to use the experience to deepen their understanding of one another. In the case study at the end of the chapter, sculpting is a powerful vehicle for helping this highly divided family understand how each person is experiencing the son's coming out as well as see the good intensions of each.

Touch

Satir (1988) used touch in therapy to initially connect with clients and to encourage and reassure them when they were practicing new ways of communicating, thereby underscoring emotional content and providing palpable support. She also used touch with children to help teach alternatives to violent behavior and to model for parents how to manage difficult children. The fact that she was an extremely nurturing female figure contributed significantly to how clients experienced her physical touch. In today's practice environments, touching is generally discouraged because it can easily be misinterpreted as sexual harassment or may make a client feel uncomfortable. Thus therapists need to carefully consider the legal and ethical issues of using touch. That said, touch may be appropriate in certain practice and cultural contexts. At minimum, therapists can learn from Satir the importance of coaching clients in touching *each other*—their children and spouses—in more loving and helpful ways. For example, rather than using a demonstration, therapists can guide a parent on how to hold children who are having a tantrum.

Interventions for Special Populations
Family Reconstruction: Group Intervention

A form of group psychodrama, the family reconstruction is used to allow clients to safely explore unresolved family issues and life events in the safety of the group setting (Satir et al., 1991). The client, who is called the "star," first identifies key events in his/her life chronology and significant sources of influence. The star then picks people from the group to reenact pivotal life experiences and relationships. The therapist facilitates the reenactment with the following three goals in mind:

- Identify the roots of old learning and their role in the present
- Develop a more realistic picture of the client's parents
- Discover unique strengths and potentials

Parts Party

A group activity similar to the family reconstruction, the parts party involves the client identifying group members to represent aspects of the self (Satir et al., 1991). The person may have members enact generic characteristics (martyr, victim, savior) or use famous figures to represent different aspects of self. In this way, the therapist facilitates a process in which the client is better able to accept different aspects of self and to identify the contexts in which they have been and continue to be useful. The language of "parts" can also be used to facilitate similar discussions and insights in individual, couple, and family therapy.

Putting It All Together

Case Conceptualization Template

Assessment of Family Functioning

- *Role of the Symptom in the System:* Describe how the symptom maintains the family's homoeostasis at an emotional level, often regulating closeness and disance.
- *Family Dynamics:* Identify salient dynamics:
  - Power struggles
  - Parental conflicts
  - Lack of validation
  - Lack of intimacy
- *Family Roles:* Identify possible family roles:
  - The martyr
  - The victim or helpless one
  - The rescuer
  - The good child or parent
  - The bad child or parent
- *Family Life Chronology:* Describe key historical events, deaths, births, divorces, major life changes.
- *Survival Triad:* Describe the emotional and nurturing relationships between each child and parents.

Assessment of Individual Functioning

- *Survival Stances:* Identify the survival stance of each person in the system:
  - Placator
  - Blamer
  - Superreasonable
  - Irreleveant
  - Congruent
- *Six Levels of Experience: The Iceberg:* For symptomatic behavior, describe:
  - Behavior
  - Coping
  - Feelings
  - Perceptions
  - Expectations
  - Yearnings
- *Self-Worth and Self-Esteem:* Describe each person's level of self-worth.
- *Mind-Body Connection:* Describe any salient mind-body connections.

Treatment Plan Template for Individual

Satir Initial Phase of Treatment for Individual

Initial Phase Counseling Tasks

1. Develop working counseling relationship. *Diversity Note: [Describe how you will adjust to respect cultured, gendered, and other styles of relationship building and emotional expression.]*
   a. *Make contact* with client's authentic self using *empathy* and *therapeutic presence*.

2. Assess individual, systemic, and broader cultural dynamics. *Diversity Note: [Describe how you will adjust assessment based on cultural, socioeconomic, sexual orientation, gender, and other relevant norms.]*
   a. Assess individual functioning by noting *survival stance, six levels of experience* related to problem areas, and sense of *self-worth*.
   b. Assess relational functioning by identifying *relational dynamics, family roles, survival triad,* and *family life chronology*.

Initial Phase Client Goals

1. Decrease use of *survival stance* in [key context where symptoms occur] to reduce depressed mood and anxiety.
   a. *Ingredients of interaction* to identify meanings, feelings, defenses, and rules for commenting in problem interactions.
   b. *Facilitate emotional expression* related to depression and anxiety feelings.

Satir Working Phase of Treatment for Individual

Working-Phase Counseling Task

1. Monitor quality of the working alliance. *Diversity Note: [Describe how you will attend to client response to interventions that indicate therapist using expressions of emotion that are not consistent with client's cultural background.]*
   a. *Assessment Intervention:* Monitor client's ability to relate from *authentic self* in relation to therapist.

Working-Phase Client Goals

1. Reduce *"shoulds"* and alter *expectations* for [specify area of life] to reduce sense of depression, hopelessness, and anxiety.
   a. *Ingredients of interaction* to identify and alter "should" and expectations related to depressive and anxious feelings.
   b. *Sculpting* of "should" and expectations to increase awareness of emotions and history related to expectations.

2. Increase realistic *perceptions, attitudes, and beliefs* that inform *sense of self* to reduce depression, hopelessness, and anxiety.
   a. *Softening of family rules* to reduce rigidity of beliefs and attitudes learned in family of origin.
   b. *Coaching* on how to act from more realistic perceptions and attitudes.

3. Increase behaviors that more effectively fulfill client's *yearnings for love and acceptance* to reduce sense of depression and anxiety.
   a. *Ingredients of interactions* intervention to identify underlying yearnings.
   b. *Role play* new behaviors and forms of communication that help client fulfill yearnings for love and acceptance.

(continued)

Satir Closing Phase of Treatment with Individual

Closing-Phase Counseling Task

1. Develop aftercare plan and maintain gains. *Diversity Note: [Describe how you will access resources in the communities of which they are a part to support them after ending therapy.]*
   a. Identify specific emotions and behaviors the return of which will signal relapse and develop a plan of action to address.

Closing-Phase Client Goals

1. Increase daily experience of *self-worth* and ability to express and relate from *authentic self* in all areas of life to reduce depression and increase sense of wellness.
   a. *Sculpting* from old vs. new self to experientially understand the difference.
   b. *Coaching* to practice making decisions from authentic self.

2. Increase *congruent communication in couple/family relations* to reduce depression and anxiety.
   a. *Sculpt* key relationships from survival stance vs. congruent stance to experience the difference.
   b. *Role play* to practice communicating with others from congruent position.

Treatment Plan Template for Couple/Family

Satir Initial Phase of Treatment with Couple/Family

Initial Phase Counseling Tasks

1. Develop working counseling relationship. *Diversity Note: [Describe how you will adjust to respect cultured, gendered, and other styles of relationship building and emotional expression.]*
   a. *Make contact* with each client's authentic self using *empathy* and *therapeutic presence*.

2. Assess individual, systemic, and broader cultural dynamics. *Diversity Note: [Describe how you will adjust assessment based on cultural, socioeconomic, sexual orientation, gender, and other relevant norms.]*
   a. Assess each individual's level of functioning by noting *survival stance, six levels of experience* related to problem areas, and sense of *self-worth*.
   b. Assess relational functioning by identifying *relational dynamics, family roles, survival triad,* and *family life chronology*.

Initial Phase Client Goals

1. Decrease use of *survival stances* in couple/family relating to reduce conflict.
   a. *Family sculpting* of each person's perception of others and their roles in family system.
   b. *Coach* family in *direct communication of emotions*.

Satir Working Phase of Treatment with Couple/Family

Working-Phase Counseling Task

1. Monitor quality of the working alliance. *Diversity Note: [Describe how you will attend to client response to interventions that indicate therapist using expressions of emotion that are not consistent with client's cultural background.]*
   a. *Assessment Intervention*: Monitor each person's ability to relate from *authentic self* in relation to therapist.

(continued)

Working-Phase Client Goals

1. Reduce *"shoulds"* and alter *expectations* for of others in system and increase acceptance to reduce conflict.
   a. *Ingredients of interaction* to identify and alter "should" and expectations of others.
   b. *Sculpting* of "should" and expectations to increase awareness of emotions and history related to expectations.

2. Increase realistic *perceptions, attitudes, and beliefs* that inform incongruent communication reduce conflict.
   a. *Softening of family rules* to reduce rigidity of beliefs and attitudes learned in family of origins.
   b. *Coaching* on how to interact with more realistic perceptions and attitudes.

3. Increase behaviors that more effectively fulfill client's *yearnings for love and acceptance* to reduce conflict.
   a. *Ingredients of interactions* intervention to identify and express underlying yearnings.
   b. *Role play* new behaviors and forms of communication that help client fulfill yearnings for love and acceptance.

Satir Closing Phase of Treatment with Couple/Family

Closing-Phase Counseling Task

1. Develop aftercare plan and maintain gains. *Diversity Note: [Describe how you will access resources in the communities of which they are a part to support them after ending therapy.]*
   a. Identify specific *emotions* and behaviors the return of which will signal relapse and develop a plan of action to address.

Closing-Phase Client Goals

1. Increase each person's daily experience of *self-worth* and ability to express and relate from *authentic self* to reduce conflict and increase sense of wellness.
   a. *Sculpting* from old vs. new selves to experientially understand the difference.
   b. *Coaching* to practice making decisions from authentic self.

2. Increase *congruent communication* in work/school/extended family relations to reduce conflict and increase sense of wellness.
   a. *Sculpt* key relationships from survival stance vs. congruent stance to experience the difference.
   b. *Role play* to practice communicating with others from congruent position.

Research and the Evidence Base

Research on Humanistic Principles

With the notable exception of emotionally focused couples therapy (Johnson, 2004), an empirically supported treatment (see Chapter 15), there has been little outcome research on the effectiveness of specific experiential family therapies. However, what has received attention is the effectiveness of the therapeutic relationship as defined by experiential and humanistic therapies. Both streams of common factors research (see Chapter 2) identify the quality of the therapeutic relationship as highly correlated with therapeutic outcome, Lambert (1992) estimating that 30% of therapeutic outcome in any form of therapy can be attributed to the therapeutic relationship as defined in the humanistic tradition: nonjudgmental, empathetic, and engaged (Miller, Duncan, & Hubble, 1997). Furthermore, the vast majority of studies over the past four decades have found that the *client's* perception of the therapeutic relationship—more so than the therapist's or neutral third party's—is correlated with positive outcome; thus, it is key that the client's perception is used to measure the quality of the therapeutic relationship (Kirschenbaum & Jourdan, 2005). To support this claim, Duncan et al. (2003) have found that they can predict therapy outcome by measuring the quality of the therapeutic alliance in the first three sessions, suggesting that clients must experience the relationship with the therapist as safe and effective early in treatment.

Furthermore, there is a significant stream of research that supports experiential therapists' claims that emotional expression promotes well-being (Stanton & Low, 2012). In particular, writing to express emotions has been found to transform meaning and increase positive emotion related to the event (Langens & Schüler, 2007). Of particular interest to family therapists, constructive expression of emotion characterizes more satisfying intimate relationships (Yoshida, 2011). Additionally, young families with a positive attitude toward emotional expression also reported a greater sense of social support, suggesting that such an attitude may strengthen relationships on multiple levels (Castle, Slade, Barranco-Wadlow, & Rogers, 2008). In sum, although the specific interventions and overall outcome have not received significant research support, the principles and practices behind the humanistic approach to therapeutic relationships and to facilitating emotional expression have strong, consistent support.

Tapestry Weaving: Working with Diverse Populations

Cultural, Ethnic, and Gender Diversity

Used with a wide range of clients, experiential approaches such as Satir's model value clear, congruent emotional expression and the willingness to be vulnerable. Therefore, when working with populations that have different attitudes toward emotional expression and/or are in a treatment context where such vulnerability feels unsafe, therapists need to proceed thoughtfully. For example, the emotional expression often promoted in Satir's communication approach may not be comfortable for some men or East Asian–Americans (Wang, 1994), populations that generally value less dramatic and more indirect emotional expression; thus, they should be used with caution. That said, one recent study found that expressive writing was found to be beneficial for most people but especially those of Asian origin (Lu & Stanton, 2010). Similarly, in another study both Asian and female college students rated experiential therapy approaches more positively than cognitive-behavioral or psychodynamic (Yu, 1998). Additionally, experiential approach may be initially too threatening for mandated clients, who often feel that they might be in a worse position with the courts or social agencies that mandated their treatment if they express too much or certain emotions. Given the fact that a report will be sent to an outside party, these clients are less trusting of the therapeutic relationship.

Regarding gender, therapists need to monitor their expectations for women's level and style of emotional expression; nontraditional female clients report feeling that their therapists expect that they express their emotions in a certain way and report feeling judged and misunderstood because they do not subscribe to stereotyped modes of female emotional expression (Gehart & Lyle, 2001). Therapists working with such clients must be very careful to (a) avoid inaccurate assessment due to different cultural and gender standards and values concerning emotional expression, (b) adjust the level of intimacy in the therapeutic alliance to fit with the client's level of comfort, and (c) choose interventions that actively engage clients at their comfortable level.

Sexual Identity Diversity

Grounded in their values of authentic self-expression, experiential therapists avoid conceptualizations of sexuality that are based on norms and performance (Kleinplatz, 1996). Thus, experiential therapies have been widely used with gay, lesbian, bisexual, transgendered, and questioning (GLBTQ) clients (Davies, 2000). When considering the experience of GLBTQ clients, experiential therapists focus on the burden of societal rejection of the client's authentic self and help clients move toward finding safe ways and contexts to express their authentic selves (Pachankis & Bernstein, 2012). Experiential therapists should also consider how their social contexts may more seriously constrain communication styles, both in their intimate relationships, with supportive vs. nonsupportive family members, as well as colleagues and strangers. Experiential practices, such as Satir's parts party and family reconstruction have been used in gay and lesbian group programs to allow them to reexperience support and their family-of-origin relationships in new and more empowering ways (Picucci, 1992).

Online Resources

Internal Family Systems:
www.selfleadership.org

Satir Global Network (formerly AVANTA). Includes links to training institutes in Asia, Europe, and South America:
www.avanta.net

Satir Institute of the Pacific (John Banmen):
http://www.satirpacific.org/

Satir Institute of the Rockies:
http://www.satirtraining.org

Satir Institute of the Southeast:
http://www.satirinstitute.org/

Satir Training Centre Ottawa:
http://www.satirottawa.ca/main/

Science and Behavior Books (Publishing Books on the Satir Model):
www.sbbks.com

References

*Asterisk indicates recommended introductory readings.

Azpeitia, L. M. (1991). The Satir model in action [course reader]. Encino, CA: California Family Study Center.

Azpeitia, L. M. (1995). Blossoms in Satir's garden: Lynne Azpeitia's work with gifted adults. *Advanced Development, Special Edition*, 127–146.

Bandler, R., Grinder, J., & Satir, V. (1976). *Changing with families: A book about further education for being human* (Vol. 1). Palo Alto, CA: Science and Behavior Books.

Banmen, J. (Guest Ed.). (2002). The Satir model: Yesterday and today (Special issue). *Contemporary Family Therapy, 24*.

Banmen, J. (2003). *Meditations of Virginia Satir*. Palo Alto, CA: Science and Behavioral Books.

Castle, H., Slade, P., Barranco-Wadlow, M., & Rogers, M. (2008). Attitudes to emotional expression, social support and postnatal adjustment in new parents. *Journal of Reproductive and Infant Psychology, 26*(3), 180–194.

Davies, D. (2000). Person-centered therapy. In D. Davies & C. Neal (Eds.), *Therapeutic perspectives on working with lesbian, gay and bisexual clients* (pp. 91–105). Maidenhead, BRK England: Open University Press.

Duncan, B. L., Miller, S. D., Sparks, J. A., Claud, D. A., Reynolds, L. R., Brown, J., & Johnson, L. D. (2003). *Journal of Brief Therapy, 3*, 3–12.

Gehart, D. R., & Lyle, R. R. (2001). Client experience of gender in therapeutic relationships: An interpretive ethnography. *Family Process, 40*, 443–458.

Gehart, D., & McCollum, E. (2008). Teaching therapeutic presence: A mindfulness-based approach. In S. Hicks (Ed.), *Mindfulness and the healing relationship*. New York: Guilford.

Gomori, Maria. (2002). *Passion for freedom*. Palo Alto, CA: Science and Behavior Books.

*Johnson, S. M. (2004). *The practice of emotionally focused marital therapy: Creating connection* (2nd ed.). New York: Brunner/Routledge.

Johnson, S. M. (2005). *Emotionally focused couple therapy with trauma survivors: Strengthening attachment bonds*. New York: Guilford.

Kirschenbaum, H., & Jourdan, A. (2005). The current status of Carl Rogers and the Person-Centered Approach. *Psychotherapy: Theory, Research, Practice, and Training, 42*, 37–51.

Kleinplatz, P. J. (1996). Transforming sex therapy: Integrating erotic potential. *The Humanistic Psychologist, 24*(2), 190–202. doi:10.1080/08873267.1996.9986850

Lambert, M. (1992). Psychotherapy outcome research: Implications for integrative and eclectic therapists. In J. C. Norcross & M. R. Goldfried (Eds.), *Handbook of psychotherapy integration* (pp. 94–129). New York: Wiley.

Langens, T. A., & Schüler, J. (2007). Effects of written emotional expression: The role of positive expectancies. *Health Psychology, 26*(2), 174–182. doi:10.1037/0278-6133.26.2.174

Lu, Q., & Stanton, A. L. (2010). How benefits of expressive writing vary as a function of writing instructions, ethnicity and ambivalence over emotional expression. *Psychology & Health, 25*(6), 669–684.

Miller, S. D., Duncan, B. L., & Hubble, M. (1997). *Escape from Babel: Toward a unifying language for psychotherapy practice*. New York: Norton.

Neff, K. (2003). Self-compassion: An alternative conceptualization of a healthy attitude toward oneself. *Self and Identity, 2*, 85–101.

Pachankis, J. E., & Bernstein, L. B. (2012). An etiological model of anxiety in young gay men: From early stress to public self-consciousness. *Psychology of Men & Masculinity, 13*(2), 107–122. doi:10.1037/a0024594

Picucci, M. (1992). Planning an experiential weekend workshop for lesbians and gay males in recovery. *Journal of Chemical Dependency Treatment, 5*(1), 119–139. doi:10.1300/J034v05n01_10

Rogers, Carl. (1961). *On becoming a person: A therapist's view of psychotherapy*. London: Constable.

Rogers, C. (1981). *Way of being*. Boston: Houghton Mifflin.

Satir, V. (1983). *Conjoint family therapy* (3rd revised ed.). Palo Alto, CA: Science and Behavior Books. (Original work published in 1983).

Satir, V. (1972). *Peoplemaking*. Palo Alto, CA: Science and Behavior Books.

Satir, V. (1988). *The new peoplemaking*. Palo Alto, CA: Science and Behavior Books.

Satir, V., & Baldwin, M. (1983). *Satir step by step: A guide to creating change in families*. Palo Alto, CA: Science and Behavior Books.

*Satir, V., Banmen, J., Gerber, J., & Gomori, M. (1991). *The Satir model: Family therapy and beyond*. Palo Alto, CA: Science and Behavior Books.

Schwartz, R. C. (1995). *Internal family systems therapy*. New York: Guilford.

Stanton, A. L., & Low, C. A. (2012). Expressing emotions in stressful contexts: Benefits, moderators, and mechanisms. *Current Directions in Psychological Science, 21*(2), 124–128. doi:10.1177/0963721411434978

Wang, L. (1994). Marriage and family therapy with people from China. *Contemporary Family Therapy: An International Journal, 16*(1), 25–37. doi:10.1007/BF02197600

*Whitaker, C. A., & Bumberry, W. M. (1988). *Dancing with the family*. New York: Brunner/Mazel.

Yoshida, T. (2011). Effects of attitudes toward emotional expression on anger regulation tactics and intimacy in close and equal relationships. *The Japanese Journal of Social Psychology, 26*(3), 211–218.

Yu, J. (1998, November). Asian students' preferences for psychotherapeutic approaches: Cognitive-behavioral, process-experiential and short-term dynamic therapies. *Dissertation Abstracts International, 59*, 2444.

Experiential Case Study: Son Comes Out in Immigrant Persian Family

Rhana (AF), a 36-year-old Persian woman is seeking family counseling along with Javid (AM), her husband of 20 years who is 46 years old, and their 16-year-old son, Jamshid (CM). The couple immigrated to the United States from Iran two years after their marriage, leaving all family and loved ones behind. With only themselves to rely on for their emotional needs, the couple worked hard to maintain their traditional expectations, while it was difficult for the husband at first, AF decided to continue her education and soon she began working as a nurse practitioner. After the birth of her son, she continued to work but adjusted to a part-time shift to accommodate for attending to the house and her son. The son has been the pride and joy of his parents, especially his mother.

The husband, a successful retail owner, works over 60 hours a week, leaving him little time to spend with his family. AF describes their family dynamic as generally peaceful. Their challenges started when recently her son shared the news about his homosexuality asking for Rhana to break the news to the father. Anxious about the son's and the family's future and worried about her husband's reaction and terrified of larger family and cultural stigma, she suffered for weeks until she found the courage to share the news with her husband, whose initial devastation was quickly overtaken with anger and rage. He refused to accept his son's sexuality and openly blamed AF for not having followed traditional role.

Since the news, the father has announced he wants to move back to Iran and expects his wife and son to follow. AF threatened the husband with separation and so the husband has agreed to come to counseling, but he is not eager to participate. The son reports feeling very sad and guilty about the pain he has caused his parents; he is anxious and worried about what the future holds for his family.

Satir Case Conceptualization

Assessment of Family Functioning

- *Role of the Symptom in the System:* The family has historically maintained peace by everyone doing their duty: the father as breadwinner, the mother as homemaker, and the son as the shining star and "glue" that keeps the parents united with a common cause; an inspiration that gives their immigrant struggle meaning. The son's revelation of his sexual orientation has threatened this family order, especially for his father, who is responding with desperate means to maintain the family he was so proud of.
- *Family Dynamic: Identity Salient Dynamics:* The couple reports little intimacy, and instead a focus on their respective roles, which is not uncommon for Iranian immigrants, with the immigration experience in some ways intensifying and in other ways relaxing these norms. Until the revelation of the son, this had helped them maintain a largely conflict-free relationship. The father's rejection of the son's sexuality has placed the mother in a difficult position between the two.
- *Family Roles: Identify Possible Family Roles:*
  Mother = Martyr (for giving up her dreams for the family) and caretaker
  Father = Martyr, for working long hours to support family; head of household
  Son = Star of the family.
- *Family Life Chronology:*
  1994: AF and AM marry
  1996: AF and AM immigrate to the United States
  1996: CM born, in the United States
  2000: Couple buys house
  2002: AM's father dies; AM returns to Iran for three months
  2006: AF mother dies; AF and CM return to Iran for two months
  2014: CM announces he is gay
- *Survival Triad:* CM has always been more emotionally connected with his mother, whereas he has had a more formal connection with his father, as is culturally typical. The father's recent rejection of his sexuality has seriously ruptured the father-son relationship, thus intensifying the relationship between the mother and son.

Assessment of Individual Functioning

- *Survival Stances:* AF and CM function from a placating stance, trying to do what it takes to keep the family together. On the other hand, AM typically takes a blamer position, and sometimes even a superreasonable stance, citing Iranian cultural tradition as rules which must be followed.
- *Six Levels of Experience: The Iceberg: For Symptomatic Behavior*

  Behavior
  AF = Talks with AM and CM to calm each down.
  AM = Works even longer hours; threatens to move back to Iran.
  CM = Avoids speaking to father and turns to mother for acceptance and asks her to help him connect with the father.
  Coping
  AF & CM use placating stance.
  AM: Uses blaming (AF's work is the cause) and superreasonable stance (son's sexuality is not natural or sanctioned by God).
  Feelings
  AF: Strong feelings of guilt, confusion, and loneliness; no one to turn to for support.
  AM: Dissatisfaction and disapproval of AF's life choices and CM's revelation that he is gay.
  CM: Feelings of sadness, guilt, and shame, which is only partially associated with his sexual orientation and partially with a sense of never being good enough to live up to his father's high expectations in general.
  Perceptions
  AF: Perceives herself as having few options besides pleasing her husband and child.

AM: Perceives himself as having performed his familial obligations and expects his wife and child to do the same.

CM: Perceives his mother as understanding him and American culture more; he sees his father as overly devoted to Iranian culture, which he does not see as fully applicable in the United States.

Expectations

AF: Has relatively realistic expectations of others, but has unrealistic expectation of self.

AM: Has strong belief that the mother is the primary responsible party in shaping the life of the son, and expects her to do so in an appropriate Iranian way.

CM: Expects his parents to be Americanized than they are and to support him the way his friends are supported by their friends.

Yearning

AF: Yearns to fulfill her traditional role as a wife and mother in order to earn validation from her family/culture and also to seek individual fulfillment as do other American women.

AM: Yearns to fulfill his culturally assigned role of a protector and provider and to enjoy the respect that comes with this role.

CM: Yearns for acceptance and validation from his parents in addition to the types of freedoms that his peers have.

Satir Treatment Plan

Satir Initial Phase of Treatment with Family

Initial Phase Counseling Tasks

1. Develop working counseling relationship taking diversity into consideration. *Diversity Notes: Honor father's role as head of household; use respectful, formal language and tone. Try to engage son with humor. Respectfully acknowledge both son and father's position on son's sexuality.*
   a. *Make contact* with each client's authentic self using *empathy* and *therapeutic presence*, validating each person's experience, creating a safe haven by authentically responding to the need of self and others.

2. *Assess individual systemic and broader cultural dynamics. Diversity Notes:* Adjust assessment to address Iranian family norms, immigration status and stresses, and son's sexuality. Pay attention to cultural difference between AM (Iranian male) and CM (GLBTQ) for their emotional expressive styles, and communicative approach; with AM use more indirect and less dramatic emotional expression and with CM focus on experiences of rejection and work toward establishing safety. Use less directive and open-ended reflections with AF, who has less cultural permission to have opinions.
   a. Assess individual functioning by noting *survival stance, six levels of experience* related to problem areas, and sense of *self-worth.*
   b. *Assess* relational function by identifying *relational dynamics, family roles, survival triad, and family life chronology.*

Initial Phase Client Goals

1. Decrease use of survival stances in family relating to reduce conflict and increase authentic contact between family members.
   a. *Family sculpting* of each person's perception of others and their roles in family system; begin with father, then mother, then son; sculpt both current and desired; wait for discussion until after everyone is done.
   b. *Coach* family in *direct communication of emotions.* Physically have AM and AF turn their chairs toward one another and ask AF to tell AM about how she felt about his blaming her about the son's homosexuality and take necessary caution

(continued)

to monitor for client's nonverbal and verbal tendencies to revert to survival stance. The same technique can be applied to everyone. Make adequate adjustment to ensure "I" statements are used.

Satir Working Phase of Treatment with Family

Working-Phase Counseling Task

1. Monitor quality of the working alliance. *Diversity Note: Facilitate necessary adjustment to accommodate for each person's style of emotional expression, respecting AM's cultural and gender mandates to seem in control, and acknowledging how AF and CM may feel the need to be more restrained in front of AM.*
   a. Monitor each client's ability to relate from *authentic self* in relations to therapist. Use self-disclosure to provide a role model for communicating congruently.

Working-Phase Client Goals

1. Reduce *"shoulds"* and alter *expectations* for others in the system while still being mindful of Iranian cultural norms to reduce conflict.
   a. *Ingredients of interaction* to identify and alter "should" and expectations of others. Revisit a troubling interaction between AM and CM using the following question to teach client to recognize their internal and relational processes: What do I hear and see? What meaning do I make of what I hear and see? What feelings do I have about the meaning I make? What feelings do I have about these feelings? What inferences do I use? What rules for commenting do I use? (Address cultural expectations.) What is my response in the situation?
   b. *Sculpting* of "should" and expectations to increase awareness of emotions and history related to expectations. Use this session for exploring how each person sculpts the family before and after.

2. Increase realistic *perceptions, attitudes, and beliefs* that inform incongruent communication to reduce conflict.
   a. *Softening of family rules* to reduce rigidity of beliefs and attitudes learned in family and culture of origin. Change rules dictating "shoulds" and "woulds" to guidelines, minimizing number of rules and being adaptive to fit the context and needs of the family. May use "ingredients of an interaction" to increase feeling about family rules. Discuss how to honor culture as well as CM's experience of his sexual orientation.
   b. *Coaching* on how to act from more realistic perceptions and attitudes continue with creating "I" statements, taking responsibility for feelings rather than blaming others, "when X happened, I felt ..."), eliminate mind reading, and recognize double binds.

3. Increase behaviors that more effectively fulfill client's *yearnings for love and acceptance* to reduce conflict and increase intimacy.
   a. *Ingredients of interactions* intervention to identify underlying yearnings. Using this to extend family's understanding of feelings about feelings to identify difficult underlying yearning for the emotional void that exists in the family.
   b. *Role play* new behaviors and forms of communication that help family fulfill yearning for love and acceptance, especially between father and son and husband and wife.

Satir Closing Phase of Treatment with Family

Closing-Phase Counseling Task

1. Develop aftercare plan and maintain gains. *Diversity Note: Identify resources and support groups for LGBT families and emphasis active participation in community*

(continued)

awareness movements. Identify resources in Iranian community, especially for supporting couple.
   a. Identify specific *emotions* and behaviors that return of which will signal relapse and develop a plan of action to address.

Closing-Phase Client Goals

1. Increase each person's daily experience of *self-worth* and ability to express and relate from *authentic self* to reduce conflict and increase sense of wellness.
   a. *Sculpting* from old vs. new selves to experientially understand the difference. May include individual sessions with son to help with his sexual identity development.
   b. *Coaching* to practice making decision from authentic self.

2. Increase *congruent communication* in work/school/extended family relations to reduce conflict and increase sense of wellness.
   a. *Sculpt* key relationships from survival stance vs. congruent stance to experience the difference; for CM this may be with peers; for AF this may be with extended family; for AM this may be with work colleagues.

CHAPTER 9

Symbolic-Experiential and Internal Family Systems

"When I meet with a family, I am absolutely certain they have within themselves the capacity to struggle and grow. There is not a need to assess or evaluate this. I know this is possible."

—Whitaker & Bumberry, 1988, p. 20

Lay of the Land

This chapter continues our exploration of experiential family therapies from Chapter 8, describing symbolic-experiential therapy and internal family systems therapy. Similar to the Satir approach, these two approaches are closely associated with their progenitors. In addition, these approaches make extensive use of imagery and experiential forms of learning, although using unique and highly distinct strategies.

Symbolic-Experiential Therapy

In a Nutshell: The Least You Need to Know

Symbolic-experiential therapy is an experiential family therapy model developed by Carl Whitaker. Whitaker referred to his work as "therapy of the absurd," highlighting the unconventional and playful wisdom he used to help transform families (Whitaker, 1975). Relying almost entirely on emotional logic rather than cognitive logic, his work is often misunderstood as nonsense, but it is more accurate to say that he worked with "heart sense." Rather than intervene on behavioral sequences like strategic-systemic therapists, Whitaker focused on the emotional process and family structure (Roberto, 1991). He intervened directly at the emotional level of the system, relying heavily on "symbolism" and real life experiences as well as humor, play, and affective confrontation.

For the astute observer, Whitaker's work embodied a deep and profound understanding of families' emotional lives; to the casual observer, he often seemed rude or inappropriate. When he was "inappropriate," it was always for the purpose of confronting or otherwise intervening on emotional dynamics that he wanted to expose, challenge, and transform. He was adamant about balancing strong emotional confrontation with warmth and support from the therapist (Napier & Whitaker, 1978). In many ways, he encouraged therapists to move beyond the rules of polite society and invite themselves and clients to be genuine and real enough to speak the whole truth.

The Juice: Significant Contributions to the Field

If you remember one thing from this chapter, it should be this:

The Battle for Structure and the Battle for Initiative

Whitaker referred to two "battles" in therapy: the battle for structure and the battle for initiative. Even if you are turned off by the war metaphor, every competent therapist should consider the principles he or she is describing. The battle for *structure* should be won by the *therapist,* who sets the boundaries and limits for therapy (Whitaker & Bumberry, 1988). Therapists need to win this battle because they are responsible for setting up a program for change and therefore need to ensure that the necessary structure for change is in place:

- That the necessary people attend therapy
- That therapy occurs frequently enough to produce progress
- That the session content and process will produce change

The "battle" occurs when the therapist must insist on these key pieces. Whitaker was quite clear that if clients were not able to meet the minimal structure requirements, he would not do therapy. He saw the therapist's personal integrity at the heart of this battle:

> The key point here is for the therapist to face the need to act with personal and professional integrity. You must act on what you believe. Betrayals help no one. The Battle for Structure is really you coming to grips with yourself and then presenting this to them. It's not a technique or power play. It's a setting of the minimum conditions you require before beginning. (Whitaker & Bumberry, 1988, p. 54)

Although it is unethical to do therapy if you do not believe that you can render successful treatment, frequently therapists "settle" for trying to do marital or family therapy without the key players attending sessions or do not direct the content or process to the areas they believe need to be addressed. Losing this battle results in stagnant therapy.

Conversely, the battle for *initiative* needs to be won by the *client.* It is the client who must have the most investment and initiative to pursue change. This insight is often summarized as: *therapists should never work harder than their clients.* This is a particularly challenging battle for new therapists, who can be overly helpful and often want to move faster than their clients are able. However, if the therapist wants change more than the client, this creates a problematic dynamic and, paradoxically, often stalls change. The therapist needs to wait and sometimes let the tension and crisis build until the client develops the incentive and motivation to make changes. If the therapist has more initiative toward change, clients feel they are being dragged or forced and then start to dig in their heels or find little ways to sabotage the therapist's efforts at change. Instead, when the client has the greater motivation for change, the process flows more smoothly. Thus therapists must be ready to follow clients' lead on how hard to work, following their flow of energy and enthusiasm.

The battle for initiative can be interpersonally uncomfortable, involving awkward silences, "I don't know" answers, or tension. Clients may feel frustrated that the therapist is not taking the lead in choosing topics of discussion and providing ready solutions. Whitaker and Bumberry (1988) explain the purpose of allowing this tension to build: "It's an issue of the family becoming somebody. They need to grapple with each other. It's an invitation to them to come alive and stop play-acting" (p. 66).

Rumor Has It: The People and Their Stories

Carl Whitaker Along with his colleagues, Thomas Malone and John Warkentin, Whitaker began seeing families in the 1940s and so was one of the earliest pioneers in the field (Roberto, 1991). A psychoanalytically trained psychiatrist, he began to shift away from conceptualizing client problems as internal conflicts toward viewing problems as part of dysfunctional interactions. In his early work with psychosis and trauma, he focused on the emotional dynamics in session and within the family system. As his work evolved, he increasingly focused on affect and here-and-now experiences within family relationships. While Whitaker was chair of the Department of Psychiatry at Emory University, Whitaker and Malone began to use cotherapy, a hallmark of the approach. Whitaker's best-known colleagues include Augustus Napier, Whitaker's coauthor on *The Family Crucible;* William Bumberry, Whitaker's coauthor on *Dancing with the Family*; David Keith (Keith, Connell, & Connell, 2001); and Gary Connell (Connell, Mitten, & Bumberry, 1999).

The Big Picture: Overview of Treatment
Therapy of the Absurd
Symbolic-experiential therapy is often referred to as the "therapy of the absurd" (Whitaker, 1975). However, absurdity in this case is not absurdity for absurdity's sake (whatever that might be); instead, symbolic-experiential therapists employ a specific form of absurdity for a specific purpose. Absurdity is used to *perturb* (shake or wake up) the system in a compassionate and caring way. Sometimes the "caring" takes the form of speaking a truth no one else has been willing to speak, but the therapist is always careful to convey the spirit of caring behind such brutally honest comments (Whitaker & Bumberry, 1988). Usually, however, therapy of the absurd involves humor, playfulness, and silliness. By being able to play with otherwise "serious matters," therapists invite themselves as well as their clients into a more resourceful position in relation to the problem. A new attitude of lightness and hope emerges from this playfulness. Therapy of the absurd also uses paradoxical techniques that take the symptom and exaggerate it 10% so that clients can see the folly of their fears and habits. Symbolic-experiential therapists almost always employ paradox in a playful way. In the case study at the end of the chapter, the therapist uses absurdity to help a Cherokee family embrace the painful truth they have been avoiding: that they lost their mother/wife to a drunk driver. The therapist, or course, must use absurdity in a respectful way that helps them see how their avoidance behaviors only exacerbate their pain rather than avoid it.

Making Connection: The Therapeutic Relationship
"Families do not fail, therapists do." —Whitaker & Ryan, 1989, p. 56

Therapist's Authentic Use of Self
Symbolic-experiential therapists strive to be authentic and genuine and, arguably, are the most authentic of family therapists in that they do not follow many of the pretenses that many would consider professional or appropriate boundaries; they are the first to point out that the emperor has no clothes (Connell et al., 1999; Napier & Whitaker, 1978; Whitaker & Bumberry, 1988). They are fully themselves and do not hide this from their clients. If they are bored, they show it; if they are annoyed, they express it. If they see an elephant in the middle of the room, they say something. This level of authenticity requires extensive supervision and training to ensure that therapists are able to maintain exceptionally clear boundaries between their personal issues and clients' issues. For this reason, therapists who do not work this way are often baffled when watching symbolic-experiential therapists in action because they use a different set of relational rules for being "professional." In the end, being authentic at this level is primarily for the benefit of the client: to model the type of authenticity the therapist wants the client to develop and to create an environment in which the client can do this. If the therapist is hiding behind professional boundaries and the "role," the client has little chance of fully developing this type of authenticity.

Personal Integrity

Whitaker insisted that the therapist maintain a clear and unwavering integrity as a person (Whitaker & Bumberry, 1988). This integrity requires a fierce adherence to personal beliefs and the willingness to stand up for them, even when they are unpopular and may make people upset. Integrity is required to push families to address the painful issues they have been avoiding.

Therapist's Responsibility

Symbolic-experiential therapists strive "to be responsive *to* the family without being responsible *for* them" (Whitaker & Bumberry, 1988, p. 44). They are careful not to take on responsibility for clients' lives, but instead are responsible for pushing clients to accept full responsibility for their own lives. The therapist's greatest responsibility is to ensure that the therapeutic process promotes change: to win the battle for structure. The therapist is *active* but not directive.

Stimulating Mutual Growth

The therapeutic process in symbolic-experiential therapy stimulates mutual growth: both the therapist and client grow together through their authentic encounter with each other (Connell et al., 1999; Napier & Whitaker, 1978). Because therapists are fully authentic—the same people they are in other relationships—they learn about their own limitations, blind spots, and weaknesses and use the encounters with clients to also grow and become more fully authentic people themselves. The encounter touches each participant—client and therapist—at a deeply profound level, leaving both transformed. In the case study at the end of the chapter, the therapist must allow herself to be touched by the enormous tragedy this family has experienced as well as each person's brave attempt to cope with it.

Use of Cotherapists

Whitaker encouraged the use of a cotherapist, recommending that one therapist be nurturing and the other more confrontational so that the family has a strong base of support as well as a process for having difficult issues raised and addressed (Napier & Whitaker, 1978). In providing a balance of support and challenge, the cotherapy team models a coparenting relationship.

The Viewing: Case Conceptualization and Assessment
Authentic Encounters and the Affective System

Case conceptualization in symbolic-experiential therapy is one of the most difficult to fully capture in words. In one sense, the therapist relies primarily on the in-the-moment *authentic encounters* with the client to directly experience who the other is in a holistic way (Connell et al., 1999; Whitaker & Bumberry, 1988). To anyone new to the practice, that statement is a bit too vague to be helpful. It is much like riding a bike: beginners need each step broken down, whereas those who are experienced say, "it's easy; just pedal," forgetting how difficult it was to get started (and how long dad pushed from behind). The little steps that allow symbolic-experiential therapists to effortlessly "roll" along and "intuit" their case conceptualization are grounded in a systemic understanding of the family: boundaries, homeostasis, triangles, and other factors. However, symbolic-experiential therapists focus primarily on the family's *emotional system* rather than their behavioral interactions. When they get a sense of boundaries or triangles, they focus on the emotional exchange between parties rather than actions. Therapists "feel" their way through the system.

Trial of Labor

The assessment of the family is accomplished through a *trial of labor,* which refers to observing how the family responds to the therapist's interventions and interactions (Whitaker & Keith, 1981). During the trial of labor, the therapist tries to understand each person's preferred family roles, beliefs about life, values within relationships, developmental and family histories, and interactional patterns. More specifically, the therapist attends to two broad patterns: (a) the structural organization of the family and (b) the emotional processes and exchanges within the family (Roberto, 1991). In assessing structure, Whitaker used many of the same criteria as structural therapists.

Assessing Structural Organization
- *Permeable Boundaries Within the Family:* Interpersonal boundaries should be permeable, not overly rigid or diffuse.
- *Clear Boundaries With Extended Family and Larger Systems:* Boundaries with larger systems should allow for the autonomy of the nuclear family as well as connection with broader systems.
- *Role Flexibility:* Family roles, including the scapegoat or good/bad child, should rotate frequently.
- *Flexible Alliances and Coalitions:* Alliances and coalitions are inevitable but should be flexible, changing with each new situation or challenge rather than always involving the same people on the same team.
- *Generation Gap:* Generations should have clear boundaries, resulting in strong marital and sibling subsystems.
- *Gender-Role Flexibility:* Gender roles should be negotiable, resisting stereotyped gender norms in favor of the ability of each parent or partner to assume a wide range of roles as necessary.
- *Transgenerational Mandates:* Transgenerational behavioral expectations and values are assessed across three to four generations; in healthy families, these are open to renegotiation.
- *"Ghosts":* Therapy identifies deceased or living extended family members who are creating cross-generational stress.

Assessing Emotional Process
- *Differentiation and Individuation:* Each family member should be able to hold unique opinions and speak for him or herself.
- *Tolerance of Conflict:* Healthy families are able to tolerate the overt and explicit expression of differences and conflict.
- *Conflict Resolution and Problem Solving:* Healthy families are able to engage in overt conflict and successfully resolve conflicts and solve problems, which may involve win-win scenarios, compromises, or acceptance of differences.
- *Sexuality:* In healthy families, couples share sexual intimacy, and sexuality is contained within generational lines.
- *Loyalty and Commitment:* Members experience a clear sense of loyalty and commitment while allowing for individual autonomy.
- *Parental Empathy:* Parents should demonstrate empathy for children's experience while still maintaining boundaries and structure; parents who were abused as children often fail to have sufficient empathy or are overly empathetic and do not set healthy boundaries.
- *Playfulness, Creativity, and Humor:* Fun and laughter are signs of healthy family functioning.
- *Cultural Adaptations:* Immigrant families are able to balance the needs of their culture of origin and their current cultural context.
- *Symbolic Process:* Each family has particular symbols and images that are "affectively loaded" and thus helpful in facilitating change.

Focus on Competency
When assessing families, symbolic-experiential therapists emphasize strengths, competencies, and resources for change (Roberto, 1991). Families are viewed as highly resilient and resourceful, and therapists focus on activating these resources (Whitaker & Bumberry, 1988). For example, in the case study at the end of the chapter, the therapist

is as interested in the older children's competence in taking care of their young sister after their mother's death as she is in their truancy, alcohol use, and sexual promiscuity; clearly they have strengths that can be used to address the more challenging behaviors.

Symptom Development

Symptoms develop when dysfunctional structures and processes persist over time (Roberto, 1991). Healthy families experience periods of dysfunction and difficulty, but these do not become chronic. The persistence of dysfunction can occur over generations, with offspring feeling obligated to adhere to family myths and legacies or to make up for prior losses.

Targeting Change: Goal Setting

Symbolic-experiential therapists have three primary long-term goals for all clients:

- *Increase Family Cohesion:* Create a sense of nurturance and confidence in problem solving (Roberto, 1991)
- *Promote Personal Growth:* Support the completion of developmental tasks for all family members (Roberto, 1991; Whitaker & Bumberry, 1988)
- *Expand the Family's Symbolic World* (Whitaker & Bumberry, 1988)

Increase Family Cohesion

The first goal involves increasing cohesion and the authenticity of family relationships, which means increasing the sense of love and meaningful connection between family members. Therapists focus on two key areas to achieve this goal:

- *Cohesion:* More than many other family therapists, symbolic-experiential therapists focus on increasing the sense of family cohesion, the emotional connection between family members. The outer expression of cohesion and connection varies across cultures and genders, but it is generally characterized by a strong sense of belonging, being loved, being wanted, and loyalty.
- *Interpersonal Boundaries:* Symbolic-experiential family therapists also use the term *boundaries* to refer to the relational rules that regulate closeness and distance. Boundaries should allow (a) each person to have enough freedom to be fully himself or herself (i.e., authentic), and (b) strong emotional connection and intimacy between family members.
- *Transgenerational Boundaries:* Transgenerational boundaries should allow for sufficient autonomy of the nuclear family while encouraging connection with extended families.

Promote Personal Growth

Symbolic-experiential family therapists aim to increase each person's level of personal growth by successfully navigating developmental tasks, a process also referred to as *self-actualization*. Because self-actualization is a lifelong process, how a therapist and client know when therapy is done can be a tricky question. At minimum, client self-actualization and growth should be promoted to the point where the client is no longer experiencing symptoms at the individual level, such as depression or anxiety, and can function in most areas of daily life, such as school or work. Beyond a base level of individual functioning, therapists aim to promote personal growth and a more authentic experience of self and self-expression. This is a more difficult quality to quantify, but it is generally easy for clients to articulate a sense of growth or being more of who they are. Therapists encourage clients to pursue specific self-growth goals that relate to their presenting problem. Examples include the following:

- Increase ability to express thoughts and feelings respectfully with others and decrease people-pleasing behaviors
- Increase ability to connect with others at a deeper emotional level and to express this connection verbally

- Increase ability to consciously respond to stress rather than react with anger or fear
- Increase proactive handling of problems and reduce procrastination and avoidance

Expand the Family's Symbolic World

Symbolic-experiential therapists believe that people filter their lives through relatively few constructs or beliefs about life; these constructs constitute a person's *symbolic world*. All experiences are filtered through this system of symbolic meaning to interpret life events as good or bad, problematic or joyful. As a growth-oriented approach, experiential therapy aims to expand the meaning of experience and broaden the client's life horizons: "If we can aid in the expansion of the symbolic world of the families we see, they can live richer lives" (Whitaker & Bumberry, 1988, p. 75). For example, if a family identifies itself with hard work and success at the expense of personal relationships, the therapist explores the source of this identity and may try to expand the meaning of success to include relational, health, and other spheres of life in addition to work. In the case study at the end of the chapter, the therapist draws from the family's Native American culture as well as their own imaginations to help them envision a life that still has meaning and a sense of family even after the loss of their mother/wife.

The Doing: Interventions

Creating Confusion and Disorganization

In the early phases of therapy, symbolic-experiential therapists create confusion and disorganization to break the family out of their rigid interaction patterns: "confusion is, by itself, one of the most potent ways to symbolically open up the infrastructure of the family" (Whitaker & Bumberry, 1988, p. 82). Confusion can be created with absurd comments (e.g., offering ridiculous solutions), role reversals (e.g., relabeling a child's correcting of the parent as the child trying to parent the parent), or appealing to universal principles that are at odds with the family's beliefs (e.g., that teen rebellion is normal or a rite of passage).

Here-and-Now Experiencing

Believing that people rarely grow emotionally by intellectual education, symbolic-experiential therapists use present-moment interactions and their own affect to promote change (Mitten & Connell, 2004; Whitaker & Bumberry, 1988). In fact, Whitaker and Bumberry (1988) go so far as to quip: "Nothing worth learning can be taught" (p. 85). Thus therapists use the immediacy of what is in the room—including their own negative emotional responses to clients—to highlight and redirect structural change and confront dysfunctional patterns and beliefs.

Redefining and Expanding Symptoms

Symbolic-experiential therapists often choose to redefine symptoms as ineffective efforts toward growth, thus pointing in the direction of needed change (Connell et al., 1999; Roberto, 1991). Although similar to positive connotation in the Milan approach, the symbolic-experiential approach uses the specific reframe of striving for *growth and authenticity*. For example, a child's refusal to do homework is redefined as fear of failing at a new level of challenge at school. In addition, therapists expand the symptom from an individual matter to a family matter, often extending this to an intergenerational problem (e.g., fear of failure kept grandfather from pursuing his dream of owning his own business; Mitten & Connell, 2004).

Spontaneity, Play, and "Craziness"

Experiential therapists use spontaneity and fun toward several ends (Mitten & Connell, 2004; Roberto, 1991; Whitaker & Bumberry, 1988). First, by being playful, therapists build a strong therapeutic relationship that allows them to directly and honestly confront clients without encountering resistance. When working alone, they position themselves to be the type of person whom the client trusts enough to listen to the raw truth;

when working as cotherapists, one therapist tends to be nurturing and the other confrontational. Playfulness also helps to reframe problems that have been unrealistically magnified, as is often the case with parents or spouses who magnify a single flaw in the other and de-emphasize the balance of good qualities. As folk wisdom teaches, laughter is often the best medicine, and symbolic-experiential therapists are skilled in using laughter to help their clients heal. The use of humor and play often goes against common stereotypes about therapists and therapy, which are based primarily on psychodynamic therapies, but most clients find laughter helpful, or at least enjoyable. Thus, if you observe symbolic-experiential therapists in action, you may see them tossing a Frisbee, singing a silly song, telling a well-chosen joke, or playing musical chairs. Even when dealing with grief, as the therapist does in the case study at the end of this chapter, the therapist uses humor and play to create a safe environment for the family to address the grief they have been avoiding for over two years.

Separating Interpersonal from Personal Distress

Symbolic-experiential therapists help clients learn how to separate personal issues from interpersonal issues (Roberto, 1991). Often people get into trouble because they do not know how to allow each person in a relationship to have personal autonomy while also being intimately connected with others. When clients are unreasonably demanding to their partners or children, the therapist helps them sort out where their spheres of influence should begin and end. For example, if a parent is demanding that a child play a certain instrument or sport that the child clearly does not enjoy, the therapist will confront the family about where the parenting ends and the child's autonomy begins. Similarly, with couples, if one partner insists that the other feel a certain way about an issue (e.g., be equally upset about a friend's comment), the therapist will invite the couple to separate out and honor each person's unique response to the situation as well as the underlying need for validation in wanting the partner to respond a certain way.

Affective Confrontation of Rigid Patterns and Roles

Symbolic-experiential family therapists use affective confrontation to interrupt rigid patterns. The goals may be (a) to raise clients' awareness when they do not know how they are contributing to the problem, (b) to raise a taboo subject that the client and others have been avoiding, or (c) to increase motivation to make changes when there is cognitive awareness but no change in action (Roberto, 1991). Whitaker explained:

> I am comfortable pushing the family because of my belief that they have unlimited potential. They have the capacity to expand and progress, if only they have the courage to try. My job is to struggle to mobilize that courage .... Not to push, under the assumption that I might make things worse, is to decide for the family that they're too sick to care and too inept to grow. (Whitaker & Bumberry, 1988, p. 37)

"When did you divorce your husband and marry your son?" and "You are aware that you have abandoned the family to advance your career" are examples of confrontations used to interrupt dysfunctional patterns. In addition, affective confrontation is useful for increasing motivation when the insight is there but no action has followed. For example, with parents who need to save their marriage but have no time for a date because the kids have too many after-school activities, a therapist can say, "What do you think would be more detrimental for your daughter: missing dance practice once a week for a few months or having her parents divorce? Do you want to ask your child what her preference is?"

Augmenting Despair and Amplifying Deviation

With clients who are unrealistically hopeless (pessimists), symbolic-experiential therapists use paradoxical techniques such as augmenting despair and amplifying deviation (Roberto, 1991). As with other paradoxical techniques, the therapist exaggerates the client's symptom, such as despair, slightly—perhaps 10%–20%—just enough to get outside the client's comfort zone (or normal range of despair) so that the client can see how out of proportion the despair and negative assumptions are with the facts. This can be done in a playful way or in a more direct and literal way, depending on what would be most useful to the client and most congruent with his/her personality. For example, the therapist may jokingly suggest to a client who feels hopeless about meeting the man of her dreams at age 25 that she perhaps consider a career that will substitute for a marriage, perhaps even join a convent; if such a comment is well timed and delivered, the client is likely to see how unrealistic her despair is.

Absurd Fantasy Alternatives

Symbolic-experiential therapists enjoy inviting clients into absurd fantasy scenarios to shake them out of their patterns and to make realistic solutions more palatable (Mitten & Connell, 2004; Roberto, 1991). Designed to increase the family's flexibility and openness to new behaviors, fantasy alternatives get clients out of their habitual ways of looking at things by playfully perturbing or shaking up the system so that new symbolic meanings, ideas, and perspectives can emerge. Therapists may suggest, "If washing dishes is such a problem, why don't you invest in paper plates?" or "If you need that much space, why don't you build a private cell in the backyard where you can really be alone? I guess that would lead to a fight with your wife as to whether or not to install a heating system."

Reinforcing Parental Hierarchy

Symbolic-experiential therapists are careful to reinforce the parental hierarchy and clearly establish generational boundaries between parents and children (Whitaker & Bumberry, 1988). This involves supporting parents when they make a request of a child, instructing parents to manage a child's behavior in session (rather than having the therapist take over), beginning with parents, and greeting parents first.

Stories, Free Associations, and Metaphors

Symbolic-experiential therapists share stories, free-associate, and offer metaphors to provide powerful images and examples that will inspire clients to change (Mitten & Connell, 2004). Clients can often receive a message more easily through a fictional story or metaphor because it is about somebody else—not them—and so there is less resistance or debate about details. When working with a family that has an established tradition of storytelling and legend, such as the Native American family in the case study at the end of the chapter, the therapist accesses these stories and metaphors to create deeply meaningful and personal sources of inspiration and transformation.

Putting It All Together

Case Conceptualization Template

- *Assessing Structural Organization*
  - *Permeable Boundaries Within the Family:* Interpersonal boundaries should be permeable, not overly rigid or diffuse.
  - *Clear Boundaries With Extended Family and Larger Systems:* Boundaries with larger systems should allow for the autonomy of the nuclear family as well as connection with broader systems.
  - *Role Flexibility:* Family roles, including the scapegoat or good/bad child, should rotate frequently.
  - *Flexible Alliances and Coalitions:* Alliances and coalitions are inevitable but should be flexible, changing with each new situation or challenge rather than always involving the same people on the same team.
  - *Generation Gap:* Generations should have clear boundaries, resulting in strong marital and sibling subsystems.
  - *Gender-Role Flexibility:* Gender roles should be negotiable, resisting stereotyped gender norms in favor of the ability of each parent or partner to assume a wide range of roles as necessary.

- *Transgenerational Mandates:* Transgenerational behavioral expectations and values are assessed across three to four generations; in healthy families, these are open to renegotiation.
- *"Ghosts":* Therapy identifies deceased or living extended family members who are creating cross-generational stress.
- Assessing Emotional Process
    - *Differentiation and Individuation:* Each family member should be able to hold unique opinions and speak for him or herself.
    - *Tolerance of Conflict:* Healthy families are able to tolerate the overt and explicit expression of differences and conflict.
    - *Conflict Resolution and Problem Solving:* Healthy families are able to engage in overt conflict and successfully resolve conflicts and solve problems, which may involve win-win scenarios, compromises, or acceptance of differences.
    - *Sexuality:* In healthy families, couples share sexual intimacy, and sexuality is contained within generational lines.
    - *Loyalty and Commitment:* Members experience a clear sense of loyalty and commitment while allowing for individual autonomy.
    - *Parental Empathy:* Parents should demonstrate empathy for children's experience while still maintaining boundaries and structure; parents who were abused as children often fail to have sufficient empathy or are overly empathetic and do not set healthy boundaries.
    - *Playfulness, Creativity, and Humor:* Fun and laughter are signs of healthy family functioning.
    - *Cultural Adaptations:* Immigrant families are able to balance the needs of their culture of origin and their current cultural context.
    - *Symbolic Process:* Each family has particular symbols and images that are "affectively loaded" and thus helpful in facilitating change.
- *Focus on Competency*
  Describe strengths and resiliencies.
- *Symptom Development*
  Describe history of symptoms, especially noting intergenerational transmission of problem.

Treatment Plan Template for Individual

Symbolic-Experiential Initial Phase of Treatment for Individual

Initial Phase Counseling Tasks

1. Develop working counseling relationship. *Diversity Note: [Describe how you will adjust to respect cultured, gendered, and other styles of relationship building and emotional expression.]*
    a. Use *therapist's authentic self* to form *mutual relationship* that stimulates each party's growth.

2. Assess individual, systemic, and broader cultural dynamics. *Diversity Note: [Describe how you will adjust assessment based on cultural, socioeconomic, sexual orientation, gender, and other relevant norms.]*
    a. Assess emotional processes, *including level of individuation, ability to tolerate conflict, problem solving, sexuality, loyalty, empathy, playfulness, cultural adaptations, and symbolic processes.*
    b. Assess family dynamics, including *boundary within family, boundaries with external systems, role flexibility, alliances, gender-role flexibly, transgenerational mandates, and family "ghosts."*

(continued)

Initial Phase Client Goals

1. Increase tolerance for and expression of *authentic emotion* to reduce depressed mood and anxiety.
   a. Use of *play and "craziness"* to invite clients into present moment experiencing of authentic emotion.
   b. *Redefining and expanding symptoms* as ineffective attempts at growth and identify alternative directions.

Symbolic-Experiential Working Phase of Treatment for Individual

Working-Phase Counseling Task

1. Monitor quality of the working alliance. *Diversity Note: [Describe how you will attend to client response to interventions that indicate therapist using expressions of emotion that are not consistent with client's cultural background.]*
   a. *Assessment Intervention:* Monitor client's response to *therapist's authentic use of self* and regularly discuss.

Working-Phase Client Goals

1. Reduce *rigid patterns* that fuel depression and anxiety behaviors [specify if possible] to reduce depressed mood and anxiety.
   a. *Affective confrontation* of rigid patterns using humor, stories, and direct confrontation.
   b. *Augment despair and amplify deviation* by exaggerating rigid patterns to the point of absurdity.

2. Increase the breadth and flexibility of *client's symbolic world* to reduce sense of helplessness and feeling trapped.
   a. *Absurd fantasy alternatives* to help client expand word view.
   b. *Stories, free association, and metaphors* to expand client sense of possibilities.

3. Increase capacity to *tolerate difference and conflict* to reduce depressed mood and anxiety.
   a. *Here-and-now experiencing* of conflict and difference to reduce fears related to conflict.
   b. *Play and "craziness"* that exaggerates fear of conflict and rejection.

Symbolic-Experiential Closing Phase of Treatment for Individual

Closing-Phase Counseling Task

1. Develop aftercare plan and maintain gains. *Diversity Note: [Describe how you will access resources in the communities of which they are a part to support them after ending therapy.]*
   a. *Challenge fears of relapse* and happy-ever-after hopes and identify *playful options* to responding to relapse.

Closing-Phase Client Goals

1. Increase ability to *experience present moment emotions* and *authentically express them* to reduce depression and anxiety.
   a. *Here-and-now experiencing* of both positive and negative emotions.
   b. Introduce client-initiated *play and creativity* into daily life activities.

2. Increase ability to form *cohesive intimate relationships* to reduce depressed mood and anxiety.
   a. *Separate personal from interpersonal distress* to help client more authentically relate.
   b. *Confront* fears of intimacy, worthiness, and rejection to enable client to more authentically relate.

Treatment Plan Template for Couple/Family

Symbolic-Experiential Initial Phase of Treatment for Couple/Family

Initial Phase Counseling Tasks

1. Develop working counseling relationship. *Diversity Note: [Describe how you will adjust to respect cultured, gendered, and other styles of relationship building and emotional expression.]*
   a. Use *therapist's authentic self* to form *mutual relationships* that stimulates each party's growth.

2. Assess individual, systemic, and broader cultural dynamics. *Diversity Note: [Describe how you will adjust assessment based on cultural, socioeconomic, sexual orientation, gender, and other relevant norms.]*
   a. Assess emotional processes, *including level of individuation, ability to tolerate conflict, problem solving, sexuality, loyalty, empathy, playfulness, cultural adaptations, and symbolic processes.*
   b. Assess family dynamics, including *boundary within family, boundaries with external systems, role flexibility, alliances, gender-role flexibly, transgenerational mandates, and family "ghosts."*

Initial Phase Client Goals

1. Increase tolerance for and expression of *authentic emotion* to reduce relational conflict.
   a. Use of *play and "craziness"* to invite couple/family into present moment experiencing of authentic emotion.
   b. *Redefining and expanding symptoms* as ineffective attempts at growth and identify alternative directions.

Symbolic-Experiential Working Phase of Treatment for Couple/Family

Working-Phase Counseling Task

1. Monitor quality of the working alliance. *Diversity Note: [Describe how you will attend to client response to interventions that indicate therapist using expressions of emotion that are not consistent with client's cultural background.]*
   a. *Assessment Intervention:* Monitor couple/family response to *therapist's authentic use of self* and regularly discuss.

Working-Phase Client Goals

1. Reduce *rigid relational patterns, clarify boundaries* and increase *role flexibility* to reduce conflict.
   a. *Affective confrontation* of rigid patterns using humor, stories, and direct confrontation.
   b. *Augment despair and amplify deviation* by exaggerating rigid patterns to the point of absurdity.

2. Increase capacity to *tolerate difference and conflict* within system to reduce conflict.
   a. *Here-and-now experiencing* of conflict and difference to reduce fears related to conflict.
   b. *Play and "craziness"* that exaggerates fear of conflict and rejection.

3. Increase the breadth and flexibility of *couple/family's symbolic world* to reduce conflict by expanding possibilities for relating.
   a. *Absurd fantasy alternatives* to help expand couple/family word view.
   b. *Stories, free association, and metaphors* to expand sense of possibilities.

(continued)

Symbolic-Experiential Closing Phase of Treatment for Couple/Family

Closing-Phase Counseling Task

1. Develop aftercare plan and maintain gains. *Diversity Note: [Describe how you will access resources in the communities of which they are a part to support them after ending therapy.]*
   a. *Challenge fears of relapse* and happy-ever-after hopes and identify *playful options* to responding to relapse.

Closing-Phase Client Goals

1. Increase *couple/family cohesion* and *authentic expression of self within system* to reduce conflict and increase sense of well-being.
   a. *Separate personal from interpersonal distress* to help client more authentically relate.
   b. *Confront* fears of intimacy, worthiness, and rejection to enable clients to more authentically relate.
2. Increase each member's ability to *experience present moment emotions* and *authentically express him/herself* to reduce conflict and increase sense of well-being.
   a. *Here-and-now experiencing* of both positive and negative emotions.
   b. Introduce client-initiated *play and creativity* into daily life activities.

Internal Family Systems Therapy

In a Nutshell: The Least You Need to Know

Internal family systems therapy shares humanistic assumptions common to all experiential approaches—that people have a healthy core self that therapy aims to restore—and the interventions are classically humanistic in that clients are invited to reflect on their inner emotional life as it unfolds in the moment. Nonetheless, it is in many respects an integrative theory, integrating elements of structural, strategic, intergenerational, and narrative therapy (Schwartz, 2001). The fundamental premise of internal family systems is that each person's inner life has multiple parts that form a system that functions like a family system: hence the name *internal* family systems. Each person's inner world is characterized as having various *parts* (a term used by other experiential therapists, such as Satir and Fritz Perls). Unlike other theorists who discuss inner parts, Schwartz theorizes that these parts interrelate as a coherent system with the same dynamics one would see in a family. Thus, the conceptualization and approaches used to help family relations can be used to help a person's various parts interrelate more effectively. Furthermore, each person has a *Self* that is its core and distinct from its parts that has vision, compassion, and confidence; the goal of therapy is to have this Self provide inner leadership to the various parts. When people experience trauma, family imbalance, or polarized relationships, their inner system of parts take on *extreme roles* to cope. Schwartz has identified three basic patterns or roles parts assume: exiles (parts that are vulnerable and therefore constrained or hidden), managers (parts that try to keep the Self from potential harm), and firefighters (the parts that come out in crisis). The goal of therapy is to help a person move from managing their life reactively with their parts to becoming led by the Self. This model has been used extensively with survivors of trauma, most notably sexual abuse survivors, and eating disorders (Schwartz, 1987).

The Juice: Significant Contributions to the Field

If you remember one thing from this chapter, it should be this:

Individuals as Systems

A unique approach, internal family systems therapists conceptualize individuals as have an *internal* system of self parts that function in ways similar to a family system.

Other therapists have conceptualized the self as having parts: id, ego, and superego in psychoanalysis, internal objects in object relations therapy, disowned parts in Gestalt, and parts in Satir's human growth model approach. But no other approach conceptualizes these parts as a system: that work together to maintain homeostasis and handle crisis (positive feedback). Thus, just as in family systems theory, each part must be understood in the *context* of other parts. For example, the part that cuts to help manage emotions is viewed in context of other parts, often the more vulnerable parts it is trying to protect. Just as in real families, parts in internal family systems have boundaries—sometimes enmeshed, sometimes disengaged—and form coalitions. The internal family systems therapists help clients learn how to identify not only their parts but how they work—or don't work—together. Similar to the nonblaming, nonpathologizing stance of systemic family therapists, internal family systems therapists do not see some parts as good and others as bad; instead, each plays a critical role in the internal family system. The therapist helps clients understand these complex inner relationships and how to gently shift those relations. In general, three types of parts are identified: exiles (the vulnerable parts that one disowns), managers (the parts that strive to keep the exiles contained), and the firefighters (the parts that are activated in crisis). Unlike family systems, Schwartz describes internal systems as having a preferred orchestrator: the Self, which is a compassionate, curious state of mind that most effectively balances the other parts.

Rumor Has It: The People and Their Stories

Richard Schwartz Richard Schwartz developed internal family systems in his clinical work with clients who described themselves as having various "parts." In 2000, he founded The Center for Self Leadership, which offers several levels of training to therapists and the general public.

Making Connection: The Therapeutic Relationship
Collaboration

Internal family systems therapists use a *collaborative* approach to building relationships with clients based on the assumption that all people have a highly competent Self at their core; the therapist's job is to remove the constraints to this Self (Schwartz, 1995). The therapist seeks to create a safe environment for the client to explore their internal worlds, especially their more vulnerable parts. The therapist elicits and respects clients' preferences in terms of when and how to intervene. In this approach, there is a shared responsibility for change: both the therapist and client need to do their parts to make the process work.

Therapist Parts

Like other experiential approaches, the self-of-the-therapist is a key ingredient to successful outcomes. In internal family systems, this translates to therapists having knowledge of their parts and having learned to how guide themselves with the wiser Self (see The Self below) rather than more reactive parts. Schwartz (1995) describes several common therapist parts that interfere with their effectiveness in session:

- *Striving Managers:* Therapist parts that want rapid change; these parts can become highly directive or coercive.
- *Approval-Seeking Managers:* Therapist parts that are worried about being liked and valued.
- *Pessimistic Managers:* These parts want to blame or give up when therapy does not proceed according to plan.
- *Caregiving Managers:* Therapist parts that want to overfunction on behalf of and rescue clients rather than let them engage in their own person struggle.
- *Angry Parts:* Therapist parts that feel burdened by the clients' needs.
- *Hurt Parts:* Therapist parts that overidentify with the client's pain.
- *Evaluating Parts:* Therapist parts that are critical of various therapist characteristics—weight, relationships, etc.—and then can't stand those parts in others.

Transference and Countertransference

Internal family systems therapists have adapted a version of the psychodynamic concepts of transference and countertransference to describe how clients may relate to therapists as if they were someone from their past (transference) or how the therapist may respond to the client based on another past relationship (countertransference; Schwartz, 1995). However, in internal family systems, the assumption is that it is not the client per se but a *part of the client* that is reacting to the therapist based on a past trauma: essentially that part of the self is frozen in time. The same can happen to the therapist, which is why it is important that therapists be well familiar with their own parts.

The Viewing: Case Conceptualization and Assessment

The Self

In internal family systems, the *Self* is the seat of consciousness and is considered the natural leader of the internal family system. The Self is present from birth and has a natural capacity for compassion, perspective, curiosity and acceptance (Schwartz, 1995, 2001). The Self is both a boundless state of consciousness—one similar to the observing mind in mindfulness practice (see Chapter 11)—and an active, compassionate inner leader. When the Self leads, a person experiences inner balance and harmony. At their most fundamental level, the parts serve to protect the Self; in cases of trauma or crisis, the parts will remove the Self from and from leadership to safeguard it.

Parts: Exiles, Managers, and Firefighters

When conceptualizing clients' internal systems that are out of balance, therapists identify three types of *parts* or subpersonalities: exiles, managers, and firefighters. Schwartz (1995, 2001) estimates that most clients identify 5–15 parts in the course of therapy. These parts may be conceptualized as inner people who have different ages, abilities, attitudes, and desires; together they form a family or tribe. This multiplicity of parts is considered a natural quality of the human mind; however, due to trauma or other burdens, these parts take on extreme roles and become imbalanced. All parts are considered valuable and having the potential to be constructive, even though some parts may be connected with distressing behaviors or thoughts. The goal of therapy is to help bring these parts into balance by enabling Self-leadership.

Exiles The exiles refer to the parts that are closeted away due to burdens of shame, guilt, or fear of not being lovable (Schwartz, 1995, 2001). This group is oppressed, kept from the conscious mind at any cost. Of course, the more one tries to repress these parts, the more they want to get out (systemic dynamics again). Exiles are often frozen in painful memories of the past that would be too overwhelming; thus, they must be locked up. These parts will do most anything for small amounts of love and acceptance, including enduring otherwise abusive relationships.

Managers Always with an eye to safety, the job of managers is to keep the exiles locked up so they do not escape and flood the person with strong, painful emotions. Managers work to protect the rest of the system from exiles and to protect the exiles themselves. The rigidity or severity of the managers depends on how serious the perceived threat. Thus, most people have several different managers, each with its own personal style. Some common management styles include:

- *The Controller:* A part that tries to stay in control of all situations and relationships to avoid even the slightest possible danger.
- *The Evaluator/Perfectionist:* A part that is perfectionist about appearance and behavior to ensure safety by pleasing others.
- *The Dependent One:* This part creates safety by playing a victim role to get others to help and protect them.
- *The Passive Pessimist:* This type of part creates safety by becoming apathetic and withdrawn and avoiding danger.

- *The Caretaker:* Especially encouraged in women, the caretaker part manages by ensuring everyone else is happy, often at the person's own expense.
- *The Worrier/Sentry:* This part helps manage by always being on guard for possible danger.
- *The Denier:* This part creates safety by distorting reality to protect the person from risky information or feedback.
- *Entitle One:* Most commonly seen in men, this part creates safety by taking what it wants regardless of how others are affected.

Firefighters When the managers fail to keep the exiles safely tucked away, the firefighters come to the rescue to contain the dangerous images, emotions, and sensations. These parts are automatically triggered when exiles are activated and they respond forcefully. The strategies of firefighters often involve extreme behavior, such as cutting, binging, substance abuse, binge eating, or risky sexual behavior. The person often engages in these activities, which can be simultaneously numbing and all-consuming. Firefighting activity may include numbing forms of rage, sensual indulgence, or suicidal thoughts. When they are triggered, the client is most likely to have their most severe "symptoms" and the therapist is most likely to get a late night crisis call. And, although firefighters and managers have the same general goal of keeping the exiles in the proper places, they have very different methods, often causing conflict between the two.

Polarization

Polarization describes what happens when parts assume leadership rather than the Self in order to protect the self (Schwartz, 1995). As one part shifts to an extreme position in response to trauma or other burdens, the other parts must oppose or take a counter role (as is natural in any system). Polarization tends to self-confirming: the more vulnerable one feels, the more one needs to take extreme protection measures; this continues in a vicious cycle. Polarization can lead to coalitions in which groups of parts unite in competition with others (e.g., the various manager parts work together to keep the firefighters at bay).

External Family System Imbalances

In addition to assessing internal family systems, internal family systems therapists also assess the external "real" family system. When assessing the family, the therapist considers the following:

Burdens and Constrained Development Families can experience several different types of burdens and constraints, all of which throw the system out of balance and require its members to take on extreme survival roles. First, *traumatic burdens* can take many forms, most commonly abuse or sudden loss. Second, *environmental burdens* that require the family to live up to social definitions of success to be valued can create the same type of burden that a child experiences when wanting approval from a parent. Third, *legacy burdens* refer to values and beliefs that have been handed down from one generation to another, such as shame, perfectionism, and high performance expectations. Fourth, *development burdens* refer to difficulties adjusting to the demands of family life stages and to unexpected events, such as a death or birth. Finally, *tangible burdens*, such as chronic illness, poverty, or disability, can constrain a family's functioning.

Imbalances Another assessed aspect of family function is the relative level of balance or imbalance in terms of influence (who makes decisions), resources (physical and emotional), responsibility (homemaking and providing), and boundaries (who has access to whom). These elements do not need to be equally distributed among members, but rather there needs to be an appropriate balance that is healthy for all members.

Family Harmony Internal family systems therapists also assess family harmony, as they would internal system harmony. Harmonious families are cohesive, flexible, communicative, and supportive to all of the members. In contrast, when families become burdened and imbalanced, they become polarized, often taking on the role of exile, manager, or firefighter at the family level.

Leadership Finally, just as the internal system needs an effective manager—namely the Self—the family also needs effective leadership. Effective family leadership results in a balance of resources and responsibilities; healthy boundaries between members; nurturance for all members; and a shared identity or vision. Ineffective leadership can take the form of *abdicated leadership* (the leader being overwhelmed by burdens and giving up), *polarized leadership* (the leaders become polarized), *discredited leadership* (the leader has lost respect), or *biased leadership* (the leader favors some members over others).

Parts Patterns Between Two People

When assessing the relationship between parts of individuals in relationships, internal family systems therapists identify patterns of polarization and enmeshment. Common patterns include:

- *Manager-Manager Polarizations:* Manager-manager conflicts result when two people frequently relate from one or more of their respective manager positions rather than relating from positions of Self-leadership or other parts. These polarizations are typically characterized by conflict or a cold war status.
- *Manager-Exile Polarizations:* Manger-exile polarizations occur when one person is overwhelmed by their painful exiles and the other responds with their managers to help caretake for and contain the other. Often this escalates as the first feels increasingly desperate and the second feels trapped or resentful of caretaking, at which point it may shift to a manger-manager polarization.
- *Manager-Firefighter Polarizations:* As many troubled families are dominated by extreme managers, even a small amount of firefighter activity results in immediate controlling behaviors from another's manager. This is typified by rebellious acting out of a teen and a parent frantically trying to manage with greater and greater punishments.
- *Enmeshment Between Parts:* Enmeshment most often occurs when one person turns to another to manage and sooth their exiled parts, and this can take four forms:
    - Person A tries to get B to take care of his/her exiles.
    - Person A tries to get desired qualities from B.
    - Person A turns to person B as a redeemer from their sense of worthlessness.
    - Person A fears loss or harm to B.

Targeting Change: Goal Setting
Self-leadership

Healthy functioning in internal family systems is defined by Self-leadership: the Self organizes the parts of the internal system. When Self-leadership is realized, the parts do not disappear, they just do not take on their extreme, polarized roles that lead to problematic symptoms, such as depression, self harm, anxiety, etc. The parts become less rigid and serve to work together to help problem solve or otherwise help out, each having a valuable role to play, such as identifying potential dangers and opportunities. When the parts do conflict—perhaps one advocating for safety and the other adventure—the Self is able to step in and effectively mediate. However, when there is Self-leadership, the parts are less noticeable to the individual because they operate harmoniously.

The Doing: Interventions
Introducing the Language of Parts

At some point early in therapy, the therapist will introduce "parts language" to the client (Schwartz, 1995). Most often this is done by summarizing and reflecting what the client is saying, such as "So, it sounds like there is a part of you that wants to stay in the relationship and a part that isn't so sure if it is what you really want." Most clients resonate with the idea of having different parts of themselves who are not always in agreement. Some even feel relieved that the "bad part of them" isn't all of who they are; others feel a renewed sense of hope because the problem is a bit more manageable: I just need to change a part of me. Some common ways to introduce parts language include:

- *Summarize and Reflect Conflicting Feelings Using Parts Language:* "So, there is one part of you that is angry at your parents for not giving you the approval you want; and another part of you is still working hard to win their approval."
- *Inquire About Inner Dialogue and Potential Conflicting Thoughts/Emotions.* "When you are feeling sad, what do you say to yourself inside? [client responds]. "Oh, so a part of you says X. Do you ever argue with yourself about this?

Schwartz (1995) notes that when the language is natural for the therapist, clients generally are comfortable too. However, if a therapist is too rigid, tentative, or early in introducing the language, clients may be resistant.

Once client is basically comfortable with the parts language, the therapist usually offers a more detailed explanation of the internal family systems model:

> You may have noticed me using the term "parts" to describe your feelings or thoughts. I do that because I believe that we all have many different personalities that fight inside and try to take over power from one another. When they are at war, it feels like you are out of control, doesn't it? And sometimes a part will take over and make you do or say things you don't want to do, right? Well, I know that even though those parts of you get extreme and destructive at times, they all want something good for you. I know how to help you get them to change into their preferred roles so they get along with each other and stop doing this to you. Are you interested? (p. 92).

Assessing Internal Relationships

After the parts language has been successfully introduced, the therapist then helps map the client's internal relationships. Before moving on, both the client and therapist should have a better sense of the client's inner ecology. In general, the therapist wants to track two basic types of relationships: (a) the relationship between the Self and parts, and (b) the relationship between parts.

Assessing the Relationship Between Self and Parts

The therapist can map the relationship between the Self and parts by asking the following:

- How do you feel toward this part of yourself?
- Why do you think this part does what it does?
- How often do you hear from this part of you?
- How much influence do you have over this part and how much influence does it have over you?
- How would you like this relationship to change?
- Where, when, and by whom is this part activated?
- How does it affect you when it is activated?
- Are you ever able to calm it or separate from it?

Assessing the Relationship Between Parts

- How do you think these two parts feel toward, influence, or activate each other?
- Why do they relate this way?
- How might they be helped to get along better?

—Adapted from Schwartz (1995), pp. 93–94

When working to change parts, Schwartz offers two guidelines:

1. When trying to change an extreme part, the polarized part must simultaneously be changed to maintain the system's homeostasis (e.g., the self-harming part may cause the biggest problem, but the therapist also needs to identify the part that it is trying to protect and change both together).
2. Anything you can do to help the parts trust the leadership of the Self will help (e.g., the parts trust that the Self will not let the other parts take over).

Controlling Blending

The manager and firefighter parts are organized around the fear that the exiles will take over the system, or in other words *blend* with the Self so that the Self is no longer in control. Thus, to work with exiled parts effectively, the therapist and client need to work to control blending so that managers and firefighters don't get activated. To do this, internal family systems therapists simply explain to the exiled parts that although they may have an urge to blend, it is really not in their best interest because then the Self is not able to protect them. Schwartz (1995) explains that this usually works. Once this arrangement is in place, the therapist asks the client's Self to not let the managers interfere and then approach the exiles and stop when the Self begins to feel overwhelmed. The therapist can then help the client's Self engage the childlike exiles to learn about why they are hurt and how to help them feel safer. In addition, the therapist periodically stops to check in with the managers to see if they are concerned.

In-sight and Imagery

Once clients are familiar with the parts language and express a desire to work more with parts, the therapist begins by helping them develop an image of the part, a process that is formally called in-sight (as in able to see inside). This visual image often helps the client further externalize and better relate to parts. Therapists help clients in identifying images for their parts:

> Focus on that part of you, however you experience it. If it is a feeling, focus on the feeling. If it's a thought pattern, or an inner voice, focus on that. If it seems to be a sensation located in a place in your body, focus on that place. As you focus on this part, see if an image for it comes to you. Don't try images on—just wait for the part to show itself to you .... If you are not getting any image, that's okay, because we can do this without your seeing the part (p. 114).

The Room Technique

The room technique is used when a client first encounters a part that is either overwhelming or elusive. In this technique, the therapist suggests that the client put the part in a separate room, close the door (lock if necessary), and observe the part from a window:

> Go ahead and put that Angry part into a room by itself and close the door. Then observe it through a window? What do you notice? Does it say anything?

This technique is a type of boundary making similar to what structural therapists do with families (see Chapter 7). With this safe distance, most clients begin to experience compassion, curiosity, or acceptance regarding the part in the room.

Direct Access

Similar to the empty-chair work in Gestalt therapy, direct access in internal family systems refers to the therapist directly talking with parts, often by having the client switch chairs as they speak from various parts (one chair per part). This approach is particularly useful with clients who have difficulty with imagery. With direct access the therapist interacts directly with the parts, whereas in in-sight work the therapist asks the client to interact with the part in imagery. One of the benefits is that the therapist and client often get a dramatic view of these inner parts and their relationships with one another; clients often demonstrate significant shifts in tone of voice and mannerisms. Direct access can be used in conjunction with in-sight or without-in-sight work.

Putting It All Together

Case Conceptualization Template

- *Self-Leadership*
  Describe client's capacity and frequency of Self-leadership, noting times and relationships where it is most prevalent.
- *Parts: Exiles, Managers, and Firefighters*
  - *Exiles*: Describe exiled parts
  - *Managers*: Describe manager parts, such as:
    - The controller
    - The evaluator/perfectionist
    - The dependent one
    - The passive pessimist
    - The caretaker
    - The worrier/sentry
    - The denier
    - Entitle one
  - *Firefighters*: Describe the firefighters and their crisis behaviors
- *Polarization*
  Describe patterns of polarization, noting contexts, relationships, etc.
- *External Family System Imbalances*
  - *Burdens and Constrained Development*
  - *Imbalances*
  - *Family Harmony*
  - *Leadership*
- *Parts Patterns Between Two People*
  Describe problematic parts patterns between people. Common patterns include:
  - Manager-manager polarizations
  - Manager-exile polarizations
  - Manager-firefighter polarizations
  - Enmeshment between parts

Individual Treatment Plan: Trauma

The following treatment plan can be used with a client recovering from trauma.

Treatment Plan Template for Individual

Initial Therapeutic Tasks

1. Develop working counseling relationship. *Diversity Note: Adjust to respect cultured, gendered, and other styles of relationship building and emotional expression.*
   Relationship building approach/intervention:
   a. Develop a *collaborative* relationship with clients, engaging clients from the position of Self rather than therapist part.

2. Assess individual, systemic, and broader cultural dynamics. *Diversity Note: Adjust assessment based on cultural, socioeconomic, sexual orientation, gender, and other relevant norms.*
   Assessment strategies:
   a. Identify client's specific *exile, managerial, and firefighter parts,* and identify *polarizations* in the internal system.
   b. Assess external family system *imbalances, burdens, level of harmony, and quality of leadership.*

(continued)

Initial Phase Client Goals

1. Decrease *polarity* between *firefighter* and *exile* parts to reduce self-harm and feelings of being overwhelmed.
   a. *Introduce parts language* to describe inner conflict related to trauma and crisis symptoms.
   b. *In-sight imagery* to calm *firefighters* and increase *exile* sense of safety by increasing trust in Self-leadership.

IFS Working Phase of Treatment with Individual

Working-Phase Counseling Task

1. Monitor quality of the working alliance. *Diversity Note: Attend to client's response to interventions that indicate therapist's approach not connecting with client's culturally informed meaning systems.*
   a. *Assessment Intervention:* Monitor client responses to ensure *parts language* and selected interventions are meaningful and appropriate.

Working-Phase Client Goals

1. Increase supportive contact between *Self* and *exiled* parts to reduce flashbacks, hypervigilence, and intrusive memories.
   a. *Control blending* to enable client to safely experience repressed/exiled thoughts, feelings, and memories.
   b. *Assess relationship between Self and exiles* to better understand how to create a sense of safety.

2. Decrease *polarity* of parts related to trauma experience to reduce feelings of hopelessness and powerlessness.
   a. *Direct access* of polarized parts to facilitate dialogue and understanding.
   b. *In-sight imagery with room technique* to help client engage polarized parts from a position of Self-leadership.

3. Decrease use of *manager parts* to maintain sense of balance to reduce anxiety and depressive symptoms.
   a. *Assess relationships between parts and between Self and parts* to decrease reliance on managers to contain exiled parts.
   b. *In-sight imagery* and/or *direct access* to reassure manager parts that they needn't work so hard to increase their trust of Self-leadership.

IFS Closing Phase of Treatment with Individual

Closing-Phase Counseling Task

1. Develop aftercare plan and maintain gains. *Diversity Note: Access resources in the communities of which they are a part to support them after ending therapy. Intervention:*
   a. Identify behaviors, relationships, and choices that support *Self-leadership*.

Closing-Phase Client Goals

1. Decrease *external family imbalances* [specify burdens, imbalances, harmony, or leadership issues] to reduce sense of powerlessness and increase sense of safe connection with supportive others.
   a. Couple/family sessions to explore how *parts of each interact* and to identify alternative behaviors that enable members to relate from position of *Self-leadership*.
   b. Identify *relational imbalances*, how parts relate to create these imbalances, and identify alternatives.

(continued)

2. Increase ability to maintain *Self-leadership* in work and social settings to reduce depression, anxiety, and potential for relapse.
   a. *Direct access* of parts to discuss how to allow for increased Self-leadership in work and social settings.
   b. *Assess relationship between parts and Self and parts* as applies in work and social contexts.

Couple Treatment Plan: Couple/Family Conflict

The following treatment plan can be used to develop treatment plans for couples reporting conflict and couple distress.

Treatment Plan Template for Couple/Family

Initial Therapeutic Tasks

1. Develop working counseling relationship. *Diversity Note: [Describe how you will adjust your approach to respect cultured, gendered, and other styles of relationship building and emotional expression.]*
   a. Develop a *collaborative* relationship with members of the couple/family, especially their *manager parts*, engaging them from the position of therapist's Self.

2. Assess individual, systemic, and broader cultural dynamics. *Diversity Note: [Describe how you will adjust assessment based on cultural, socioeconomic, sexual orientation, gender, and other relevant norms.]*
   a. Identify each person's specific *exile, managerial, and firefighter parts,* and identify *polarizations* that are most closely related to the couple/family conflict.
   b. Track *part sequences across partners/family members* as well as relational *imbalances, burdens, level of harmony, and quality of leadership.*

Initial Phase Client Goals

1. Decrease the impact of *couple/family burdens,* including those related to development, trauma, legacy, or constraining environments, to reduce couple conflict.
   a. Identify source of *burdens* and the associated relational *imbalances*.
   b. Introduce *parts language* to begin mapping each person's inner ecology and how they relate to relational conflict.

IFS Working Phase of Treatment with Couple/Family

Working-Phase Counseling Task

1. Monitor quality of the working alliance. *Diversity Note: [Describe how you will attend to each person's response to interventions that indicate therapist's approach not connecting with client's culturally informed meaning systems.]*
   a. Monitor each person's responses to ensure *parts language* and selected interventions are meaningful and appropriate.

Working-Phase Client Goals

1. Decrease *polarized patterns [specify type]* and *enmeshment [specify between whom]* in the relational system to reduce conflict.
   a. Describe how *inner conflicts* and parts relate to the external conflict the couple/family is experiencing and *negotiate a truce between parts*.
   b. *In-sight imagery* to reassure each party's *managers* that they can trust the therapy process and increase *exile* sense of safety to increase safety in the relationship.

(continued)

2. Increase each partner/family member's ability to engage the other from a *position of Self-leadership* to reduce conflict and increase relational satisfaction.
   a. *Control blending* to enable each person to safely experience and communicate exiled thoughts and emotions.
   b. Identify how *inner dynamics of each person* contributes to negative interaction patterns and help couple to alter these by relating from position of Self-leadership.
3. Increase effective *leadership* within the couple/family system to reduce conflict.
   a. *Direct access* of polarized parts to facilitate dialogue and understanding or how these parts have contributed to ineffective leadership.
   b. *In-sight imagery with room technique* to help engage polarized parts from a position of Self-leadership.

IFS Closing Phase of Treatment with Couple/Family

Closing-Phase Counseling Task

1. Develop aftercare plan and maintain gains. *Diversity Note: [Describe how you will access resources in the communities of which clients are a part to support them after ending therapy.]*
   a. Identify relationships, communities, and choices that support *Self-leadership*.

Closing-Phase Client Goals

1. Decrease *imbalances* in extended family and relational system [specify burdens, imbalances, harmony, or leadership issues] to reduce conflict and increase confidence.
   a. Couple/family sessions to explore how *parts of each interact* and to identify alternative behaviors that enable members to relate from position of *Self-leadership*.
   b. Identify *relational imbalances,* how parts relate to create these imbalances, and identify alternatives.
2. Increase ability to maintain *Self-leadership* in work/school and social settings to reduce conflict and increase wellness.
   a. *Direct access* of parts to discuss how to allow for increased Self-leadership in work/school and social settings.
   b. *Assess relationship between parts and Self and parts* as applies in work/school and social contexts.

Research and the Evidence Base

Similar to the Satir model (Chapter 8), symbolic experiential and internal family systems approaches find their best empirical support from the common factors research (Chapter 2), specifically support for a warm, empathic therapeutic relationship, a hallmark of all experiential therapies. In addition, symbolic experiential and internal family systems also have a rich history of case studies that form their evidence base. Symbolic experiential therapy was developed through Whitaker's innovative study of families, which he reports in numerous publications (Connell, Mitten, & Whitaker, 1993; Napier & Whitaker, 1978; Whitaker & Bumberry, 1988). Internal family systems have several reported case studies with individuals and families with sexual abuse histories and eating disorders (Holmes, 1994; Schwartz, 1987; Schwartz & Grace, 1989; Wilkins, 2007).

A two-part study examined the correlation of Self-leadership to psychological, health, and work outcomes (Dolbier, Soderstrom, & Steinhardt, 2001). Self-leadership in a sample of college students was found to be significantly correlated with psychological functioning, including effective coping abilities, greater optimism and hardiness, and better physical health; it was negatively correlated with ineffectiveness and interpersonal distrust. In a sample of corporate employees, Self-leadership was correlated with greater work satisfaction, enhanced communication, management skills, effective work relationships, less stress, and greater physical health.

Finally, Mitten and Connell (2004) designed a qualitative study to identify the core variables of symbolic experiential therapy. Based on the analysis of Whitaker's work by several experts in the approach, the following six variables were identified as characterizing symbolic experiential work:

1. *Generating an Interpersonal Set*: Shifting the focus from the identified patient to the family system.
2. *Creating a Suprasystem*: Joining the family system to create a therapeutic suprasystem.
3. *Stimulating a Symbolic Context*: Shifting focus from the content to attend to the family's use of symbolism.
4. *Activating Stress Within the Family*: Using anxiety to foster growth, especially when therapy was stuck.
5. *Creating Symbolic Experience*: Amplifying the family's symbolic experience, often by exaggerating the roles of each member.
6. *Moving Out of the System*: Once the family makes it clear that they are no longer experiencing a sense of urgency and are managing the problem, the therapist finds graceful ways to "exit" the system.

Tapestry Weaving: Working with Diverse Populations

Cultural, Ethnic, and Gender Diversity

Chapter 8 covers key issues when working with diverse populations using an experiential approach. In addition, therapists should consider unique adaptations for diverse populations when using each of the theories in this chapter. For example, symbolic-experiential therapists attend closely to symbols and language, both of which are highly informed by ethnic, cultural, religious, and linguistic backgrounds (Connell, Mitten, & Whitaker, 1993). In this approach, therapists observe "the family's process and listen for symbolic meaning embedded within the family's verbal accounts of their life together" (p. 244). Accurately interpreting any person's symbolic world is challenging; to do so with a person for another cultural background is even more difficult. When clients are not using their native or preferred language, then the translation of symbolic meaning is even more likely to fail. Thus, when using this approach with persons from another background, therapists must inquire humbly about the family's symbolic meaning to ensure understanding. Ideally, therapy should be conducted in the client's preferred language so the therapist can identify culture-specific meanings. When approached with awareness, symbolic-experiential therapy can be used to sensitively and meaningfully access diverse clients' experiences.

African Americans and Biracial Individuals

Internal family systems have been used with African Americans and biracial individuals to explore internalized parts that are informed by their experience of race. Cooper (2000) developed an eight-week group using internal family systems to help biracial individuals increase their awareness of their internal experience of biracial identity. Clients reported that the parts language of internal family systems as well as the concept of Self helped them to develop a greater sense of ease with their identity and reduced symptoms and difficulties related to solidifying such an identity.

Similarly, Wilkins (2007) reports how internal family systems can be useful to helping African American families surviving sexual abuse to address issues of power, privilege, and oppression. First, it is helpful to remember that sexual abuse is often not revealed in African American families due to fear of further stigmatization and a sense of powerlessness. Once the abuse is revealed, the parts language of internal family systems can be used to explore the multiple, contradictory parts of clients that struggle with not just the present episode of victimization but also with a residual yet clearly present cultural legacy of sexual trauma during slavery. Slaves who were often brutally sexually abused had few options for coping: suffer silently, display outward strength, accept one's helpless role, and detach from one's feelings. These strategies have been passed on over the generations, taking the form of both resilience and maladaptive behaviors. Internal family systems therapy enables clients to identify the manager parts, often taking the form of needing to have a never-failing rock-solid strength no matter the circumstances and/or a bottomless rage and anger that keeps others at a distance, and safely connect with exiled vulnerable parts that were forced to hide due to their experience of marginalization as well as the residual effects of trauma over generations. Additionally, therapists can help individuals and families explore the cultural burdens that are related to the development of particular parts.

Sexual Identity Diversity

As covered in Chapter 8, the general experiential issues for working with gay, lesbian, bisexual, transgendered, and questioning (GLBTQ) clients can apply with the approaches in this chapter. Little specifically has been written on using specifically using symbolic-experiential or internal family systems therapy with GLBTQ; however, similar principles for working with ethnic and racial diversity apply. Specifically, when using symbolic experiential approaches, therapists should be particularly mindful of the unique symbols and meanings related to marginalization and being an "invisible minority" (Bell, Özbilgin, Beauregard, & Sürgevil, 2011) that many GLBTQ clients experience. Similarly, using parts language to identify the multiple and often contradictory parts of GLBTQ individuals—including parts that would prefer to not be—can help clients develop a greater sense of understanding and acceptance of disparate parts of the Self. Similar to working with diverse clients, identifying the social burdens related to the client's sexual identity can also help clients to better understand how these parts have developed and how best to relate to them.

Online Resource

Internal Family Systems: www.selfleadership.org

References

*Asterisk indicates recommended introductory readings.

Bell, M. P., Özbilgin, M. F., Beauregard, T., & Sürgevil, O. (2011). Voice, silence, and diversity in 21st century organizations: Strategies for inclusion of gay, lesbian, bisexual, and transgender employees. *Human Resource Management, 50*(1), 131–146. doi:10.1002/hrm.20401

Connell, G. M., Mitten, T. J., & Whitaker, C. A. (1993). Reshaping family symbols: A symbolic-experiential perspective. *Journal of Marital and Family Therapy, 19*(3), 243–251. doi:10.1111/j.1752-0606.1993.tb00985.x

Connell, G., Mitten, T., & Bumberry, W. (1999). *Reshaping family relationships: The symbolic-experiential therapy of Carl Whitaker*. Philadelphia, PA: Brunner/Mazel.

Cooper, B. (2000, May). The use of internal family systems therapy to treat issues of biracial identity development. *Dissertation Abstracts International, 60,* 5767.

Dolbier, C. L., Soderstrom, M., & Steinhardt, M. A. (2001). The relationships between self-leadership and enhanced psychological, health, and work outcomes. *Journal of Psychology: Interdisciplinary and Applied, 135*(5), 469–485. doi:10.1080/00223980109603713

Holmes, T. (1994). Spirituality in systemic practice: An internal family systems perspective. *Journal of Systemic Therapies, 13*(3), 26–35.

Keith, D., Connell, G., & Connell, L. (2001). *Defiance in the family: Finding hope in therapy*. New York: Routledge.

Mitten, T. J., & Connel, G. M. (2004). The core variables of symbolic-experiential therapy: A qualitative study. *Journal of Marital and Family Therapy, 30,* 467–478.

*Napier, A. Y., & Whitaker, C. (1978). *The family crucible: The intense experience of family therapy*. New York: Harper.

*Roberto, L. G. (1991). Symbolic-experiential family therapy. In A. S. Gurman & D. P. Kniskern (Eds.), *Handbook of family therapy* (Vol. 2, pp. 444–476). New York: Brunner/Mazel.

Schwartz, R. C. (1987). Working with "internal and external" families in the treatment of bulimia. *Family Relations: An*

Interdisciplinary Journal of Applied Family Studies, 36(3), 242–245.
Schwartz, R. C. (1995). *Internal family systems therapy.* New York: Guilford.
Schwartz, R. C. (2001). *Introduction to the internal family systems model.* Fort Collins, CO: Trailhead Publications.
Schwartz, R. C., & Grace, P. (1989). The systemic treatment of bulimia. *Journal of Psychotherapy & the Family,* 6(3–4), 89–105. doi:10.1300/j085V06N01_09
Whitaker, C. A. (1975). Psychotherapy of the absurd: With a special emphasis on the psychotherapy of aggression. *Family Process,* 14, 1–15.
*Whitaker, C. A., & Bumberry, W. M. (1988). *Dancing with the family.* New York: Brunner/Mazel.
*Whitaker, C. A., & Keith, D. V. (1981). Symbolic-experiential family therapy. In A. S. Gurman & D. P. Kniskern (Eds.), *Handbook of family therapy* (pp. 187–224). New York: Brunner/Mazel.
Whitaker, C. A., & Ryan, M. C. (1989). *Midnight musings of a family therapist.* New York: Norton.
Wilkins, E. J. (2007). Using an IFS informed intervention to treat African American families surviving sexual abuse: One family's story. *Journal of Feminist Family Therapy: An International Forum,* 19(3), 37–53.

Symbolic Experiential Case Study: Cherokee Family Grieves Mother

Rick Powell (AM38), who works as a bus driver and identifies with his Cherokee heritage, has three children: Nathaniel (CM15), Aiyana (CF13), and Nadia (CF8). The family was referred to counseling after allegations of neglect arose when Nathaniel missed over thirty days of school in a three-month period. Anna Powell (AF), Rick's wife and the children's mother, died two years ago unexpectedly when she was hit by a drunk driver. Since that time, Rick has been struggling to raise the three children on his own and works two jobs in order to provide for the family. Rick is very concerned about his son's refusal to attend school and has recently discovered that he has started to drink alcohol. Rick is a recovering alcoholic and stopped drinking after his wife was killed. He is also concerned about his 13-year-old daughter who he fears may be at risk for teen pregnancy and doesn't know how to get through to her. Rick describes his youngest daughter Nadia as quiet, curious, and smart; she reminds him of his wife. Anna's parents help to care for the children; however, they both still work and are not always able to be there.

Symbolic-Experiential Case Conceptualization
Authentic Encounters and Affective System

Family members are resistant toward expressing emotions with one another, especially after the death of the mother. However, emotional exchanges between family members are quick to escalate, when the topic of their mother's death is brought up, especially between AM and CM15. When the family feels safe enough to acknowledge their loss, they express intense pain, sadness, and hurt as anger. But most of the time, they maintain the Cherokee tradition of not invoking the name of the deceased, although they do not generally practice Native American religion.

Assessing Structural Organization

Boundaries History of enmeshed boundaries in this family system have intensified since the death of Anna. There are minimal boundaries between the nuclear family and extended family. AM has attempted to create boundaries since his wife's passing by attempting to keep himself and his children away from those extended family members who have a history of alcoholism. CM15 and CF13 are angry with their father for this and feel as though he is trying to cut them off from their family.

Alliances CM15 and CF13 have become allies and rebel against AM. Both children have been parentified since death of mother and unwillingly have been thrust into a caregiver role for CF8. Almost all family members highly respect AM's grandmother, who is respected for her quiet wisdom and kindness.

Role Flexibility Since the death of AF, the kids have taken on increasingly rigid roles. CF8 takes on the role of the "good" child and often acts as the peacekeeper in the family. While CM15 and CF13 both engage in inappropriate behaviors, apparently to compensate for having to be responsible for taking care of their little sister and many of the household tasks their mother had taken care of.

Transgenerational Mandates   AM was reluctant to seek help for himself and his children due to a belief instilled in him by his family of origin that problems that arise in the family should stay within the family. It is considered taboo to seek outside help in AM's family of origin.

Work is valued over education, which has contributed to CM15's truancy.

A history of alcoholism within this family system over generations accounts for a casual attitude about drinking among extended family members and the promotion of drinking to cope with the challenges of life.

"Ghosts"   AF still has a strong presence in the family, with the older children rejecting AM's new roles since her death, in part to help cope with her loss. The children state that "everything would be different" if their mother were still alive.

Assessing Emotional Process

Tolerance of Conflict   AM struggles to tolerate the conflicts between himself and his children. He feels like a failure and is angry about his wife's sudden death; he has never fully grieved her death because he has been too busy trying keep the house running, earning money, and taking care of the children. The children also struggle to tolerate conflict and engage in maladaptive behaviors to cope with unexpressed feelings.

Conflict Resolution and Problem Solving   Prior to her death, AF was the primary problem solver and arbiter. Now, the family has become quite fragmented, unable to solve any small problem because they have not yet faced the big problem: the loss of AF.

Loyalty and Commitment   AM feels his family of origin has been unfaithful to him in their continued use of alcohol since his wife's death and that they simply fall into stereotypes of Native Americans. CM15 and CF13 remain loyal to their extended family and are committed to caring for CF8, stating: "Our mom wants us to continue caring for her."

Symbolic Process   AM states that Anna was the "glue" that held the family together. Since her death it seems as if "everything has fallen apart."

Since their mother's tragic death, alcohol has become a symbol of betrayal for AM. AM is furious that CM15 has been drinking alcohol and views this as the ultimate betrayal. AM also feels abandoned by his extended family whom he feels has not acknowledged the death of his wife in the way he would want.

Focus on Competency

Rick's commitment to become sober since the death of his wife is a significant strength. He has taken on another job to support his family financially and is committed to providing for his children. CM15 and CF13's love and dedication to taking care of CF8 is another notable family strength. Both CF13 and CF8 do well in school. The children are committed to staying connected with their extended family despite their father's efforts to keep them at a distance.

Symptom Development

The death of Anna was a huge loss to this family. The family has never really grieved the death of their mother, individually or collectively, as emotions are not typically expressed in this family, especially without the mother's gentle peacemaker influence. Each individual family member has developed a maladaptive way of coping with feelings of loss. As there is a history of alcoholism in this family, the tragic way in which their mother was killed has forced this family to examine the role alcohol plays in their lives.

Symbolic-Experiential Treatment Plan

Symbolic-Experiential Initial Phase for Family

Initial Phase Therapeutic Tasks

1. *Develop working counseling relationship.* Diversity Note: Be considerate of family's level of comfort with emotional expression with respect to their Cherokee culture; respect a desire for less eye contact and native spiritual beliefs.
   a. Use *therapist's authentic self* to form *mutual relationships* that stimulates each party's growth.

2. *Assess individual, systemic, and broader cultural dynamics.* Diversity Note: Adapt assessment of emotions, emotional expression, and authenticity to family and cultural norms. Assess role of tribe and elders in family system. Consider any tribal affiliation, language(s) spoken, levels of self-identity, where individual family members grew up, and any current relationship to a tribe or culture.
   a. Assess emotional processes, *including level of individuation, ability to tolerate conflict, problem solving, sexuality, loyalty, empathy, playfulness, cultural adaptations and symbolic processes.*
   b. Assess family dynamics, *including boundary within family, boundaries with external systems, role flexibility, alliances, gender-role flexibility, transgenerational mandates, and family "ghosts"* especially as related to death of AF.

Initial Phase Client Goals

1. Increase tolerance for and expression of *authentic emotion* while respecting cultural traditions of handling emotions privately to reduce relational conflict and facilitate grieving process.
   a. Use of *play and "craziness"* to invite family into present moment experiencing of authentic emotion while also acknowledging their grief.
   b. *Reframe Symptoms:* Reframe CM15's alcohol use and truancy and CF13's promiscuity as their way of helping distract their family/father from their overwhelming grief; thank children for their sacrifice.

Symbolic-Experiential Working Phase of Treatment for Family

Working-Phase Counseling Task

1. Monitor quality of the working alliance. *Diversity Note:* Attend to and respect client's Cherokee norms for expression of emotion and respect for elders/professionals; allow sufficient space and create sense of safety.
   a. Monitor family's response to *therapist's authentic use of self* and regularly discuss.

Working-Phase Client Goals

1. Reduce *rigid relational patterns, clarify boundaries and increase role flexibility* to reduce conflict.
   a. *Affective confrontation* of rigid patterns using humor, stories, and direct confrontation, with particular attention given to the subject of Anna's passing which the family has been avoiding. Integrate native stories as much as possible; inquire about their favorite tales from grandmother.
   b. *Augment despair and amplify deviation* by exaggerating CM15 and CF13's problem behaviors to the point of absurdity.

2. Increase capacity to *tolerate authentic emotional expression as well as difference and conflict* within system to reduce conflict and facilitate grieving process.

(continued)

a. *Here-and-now experiencing* of grief, conflict and difference to reduce fears related to experiencing intense emotions; private sessions for father and children to discuss grief.
   b. *Play and "craziness"* that allows them to safely experience difficult or taboo emotions.

3. Increase the breadth and flexibility of *family's symbolic world* to reduce conflict by expanding possibilities for relating.
   a. *Absurd fantasy alternatives* to help expand view of death and their possibilities for continuing as a "real" family.
   b. *Stories and metaphors* that draw from native and family traditions to expand sense of possibilities.

Symbolic-Experiential Closing Phase of Treatment for Family

Closing-Phase Counseling Task

1. Develop aftercare plan and maintain gains. *Diversity Note: Possibly access culturally congruent grief groups, including more "mainstream" resources as well as tribal resources. Provide culturally appropriate resources for overcoming and/or coping with alcoholism within families.*
   a. *Challenge fears of relapse* and happy-ever-after hopes and identify *playful options* to responding to relapse.

Closing-Phase Client Goals

1. Increase *family cohesion* and *authentic expression of self within system* to facilitate grieving and increase sense of well-being.
   a. *Separate personal from interpersonal distress* to help client more authentically relate.
   b. *Confront* fears of intimacy, worthiness, and loss to enable clients to more authentically relate and to accept the reality of their loss as well as possibilities for moving forward.

2. Increase each member's ability to *experience present moment emotions* and *authentically express him/herself* to facilitate grief and increase sense of well-being.
   a. *Here-and-now experiencing* of both positive and negative emotions, especially related to loss of mother; adapt for culture.
   b. Introduce client-initiated *play and creativity* into daily life activities to create new sense of family identity and ritual.

CHAPTER 10

Intergenerational and Psychoanalytic Family Therapies

"Bowen theory is really not about families per se, but about life."
—Friedman, 1991, p. 134

Lay of the Land

Although distinct from each other, Bowenian intergenerational therapy and psychoanalytic family therapy share the common roots of (a) psychoanalytic theory and (b) systemic theory. A psychoanalytically trained psychiatrist, Bowen (1985) developed a highly influential and unique approach to therapy that is called Bowen intergenerational therapy. Drawing heavily from object relations theory, psychoanalytic and psychodynamic family therapies have developed several unique approaches, including *object relations family therapy* (Scharff & Scharff, 1987), *family-of-origin therapy* (Framo, 1992), and *contextual therapy* (Boszormenyi-Nagy & Krasner, 1986). These therapies share several key concepts and practices:

- Examining a client's early relationships to understand present functioning
- Tracing transgenerational and extended family dynamics to understand a client's complaints
- Promoting insight into extended family dynamics to facilitate change
- Identifying and altering destructive beliefs and patterns of behavior that were learned early in life in one's family of origin

Bowen Intergenerational Therapy

In a Nutshell: The Least You Need to Know

Bowen intergenerational theory is more about the nature of being human than it is about families or family therapy (Friedman, 1991). The Bowen approach requires therapists to work from a broad perspective that considers the evolution of the human species and the characteristics of all living systems. Therapists use this broad perspective to conceptualize client problems and then rely primarily on the therapist's use of self to effect change. As part of this broad perspective, therapists routinely consider the *three-generational emotional process* to better understand the current presenting symptoms. The process of therapy involves increasing clients' awareness of how their current behavior is connected to multigenerational processes and the resulting family dynamics. The therapist's primary tool for promoting client change is the therapist's personal level of *differentiation*, the ability to distinguish self from other and manage interpersonal anxiety.

The Juice: Significant Contributions to the Field

If you remember a couple of things from this chapter, they should be:

Differentiation

Differentiation is one of the most useful concepts for understanding interpersonal relationships, although it can be difficult to grasp at first (Friedman, 1991). An *emotional or affective* concept, differentiation refers to a person's ability to separate intrapersonal and interpersonal distress:

- *Intrapersonal*: Separate thoughts from feelings in order to *respond* rather than *react*.
- *Interpersonal*: Know where oneself ends and another begins without loss of self.

Bowen (1985) also described differentiation as the ability to balance two life forces: the need for *togetherness* and the need for *autonomy*. Differentiation is conceptualized on a *continuum* (Bowen, 1985): a person is more or less differentiated rather than differentiated or not differentiated. Becoming more differentiated is a lifelong journey that is colloquially referred to as "maturity" in the broadest sense.

A person who is more differentiated is better able to handle the ups and downs of life and, more importantly, the vicissitudes of intimate relationships. The ability to clearly separate thoughts from feelings and self from others allows one to more successfully negotiate the tension and challenges that come with increasing levels of intimacy. For example, when one's partner expresses disapproval or disinterest, this does not cause a differentiated person's world to collapse or inspire hostility. Of course, feelings may be hurt, and the person experiences that pain. However, he/she doesn't immediately *act on* or *act out* that pain. Differentiated people are able to reflect on the pain: clearly separate out what is their part and what is their partner's part and identify a respectful way to move forward. In contrast, less differentiated people feel compelled to immediately react and express their feelings before thinking or reflecting on what belongs to whom in the situation. Partners with greater levels of differentiation are able to tolerate difference between themselves and others, allowing for greater freedom and acceptance in all relationships.

Because differentiated people do not immediately react in emotional situations, a common misunderstanding is that differentiation implies lack of emotion or emotional expression (Friedman, 1991). In reality, highly differentiated people are actually able to engage *more* difficult and intense emotions because they do not overreact and instead can thoughtfully reflect on and tolerate the ambiguity of their emotional lives.

It can be difficult to assess a client's level of differentiation because it is expressed differently depending on the person's culture, gender, age, and personality (Bowen, 1985). For example, to the untrained eye, emotionally expressive cultures and genders may look more undifferentiated, and emotionally restricted people and cultures may appear more differentiated. However, emotional coolness often is a result of *emotional cutoff* (see later section on Emotional Cutoff), which is how a less differentiated person manages intense emotions. Therapists need to assess the actual functioning intrapersonally (ability to separate thought from feeling) and interpersonally (ability to separate self from other) to sift through the diverse expressions of differentiation. In the case study at the end of the chapter, the therapist helps a mother and her 17-year-old daughter, her youngest child, increase their level of differentiation as the family prepares to launch their last child.

Genograms

The genogram has become one of the most commonly used family assessment instruments (McGoldrick, Gerson, & Petry, 2008). At its most basic level, a genogram is a type of family tree or genealogy that specifically maps key multigenerational processes that illuminate for both therapist and client the emotional dynamics that contribute to the reported symptoms. The case study at the end of the chapter includes an example for you to get a sense of what it looks like.

New therapists are often reluctant to do genograms. When I ask students to do their own, most are enthusiastic. However, when I ask them to do one with a client, most are reluctant. They may say, "I don't have time" or "I don't think these clients are the type who would want to do a genogram." Yet after completing their first genogram with a client, they almost always come out saying, "That was more helpful than I thought it was going to be." Especially for newer therapists—and even for seasoned clinicians—genograms are always helpful in some way. Although originally developed for the intergenerational work in Bowen's approach, the genogram is so universally helpful that many therapists from other schools adapt it for their approach, creating solution-focused

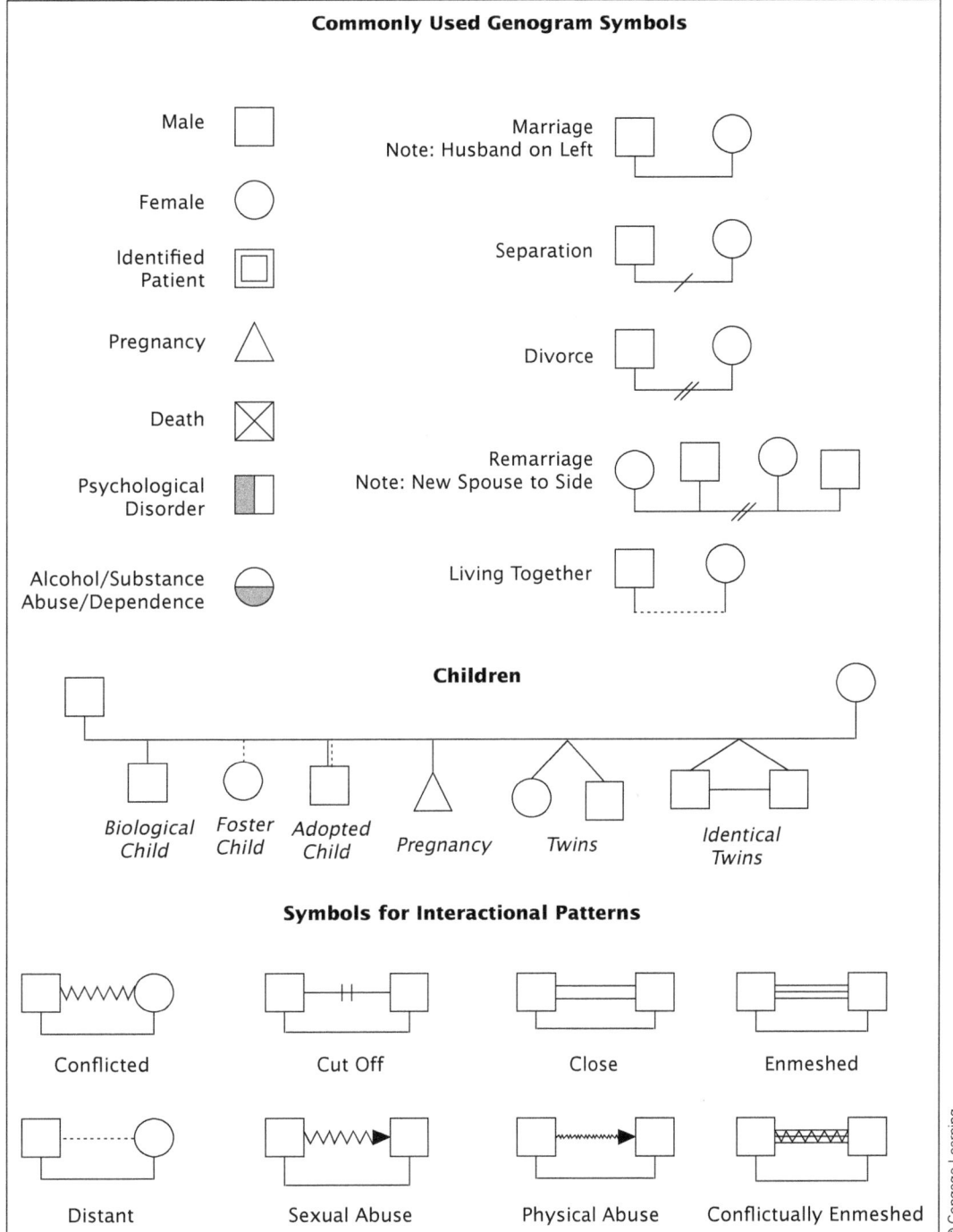

genograms (Kuehl, 1995) or culturally focused genograms (Hardy & Laszloffy, 1995; Rubalcava & Waldman, 2004).

The genogram is simultaneously (a) an assessment instrument and (b) an intervention, especially in the hands of an intergenerational therapist. As an assessment instrument, the genogram helps the therapist identify intergenerational patterns that surround the problem, such as patterns of parenting, managing conflict, and balancing autonomy with togetherness. As an intervention, genograms can help clients see their patterns more clearly and how they may be living out family patterns, rules, and legacies without conscious awareness. As a trainee, I worked with one client who had never spoken to her parents about how her grandfather had sexually abused her and had no intention of doing so because she believed it would tear the family apart. This changed the day we constructed her genogram. I had her color in each person she knew he had also abused. When she was done, the three-generation genogram had over 12 victims colored in red; she went home and spoke to her mother that night and began a multigenerational process of healing for her family.

To create a genogram, most therapists use the conventions created by McGoldrick et al. (2008). The following figure depicts the symbols commonly used in creating a genogram, with specific symbols for gender, marriage, divorce, partnership without marriage, children, conflict, distance, household membership, etc.

In addition to basic family configuration, genograms often also track key multigenerational patterns, including:

- Family strengths and resources
- Substance and alcohol abuse and dependence
- Sexual, physical, and emotional abuse
- Personal qualities and/or family roles; complementary roles (e.g., black sheep, rebellious one, overachiever/underachiever)
- Physical and mental health issues (e.g., diabetes, cancer, depression, psychosis)
- Historical incidents of the presenting problem, either with the same people or how other generations and family members have managed this problem

Rumor Has It: The People and Their Stories

Murray Bowen A psychoanalytically trained psychiatrist, Bowen (1966, 1972, 1976, 1985) began working with people diagnosed with schizophrenia at the Menninger Clinic in the 1940s and continued his research in the 1950s at the National Institute for Mental Health (NIMH), where he hospitalized entire families with schizophrenic members to study their emotional processes. He then spent the next 30 years at Georgetown University developing one of the most influential theories of family and natural systems, which has influenced generations of family therapists.

Georgetown Family Center: Michael Kerr A longtime student of Bowen, Michael Kerr has also been one of his most influential students and has served as director of the Georgetown Family Center where Bowen refined his clinical approach.

The Center for Family Learning: Philip Guerin and Thomas Fogarty Guerin and Fogarty cofounded the Center for Family Learning in New York, one of the premier training centers for family therapy. Both Guerin and Fogarty have written extensively on the clinical applications of Bowen's model.

Monica McGoldrick and Betty Carter Betty Carter and Monica McGoldrick (1999) used Bowen's theory to develop their highly influential model of the *family life cycle,* which uses the Bowenian concept of balancing the need for togetherness and independence to understand how families develop. McGoldrick's work with genograms is the definitive work on this tool subject (McGoldrick, Gerson, & Petry, 2008).

David Schnarch Grounded in Bowen's intergenerational approach, Schnarch developed a unique approach to working with couples, the sexual crucible model, which is designed to increase a couple's capacity for intimacy by increasing their level of differentiation. One of the hallmarks of this approach is harnessing the intensity in the couple's sexual relationship to promote the differentiation process.

The Big Picture: Overview of Treatment

Much like other approaches that have psychodynamic roots, intergenerational therapy is a *process*-oriented therapy that relies heavily on the *self-of-the-therapist*, most specifically the therapist's level of differentiation, to promote client change (Kerr & Bowen, 1988). This therapy does not emphasize techniques and interventions. Instead, therapists use genograms and assessment to promote insight and then intervene as differentiated persons. For example, when one partner tries to get the therapist to take his/her side in an argument, the therapist responds by simultaneously modeling differentiation and gently promoting it in the couple. By refusing to take sides and also helping the couple tolerate their resulting anxiety (their problem is still not fixed, and neither partner has been "validated" by the therapist), the therapist creates a situation in which the couple can increase their level of differentiation: they can use self-validation to soothe their feelings and learn how to tolerate the tension of difference between them. Change is achieved through alternately using insight and the therapeutic relationship to increase clients' levels of differentiation and tolerance for anxiety and ambiguity.

Making Connection: The Therapeutic Relationship
Differentiation and the Emotional Being of the Therapist

More than in any other family therapy approach, in intergenerational therapy the therapist's level of differentiation (Bowen, 1985; Kerr & Bowen, 1988) and emotional being (Friedman, 1991) are central to the change process. Intergenerational therapists focus on developing a therapeutic relationship that encourages all parties to further their differentiation process: "*the differentiation of the therapist is technique*" (Friedman, 1991, p. 138; italics in original). Intergenerational therapists believe that clients can only differentiate as much as their therapists have differentiated (Bowen, 1985). For this reason, the therapist's level of differentiation is often the focus of supervision early in training, and therapists are expected to continually monitor and develop themselves so that they can be of maximum assistance to their clients. Bowen therapists assert that the theory cannot be learned through books (such as this one) but can only be learned through a relationship with a supervisor or teacher who uses these ideas to interact with the student (Friedman, 1991).

A Nonanxious Presence

The greater a therapist's level of differentiation, the more the therapist can maintain a nonanxious presence with clients (Kerr & Bowen, 1988). This is not a cold, detached stance but rather an emotionally engaged stance that is *nonreactive*, meaning that the therapist does not react to attacks, "bad" news, and so forth without careful reflection. The therapist does not rush in to rescue clients from anxiety every time they feel overwhelmed by anger, sadness, or another strong emotion; instead, the therapist calmly wades right into the muck the client is trying to avoid and guides the client through the process of separating self from other and thought from feelings (Friedman, 1991). The therapist's calm center is used to help clients move through the differentiation process in a safe, contained environment in which differentiation is modeled. When clients are upset, the "easiest" thing to do is to soothe and calm their anxieties, fears, and strong emotions; this makes everyone calmer sooner, but nothing is learned. The intergenerational therapist instead shepherds a more difficult process of slowly coaching clients through that which they fear or detest in order to facilitate growth. For example, in the case study at the end of the chapter, the therapist uses nonanxious presence to help the family learn new ways to negotiate the constant conflict between the mother and teenage daughter.

The Viewing: Case Conceptualization and Assessment

Viewing is the primary "intervention" in intergenerational therapy because the approach's effectiveness relies on the therapist's ability to accurately assess the family dynamics and thereby guide the healing process (Bowen, 1985). Although this is true with all therapies, it is truer with intergenerational therapies because the therapist's level of differentiation is critical to the ability to accurately "see" what is going on.

Emotional Systems

Bowen viewed families, organizations, and clubs as emotional systems that have the same processes as those found in all natural systems: "Bowen has constantly emphasized over the years that we have more in common with other forms of protoplasm (i.e., life) than we differ from them" (Friedman, 1991, p. 135). He viewed humans as part of an *evolutionary emotional process* that goes back to the first cell that had a nucleus and was able to differentiate its functions from other cells (i.e., human life begins with one cell that divides to create new cells, which then differentiate to create the different systems and structures of the body: blood, muscle, neurons, etc.). This process of differentiating yet remaining part of a single living organism (system) is a primary organizing concept in Bowen's work, and the family's emotional processes are viewed as an extension (not just a metaphor) of the differentiation process of cells. Thus Bowen's theory of natural systems focuses on the relationship between the human species and all life past and present.

Of particular interest in family therapy are natural systems that have developed emotional interdependence (e.g., flocks of birds, herds of cattle, and human families; Friedman, 1991). The resulting system or emotional field profoundly influences all of its members, defining what is valued and what is not. When a family lacks sufficient differentiation, it may become emotionally fused, an undifferentiated family "ego mass." Intergenerational therapists focus squarely on a family's unique emotional system rather than on environmental or general cultural factors, and they seek to identify the rules that structure the particular system.

This approach is similar to other systemic conceptualizations of the family as a single organism or system; however, Bowen emphasizes that it is fundamentally an *emotional system*. Because this system has significant impact on a person's behavior, emotions, and symptoms, one must always assess this context to understand a person's problems.

Chronic Anxiety

Bowen viewed chronic anxiety as a biological phenomenon that is present in all natural systems. Chronic anxiety involves automatic physical and emotional reactions that are not mediated through conscious, logical processes (Friedman, 1991). Families exhibit chronic anxiety in their responses to crises, loss, conflict, and difficulties. The process of differentiation creates a clearheadedness that allows individuals and families to reduce the reactivity and anxiety associated with survival in natural systems and instead make conscious choices about how to respond. For example, chronic anxiety in a family may result from a mother feeling guilty about a child's lack of success, in which case it is the therapist's job to help the mother increase her level of differentiation so that she can respond to the child's situation from a clear, reasoned position rather than with a blind emotional reactivity that rarely helps the situation.

The Multigenerational Transmission Process

The multigenerational transmission process is based on the premise that emotional processes from prior generations are present and "alive" in the current family emotional system (Friedman, 1991). In this process, children may emerge with higher, equal, or lower levels of differentiation than their parents (Bowen, 1985). Families with severe emotional problems result from a multigenerational process in which the level of differentiation has become lower and lower with each generation. Bowen's approach is designed to help an individual create enough distance from these processes to comprehend the more universal processes that shape human relationships and individual identities (Friedman, 1991).

Multigenerational Patterns

Intergenerational therapists assess multigenerational patterns, specifically those related to the presenting problem. Using a genogram or oral interview, the therapist identifies patterns of depression, substance use, anger, conflict, parent-child relationships, couple relationships, or whatever issues are most salient for the client. The therapist then identifies how the current situation fits with these patterns. Is the client replicating or rebelling against the pattern? How has the pattern evolved with this generation? The therapist thereby gains greater clarity into the dynamics that are feeding the problem. In cases of immigration, such as that at the end of this chapter, the historic family patterns may change because of different cultural contexts (e.g., the family attempts or is forced to blend and adapt), may be rigidly the same (e.g., the family wants to adhere to traditions), or may be radically different (e.g., the family wants to "break" from the past). In the case study at the end of the chapter, the family has intergenerational patterns that include a rebellious child, a family peacemaker, and a martyr.

Level of Differentiation (see Juice)

When differentiation is used as part of case conceptualization, the therapist assesses the client's level of differentiation along a continuum, which Bowen developed into a differentiation scale that ranges from 1 to 100, with lower levels of differentiation represented by lower numbers (Bowen, 1985). Bowen maintained that people rarely reach higher than 70 on this scale.

Although there are pen-and-paper measures such as the Chabot Emotional Differentiation Scale (Licht & Chabot, 2006), most therapists simply note patterns of where and how a person is able or unable to separate self from other and thought from emotion. What is most useful for treatment is not some overall score or general assessment of differentiation, but the specific places where clients need to increase their level of differentiation to resolve the presenting problem. For example, a couple may need to increase their ability to differentiate self from other in the area of sex so that they can create a better sexual relationship that allows each person to have preferences, discuss them, and find ways to honor these preferences without becoming emotionally overwhelmed.

Emotional Triangles

Bowen identified triangles as one of the most important dynamics to assess because they are the basic building block of families (Bowen, 1985; Friedman, 1991; Kerr & Bowen, 1988). A triangle is a process in which a dyad draws in a third person (or some thing, topic, or activity) to stabilize the primary dyad, especially when there is tension in the dyad. Because triangles use a third person or topic to alleviate tension, the more you try to change the relationship with the third entity, the more you ironically reinforce the aspects you want to change. Thus, therapists assess triangles to identify the primary relationship that needs to be targeted for change.

Bowen maintained that triangulation is a fundamental process in natural systems (Bowen, 1985). Everyone triangulates to some degree: going down the hall to complain about your boss or coworker is triangulation. However, when this becomes the primary means for dealing with dyadic tension and the members of the dyad never actually resolve the tension themselves, then pathological patterns emerge. The more rigid the triangle, the greater the problems.

The classic family example of a triangle is a mother who becomes overinvolved with her children to reduce unresolved tension in the marriage; the case study at the end of the chapter includes such a situation. This overinvolvement can take the form of positive interactions (overinvolvement in school and social activities, emotional intimacy, constant errands or time devoted to the child) or negative interactions (nagging and worrying about the child). Another common form of triangulation is seen in divorced families, in which both parents often triangulate the child, trying to convince the child to take their side against the other parent. Triangulation can also involve using alcohol or drugs to create dyadic stability, complaining or siding with friends or family of origin against one's spouse, or two siblings siding against a third.

The Family Projection Process

The family projection process describes how parents "project" their immaturity onto one or more children (Bowen, 1985), causing decreased differentiation in subsequent generations. The most common pattern is for a mother to project her anxiety onto one child, focusing all her attention on this child to soothe her anxiety, perhaps becoming overly invested in the child's academic or sporting activities. The child or children who are the focus of the parent's anxiety will be less differentiated than the siblings who are not involved in this projection process.

Emotional Cutoff

A particularly important process to assess is emotional cutoff, which refers to situations in which a person no longer emotionally engages with another in order to manage anxiety; this usually occurs between children and parents. Emotional cutoff can take the form of no longer seeing or speaking to the other or, alternatively, being willing to be at the same family event with virtually no interaction. Often people who display cutoff from their family believe that doing so is a sign of mental health (e.g., "I have set good boundaries") or even a sign of superiority (e.g., "It makes no sense for me to spend time with *that type* of person"). They may even report that this solution helps them manage their emotional reactivity. However, cutoff is almost always a sign of lower levels of differentiation (Bowen, 1985). Essentially, the person is so emotionally fused with the other that he/she must physically separate to be comfortable. The higher a person's level of differentiation, the less need there is for emotional cutoff. This does not mean that a highly differentiated person does not establish boundaries. However, when differentiated people set boundaries and limit contact with family, they do so in a way that is respectful and preserves emotional connection, and not out of emotional reactivity (e.g., after an argument).

Emotional cutoff requires a little more attention in assessment because it can "throw off" an overall assessment of differentiation and family dynamics. People who emotionally cut themselves off as a means of coping often appear more differentiated than they are; it may also be harder to detect certain family patterns because in some cases the client "forgets" or honestly does not know the family history. However, at some times and in certain families, more cutoff is necessary because of extreme patterns of verbal, emotional, or childhood abuse. In such cases, where contact is not appropriate or possible, the therapist still needs to assess the *emotional* part of the cutoff. The more people can stay emotionally engaged (e.g., have empathy and cognitive understanding of the relational dynamics) without harboring anger, resentment, or fear, the healthier they will be, and this should be a therapeutic goal.

Sibling Position

Intergenerational therapists also look at sibling position as an indicator of the family's level of differentiation; all things being equal, the more the family members exhibit the expected characteristics of their sibling position, the higher the level of differentiation (Bowen, 1985; Kerr & Bowen, 1988). The more intense the family projection process is on a child, the more that child will exhibit characteristics of an infantile younger child. The roles associated with sibling positions are informed by a person's cultural background, with immigrants generally adhering to more traditional standards than later generations. Most often, older children identify with responsibility and authority, and later-born children respond to this domination by identifying with underdogs and questioning the status quo. The youngest child is generally the most likely to avoid responsibility in favor of freedom.

Societal Regression

When a society experiences sustained chronic anxiety because of war, natural disaster, economic pressures, and other traumas, it responds with emotionally based reactive decisions rather than rational decisions (Bowen, 1985) and regresses to lower levels of functioning, just like families. These Band-Aid solutions to social problems generate a vicious cycle of increased problems and symptoms. Societies can go through cycles in which their level of differentiation rises and falls.

Targeting Change: Goal Setting

Two Basic Goals

Like any theory with a definition of health, intergenerational therapy has clearly defined long-term therapeutic goals that can be used with all clients:

1. To increase each person's level of differentiation (in specific contexts)
2. To decrease emotional reactivity to chronic anxiety in the system

Increasing Differentiation Increasing differentiation is a general goal that should be operationally defined for each client. For example, "increase AF's and AM's level of differentiation in the marital relationship by increasing the tolerance of difference while increasing intimacy" is a better goal than "increase differentiation."

Decreasing Emotional Reactivity to Chronic Anxiety Decreasing anxiety and emotional reactivity is closely correlated with the increasing differentiation. As *differentiation increases, anxiety decreases*. Nonetheless, it can be helpful to include these as separate goals to break the process down into smaller steps. Decreasing anxiety generally precedes increasing differentiation and therefore may be included in the working rather than the termination phase of therapy. As with the general goal of increasing differentiation, it is clinically helpful to tailor this to an individual client. Rather than stating the general goal of "decrease anxiety," which can easily be confused with treating an anxiety disorder (as may or may not be the case), a more useful clinical goal would address a client's specific dynamic: "decrease emotional reactivity to child's defiance" or "decrease emotional reactivity to partner in conversations about division of chores and parenting."

The Doing: Interventions

Theory Versus Technique

The primary "technique" in Bowen intergenerational theory is the therapist's ability to embody the theory. The premise is that if therapists understand Bowen's theory of natural systems and work on their personal level of differentiation, they will naturally interact with clients in a way that promotes the clients' level of differentiation (Friedman, 1991). Thus understanding—"living" the theory—is the primary technique for facilitating client change.

Process Questions

Intergenerational therapists' embodiment of the theory most frequently expresses itself through *process questions*, questions that help clients see the systemic process or the dynamics that they are enacting. For example, a therapist can use process questions to help clients see how the conflict they are experiencing with their spouse is related to patterns they observed in the parents' relationship: "How do the struggles you are experiencing with your spouse now compare with those of each of your parents? Similar or different? Is the role you are playing now similar to that of one of your parents in their marriage? Is it similar to the type of conflict you had with your parents when you were younger? Who are you most like? Least like?" These questions are generated naturally from the therapist's use of the theory to conceptualize the client's situation. The therapist in the case study at the end of the chapter uses process questions throughout to help the parents develop a stronger relationship, increase the independence of the daughter, and decrease the mother's overinvolvement in the teen's life.

Encouraging Differentiation of Self

According to Bowenian theory, families naturally tend toward togetherness and relationship as part of survival. Thus therapeutic interventions generally target the counterbalancing force of differentiation (Friedman, 1991) by encouraging clients to use *"I" positions* to maintain individual opinions and mood states while in relationships with others. For example, if spouses are over-reactive to the moods of the other, every time one person is in an angry or unhappy mood, the other feels there is no other choice but to also be in that mood state. Therapists promote differentiation by coaching the second spouse to maintain his/her emotional state without undue influence from the other.

Genograms

The genogram is used both as an assessment tool and an intervention (McGoldrick, Gerson, & Petry, 2008). As an intervention, the genogram identifies not only problematic intergeneration patterns but also alternative ways for relating and handling problems. For example, if a person comes from a family in which one or more children in each generation have strongly rebelled against their parents, the genogram can be used to identify this pattern, note exceptions in the larger family, and identify ways to prevent or intervene on this dynamic. The genogram's visual depiction of the pattern across generations often inspires a greater sense of urgency and commitment to change than when the dynamics are only discussed in session. Constructing the genogram often generates a much greater sense of urgency and willingness to take action compared to relying strictly on process questions and a discussion of the dynamics. Chapter 17 includes a brief description of how to construct a genogram and use it in session, but I suggest you use McGoldrick, Gerson, and Petry (2008) as your primarily genogram reference.

Detriangulation

Detriangulation involves the therapist maintaining therapeutic neutrality (differentiation) in order to interrupt a client's attempt to involve the therapist or someone else in a triangle (Friedman, 1991). Whether working with an individual, couple, or family, most therapists at some point will be "invited" by clients to triangulate with them against a third party who may or may not be present in the room. When this occurs, the therapist "detriangulates" by refusing to take a side, whether literally or more subtly. For example, if a client says, "Don't you think it is inappropriate for a child to talk back?" or "Isn't it inappropriate for a husband to go to lunch with a single woman who is attracted to him?," the quickest way to relieve the client's anxiety is to agree: this makes the relationship between the client and therapist comfortable, allowing the client to immediately feel "better," "understood," and "empathized with." However, by validating the client's position and taking the client's side against another, the therapist undermines the long-term goal of promoting differentiation. Thus, if therapy becomes "stuck," therapists must first examine their role in a potential triangle (Friedman, 1991).

Rather than take a side, the therapist invites clients to validate *themselves*, examine their own part in the problem dynamic, and take responsibility for their needs and wants. There is often significant confusion in the therapeutic community about "validating" a client's feelings. *Validation* implies approval; however, approval from the therapist undermines a client's sense of autonomy. Intergenerational therapists emphasize that when therapists "validate" by saying "It is normal to feel this way," "It sounds like he really hurt you," or in some way imply "You are entitled to feel this way," they close down the opportunity for differentiation. Instead, clients are coached to approve or disapprove of their own thoughts and feelings and then take responsibility and action as needed.

Relational Experiments

Relational experiments are behavioral homework assignments that are designed to reveal and change unproductive relational processes in families (Guerin, Fogarty, Fay, & Kautto, 1996). These experiments interrupt triangulation processes by increasing direct communication between a dyad or by reversing pursuer/distancer dynamics that are fueled by lack of differentiation.

Going Home Again

Most adults are familiar with this paradox: you seem to be a balanced person who can manage a demanding career, an educational program, and a complex household; yet, when you go home to visit family for the holidays, you find yourself suddenly acting like a teenager—or worse. Intergenerational therapists see this difference in functioning as the result of unresolved issues with the family of origin that can be improved by increasing differentiation. Even if you cannot change a parent's critical comments or a sibling's arrogance, you can be in the presence of these "old irritants" and not regress to past behaviors but instead keep a clear sense of self. As clients' level of differentiation grows, they are able to maintain a stronger and clearer sense of self in the family nuclear system. The technique of "going home" refers to when therapists encourage clients to interact with family members while maintaining a clearer boundary between self and other and to practice and/or experience the reduced emotional reactivity that characterizes increases in differentiation (Friedman, 1991).

Interventions for Special Populations
The Sexual Crucible Model

One of the most influential applications of Bowen intergenerational theory is the sexual crucible model developed by David Schnarch (1991). The model proposes that marriage functions as a "crucible," a vessel that physically contains a volatile transformational process. In the case of marriage, the therapist achieves transformation by helping both partners differentiate (or more simply, forcing them to "grow up"). As with all crucibles, the contents of marriage must be contained because they are unstable and explosive.

Schnarch views sexual and emotional intimacy as inherently intertwined in the process of differentiation. He directs partners to take responsibility for their individual needs rather than demand that the other change to accommodate their needs, wants, and desires. To remain calm, each person learns to self-soothe rather than demand that the other change. Schnarch also includes exercises like "hugging to relax" in which he helps couples develop a greater sense of physical intimacy and increased comfort with being "seen" by the other. He has developed this model for therapists to use with clients and has also made it accessible to general audiences (Schnarch, 1998).

Schnarch has developed a comprehensive and detailed model for helping couples create the type of relationship most couples today expect: a harmonious balance of emotional, sexual, intellectual, professional, financial, parenting, household, health, and social partnerships. However, Schnarch points out that this multifaceted intimacy has never been the norm in human relationships. His model is most appropriate for psychologically minded clients who are motivated to increase intimacy.

Putting It All Together

Case Conceptualization Template

- *Chronic Anxiety*
  Describe patterns of chronic anxiety within the family: each person's role, how it relates to symptoms, etc.
- *Multigenerational Patterns*
  Based on genogram, identify multigenerational patterns, attending to the following themes:
  - Family strengths
  - Substance/alcohol abuse
  - Sexual/physical/emotional abuse
  - Parent/child relations
  - Physical/mental disorders
  - Historical incidents of presenting problem
  - Roles within the family: martyr, hero, rebel, helpless one, etc.
- *The Multigenerational Transmission Process*
  Describe multigenerational transmission of functioning, attending to acculturation issues, residual effects of trauma and loss, significant legacies, etc.
- *Level of Differentiation*
  Describe each person's relative level of differentiation and provide examples for how it is expressed.

- *Emotional Triangles*
  Identify patterns of triangulation in the family.
- *The Family Projection Process*
  Describe patterns of parents projecting their anxiety onto one more child who becomes the focus of attention.
- *Emotional Cutoff*
  Describe any cutoffs in the family.
- *Sibling Position*
  Describe sibling position patterns that seem to be relevant for the family.

Treatment Plan Template for Individual

Intergenerational Initial Phase of Treatment with Individual

Initial Phase Counseling Tasks

1. Develop working counseling relationship. *Diversity Note: [Describe how you will adjust to respect cultured, gendered, and other styles of relationship building and emotional expression.]*
   a. Engage with client from a differentiated position conveying a nonanxious presence.

2. Assess individual, systemic, and broader cultural dynamics. *Diversity Note: [Describe how you will adjust assessment based on cultural, socioeconomic, sexual orientation, gender, and other relevant norms.]*
   a. Use a three-generation genogram to identify multigenerational patterns, chronic anxiety, triangles, emotional cutoff, family projection process, and sibling position.
   b. Assess client and significant other's levels of differentiation in current crisis/ problem situation and in the past.

Initial Phase Client Goals

1. Reduce triangulation between client and [specify] to reduce depression and anxiety.
   a. Detriangulate by maintaining therapeutic neutrality and refocusing client on his/her half of problem interactions.
   b. Relational experiments to go home and practice relating directly rather than triangulating.

Intergenerational Working Phase of Treatment with Individual

Working-Phase Counseling Task

1. Monitor quality of the working alliance. *Diversity Note: [Describe how you will attend to client response to interventions that indicate therapist using expressions of emotion that are not consistent with client's cultural background.]*
   a. *Assessment Intervention:* Monitor therapist responses (both verbal and nonverbal) to ensure relating from differentiated position and avoiding triangulation.

Working-Phase Client Goals

1. Decrease chronic anxiety and reactivity to stressors to reduce anxiety.
   a. Encourage differentiated responses to common anxieties and triggers.
   b. Relational experiments to practice responding rather than simply reacting to perceived anxieties and stressors.

2. Decrease mindless repetition of unproductive multigenerational patterns and increase consciously chosen responses to stressors to reduce depression and hopelessness.

(continued)

a. Use genogram to identify multigenerational patterns and intergenerational transmissions related to presenting problems.
b. Process questions to help clients see the multigenerational processes and make differentiated choices instead of mindlessly repeating pattern.

3. Decrease emotional cutoffs and reengage in difficult relationships from a differentiated position to reduce anxiety.
    a. Process questions to identify the fusion underlying cutoffs.
    b. Going home again to help client reengage in cutoff relationships from a differentiated position.

Intergenerational Closing Phase of Treatment with Individual

Closing-Phase Counseling Task

1. Develop aftercare plan and maintain gains. *Diversity Note: [Describe how you will access resources in the communities of which they are a part to support them after ending therapy.]*
    a. Identify relationships and practices that help client maintain differentiation in key relationships.

Closing-Phase Client Goals

1. Increase client's ability to balance need for togetherness and autonomy in intimate relationships to reduce depression and anxiety.
    a. Process questions to explore how togetherness and autonomy can both be honored.
    b. Relational experiments to practice relating to others from a differentiated position.
2. Increase ability to respond to family-of-origin interactions from a position of engaged differentiation to reduce to depression and sense of helplessness.
    a. Encouraging differentiated responses when engaging family of origin.
    b. Going home again exercises to redefine relationship with family of origin.

Treatment Plan Template for Couple/Family

Intergenerational Initial Phase of Treatment with Couple/Family

Initial Phase Counseling Tasks

1. Develop working counseling relationship. *Diversity Note: [Describe how you will adjust to respect cultured, gendered, and other styles of relationship building and emotional expression.]*
    a. Engage with each client from a differentiated position conveying a nonanxious presence.
2. Assess individual, systemic, and broader cultural dynamics. *Diversity Note: [Describe how you will adjust assessment based on cultural, socioeconomic, sexual orientation, gender, and other relevant norms.]*
    a. Use a three to four generation genogram to identify multigenerational patterns, chronic anxiety, triangles, emotional cutoff, family projection process, and sibling position.
    b. Assess client and significant other's levels of differentiation in current crisis/problem situation and in the past.

(continued)

Initial Phase Client Goals

1. Reduce triangulation between [specify] and [specify] to reduce conflict.
   a. Detriangulate in session by maintaining therapeutic neutrality and refocusing each person on his/her half of problem interactions.
   b. Process questions to increase awareness of how triangulation is used to unsuccessfully manage conflict.

Intergenerational Working Phase of Treatment with Couple/Family

Working-Phase Counseling Task

1. Monitor quality of the working alliance. *Diversity Note: [Describe how you will attend to client response to interventions that indicate therapist using expressions of emotion that are not consistent with client's cultural background.]*
   a. *Assessment Intervention:* Monitor therapist responses (both verbal and nonverbal) to ensure relating from differentiated position and avoiding triangulation.

Working-Phase Client Goals

1. Decrease chronic anxiety in system and reactivity to stressors to reduce conflict.
   a. Encourage differentiated responses to common anxieties and triggers.
   b. Relational experiments to practice responding rather than simply reacting to perceived anxieties and stressors.

2. Decrease mindless repetition of unproductive multigenerational patterns and increase consciously chosen responses to stressors to reduce conflict.
   a. Use genogram to identify multigenerational patterns and intergenerational transmissions related to presenting problems.
   b. Process questions to help clients see the multigenerational processes and make differentiated choices instead of mindlessly repeating pattern.

3. Decrease emotional cutoffs and reengage in difficult relationships from a differentiated position to reduce conflict.
   a. Process questions to identify the fusion underlying cutoffs.
   b. Going home again to help clients reengage in cutoff relationships from a differentiated position.

Intergenerational Closing Phase of Treatment with Couple/Family

Closing-Phase Counseling Task

1. Develop aftercare plan and maintain gains. *Diversity Note: [Describe how you will access resources in the communities of which they are a part to support them after ending therapy.]*
   a. Identify relationships and practices that help client maintain differentiation in key relationships.

Closing-Phase Client Goals

1. Increase each person's ability to balance need for togetherness and autonomy in intimate relationships to reduce conflict and increase intimacy.
   a. Process questions to explore how togetherness and autonomy can both be honored within the relationship; discuss needs of each person and how they may differ and be accommodated.
   b. Relational experiments to practice relating to others from a differentiated position.

2. Increase ability to respond to family-of-origin interactions from a position of engaged differentiation to reduce conflict and increase intimacy.
   a. Encouraging differentiated responses when engaging family of origins.
   b. Going home again exercises to redefine relationships with family of origins.

Psychoanalytic Family Therapies

In a Nutshell: The Least You Need to Know

Many of the founders of family therapy were psychoanalytically trained, including Don Jackson, Carl Whitaker, Salvador Minuchin, Nathan Ackerman, and Boszormenyi-Nagy. Although some disowned their academic roots as they developed methods for working with families, others, such as Nathan Ackerman and Boszormenyi-Nagy, did not. In the 1980s, renewed interest in object relations therapies led to the development of *object relations family therapy* (Scharff & Scharff, 1987).

These therapies use traditional psychoanalytic and psychodynamic principles that describe inner conflicts and extend these principles to external relationships. In contrast to individual psychoanalysts, psychoanalytic family therapists focus on the family as a nexus of relationships that either support or impede the development and functioning of its members. As in traditional psychoanalytic approaches, the process of therapy involves analyzing intrapsychic and interpersonal dynamics, promoting client insight, and working through these insights to develop new ways of relating to self and others. Some of the more influential approaches are *contextual therapy* (Boszormenyi-Nagy & Krasner, 1986), *family-of-origin therapy* (Framo, 1992), and *object relations family therapy* (Scharff & Scharff, 1987).

The Juice: Significant Contributions to the Field

If you remember one thing from this chapter, it should be this:

Ethical Systems and Relational Ethics

Boszormenyi-Nagy (Boszormenyi-Nagy & Krasner, 1986) introduced the idea of an *ethical system* at the heart of families that, like a *ledger*, keeps track of *entitlement* and *indebtedness*. Families use this system to maintain trustworthiness, fairness, and loyalty between family members; its breakdown results in individual and/or relational symptoms. Thus the goal of therapy is to reestablish an ethical system in which family members are able to trust one another and to treat one another with fairness.

Clients often present in therapy with a semiconscious awareness of this ethical accounting system. Their presenting complaint may be that things are no longer fair in the relationship; parents are not sharing their duties equitably or one child is being treated differently than another. In these cases, an explicit dialogue about the family's ethical accounting system—what they are counting as their entitlement and what they believe is owed them—can be helpful in increasing empathy and understanding among family members.

Rumor Has It: The People and Their Stories

Nathan Ackerman and the Ackerman Institute A child psychiatrist, Nathan Ackerman (1958, 1966) was one of the earliest pioneers in working with entire families, which he posited were split into factions, much the way an individual's psyche is divided into conflicting aspects of self. After developing his family approach at the Menninger Clinic in the 1930s and at Jewish Family Services in New York in the 1950s, he opened his own clinic in 1960, now known as the Ackerman Institute, which has remained one of the most influential family therapy institutes in the country. With Don Jackson, he cofounded the field's first journal, *Family Process*.

Ivan Boszormenyi-Nagy With one of the most difficult names to pronounce in the field (Bo-zor-ma-nee Naj), Boszormenyi-Nagy was an early pioneer in psychoanalytic family therapy. His most unique contribution was his idea that families had an ethical system, which he conceptualized as a *ledger of entitlement and indebtedness* (Boszormenyi-Nagy & Krasner, 1986).

James Framo A student of Boszormenyi-Nagy, James Framo is best known for developing *family-of-origin therapy;* as part of treatment with individuals, couples, and families, he invited a client's entire family of origin in for extended sessions (Boszormenyi-Nagy & Framo, 1965/1985; Framo, 1992). Framo located the primary problem not only in the family unit but also in the larger extended family system.

David and Jill Scharff A husband-and-wife team, David and Jill Scharff (1987) developed a comprehensive model for object relations family therapy. Rather than focusing on individuals, they apply principles from traditional object relations therapy to the family as a unit.

The Women's Project Bowenian trained social workers Marianne Walters, Betty Carter, Peggy Papp, and Olga Silverstein (1988) reformulated many foundational family therapy concepts through a feminist lens. Their work challenged the field to examine gender stereotypes that were being reinforced in family therapy theory and practice, within and beyond the practice of Bowen family therapy.

The Big Picture: Overview of Treatment

The psychodynamic tradition includes a number of different schools that share the same therapeutic process. The first task is to create a caring therapeutic relationship, or *holding environment* (Scharff & Scharff, 1987), between the therapist and client. Then the therapist analyzes the intrapsychic and interpersonal dynamics—both conscious and unconscious, current and transgenerational—that are the source of symptoms (Boszormenyi-Nagy & Krasner, 1986; Scharff & Scharff, 1987). The therapist's next task is to promote client insight into these dynamics, which requires getting through client defenses. Once clients have achieved insights into the intrapsychic and interpersonal dynamics that fuel the problem, the therapist facilitates *working through* these insights to translate them into action in clients' daily lives.

Making Connection: The Therapeutic Relationship
Transference and Countertransference

A classic psychoanalytic concept, *transference* refers to when a client projects onto the therapist attributes that stem from unresolved issues with primary caregivers; therapists use the immediacy of these interactions to promote client insight (Scharff & Scharff, 1987). *Countertransference* refers to when therapists project back onto clients, losing their therapeutic neutrality and having strong emotional reactions to the client; these moments are used to help the therapist and client better understand the reactions the client brings out in others. In therapy with couples and families, the processes of transference and countertransference vacillate more than in individual therapy because of the complex web of multiple relationships.

Contextual and Centered Holding

In contrast to traditional psychoanalysts, who are viewed as neutral "blank screens," object relations family therapists are more relationally focused, creating a nurturing relationship they call a *holding environment*. They distinguish between two aspects of holding in family therapy: contextual and centered (Scharff & Scharff, 1987). *Contextual holding* refers to the therapist's handling of therapy arrangements: conducting sessions competently, expressing concern for the family, and being willing to see the entire family. *Centered holding* refers to connecting with the family at a deeper level by expressing empathetic understanding to create a safe emotional space.

Multidirected Partiality

The guiding principle for relating to clients in contextual family therapy is *multidirectional partiality*, that is, being "partial" with all members of the family (Boszormenyi-Nagy & Krasner, 1986). Therapists must be accountable to everyone who is potentially affected by the interventions, including those not immediately present in the room, such as extended family members. This principle of inclusiveness means that the therapist must bring out the humanity of each member of the family, even the "monster member"

(Boszormenyi-Nagy & Krasner, 1986). In practice, multidirectional partiality generally involves *sequential siding* with each member by empathizing with each person's position in turn.

The Viewing: Case Conceptualization and Assessment
Interlocking Pathologies

Expanding the classic psychodynamic view of symptomology, Ackerman (1956) held that the constant exchange of unconscious processes within families creates interlocking or interdependent pathologies and that any individual's pathology reflects those family distortions and dynamics, a position similar to that of systemic therapies. Thus, when working with a family, the therapist seeks to identify *how* the identified patient's symptoms relate to the less overt pathologies within the family.

Self-Object Relations Patterns

Object relations therapists emphasize the basic human need for relationship and attachment to others. Thus they assess *self-object relations*: how people relate to others based on expectations developed by early experiences with primary attachment objects, particularly mothers (Scharff & Scharff, 1987). As a result of these experiences, external objects are experienced as ideal, rejecting, or exciting:

- *Ideal Object*: An internal mental representation of the primary caretaker that is desexualized and deaggressivized and maintained as distinct from its rejecting and exciting elements.
- *Rejecting Object*: An internal mental representation of the caregiver when the child's needs for attachment were rejected, leading to anger
- *Exciting Object*: An internal mental representation of the caretaker formed when the child's needs for attachment were overstimulated, leading to longing for an unattainable but tempting object

Splitting The more intense the anxiety resulting from frustration related to the primary caregiver, the greater the person's need to split these objects, separating good from bad objects by repressing the rejecting and/or exciting objects, thus leaving less of the *ego*, or conscious self, to relate freely. To the degree that splitting is not resolved, there is an "all good" or "all bad" quality to evaluating relationships. In couples, splitting often results in seeing the partner as "perfect" (all good) in the early phases of the relationship, but when the partner no longer conforms to expectations, the partner becomes the enemy (all bad). In families, splitting can also take the form of the perfect versus the problem child.

Projective Identification In couples and other intimate relationships, clients defend against anxiety by projecting certain split-off or unwanted parts of themselves onto the other person, who is then manipulated to act according to these projections (Scharff & Scharff, 1987). For example, a husband may project his interest in other women onto his wife in the form of jealousy and accusations of infidelity; the wife then decides to hide innocent information that may feed the husband's fear, but the more she tries to calm his fears by hiding information, the more suspicious and jealous he becomes.

Repression Object relations therapists maintain that children must *repress* anxiety when they experience separation with their primary caregiver (attachment object), which results in less of the ego being available for contact with the outside world. Until this repressed material is made conscious, the adult unconsciously replicates these repressed object relationships. One of the primary aims of psychoanalytic therapy is to bring repressed material to the surface.

Parental Interjects Framo (1976) believes that the most significant dynamic affecting individual and family functioning is parental introjects, the internalized negative aspects of parents. People internalize these attributes and unconsciously strive to make all future intimate relationships conform to them, such as when they hear a parent's critical comments in the neutral comments of a partner. Therapists help clients become conscious of these introjects to increase their autonomy in intimate relationships.

Transference Between Family Members

Similar to the way they assess transference from client to therapist, object relations therapists assess for transference from one family member onto another (Scharff & Scharff, 1987). Transference between family members involves one person projecting onto other members' introjects and repressed material. The therapist's job is to help the family disentangle their transference, using interpretation to promote insight into intrapsychic and interpersonal dynamics. It is often easier to promote insight into transference patterns in family therapy than in individual therapy because these patterns happen "live" in the room with the therapist, thus reducing the potential for a client to rationalize or minimize.

Ledger of Entitlement and Indebtedness

Boszormenyi-Nagy (Boszormenyi-Nagy & Krasner, 1986) conceptualized the moral and ethical system within the family as a *ledger of entitlements and indebtedness*, or more simply a *ledger of merits*, an internal accounting of what one believes is due and what one owes others. Of course, because in families each person has his/her own internal accounting system that has a different bottom line, tensions arise over who is entitled to what, especially if there is no consensus on what is fair and how give-and-take should be balanced in the family.

- *Justice and Fairness:* The pursuit of justice and fairness is viewed as one of the foundational premises of intimate relationships. Monitoring fairness is an ongoing process that keeps the relationship *trustworthy*. A "just" relationship is an ideal, and all relationships strive to achieve this never fully attainable goal.
- *Entitlements:* Entitlements are "ethical guarantees" to merits that are earned in the context of relationships, such as the freedom that parents are entitled to because of the care they extend to children. The person's sense of entitlement may only be evident in a crisis or extreme situation, such as a parent becoming suddenly ill. *Destructive entitlements* result when children do not receive the nurturing to which they are entitled and later project this loss onto the world, which they see as their "debtors."
- *Invisible Loyalties:* Family ledgers extend across generations, fostering invisible loyalties. For example, new couples may have unconscious commitments to their family of origin when starting their partnership. Invisible loyalties may manifest as indifferences, avoidance, or indecisiveness in relation to the object of loyalty, blocking commitment in a current relationship.
- *Revolving Slate:* This is a destructive relational process in which one person takes revenge (or insists on entitlements) in one relationship based on the relational transactions in another relationship. Instead of reconciling the "slate" or account in the relationship in which the debt was accrued, the person treats an innocent person as if he or she was the original debtor.
- *Split Loyalties:* This term refers to when a child feels forced to choose one parent (or significant caregiver) over another because of mistrust between the caregivers. Common in divorces, this highly destructive dynamic results in pathology in the child.
- *Legacy:* Each person inherits a legacy, a transgenerational mandate that links the endowments of the current generation to its obligations to future generations. "Legacy is the present generation's ethical imperative to sort out what in life is beneficial for posterity's quality of survival" (Boszormenyi-Nagy & Krasner, 1986, p. 418). Legacy is a positive force in the chain of survival.

Mature Love: Dialogue Versus Fusion

Boszormenyi-Nagy and Krasner (1986) describes mature love as a form of dialogue between two people who are conscious of the family dynamics that have shaped their lives. This type of love is quite different from fusion, experienced as an amorphous "we" similar to an infant and its caregiver. Thus clients are encouraged to make invisible loyalties overt so that they can be critically examined, allowing for conscious choice and action rather than the fear and anxiety that characterize fused relationships.

Targeting Change: Goal Setting

Goals in psychoanalytic therapies include several long-term changes in both individual and relational functioning (Boszormenyi-Nagy & Krasner, 1986; Scharff & Scharff, 1987). General goals include the following:

- Increase autonomy and ego-directed action by making unconscious processes conscious.
- Decrease interactions based on projections or a revolving slate of entitlements.
- Increase capacity for intimacy without loss of self (fusion with object).
- Develop reciprocal commitments that include a fair balance of entitlements and indebtedness.

The Doing: Interventions

Listening, Interpreting, and Working Through

In general, psychoanalytic therapies use three generic interventions:

- *Listening and Empathy:* The primary tool of psychoanalytic therapists is listening objectively to the client's story without offering advice, reassurance, validation, or confrontation. Empathy may be used to help the family to nondefensively hear the therapist's interpretation of their unconscious dynamics.
- *Interpretation and Promoting Insight:* Like other psychoanalytic therapists, family psychoanalytic therapists encourage insights into interpersonal dynamics by offering interpretations to the client, such as by analyzing self-object relations or analyzing ledgers of entitlement and indebtedness.
- *Working Through:* Working through is the process of translating insight into new action in family and other relationships. Changing one's behavior on the basis of new insight is often the most difficult part of therapy. Understanding that you are projecting onto your partner feelings and expectations that really belong in your relationship with your mother is not too difficult; changing how you respond to your partner when you feel rejected and uncared for is more challenging.

Eliciting

In contextual therapy, *eliciting* uses clients' spontaneous motives to move the family in a direction that is mutually beneficial and dialogical (Boszormenyi-Nagy & Krasner, 1986). The therapist facilitates this process by integrating the facts of the situation, each person's individual psychology, and interactive transitions to help the family rework the balances of entitlement and indebtedness, helping each member to reinterpret past interactions and identify new ways to move forward.

Detriangulating

Like other systemic therapists, psychoanalytic therapists identify situations in which the parents have triangulated a symptomatic child into the relationship to deflect attention from their couple distress (Framo, 1992). Once the child's role is made clear, the therapist dismisses the symptomatic child from therapy and proceeds to work with the couple to address the issues that created the need for the child's symptoms.

Family-of-Origin Therapy

Framo (1992) developed a three-stage model for working with couples that involved couples therapy, couples group therapy, and family-of-origin therapy. Therapists begin working with the couple alone to increase insight into their personal and relational dynamics. Next, the couple join a couples group, where they receive feedback from other couples and also view their dynamics; for many couples, insight comes more quickly when they see their problem dynamic acted out in another couple. Finally, each individual member of the couple is invited to have a 4-hour-long session with his/her family of origin without the other partner present. These extended family-of-origin sessions are used to clarify and work through past and present issues, thereby freeing individuals to respond to their partners and children without the "ghosts" of these past attachments.

Putting It All Together

Case Conceptualization Template

- *Interlocking Pathologies*
  Describe how the presenting symptoms relate to interlocking pathologies within the system.
- *Self-Object Relations Patterns*
  Identify self-object relation patterns for each person in the family:
    - Ideal object
    - Rejecting object
    - Exciting object
- *Splitting*
  Describe patterns of splitting in the system.
- *Projective Identification*
  Describe patterns of projective identification within the system.
- *Repression*
  Describe patterns of repression within the system.
- *Parental Introjects*
  Describe patterns of negative parental introjects within the system.
- *Transference Between Family Members*
  Describe transference of parental introjects and repressed material onto others in the family.
- *Ledger of Entitlement and Indebtedness*
  Describe key elements of the family's ledger:
    - *Entitlements:* Describe themes of entitlements within the family and destructive entitlements across generations.
    - *Invisible Loyalties:* Describe invisible loyalties across generations.
    - *Revolving Slate:* Describe any patterns of revolving slate.
    - *Split Loyalties:* Describe instances of children feeling pressure to choose one parent over the other.
    - *Legacy:* Describe key themes in intergenerational family legacies.
- *Mature Love: Dialogue Versus Fusion*

Describe to what degree adults have love based on dialogical exchange between two equals versus emotional fusion.

Hyman Spotnitz

Treatment Plan Template for Individual

Psychodynamic Initial Phase of Treatment with Individual

Initial Phase Counseling Tasks

1. Develop working counseling relationship. *Diversity Note: [Describe how you will adjust to respect cultured, gendered, and other styles of relationship building and emotional expression.]*
   a. Create a *holding environment* that includes *contextual* issues as well as client's dynamics.
   b. Work through client *transference* and monitor therapist *countertransference*.

(continued)

2. Assess individual, systemic, and broader cultural dynamics. *Diversity Note: [Describe how you will adjust assessment based on cultural, socioeconomic, sexual orientation, gender, and other relevant norms.]*
   a. Identify *self-object relation patterns, splitting, projective identification, repression, parental interjects, and defense patterns.*
   b. Identify *interlocking pathologies, transference with partner/family, ledger of entitlements and indebtedness, and capacity for mature love.*

Initial Phase Client Goal

1. Increase awareness of *self-object patterns* and reduce *splitting, idealizing, or other defense strategies* to reduce depressed mood and anxiety.
   a. *Listen to and interpret* for client *self-object patterns* and *defense patterns* related to depressed mood and anxiety.
   b. Identify one relationship/area of life in which the client can begin to *work through* the assessed patterns.

Psychodynamic Working Phase of Treatment with Individual

Working-Phase Counseling Task

1. Monitor quality of the working alliance. *Diversity Note: [Describe how you will attend to client response to interventions that indicate therapist using expressions of emotion that are not consistent with client's cultural background.]*
   a. *Assessment Intervention:* Continuously monitor relationship for *transference* and *countertransference*; seek consultation/supervision as necessary.

Working-Phase Client Goals

1. Decrease interactions based on projections and/or a *revolving slate of entitlements* to reduce depressed mood/anxiety.
   a. Offer *interpretations* of *projection patterns* and *revolving slate issues* to increase client awareness.
   b. Use in session examples of *transference* to help client *work through* projection patterns.

2. Reduce influence of *negative parental introjects* to enable authentic relating to reduce hopelessness and depressed mood.
   a. *Detriangulation* to help client separate negative parental interjects from interpretations and assumptions in current relationships.
   b. Identify one to two relationships in which client can *work through negative parental interjects.*

3. Increase *autonomy* and *ego-directed action* by making unconscious processes conscious to reduce depression and anxiety.
   a. *Eliciting* to develop client motivation to work in productive directions in relationships.
   b. Identify one to two relationships/areas of life in which client can *work through* unconscious dynamics to increase autonomy and goal-directed action.

Psychodynamic Closing Phase of Treatment with Individual

Closing-Phase Counseling Task

1. Develop aftercare plan and maintain gains. *Diversity Note: [Describe how you will access resources in the communities of which they are a part to support them after ending therapy.]*
   a. Identify strategies for managing *entitlements and indebtedness* as well as monitor use of *defenses.*

(continued)

Closing-Phase Client Goals

1. Increase capacity for *intimacy* and *mature love* without loss of self to reduce depression and anxiety.
    a. *Interpret defenses and projections* that hinder capacity of mature love.
    b. Identify one to two opportunities to *work through* issues that block capacity for intimacy.

2. Develop *reciprocal commitments* that include a *fair balance of entitlements and indebtedness* to increase capacity for intimacy.
    a. Identify *legacies, loyalties, and revolving slate* patterns that have imbalanced current relationships.
    b. Examine *the ledger of entitlements/indebtedness* to identify more appropriate and balanced calculations of what is due and what is owed.

Treatment Plan Template for Couple/Family

Psychodynamic Initial Phase of Treatment with Couple/Family

Initial Phase Counseling Tasks

1. Develop working counseling relationship. *Diversity Note: [Describe how you will adjust to respect cultured, gendered, and other styles of relationship building and emotional expression.]*
    a. Create a *holding environment* for all members that includes *contextual* issues as well as client's dynamics.
    b. Work through client *transference* and monitor therapist *countertransference* with each member of the system.

2. Assess individual, systemic, and broader cultural dynamics. *Diversity Note: [Describe how you will adjust assessment based on cultural, socioeconomic, sexual orientation, gender, and other relevant norms.]*
    a. Identify each client's *self-object relation patterns, splitting, projective identification, repression, parental interjects, and defense patterns*.
    b. Identify *interlocking pathologies, transference within couple/family system, ledger of entitlements and indebtedness*, and each person's capacity for mature love.

Initial Phase Client Goal

1. Increase awareness of *self-object patterns* and *transference between couple/family members* and reduce *splitting, idealizing, or other defense strategies* to reduce conflict.
    a. *Listen to and interpret* for client *self-object patterns, transference within system*, and *defense patterns* related to conflict in couple/family.
    b. Identify one aspect of relationship in which each person can take action to *work through* the assessed patterns.

Psychodynamic Working Phase of Treatment with Couple/Family

Working-Phase Counseling Task

1. Monitor quality of the working alliance. *Diversity Note: [Describe how you will attend to client response to interventions that indicate therapist is using expressions of emotion that are not consistent with client's cultural background.]*

(continued)

a. *Assessment Intervention:* Continuously monitor relationship for *transference* and *countertransference,* especially if therapist begins to take sides with one member; seek consultation/supervision as necessary.

Working-Phase Client Goals

1. Decrease couple/family interactions based on *projections* and/or a *revolving slate of entitlements* to reduce conflict.
    a. Offer *interpretations* of *projection patterns* and *revolving slate issues* to increase each person's awareness of dynamics.
    b. Use in session examples of *transference* both between members and with therapist to help clients *work through* projection patterns.

2. Reduce influence of *negative parental introjects* to enable authentic relating to reduce hopelessness and depressed mood.
    a. *Detriangulation* to help client separate negative parental interjects from interpretations and assumptions in current relationships.
    b. Identify one to two relationships in which client can *work through negative parental interjects.*

3. Increase *autonomy* and *ego-directed action* by making unconscious processes conscious to reduce conflict.
    a. *Eliciting* to develop client motivation to work in productive directions in relationship.
    b. Identify areas of relationship in which each member can *work through* dynamics and increase autonomy and goal-directed action.

Psychodynamic Closing Phase of Treatment with Couple/Family

Closing-Phase Counseling Task

1. Develop aftercare plan and maintain gains. *Diversity Note: [Describe how you will access resources in the communities of which they are a part to support them after ending therapy.]*
    a. Identify strategies for managing *entitlements and indebtedness* as well as monitor use of *defenses.*

Closing-Phase Client Goals

1. Increase each member's capacity for *intimacy* and *mature love* without loss of self to reduce conflict and increase intimacy.
    a. *Interpret defenses and projections* that hinder capacity of mature love.
    b. Identify opportunities for each member to *work through* issues that block capacity for intimacy.

2. Develop *reciprocal commitments* that include a *fair balance of entitlements and indebtedness* to increase capacity for intimacy.
    a. Identify *legacies, loyalties, and revolving slate patterns* that have imbalanced current relationships.
    b. Examine *the ledger of entitlements/indebtedness* to identify more appropriate and balanced calculations of what is due and what is owed.

Research and the Evidence Base

The focus of research on Bowenian and psychoanalytic therapies has not been on outcome, as is required to be labeled as empirically validated studies (Chapter 2); instead the focus of research has been on the validity of the concepts. Miller, Anderson, and

Keala (2004) provide an overview of the research on the validity of the intergenerational theoretical constructs. They found that research supports the relation between differentiation and (a) chronic anxiety, (b) marital satisfaction, and (c) psychological distress. However, there was little support for Bowen's assumption that people marry a person with a similar level of differentiation or his theories on sibling position; his concept of triangulation received partial empirical support.

Of particular interest to researchers is Bowen's concept of differentiation of self, which has been the focus of scores of research studies on topics such as client perceptions of the therapeutic alliance (Lambert, 2008), adolescent risk-taking behaviors (Knauth, Skowron, & Escobar, 2006), parenting outcomes in low-income urban families (Skowron, 2005), and adult well-being (Skowron, Holmes, & Sabatelli, 2003). Lawson and Brossart (2003) conducted a study that predicted therapeutic alliance and therapeutic outcome from the therapist's relationship with his or her parents, providing support for the Bowenian emphasis on the self-of-the-therapist. Another study considering a psychometric measure of differentiation identifies two aspects of differentiation: (a) affect regulation (the ability to regulate one's expressed mood), and (b) the ability to negotiate interpersonal togetherness with separateness (Jankowski & Hooper, 2012).

In regards to psychoanalytic family therapies, significant research has been conducted on the nature of attachment in problem formation (Wood, 2002). The concept of attachment is also central to two empirically supported family therapies: emotionally focused therapy (Chapter 15; Johnson, 2004) and multidimensional therapy (Chapter 5; Liddle, Dakof, Parker, Diamond, Barrett, & Tejeda, 2001). Research is needed on the outcomes and effectiveness of Bowen and psychoanalytic family therapies so that these models can be refined and further developed.

Tapestry Weaving: Working with Diverse Populations

Gender Diversity: The Women's Project

Trained as social workers, Betty Carter, Olga Silverstein, Peggy Papp, and Marianne Walters (Walters et al., 1988) joined together to promote a greater awareness of women's issues in the field of family therapy. They raised the issue of gender power dynamics within traditional families and identified how family therapists were reinforcing stereotypes that were detrimental to women. In particular, they explicated how the misuse of power and control in abusive and violent relationships made it impossible for women to end or escape their victimization, a perspective that is now accepted by most therapists and the public at large. They also asserted that therapists should be *agents of social change*, challenging sexist attitudes and beliefs in families.

Walters et al. (1988) made several suggestions for how family therapists can reduce sexism in their work with couples and families:

- Openly discuss the *gender role expectations* of each partner and parent and point out areas where the couple or family hold beliefs that are unfair or unrealistic.
- Encourage women to take *private time* for themselves to avoid losing their individual identity to the roles of wife and mother.
- Use the self-of-the-therapist to model an *attitude of gender equality*.
- Push men to take on equal responsibility both in family relationships and in the household, as well as for scheduling therapy, attending therapy with children, and/or arranging for babysitting for couples sessions.

Ethnicity and Culture Diversity

Apart from the work of the Women's Project (Walters et al., 1988; see Women's Project above), the application of Bowen intergenerational and psychoanalytic therapies to diverse populations has not been widely explored or studied. In general, these therapies are aimed at "thinking" or psychologically minded clients (Friedman, 1991). Thus, minority groups who prefer action and concrete suggestions from therapists may have difficulty with these approaches. However, the therapist's stance as an expert fits with the expectations of many immigrant and marginalized populations. The work of Bowen, Framo, and Boszormenyi-Nagy that emphasizes the role of extended family members and intergenerational patterns may be particularly useful with diverse clients whose cultural norms value the primacy of extended family over the nuclear family system. In these families, it is expected that the nuclear family subordinate their will to that of the larger family system. In addition, research on the concept of differentiation of self provides initial support for its cross-cultural validity (Skowron, 2004).

In general, the greatest potential danger in using Bowenian or psychoanalytic therapies with diverse clients is that the therapist will use inappropriate cultural norms to analyze family dynamics, thereby imposing a set of values and beliefs that are at odds with the clients' culture. For example, if a therapist, without reflection, proceeds on the Bowenian premise that the nuclear family should be autonomous and develops therapeutic goals to move an immigrant family in that direction, the therapist could put the client in the difficult situation of being caught between the therapist's goals and the extended family's expectations. Similarly, if the therapist assumes that attachment in all cultures looks the same, the client may be inaccurately and unfairly evaluated, resulting in a therapy that is ineffective at best and destructive at worst. Because these theories have highly developed systems of assessing "normal" behavior, therapists must be mindful when working with clients who do not conform to common cultural norms.

Sexual Identity Diversity

Because the issue of a child's sexual orientation and gender identity has implications for the entire family system, Bowenian therapists working gay, lesbian, bisexual, transgendered and questioning (GLBTQ) clients should pay particular attention to intergenerational relationships. One study found that gay and lesbian parents lived closer to and received more support from their own parents (Koller, 2009). In contrast, gays and lesbians who were not parents reported stronger connections with their friend networks, sometimes referred to as families of choice (Koller, 2009). Thus, therapists should pay particular attention to the role of these friendship relationships with GLBTQ clients. Another recent study considered the effects of parental disapproval on lesbian relationships, which was found to have both positive and negative effects on the relationship (Levy, 2011). The negative effects included amount and quality of time spent as a couple, stress on the couple relationship, emotional impact on couple, fear/uncertainty, communication problems, and sexual effects. The positive effects of parental disapproval of the relationship included increased couple closeness, communication, patience, maturity, and valuing of the relationship. In a study that compared three-generation genograms of heterosexual and homosexual males found that overall there were more similarities than differences but that twice as many parents of gay/lesbian children had significant marital issues and twice as many heterosexual men had more distant relationships with their fathers than gay men (Feinberg & Bakeman, 1994).

Psychodynamic therapy has long been criticized for its pathologizing of same-sex attraction, and thus psychodynamic family therapists working with gay and lesbian couples should consider using gay-affirmative psychodynamic approaches (Rubinstein, 2003). Rubinstein recommends that psychodynamic therapists working with GLBTQ clients consider a multifaceted identity formation that includes biological sex, gender identity, social sex-role, and sexual orientation. He suggests that social sex-role confusion is often the most salient issue for gay and lesbian clients, who often feel conflicted over conforming to culturally approved behaviors for maleness and femaleness. In addition, psychoanalytic therapy can be used to help GLTBQ clients address their internalized homophobia by exploring their personal meaning of being attracted to same-sex partners.

Online Resources

Ackerman Institute (Psychoanalytic Therapy and Family Therapy Training): www.ackerman.org

The Bowen Center: thebowencenter.org

Center for Family Learning (Bowen Intergenerational Therapy): jrlobdellwebdesign.com/centerforfamilylearning/index.html

Georgetown Family Center (Bowen Center for the Study of the Family): www.thebowencenter.org/

Family Process (Journal): www.familyprocess.org

Sexual Crucible Model: www.passionatemarriage.com

References

*Asterisk indicates recommended introductory readings.

Ackerman, N. W. (1956). Interlocking pathology in family relationships. In S. Rado & B. G. Daniels (Eds.), *Changing conceptions of psychoanalytic medicine* (pp. 135–150). New York: Grune & Stratton.

Ackerman, N. W. (1958). *The psychodynamics of family life*. New York: Basic Books.

Ackerman, N. W. (1966). *Treating the troubled family*. New York: Basic Books.

Boszormenyi-Nagy, I., & Framo, J. L. (1965/1985). *Intensive family therapy: Theoretical and practical aspects*. New York: Brunner/Mazel.

*Boszormenyi-Nagy, I., & Krasner, B. R. (1986). *Between give and take: A clinical guide to contextual therapy*. New York: Brunner/Mazel.

Bowen, M. (1966). The use of family theory in clinical practice. *Comprehensive Psychiatry 7*, 345–374.

Bowen, M. (1972). Being and becoming a family therapist. In A. Ferber, M. Mendelsohn, & A. Napier (Eds.), *The book of family therapy*. New York: Science House.

Bowen, M. (1976). Theory in practice of psychotherapy. In P. J. Guerin (Ed.), *Family therapy: Theory and practice*. New York: Gardner Press.

*Bowen, M. (1985). *Family therapy in clinical practice*. New York: Jason Aronson.

*Carter, B., & McGoldrick, M. (1999). *The expanded family life cycle: Individual, family, and social perspectives* (3rd ed.). Boston, MA: Allyn & Bacon.

Feinberg, J., & Bakeman, R. (1994). Sexual orientation and three generational family patterns in a clinical sample of heterosexual and homosexual men. *Journal of Gay & Lesbian Psychotherapy, 2*(2), 65–76. doi:10.1300/J236v02n02_04

Framo, J. L. (1976). Family of origin as a therapeutic resource for adults in marital and family therapy: You can and should go home again. *Family Process 15*(2), 193–210.

*Framo, J. L. (1992). *Family-of-origin therapy: An intergenerational approach*. New York: Brunner/Mazel.

Friedman, E. H. (1991). Bowen theory and therapy. In A. S. Gurman and D. P. Kniskern (Eds.), *Handbook of family therapy* (Vol. 2, pp. 134–170). Philadelphia, PA: Brunner/Mazel.

Guerin, P. J., Fogarty, T. F., Fay, L. F., & Kautto, J. G. (1996). *Working with relationship triangles: The one-two-three of psychotherapy*. New York: Guilford.

Hardy, K. V., & Laszloffy, T. A. (1995). The cultural genogram: Key to training culturally competent family therapists. *Journal of Marital and Family Therapy, 21*, 227–237.

Jankowski, P. J., & Hooper, L. M. (2012). Differentiation of self: A validation study of the Bowen theory construct. *Couple and Family Psychology: Research and Practice*, doi:10.1037/a0027469

Johnson, S. M. (2004). *The practice of emotionally focused marital therapy: Creating connection* (2nd ed.). New York: Brunner/Routledge.

*Kerr, M., & Bowen, M. (1988). *Family evaluation*. New York: Norton.

Knauth, D. G., Skowron, E. A., & Escobar, M. (2006). Effect of differentiation of self on adolescent risk behavior. *Nursing Research, 55*, 336–345.

Koller, J. (2009). A study on gay and lesbian intergenerational relationships: A test of the solidarity model. *Dissertation Abstracts International Section A, 70*, 1032.

Kuehl, B. P. (1995). The solution-oriented genogram: A collaborative approach. *Journal of Marital and Family Therapy, 21*, 239–250.

Lambert, J. (2008). Relationship of differentiation of self to adult clients' perceptions of the alliance in brief family therapy. *Psychotherapy Research, 18*, 160–166.

Lawson, D. M., & Brossart, D. F. (2003). Link among therapist and parent relationship, working alliance, and therapy outcome. *Psychotherapy Research, 13*, 383–394.

Levy, A. (2011). The effect of parental homo-negativity on the lesbian couple. *Dissertation Abstracts International, 71*, 5132.

Licht, C., & Chabot, D. (2006). The Chabot Emotional Differentiation Scale: A theoretically and psychometrically sound instrument for measuring Bowen's intrapsychic aspect of differentiation. *Journal of Marital and Family Therapy, 32*(2), 167–180.

Liddle, H. A., Dakof, G. A., Parker, K., Diamond, G. S., Barrett, K., & Tejeda, M. (2001). Multidimensional family therapy for adolescent drug abuse: Results of a randomized clinical trial. *American Journal of Drug and Alcohol Abuse, 27*, 651–688.

*McGoldrick, M., Gerson, R., & Petry, S. (2008). *Genograms: Assessment and intervention* (3rd ed.). New York: Norton.

Miller, R. B., Anderson, S., & Keala, D. K. (2004). Is Bowen theory valid? A review of basic research. *Journal of Marital and Family Therapy, 30*, 453–466.

Rubalcava, L. A., & Waldman, K. M. (2004). Working with intercultural couples: An intersubjective-constructivist perspective. *Progress in Self Psychology, 20*, 127–149.

Rubinstein, G. (2003). Does psychoanalysis really mean oppression? Harnessing psychodynamic approaches to affirmative therapy with gay men. *American Journal of Psychotherapy, 57*(2), 206–218.

*Scharff, D., & Scharff, J. (1987). *Object relations family therapy*. New York: Aronson.

Schnarch, D. M. (1991). *Constructing the sexual crucible: An integration of sexual and marital therapy*. New York: Norton.

Schnarch, D. M. (1998). *Passionate marriage: Keeping love and intimacy alive in committed relationships*. New York: Holt.

Skowron, E. A. (2004). Differentiation of self, personal adjustment, problem solving, and ethnic group belonging among persons of color. *Journal of Counseling and Development, 82*, 447–456.

Skowron, E. A. (2005). Parental differentiation of self and child competence in low-income urban families. *Journal of Counseling Psychology, 52*, 337–346.

Skowron, E. A., Holmes, S. E., & Sabatelli, R. M. (2003). Deconstructing differentiation: Self regulation, interdependent relating, and well-being in adulthood. *Contemporary Family Therapy, 25*, 111–129.

*Walters, M., Carter, B., Papp, P., & Silverstein, O. (1988). *The invisible web: Gender patterns in family relationships*. New York: Guilford.

*Wood, B. L. (2002). Attachment and family systems (Special issue). *Family Process*, 41.

Intergenerational Case Study: African American Family Launching Children

Mark (AM54) and Darla (AF53), an African American couple, come to therapy with their youngest daughter, Tasha (CF17), because Darla and Tasha's arguing has gotten extreme. Mark travels frequently with his sales job, and Darla has worked part-time but mostly devotes herself to taking care of the children, with Tasha being the last one at home. Her two older sisters are out of the house; one is married and the other is finishing college. Darla says it was easy to parent her two older girls but Tasha is different: she says she always is irritable, never thinks of the needs of others, isn't applying herself in school, and hangs out with questionable friends. Tasha complains that her mother is too hard on her, was nicer to her sisters, and makes too many demands of her. Mark tends to play the peacemaker role for the two, but says he is tired of always being in the middle. He does not understand why they just can't get along.

Intergenerational Case Conceptualization

- *Chronic Anxiety*
  The family is nearing a significant shift in the family life cycle, requiring each party to make significant shifts in their balance of independence and interdependence. AF is about to launch her youngest child, and CF, who has always been the "baby" of the family, is now about to become more independent.
- *Multigenerational Patterns*
  Based on genogram, identify multigenerational patterns, attending to the following themes:
  - *Family Strengths*: Value education; productive; caring; devoted to family.
  - *Substance/Alcohol Abuse*: Father and brother on mother's side have been involved in substance abuse.
  - *Sexual/Physical/Emotional Abuse*: AF was sexually abused by a brother when she was young; it stopped once she told her parents.
  - *Parent/Child Relations*: AF's side has tradition of dedicated mothers who are highly involved in ensuring their children's success; AM plays the peacemaker role that his mother played in his family; AM is closest to CF26, and AF is closest to CF22.
  - *Physical/Mental Disorders*: Depression in men on father's side.

- *Historical Incidents of Presenting Problem*: AF had conflict with her mother as teen also. AM's sister and father also had chronic conflict.
- *Roles Within the Family: Martyr, Hero, Rebel, Helpless One, Etc.*: AF has patterns of mother's being the family martyr; both families have patterns of rebellious youngest child.

- *The Multigenerational Transmission Process*
  Both AM and AF seem to be more stable and better functioning than their parents, claiming that their education is what has enabled them to improve their lives. They also believe that they have experienced far less racial prejudice than their parents and feel that they have also made life choices to reduce their experience of discrimination, such as moving to a neighborhood where they feel welcomed.

- *Level of Differentiation*
  AM demonstrates the most differentiation in that he is able to help AF and CF calmly resolve their differences. AF and CF experience the most emotional fusion in the family as evidenced by their frequent arguments over "little things." CF26 and CF22 are more differentiated than CF17, as they are less the center of their parent's attention.

- *Emotional Triangles*
  AF and CF triangulate both AM and the two older sisters, each complaining to the third parties about how "unreasonable" the other is. CF17 gets along better with CF26.

- *The Family Projection Process*
  AF and AM have become distant over the years as they focused on raising children and maintaining AM's career; their focus on CF17 seems to be the main way they maintain connection in the little time they have together. AF, and perhaps AM also, seems to be projecting fears of being "alone" onto CF; F is also afraid to be alone and on her own. Both parents may also be projecting a family dynamic of the youngest child as a troubled child onto CF17.

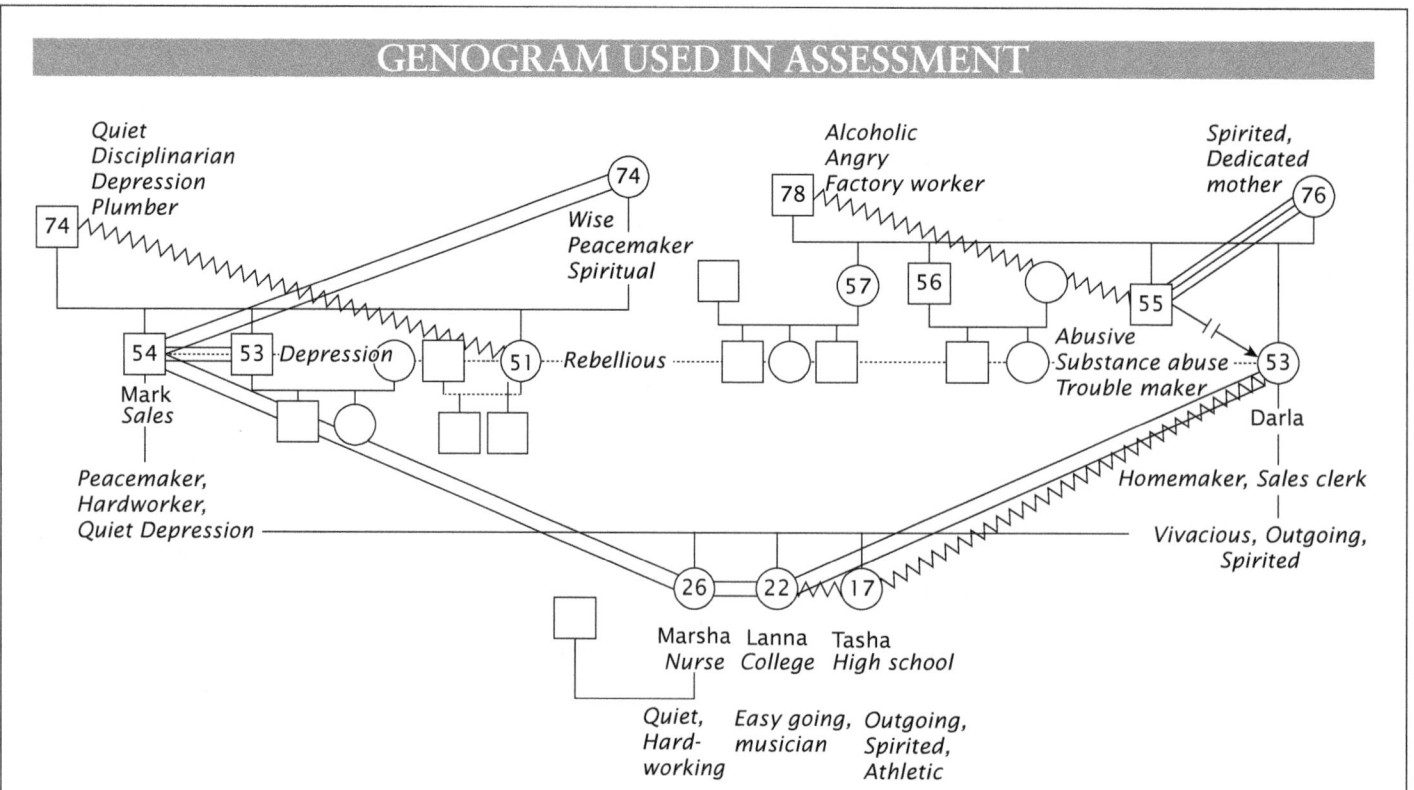

- *Emotional Cutoff*
  AF is cut off from her brother who abuses substances.
- *Sibling Position*
  - CF seems to be exhibiting characteristics of the youngest child: desiring pampering, wanting attention, and insisting on things going her way.
  - AM demonstrates characteristics of the oldest child, highly responsible and feeling obligated to make peace.
  - AF demonstrates some of the characteristics of the youngest child by not being flexible and often insisting on her way.

Intergenerational Treatment Plan

Intergenerational Initial Phase of Treatment

Initial Phase Counseling Tasks

1. Develop working counseling relationship. *Diversity Note: Use humor and relaxed tenor to engage teen. Create safe, supportive context, addressing therapist-client cultural differences as needed.*
   a. Engage with each client from a differentiated position conveying a nonanxious presence.

2. Assess individual, systemic, and broader cultural dynamics. *Diversity Note: Consider urban African American family patterns and norms when assessing family.*
   a. Use three to four generation genogram to identify multigenerational patterns, chronic anxiety, triangles, emotional cutoff, family projection process, and sibling position.
   b. Assess family member's levels of differentiation in current crisis/problem situation and in the past.

Initial Phase Client Goals

1. Reduce triangulation between AF, CF17, and AM to reduce conflict.
   a. Detriangulate in session by maintaining therapeutic neutrality and refocusing each person on his/her half of problem interactions; help AF and CF17 to resolve differences directly.
   b. Process questions to increase awareness of how triangulation is used to unsuccessfully manage conflict.

Intergenerational Working Phase of Treatment

Working-Phase Counseling Task

1. Monitor quality of the working alliance. *Diversity Note: Ensure all family members are engaged and feel safe in session; adjust therapist style and engagement to meet family's comfort with emotional expression.*
   a. *Assessment Intervention:* Monitor therapist responses (both verbal and nonverbal) to ensure relating from *differentiated* position and avoiding *triangulation*. Create opportunities for family to share their concerns with therapy process and frequently ask to ensure family agrees with goals and interventions.

Working-Phase Client Goals

1. Decrease *chronic anxiety* in system and *reactivity* to stressors to reduce conflict.
   a. Encourage *differentiated responses* to common anxieties and triggers, *coaching* CF17 to assume more responsibility and independence for self, AF to take on less responsibility for CF17, and AM and AF to focus more on their marriage.

(continued)

b. *Relational experiments* to practice responding rather than simply reacting to perceived anxieties and stressors, such as AF warmly encouraging CF17 to solve her own problems at school and socially.
c. *Relational experiments* that encourage older sisters and father to mentor CF17 rather than have AF to majority of parenting and interaction with CF.

2. Decrease mindless repetition of *unproductive multigenerational patterns* and increase consciously chosen responses to stressors to reduce conflict.
    a. Use *genogram* to identify multigenerational patterns and intergenerational transmissions related to presenting problems, focusing on martyr, rebel, and peacemaker roles in the family.
    b. *Process questions* to help clients see the multigenerational processes and make differentiated choices instead of mindlessly repeating pattern.

3. Increase intimacy and engagement between AM and AF to reduce distance between them.
    a. *Process questions* to help couple identify dynamics of their relationship, including need to readjust marriage to upcoming launch of children.
    b. *Relational experiments* to encourage couple in redefining their relationship as a couple without children as their focus.

Intergenerational Closing Phase of Treatment

Closing-Phase Counseling Task

1. Develop aftercare plan and maintain gains. *Diversity Note: Consider family supports in African American community for couple and family, including involvement in their local church.*
    a. Identify relationships and practices that help client maintain *differentiation* in key relationships.

Closing-Phase Client Goals

1. Increase couple's and CF's ability to *balance need for togetherness and autonomy* for their new stage of development to reduce conflict and increase intimacy.
    a. *Process questions* to explore how togetherness and autonomy can both be honored within the couple and family relationships; discuss needs of each person and how they may differ and be accommodated.
    b. *Relational experiments* to practice relating to others from a differentiated position, such as AF relating to CF17 as independent young adult, and for AF and AM to engage each other as their primary emotional connection.

2. Increase ability to respond to *family-of-origin interactions* from a position of engaged *differentiation* to reduce conflict and increase intimacy.
    a. *Encouraging differentiated responses* when AF and AM engage family of origins.
    b. *Going home again* exercises to redefine relationships with family of origins, especially between AF and father and brother.

CHAPTER 11

Cognitive-Behavioral and Mindfulness-Based Couple and Family Therapies

"The key thread that binds these diverse perspectives [behavioral family therapies] is a demand for continual empirical challenge. Every strategy, and every case, is subjected to empirical scrutiny that aims to define the specific therapeutic ingredients that facilitate the achievement of the specific benefits desired by the family. In other words, every family presents a new experiment with the potential to advance therapeutic frontiers."

—Falloon, 1991, p. 65, italics in original

Lay of the Land

Behavioral and cognitive-behavioral couple and family therapies (in this book called cognitive-behavioral family therapy, or CBFT—to save trees) are a group of related therapies based on the behavioral and cognitive-behavioral approaches originally developed for working with individuals. The most influential of these therapies are the following:

- *Behavioral Family Therapy:* This therapy focuses on parent training (Patterson & Forgatch, 1987).
- *Cognitive-Behavioral Family Therapy:* This therapy was developed by several therapists to integrate cognitive elements into therapy with couples and families (Dattilio, 2005; Epstein & Baucom, 2002).
- *Integrative Behavioral Couples Therapy:* This is an enhanced version of *behavioral couples therapy*, which was demonstrated to be effective in the short term but not in the long term; a humanistic component emphasizing *acceptance* of one's partner was added in an attempt to improve long-term outcomes (Jacobson & Christensen, 1996).
- *Gottman Method Couples Therapy:* This is a scientifically based approach to couples therapy based on Gottman's 30 years of research on the key differences between happy and unhappy marriages (Gottman, 1999). Although Gottman's approach is not an empirically supported treatment, its treatment goals are well supported by research data.
- *Mindfulness-Based and Mindfulness-Informed Therapies:* These approaches integrated mindfulness and acceptance practices to working with couples and families.

Behavioral and Cognitive-Behavioral Family Therapies

In a Nutshell: The Least You Need to Know

In the general mental health field, cognitive-behavioral therapies (CBTs) are some of the most commonly used therapeutic approaches. They have their roots in behaviorism—Pavlov's research on stimulus-response pairings with dogs and Skinner's research on rewards and punishments with cats—the premises of which are still widely used with phobias, anxiety, and parenting. Until the 1980s, most of the cognitive-behavioral family therapies were primarily behavioral: behavioral family therapy (Falloon, 1991) and behavioral couples therapy (Holtzworth-Munroe & Jacobson, 1991). In recent years, approaches that more directly incorporate cognitive components have developed: cognitive-behavioral family therapy (Dattilio, 2005; Epstein & Baucom, 2002) and Gottman (1999) method couples therapy approach.

Cognitive-behavioral family therapies integrate systemic concepts into standard cognitive-behavioral techniques by examining how family members—or any two people in a relationship—*reinforce* one another's behaviors to maintain symptoms and relational patterns. Therapists generally assume a directive, "teaching," or "coaching" relationship with clients, which is quite different from other approaches of "joining" or "empathizing" with clients to form a relationship. Because this approach is rooted in experimental psychology, research is central to its practice and evolution, resulting in a substantial evidence base.

The Juice: Significant Contributions to the Field

If you remember one thing from this chapter, it should be this:

Parent Training

Arguably, CBFTs greatest influence has been in the area of parenting (Patterson & Forgatch, 1987). Most therapists who work with families with young children, regardless of their primary orientation, use classic behavioral concepts of reinforcement and consistency to help improve parental efficiency (Dattilio, 2005; Patterson & Forgatch, 1987). The basic behavioral principle of *reinforcement*—that the positive or negative responses from the environment shape future behavior—is hardwired into reptilian and mammalian nervous systems, and to a large extent all living creatures. Therefore, if a dog is given a treat each time it sits on command, it learns the command by pairing compliance with the positive reinforcer, the treat. *Consistency*—reinforcing every time—is the key, especially in the beginning. Kids, employees, graduate students, and to a certain extent spouses work essentially on the same principle.

Patterson and Forgatch (1987) developed one of the most prominent approaches to parent training. Their approach is based on the following key concepts and techniques:

- *Teaching Compliance and Socialization:* Therapists aim to teach children to comply with parental requests with the broader goal of socializing children to function in society.
- *Improving Parental Requests:* Parental requests should be (a) few in number, (b) polite, (c) statements rather than questions, (d) made only once before enforcing a consequence, (e) specific, and (f) well timed.
- *Monitoring and Tracking:* Parents must monitor their children's behavior away from home by always asking four basic questions: Who? Where? What? When?
- *Creating a Contingent Environment:* Parents are encouraged to develop positive contingencies (rewards) to encourage desired behavior in children using *point charts.*
- *Five-Minute Work Chore:* Parents are taught to assign a lesser punishment, such as the 5-minute work chore, with initial infractions before removing privileges or using harsher punishments.

Being consistent and reinforcing behaviors are not the only skills parents need to master, but they are essential and therefore used widely when child behavior is a presenting concern. In the case study at the end of the chapter, an immigrant Korean couple brings their son to therapy because of conflicts with the father over grades. However, after the assessment it is clear the parents do not agree on basic expectations for the son, so the therapist starts with parent training to get the parents in agreement on household rules in the initial phase of therapy and then moves on to family communication and reducing conflict in the working phase of therapy.

Rumor Has It: The People and Their Stories

Gerald Patterson and Marion Forgatch Researchers at the Oregon Social Learning Center, Patterson and Forgatch (1987) have developed one of the most influential behavioral parent training programs.

Neil Jacobson and Andrew Christensen In the early 1970s, Neil Jacobson developed behavioral couples therapy, which was later recognized as an empirically validated treatment (Jacobson & Addis, 1993; Jacobson & Christensen, 1996). Although effective in improving couple functioning in the short term, therapeutic gains were generally lost at the two-year follow-up. Therefore, Jacobson decided to add a more affective, humanistic focus that emphasizes *acceptance* of one's partner, calling this new model "integrative behavioral couples therapy." Christensen has carried on the development of this approach since Jacobson's death in 1999.

Norman Epstein A professor at the University of Maryland, Norman Epstein has been a leader in developing cognitive and cognitive-behavioral approaches for working with couples (Epstein, 1982; Epstein & Baucom, 2002) and families (Epstein, Schlesinger, & Dryden, 1988; Freeman, Epstein, & Simon, 1987).

John Gottman For over 30 years, John Gottman (1999) has studied key factors in couples communication, divorce, and marital satisfaction. On the basis of this research, he has developed a scientifically based couples therapy, originally called the marriage clinic approach, that reduces behaviors that predict divorce, and increases those that predict long-term marital satisfaction. He has written several books for the general public, including *The Seven Principles for Making Marriages Work* (Gottman, 2002) and *And Baby Makes Three* (Gottman & Gottman, 2008).

Frank Dattilio In recent years, Frank Dattilio (2005) developed *cognitive-behavioral family therapy,* adopting traditional cognitive therapy. Specifically, work adapts Aaron Beck's (1976, 1988) concept of schemas—underlying core beliefs—for working with families.

The Big Picture: Overview of Treatment

The therapy process for most CBFTs includes the following steps:

- *Step 1. Assessment:* Obtain a detailed behavioral and/or cognitive assessment of *baseline functioning,* including the frequency, duration, and context of problem behaviors and thoughts.
- *Step 2. Target Behaviors and Thoughts for Change:* Cognitive-behavioral therapists identify *specific* behaviors and thoughts for intervention (e.g., rather than using the general goal of "improve communication," the therapist targets tantrum frequency, name calling, curfew compliance, and other problem behaviors).
- *Step 3. Educate:* Therapists educate clients on their irrational thoughts and dysfunctional patterns.
- *Step 4. Replace and Retrain:* Interventions are designed to replace dysfunctional behaviors and thoughts with more productive ones.

Making Connection: The Therapeutic Relationship

Directive Educator and Expert

Although the affective quality of connection exhibited by CBFT therapists varies greatly (i.e., from cool and detached to warm and friendly), the primary role of the therapist is the same: to serve as an expert who *directs* and *educates* the client and family on how to better manage their problems (Falloon, 1991). Following the classic medical model, traditional CBFT therapists maintained a somewhat distant stance from clients, much like medical doctors today, with their focus on diagnosing and prescribing interventions without achieving much emotional connection. Influenced by recent research such as the common factors model, which indicates that the affective quality of the therapeutic relationship is a strong predictor of positive outcomes (see Chapter 2), many CBFT therapists are increasing their use of empathy and warmth to create connection with their clients. That said, many clients, especially immigrants and those from cultures that view experts as in a position of significant authority, respond well to the expert position of the therapy. This is true of the immigrant Korean family in the case study at the end of the chapter.

Empathy in Cognitive-Behavioral Counseling

True to their research foundations, cognitive-behavioral therapists increasingly use empathy, warmth, and a nonjudgmental stance to build a counseling alliance based on research results that indicate these therapist qualities predict positive outcomes (Meichenbaum, 1997; also see common factors in Chapter 2). However, *the reason a cognitive-behavioral therapist uses empathy is quite different from the reason a humanistic therapist uses empathy*. This is frequently misunderstood, so I am going to say it again in case your mind was wandering: *cognitive-behavioral therapists use empathy for entirely different reasons than humanistic therapists.* Cognitive-behavioral therapists use empathy to create *rapport*, which then allows them to get to the "real" interventions that will change client's behaviors, thoughts, and emotions. In dramatic contrast, for experiential therapists, empathy *is* the intervention: they maintain that empathy is a curative process in and of itself (Rogers, 1961). The cognitive-behavioral therapist who uses empathy should not be seen as "manipulative" because in most cases the idea is to make the client feel more comfortable with the process, not to trick them. They should also not be seen as "integrating" experiential concepts, because they are not using an experiential concept in the way an experiential therapist would use it. Instead, they are *adapting* it to work within their philosophical assumptions about counseling and the change process.

Contemporary Cognitive-Behavioral Alliance

Judith Beck (2005) describes five practices for fostering the counseling alliance that reflect more contemporary sensibilities:

- *Actively Collaborate With the Patient:* Decisions about counseling should be jointly made with the client.
- *Demonstrate Empathy, Caring, and Understanding:* Expressing empathy helps clients trust the therapist.
- *Adapt One's Counseling Style:* Interventions, self-disclosure, and directiveness should be adjusted for each client based on personality, presenting problem, etc.
- *Alleviate Distress:* Demonstrating clinical effectiveness by helping clients solve problems and improving their moods enhances the counseling relationship.
- *Elicit Feedback at the End of the Session:* By asking clients "how did it go?" at the end of each session, therapists can intervene early in cases of alliance rupture.

By building a collaborative and engaged alliance, therapists are more likely to be effective and work through possible dysfunctional beliefs the client may have about counseling or the therapist that interfere with the process, such as "My therapist doesn't understand me" or "This will never work."

Written Contracts

CBFT therapists are perhaps the most businesslike of all therapists in their relationship with clients (Holtzworth-Munroe & Jacobson, 1991), at least in their written descriptions—and more than other therapists, they put this relationship on paper. CBFT therapists frequently use *written contracts* spelling out goals and expectations to help structure the relationship and to increase clients' motivation and dedication. Putting goals and agreements in writing and having clients sign that they agree can be a very motivating experience that creates commitment to the process.

The Viewing: Case Conceptualization and Assessment
Problem Analysis

From a CBFT therapist's perspective, most families do not come in with usefully defined problems. When a CBFT therapist hears, "We don't communicate anymore" or "My son is defiant," the therapist still has not heard a problem description. Problem analysis is the process of taking these vague descriptions and developing them into a clear description of behavioral interactions and their emotional consequences.

Problem analysis focuses on *present-day* behaviors, emotions, and cognitions. When clients tell their story about relationship distress or personal disappointments, CBFT therapists listen for (a) the behaviors, (b) emotions, and (c) thoughts that make the situation a problem. For example, if a client says that she is feeling depressed that her husband has left her, the therapist focuses on specific problematic *behaviors* (e.g., she no longer wants to see her friends), *feelings* (e.g., feeling worthless and hopeless), and *thoughts* (e.g., I will never find anyone again). These concrete, definable symptoms are the "problem," not that she is divorced. The focus of treatment is on reducing these undesirable thoughts, feelings, and behaviors and increasing more desirable ones.

Assessment of Baseline Functioning: Monitoring and Tracking

At the beginning of treatment, sometimes even before the first session, CBFT therapists conduct a baseline assessment of functioning, which provides a starting point for measuring change. They ask clients to log the (a) frequency, (b) duration, and (c) severity of specific behavioral symptoms, such as tantrums, anger, social withdrawal, or conflict. Therapists may also identify antecedent events that may have triggered the symptoms. Patterson and Forgatch (1987) use monitoring and tracking charts to help parents obtain a baseline description of their child's behavior. Although clients' verbal recall of their symptoms may seem sufficient, usually a baseline assessment provides more detailed and accurate information than recall alone, especially when remembering a child's behavior. A baseline log may look like this:

Sample Baseline Log

Problem Behavior	When?	How Long?	How Severe?	Events Before?	Events After?
Tantrum when did not get candy at grocery store; cried, said, "I hate you" to mother; refused to follow instructions	Before lunch and after doing several other errands	5 minutes of crying; 1 hour pouting afterwards	Moderate; stopped crying when left store	Very hot day; slept poorly night before; fight with brother earlier; mother did not give in to his first two requests	Another fight with brother in afternoon; mother yelled at him once got in car; refused to go to bed on time

Functional Analysis and Mutual Reinforcement

Originating in systemic theory, functional analysis identifies the precise contexts, antecedents, and consequences of the problem behavior (Falloon, 1991). However, family interactions are rarely as simple as the basic formula that works with laboratory rats: antecedent → behavior → consequence. Thus, Falloon (1991, p. 76) recommends the following questions to facilitate functional analysis in families:

> ### Functional Analysis Questions for Families
>
> - How does this specific problem handicap this person (and/or the family) in everyday life?
> - What would happen if the problem were reduced in frequency?
> - What would this person (and his or her family) gain if the problem were resolved?
> - Who (or what) reinforces the problem with attention, sympathy, and support?
> - Under what circumstances is the specific problem reduced in intensity?
> - Under what circumstances is the specific problem increased in intensity?
> - What do family members currently do to cope with the problem?
> - What are the assets and deficits of the family as a problem-solving unit?

These questions are used with the family in the case study at the end of the chapter and help illuminate the subtle nuances of the family's conflict and the meaning it holds for them.

When assessing couples, Holtzworth-Munroe and Jacobson (1991) recommend the following:

> ### Functional Analysis Questions for Couples
>
> 1. *Strengths and Skills of the Relationship*
>   - What are the major strengths of the relationship?
>   - What is each spouse's capacity to reinforce the other?
>   - What behaviors are highly valued by the other?
>   - What activities and interests do the couple currently share?
>   - What relational competencies does each possess?
>
> 2. *Presenting Problems*
>   - What are the primary complaints (defined behaviorally)?
>   - What behaviors occur too frequently? Under what circumstances and with what reinforcements do they occur?
>   - What behaviors occur too infrequently? Under what circumstances and with what reinforcements do they occur?
>   - How did these problems develop over time?
>   - Is there consensus about what needs to change?
>
> 3. *Sex and Affection*
>   - Is either unsatisfied with the frequency or quality of their sex life? What behaviors are associated with the dissatisfaction?
>   - Is either unsatisfied with the frequency or quality of nonsexual physical affection? What behaviors are associated with the dissatisfaction?
>   - Is either in an extramarital affair? Is there a history of affairs?
>
> 4. *Future Prospects*
>   - Are both seeking to improve the relationship, or are one or both contemplating separation?
>   - Have steps been taken toward separation or divorce?

(continued)

> 5. *Social Environment*
>   - What are the alternatives to this relationship, and how attractive are they?
>   - Is their social network supportive of separation?
>   - If there are children, what are the current effects, and what might be the effects of divorce?
> 6. *Individual Functioning*
>   - Does either have a significant mental or physical health disorder?
>   - What is the relationship history of each, and how does it affect the present relationship?
>
> —Adapted from Holtzworth-Munroe & Jacobson, 1991, pp. 106–107

When conducting a functional analysis, the therapist also looks for mutually reinforcing behaviors between parties and examines how these patterns are maintaining the symptom. This concept is similar to the systemic therapy concept of seeing the family's interactional patterns as interlocking steps in a dance. For example, if a parent inconsistently reinforces a child's problematic behavior (e.g., sometimes sets a consequence for talking back and sometimes does not), the behavior is likely to continue (e.g., because of an intermittent schedule of reinforcement). For example, the therapist is equally interested in assessing *how* the parents respond to their son's hyperactivity as she is in assessing the child's behaviors because the parents' responses reinforce—for better or worse—the son's behavior. Similarly, if a wife's depression results in less conflict in the marriage, this creates a positive reinforcement for the depression.

A-B-C Theory

Originally developed by Albert Ellis (1962) to analyze irrational thinking with individuals, the A-B-C theory has also been applied to working with families (Ellis, 1978, 1994). In this model, A is the "activating event," B is the "belief" about the meaning of that event, and C is the emotional or behavioral "consequence" based on the belief.

Ellis's A-B-C Theory

A = Activating event → B = Belief about A → C = Emotional and behavioral consequence

Most clients come in able to see only the connection between A and C, and thus report that A *causes* C: "I am depressed *because* my husband does not help out with the kids" or "I am *angry* because my son doesn't listen to me." The therapist's job is to help the client identify the "B" belief that the client does not put into the equation, such as "If he does not help with the children in the ways I want him to, he really does not care about me" or "Good kids follow through on parental requests without questioning the parent." These irrational beliefs are identified in assessment and targeted for change in the intervention phase.

Family Schemas and Core Beliefs

The cognitive therapy of Aaron Beck (1976) focused on identifying and changing the schemas or core beliefs that were fueling the problems in an individual's life, such as the belief that one must be perfect or that life should be fair. More recently, Dattilio (2005) has developed a system for assessing family schemas, of which we all have two sets: (a) beliefs about our family of origin, and (b) beliefs about families in general. Drawing on Beck's work, he identifies eight types of common cognitive distortions about families:

1. *Arbitrary Inference:* A belief based on little evidence (e.g., assuming your child is trying to hide something because he did not answer the cell phone immediately)
2. *Selective Abstraction:* Focusing on one detail while ignoring the context and other obvious details (e.g., believing you've failed as a parent because your child is doing poorly in school)

3. *Overgeneralization:* Just like it sounds, generalizing one or two incidents to make a broad sweeping judgment about another's essential character (e.g., believing that because your son listens to acid rock he is not going to college and will end up on drugs)
4. *Magnification and Minimization:* Going to either extreme of overemphasizing or underemphasizing based on the facts (e.g., ignoring two semesters of your child's poor grades is minimizing; hiring a tutor for one low test score is magnification)
5. *Personalization:* A particular form of arbitrary influence in which external events are attributed to oneself; especially common in intimate relationships (e.g., my spouse has lost interest in me because she did not want to have sex tonight)
6. *Dichotomous Thinking:* All-or-nothing thinking: always/never, success/failure, or good/bad (e.g., if my husband isn't "madly in love" with me, he really doesn't love me at all)
7. *Mislabeling:* Assigning a personality trait to someone based on a handful of incidents, often ignoring exceptions (e.g., saying one's husband is lazy because he does not help immediately upon being asked)
8. *Mind Reading:* A favorite in family and couple relationships: believing you know what the other is thinking or will do based on assumptions and generalizations; becomes a significant barrier to communication, especially when related to disagreements and hot topics such as sex, religion, money, and housework (e.g., before your spouse says a word, you are defending yourself)

In diverse families, such as the one at the case study at the end of the chapter, many of these beliefs may be rooted in experiences of immigration or their culture of origin.

Couple Cognition Types

Epstein and colleagues (Baucom, Epstein, Sayers, & Sher, 1989; Epstein, Chen, & Beyder-Kamjou, 2005) have identified five major types of cognitions that influence how couples emotionally and behaviorally respond to one another:

- *Selective Perceptions:* Focusing on certain events or information to the exclusion of others
- *Attributions:* Inferences about the causes of positive and negative aspects of the relationship
- *Expectancies:* Predictions about the likelihood of certain events in the relationship
- *Assumptions:* Basic beliefs or assumptions about the characteristics of the partner and/or the relationship
- *Standards:* Beliefs about the characteristics that the relationship and each partner "should" have

Targeting Change: Goal Setting

Specific treatment goals are identified through the previously discussed assessment procedures. Goals are stated in behavioral and measurable terms, such as "reduce arguments to no more than one per month." When working with couples and families, therapists use their authoritative role to identify goals that are agreeable to all (Holtzworth-Munroe & Jacobson, 1991). Immediately after clear goals are agreed upon, therapists also obtain a commitment from the couple or family to follow instructions and complete out-of-therapy assignments, often with a written contract. Getting clients to promise to complete assignments greatly increases the likelihood that clients will follow through.

Examples of Middle-Phase CBFT Goals
- Reduce CM tantrums by replacing positive and inconsistent reinforcements of tantrums with effective schedule of consistent consequences designed to extinguish the behaviors.
- Replace perfectionist beliefs about child school performance with more realistic expectations.
- Reduce generalizations and mind reading between mother and father in their parenting discussions.

Examples of Late-Phase CBFT Goals
- Develop positive mutual reinforcement cycle to reduce negativity and labeling.
- Redefine family schemas to increase tolerance of difference between members.
- Redefine couple schemas to reduce pressure for perfection and increase tolerance of weaknesses in other.

The Doing: Interventions

Classical Conditioning: Pavlov's Dogs

Used primarily to treat anxiety disorders, classical conditioning was developed by Ivan Pavlov (1932) in his famous experiments with salivating dogs. Pavlov was able to train dogs to salivate at the sound of a bell by pairing the dog's natural response to salivate at the sight of food with a bell. When the bell was rung each time food was presented, the dog learned that the bell signaled that food was coming and began salivating. After enough repetition, the dog began to salivate with just the sound of the bell (as anyone who has owned more than one dog knows, the speed at which the dog learns this is highly breed specific). This procedure is technically described as *conditioned and unconditioned stimuli and responses*.

How Classical Conditioning Works

1. *The Natural State of Affairs*

   Food (unconditioned stimulus; UCS) →
   Salivation (unconditioned response; UCR)

2. *Process of Pairing Conditional Stimulus with Response*

   Food (UCS) + *Bell* (conditioned stimulus; CS) →
   Salivation (conditioned response: CR)

3. *Resulting Pairing*

   Bell (conditioned stimulus; CS) →
   Salivation (conditioned response: CR)

Operant Conditioning and Reinforcement Techniques: Skinner's Cats

The bread and butter of CBFT interventions—especially parent training—is *operant conditioning*. Interventions based on operant conditioning use the principles identified by B. F. Skinner (1953) to modify human behavior, whether one's own or another's. The essential principle is to reward behavior in the direction of the desired behavior using small, incremental steps, a process called *shaping behavior*. Once a certain set of skills has been mastered, the bar is raised for which behavior will be reinforced (positively and/or negatively), with ever closer approximations to the desired behavior. Thus, if parents are trying to teach a child how to complete homework independently, they may begin by overseeing when, where, and how the child completes homework and reinforce success and failure under these conditions. Once the child regularly succeeds with full oversight, the child is given an area of responsibility to master—perhaps when the homework is done—and reinforced for success in this area. Next, the child may be rewarded for managing the list of homework assignments without oversight. This process continues until the child completes homework independently, much to the parents' delight.

Forms of Reinforcement and Punishment

In operant conditioning, desired behaviors can be positively or negatively reinforced or punished, depending on the behavior. The following four options are used alone or in combination to shape desired behavior.

> ### Four Options for Shaping Behavior
>
> - *Positive Reinforcement or Reward:* Rewards desired behaviors by *adding* something desirable (e.g., a treat)
> - *Negative Reinforcement:* Rewards desired behaviors by *removing* something *un*desirable (e.g., relaxing curfew)
> - *Positive Punishment:* Reduces undesirable behavior by *adding* something *un*desirable (e.g., assigning extra chores)
> - *Negative Punishment:* Reduces undesirable behavior by *removing* something desirable (e.g., grounding)

Summary of Operant Conditioning

	Increase Desired Behavior	Decrease Undesirable Behavior
Add something	Positive reinforcement; reward	Positive punishment
Remove something	Negative reinforcement	Negative punishment

Frequency of Reinforcement and Punishment

The frequency of reinforcement and punishment is key to increasing or decreasing behavior.

- *Immediacy:* The more immediate the reinforcement or punishment, the quicker the learning, especially with young children.
- *Consistency:* The more consistent the reinforcement or punishment, the quicker the learning. Consistency involves rewarding or punishing a behavior every time it occurs or on a consistent schedule (e.g., every other time) to create predictability.
- *Intermittent Reinforcement:* Random and unpredictable reinforcement increases the likelihood of a behavior, but not always the behavior you want. Inconsistent reinforcement of desired behaviors often *increases undesired behaviors;* thus, if a parent inconsistently reinforces curfew, the child is more likely to break it. However, random positive reinforcement of well-established desired behaviors helps sustain them (e.g., randomly reinforcing positive grades with periodic privileges).

The principles of positive and negative reinforcement and reward are incorporated into the following interventions.

Encouragement and Compliments Patterson and Forgatch (1987) strongly encourage *positive reinforcement* to increase desired behavior with children. When working with distressed relationships, they coach families to increase compliments and expressions of appreciation to increase positive reinforcement.

Contingency Contracting Contingency contracting can be used to promote new behaviors by creating a contingency that must be met to receive a desired reward. Parents can use contingency contracting with children that detail how privileges will be earned and lost (Falloon, 1988, 1991; Patterson & Forgatch, 1987). For example, if a child's grade point average (GPA) is above 3.0, the parents agree to an 11:00 P.M. curfew on Friday and Saturday. In the case study at the end of the chapter, once the therapist gets the parents to set up an agreed upon set of expectations, they work on age-appropriate contingency contracting that not only reinforces positive behavior and good grades but also rewards the teen's independent and self-reliance in accomplishing these goals.

Point Charts and Token Economies Generally used with younger children, point charts (Patterson & Forgatch, 1987) or token economies (Falloon, 1991) are used to shape and reward positive behaviors by allowing children to build up points that they can apply to privileges, treats, or purchases. Because the rewards must be motivating for each particular child, siblings may have different rewards. In addition, the rewards should be appropriate and readily approved by the parent. For example, if parents offer a reward that is too expensive or takes too much time from their schedule, they will have difficulty keeping up their half of the bargain. In most cases, punishment is added to a token economy by having the child lose points for poor behavior.

Behavior Exchange and Quid Pro Quo

When working with couples, mutual behavior exchanges—called *quid pro quo* ("this for that") arrangements—can be useful to help the partners negotiate relational rules (e.g., "If you make dinner, I will do the dishes"; Holtzworth-Munroe & Jacobson, 1991). However, research indicates that couples who rely primarily on quid pro quo arrangements tend to have lower levels of marital satisfaction (Gottman, 1999). Does that mean that using this technique is harmful when working with couples? Although a well-designed research study would best answer this question, it is wise to use behavior exchange judiciously with couples, balancing it with more affective techniques to increase understanding and acceptance and to avoid framing marriage as a business deal. To this end, Holtzworth-Munroe and Jacobson (1991) recommend having each partner select a behavior to "give" rather than have each "ask" for what he/she wants.

Communication and Problem-Solving Training

To help couples and families solve their problems, CBFT therapists also provide training in communication using the following guidelines (Falloon, 1991; Holtzworth & Jacobson, 1991):

- *Begin With the Positive:* When introducing a problem, each is instructed to begin with a statement of appreciation or a compliment.
- *Single Subject:* The communication training begins by identifying only one problem for the problem-solving session.
- *Specific, Behavioral Problems:* Problems are defined in specific behavioral terms rather than in global statements of feelings, characteristics, or attitudes (e.g., "he doesn't care" or "she's a nag").
- *Describe Impact:* When describing a complaint, the partner is encouraged to share the emotional impact of the behavior.
- *Take Responsibility:* Partners are encouraged to take responsibility for their half of the problem interaction.
- *Paraphrase:* After one person has spoken, the other summarizes what was heard so that misunderstandings can be immediately clarified.
- *Avoid Mind Reading:* Clients are to avoid making inferences about the other's motivations, attitudes, or feelings.
- *Disallow Verbal Abuse:* Insults, threats, and other forms of verbal abuse are not allowed; the therapist redirects the couple in appropriate directions.

Psychoeducation

A hallmark of CBFT therapy, psychoeducation involves teaching clients psychological and relational principles about their problems and how best to handle them (Falloon, 1988, 1991; Patterson & Forgatch, 1987). Psychoeducation can be done in individual or group sessions. The content typically falls into three categories:

- *Problem-Oriented:* Information about the patient's diagnosis or situation, such as ADHD, divorced, alcohol dependence, or depression. Therapists use this type of education to motivate clients to take new action.
- *Change-Oriented:* Information about how to reduce problem symptoms, such as by improving communication, reducing anger, or decreasing depression. Therapists use this type of education to help clients actively solve their problems. For such education to be successful, clients need to be highly motivated, and therapists need to introduce the new behavior in small, practical steps using everyday language.

- *Bibliotherapy: Bibliotherapy* is a fancy term for assigning clients readings that will be (a) motivating and (b) instructional for dealing with their presenting problem. Typically, therapists assign a self-help or popular psychology book, but they may also assign fiction or professional literature.
- *Cinema Therapy:* Similar to bibliotherapy, cinema therapy involves assigning clients to watch a movie that will speak to the problem issues (Berg-Cross, Jennings, & Baruch, 1990).

> ### Tips for Effectively Providing Psychoeducation
>
> - *Practice!* Yes, I'm serious. There aren't too many skills I recommend new therapists "practice" on family and friends, but psychoeducation is the major exception. Try explaining concepts and research outcome to people who haven't read books like this one. Notice the types of questions they ask after you explain a concept. That will help you learn what you might be leaving out, what type of jargon needs defining, and what people actually find useful.
> - *Ask First:* Perhaps the single greatest secret to making psychoeducation work is *timing*: providing information when the client is in a receptive state. How do you know when they are ready: ask them. "Would you be open to learning more about X?" If you fail to ask, you may not find a receptive audience. Alternatively, master the REBT style of confrontation and forget about timing.
> - *Keep It Very, Very Brief:* During a 50-minute session, I recommend keeping total psychoeducation time to 1–2 minutes—that's the max—and I am not exaggerating. Any other brilliant information you have to share should be saved for the following week because most clients cannot meaningfully integrate and act upon more than a single principle at once.
> - *Make One point—and One Point Only:* Only try to teach one concept, point, or skill in a session. Anything else is too much to be *practically* useful.
> - *Ensuring Understanding and Acceptance:* After briefly providing information, *directly ask* if clients understand and if they believe it is useful and realistic for their life.
> - *Apply It Immediately:* After you provide information, immediately identify how it can be practically applied in the client's life to address a problem that occurred in the past week or upcoming week.
> - *Step-by-Step Tasks:* After offering 2 minutes of psychoeducation, the following 48 minutes involve step-by-step instructions on how to apply the information to solve a current problem. Get specific: who does what, when, and where and how often, and is there any resistance or potential roadblocks?
> - *Follow Up on Tasks:* The next time you meet, ask the client if they used the information to any extent; if not, why?; and if so, what happened?

Challenging Irrational Beliefs

Challenging irrational beliefs involves confronting unhelpful beliefs that are creating or sustaining the problem (Ellis, 1994). This can be done in session by the therapist or out of session with a thought record (see next section). A therapist challenges a client's irrational belief in two ways:

- *Direct Confrontation:* The client is explicitly told that the belief is irrational.
- *Indirect Confrontation:* The therapist uses a series of questions to help the client see how the belief or idea is irrational and/or contributing to the creation of the problem.

The decision to use a direct or indirect approach depends on the therapeutic relationship, the therapist's style, and the client's receptiveness to a particular approach; client and therapist cultural and gender issues also significantly affect this dynamic. The direct approach generally requires that the therapist use and the client accept a more hierarchical, expert stance, whereas the indirect approach typically is more appropriate when the therapeutic relationship is less hierarchical and the client has a greater need for autonomy.

Socratic Method and Guided Discovery

Using the *Socratic method,* sometimes referred to as *guided discovery* or *inductive reasoning,* to gently encourage clients to question their own beliefs, cognitive therapists use open-ended questions that help clients to "discover" for themselves that their beliefs are either illogical (i.e., contrary to obvious evidence) or dysfunctional (i.e., not working for them; Beck, 2005). A less confrontational approach than other techniques, when questioning the validity of belief therapists generally take a relatively neutral stance, allowing the client's own logic, evidence, and reason to do the majority of convincing. Although the term *change* is used to describe what happens, in actuality tightly held beliefs are slowly eroded over time by the client questioning and requestioning their validity in different situations. In the case study at the end of the chapter, the therapist uses Socratic dialogue to help the family confront their own irrational beliefs rather than create more potential conflict by having the therapist do so directly.

Questions for Evaluating the Validity of Beliefs

- What evidence do you have to support your belief? What evidence is there to the contrary? So, what might be a realistic middle ground?
- What does respected person X (Y, and Z) say about your situation? How could they all be wrong?
- If your child [or another significant person] were to say the same thing, how would you respond?
- What is the realistic likelihood that things will really go *that* badly? What is a more realistic outcome?
- You bring up one possible reason for X. Have you considered another explanation? Perhaps ...?
- How likely is it that person X's behavior was 100% directed at you? What else might have played a role in his/her behavior?

Thought Records

Using Ellis's A-B-C theory, therapists often ask clients to confront their own irrational thinking and problem behaviors by assigning "thought records" (Datillio, 2005). A type of structured journaling, thought records provide a means for clients to analyze their own cognitions and behaviors and develop more adaptive responses. Before assigning these as homework, therapists usually practice in session on a whiteboard to demonstrate the process. Thought records generally include the following information:

- *Trigger Situation* (e.g., argument with spouse)
- *"Automatic" or Negative Thoughts* (e.g., "he/she will never change"; "he is so selfish")
- *Emotional Response:* How automatic thoughts made person feel (e.g., hurt, rejected, betrayed)
- *Evidence For:* Evidence that supports the automatic or negative thoughts and interpretations (e.g., "he has done this before"; "he has not changed and does not seem to be trying")
- *Evidence Against:* Evidence that counters the automatic thoughts (e.g., "he did seem genuinely sorry"; "he has been really trying in other areas")
- *Cognitive Distortions:* Depending on the evidence, different types of cognitive distortions (e.g., arbitrary inference, selective abstraction, overgeneralization, magnification, minimization, personalization, dichotomous thinking, mislabeling, mind reading)
- *Alternative Thought:* A more balanced perspective that incorporates both forms of "evidence" and corrects the cognitive distortion (e.g., "this is an area where we really seem to have differences; but we really do get along in so many other ways that are important to me; we are both trying to make this better")

Sample Thought Record

Trigger Situation	Automatic Thought	Emotional Response	Evidence For	Evidence Against	Cognitive Distortions	Realistic Alternative
Argument over household chores	He'll never change; he's selfish and lazy	Hurt, anger, betrayal	Has done this before; not changing	Genuinely sorry; made progress in other areas; problem less frequent	Magnification; overgeneralization; selective abstraction	We have differences in this area; we are both trying to adjust to the other's need; he will probably never be exactly what I want in this area

With *relational issues* it is helpful to add the following:

- What did I do to contribute to this problem interaction? (e.g., I came on strong and blaming in the beginning; would not accept apology)
- What can I do differently next time? (e.g., be gentler when presenting my concern; listen with an open mind)

With *behavioral issues* the following can be added:

- Problem behaviors (e.g., yelling)
- Alternative behaviors (e.g., take time out, count to 10, take a deep breath)

Homework Tasks

CBFT therapists often assign homework tasks that are designed to solve the client's problem (Datillio, 2005; Falloon, 1991; Holtzworth-Munroe & Jacobson, 1991). For example, to reduce a couple's conflict, therapists may assign communication tasks, such as using a timer to take turns listening to and summarizing what the other is saying. They may also develop tasks for reducing depression, such as journaling positive thoughts or increasing recreational and social activities. In CBFT, the tasks are logical solutions to reported problems.

Homework Compared with Strategic Directives CBFT tasks are linear and literal. In contrast, strategic and other systemic therapists assign tasks designed to (a) metaphorically make the covert overt (metaphorical task), (b) interrupt the problem interaction or behavioral pattern enough to allow the system to develop a new pattern (directive), or (c) make the uncontrollable controllable (paradox).

Homework Compared with Solution-Focused Tasks In contrast to CBFT tasks, solution-focused tasks are designed to enact the *solution* rather than reduce the problem (see Chapter 12). In addition, solution-focused tasks are (a) broken into small steps, (b) developed from clients' ideas and past successes, and (c) designed to increase motivation and hope as much as to solve the problem.

Putting It All Together

Case Conceptualization Template

Problem Definition: Define the problem in concrete, measurable behavioral terms.

Baseline Assessment: Track duration, frequency, and severity of symptoms over time; identify events that precede and follow symptom.

- *Functional Analysis:* Identify mutually reinforcing behaviors that sustain symptom.
  - How does this specific problem handicap this person (and/or the family) in everyday life?
  - What would happen if the problem were reduced in frequency?

- What would this person (and his or her family) gain if the problem were resolved?
- Who (or what) reinforces the problem with attention, sympathy, and support?
- Under what circumstances is the specific problem reduced in intensity?
- Under what circumstances is the specific problem increased in intensity?
- What do family members currently do to cope with the problem?
- What are the assets and deficits of the family as a problem-solving unit?
- What behaviors need to increase? Decrease?
- *A-B-C Theory: Identify Irrational Beliefs*
  - Identify A (Action/trigger event); C (Consequence: behavior, mood); and B (irrational belief that leads to C).
- *Cognitive Schemas and Core Beliefs*
  Identify which of the following schemas/core beliefs support the problem.
  1. Arbitrary inference
  2. Selective abstraction
  3. Overgeneralization
  4. Magnification and minimization
  5. Personalization
  6. Dichotomous thinking
  7. Mislabeling
  8. Mind reading
- *Relational Cognition Patterns*
  - Selective perceptions
  - Attributions
  - Expectancies
  - Assumptions
  - Standards

Treatment Plan Template for Individual

CBT Initial Phase of Treatment with Individual

Initial Phase Counseling Tasks

1. Develop working counseling relationship. *Diversity Note: [Describe how you will adjust to respect cultured, gendered, and other styles of relationship building and emotional expression.]*
   a. Develop an *empathic, supportive* working relationship with family in which therapist is able to effectively provide *education* from a position of *expertise*.

2. Assess individual, systemic, and broader cultural dynamics. *Diversity Note: [Describe how you will adjust assessment based on cultural, socioeconomic, sexual orientation, gender, and other relevant norms.]*
   a. *Baseline assessment* of depressive and anxiety symptoms and *functional analysis* of the role of depression/anxiety in significant relationships.
   b. Identify *irrational beliefs, family schemas, and core beliefs* that are the source of depressive and anxious thinking.

Initial Phase Client Goals

1. Reduce *irrational beliefs* related to anxiety to reduce anxiety.
   a. Use A-B-C theory to *challenge irrational beliefs.*
   b. *Psychoeducation* about the nature of anxiety and how to change it.
   c. *Thought records (homework)* for developing more realistic beliefs.

(continued)

CBT Working Phase of Treatment with Individual

Working-Phase Counseling Task

1. Monitor quality of the working alliance. *Diversity Note:* [Describe how you will attend to client response to interventions that indicate therapist using expressions of emotion that are not consistent with client's cultural background.]
   a. *Assessment Intervention:* Check in with client at regular intervals about satisfaction with therapy process and goals.

Working-Phase Client Goals

1. Reduce *irrational beliefs* related to depression to reduce depressed mood.
   a. *Psychoeducation* about the nature of depression and how to change it.
   b. Identify and *challenge irrational beliefs* underlying depression.
   c. *Thought records (homework)* to continue process of developing realistic schemas and beliefs.

2. Reduce use of *negative schemas* to interpret daily events to reduce depressed mood and anxiety.
   a. *Socratic dialogue* to allow client to discover the irrational beliefs that fuel depression and anxiety.
   b. *Homework tasks* that require action that is based on more realistic schemas about life and relationships.

3. Reduce *mutually reinforcing negative patterns* that trigger negative moods to reduce depression and anxiety.
   a. *Communication training* to improve client's half of negative interactions with others.
   b. Change consistency and frequency of reinforcement patterns for both desired and undesired relational interactions.

CBT Closing Phase of Treatment with Individual

Closing-Phase Counseling Task

1. Develop aftercare plan and maintain gains. *Diversity Note:* [Describe how you will access resources in the communities of which they are a part to support them after ending therapy.]
   a. Develop relapse prevention plan that identifies warning signs that depressive or anxious thinking is returning and using *thought records* (or similar strategy) for client to counter negative thinking.
   b. Refer to *mindfulness-based cognitive therapy group* or similar group to reduce potential for relapse (if not done earlier).

Closing-Phase Client Goals

1. Increase ability to make *realistic interpretations* of daily life and identify when thoughts are irrational to reduce depression and anxiety and increase sense of wellness.
   a. *Problem-solving training* to help client manage ongoing daily challenges.
   b. *Thought records* and *Socratic dialogue* with self to continue identifying and debunking unrealistic beliefs.

2. Increase ability to *relate to others using realistic relational schemas* to reduce depression and increase capacity for intimacy.
   a. *Thought records* to identify and counter unrealistic expectations of significant relationships.
   b. *Bibliotherapy* to develop more realistic expectations of relationships: Gottman's *7 Principles That Make a Marriage Work.*

Treatment Plan Template for Couple/Family

CBT Initial Phase of Treatment with Couple/Family

Initial Phase Counseling Tasks

1. Develop working counseling relationship. *Diversity Note: [Describe how you will adjust to respect cultured, gendered, and other styles of relationship building and emotional expression.]*
   a. Develop an *empathic, supportive* working relationship with all members of system in which therapist is able to effectively provide *education* from a position of *expertise*.

2. Assess individual, systemic, and broader cultural dynamics. *Diversity Note: [Describe how you will adjust assessment based on cultural, socioeconomic, sexual orientation, gender, and other relevant norms.]*
   a. *Baseline assessment* of any individual symptoms and *functional analysis* of the presenting complaints in significant relationships.
   b. Identify each person's and jointly held *irrational beliefs, family schemas, and core beliefs* that are the source of relational conflict.

Initial Phase Client Goals

1. Reduce *mutually reinforcing negative patterns* that trigger negative moods to reduce conflict.
   a. *Psychoeducation* about the nature of couple/family conflict and how to change it.
   b. Identify *quid pro quo* agreements on issues couple/family is willing to negotiate.
   c. Identify *unrealistic expectations* members have of each other.

CBT Working Phase of Treatment with Couple/Family

Working-Phase Counseling Task

1. Monitor quality of the working alliance. *Diversity Note: [Describe how you will attend to client response to interventions that indicate therapist using expressions of emotion that are not consistent with client's cultural background.]*
   a. *Assessment Intervention:* Check in with client at regular intervals about satisfaction with therapy process and goals.

Working-Phase Client Goals

1. Reduce *unrealistic expectations of partner/child/parents* and increase acceptance to reduce conflict.
   a. *Psychoeducation* about relationships and realistic expectations.
   b. Identify and *challenge irrational beliefs* underlying expectations of others and relationships in general.

2. Increase ability to *communicate* respectfully and effectively to reduce conflict.
   a. *Psychoeducation and communication training* to increase skills and knowledge.
   b. *Homework tasks* that help family to practice new communication skills.

3. Increase ability to effectively *problem solve* to reduce conflict.
   a. *Problem-solving training* to help couple/family learn skills and strategies for effectively solving problems.
   b. *Homework tasks* to practice problem solving with initially minor and then increasingly difficult issues.

(continued)

CBT Closing Phase of Treatment with Couple/Family

Closing-Phase Counseling Task

1. Develop aftercare plan and maintain gains. *Diversity Note: [Describe how you will access resources in the communities of which they are a part to support them after ending therapy.]*
   a. Develop relapse prevention plan that identifies warning signs that unrealistic expectations and/or conflictual patterns are returning and identify remediation strategies.
   b. Refer to *mindfulness-based couple/parenting group* or similar group to reduce potential for relapse (if not done earlier).

Closing-Phase Client Goals

1. Increase ability to *relate to partner/family members using realistic relational schemas* to reduce conflict and increase capacity for intimacy.
   a. *Thought records* to identify and counter unrealistic expectations of significant relationships.
   b. *Bibliotherapy* to develop more realistic expectations of relationships, such as John Gottman's *7 Principles That Make a Marriage Work* or Dan Siegel's *The Whole-Brain Child*.

2. Increase each member's ability to make *realistic interpretations* of daily life and identify when thoughts are irrational to reduce conflict and increase sense of wellness.
   a. *Problem-solving training* to help client manage ongoing daily challenges.
   b. *Thought records* and *Socratic dialogue* with self to continue identifying and debunking unrealistic beliefs.

Mindfulness-Based Therapies

In a Nutshell: The Least You Need to Know

Described as the third wave of behavioral therapy (with "pure" behavioral therapy the first wave and cognitive-behavioral therapy as the second), mindfulness-based approaches add a paradoxical twist to cognitive-behavioral approaches: *accepting* difficult thoughts and emotions in order to transform them (Hayes, Strohsal, & Wilson, 1999, 2011). Therapists using mindfulness-based approaches encourage clients to curiously and compassionately observe difficult thoughts and feelings *without the intention to change them*. By changing *how* clients relate to their problems—with curiosity and acceptance rather than avoidance—they experience new thoughts, emotions, and behaviors in relation to the problem, and thus have many new options for coping and resolving issues.

In couple and family therapy, mindfulness has been used to help couples and families become more accepting of one another, improve their communication, and increase intimacy (Carson, Carson, Gil, & Baucom, 2004; Duncan, Coatsworth, & Greenberg, 2009; Gehart, 2012). Mindfulness-based practices are rooted in Buddhist psychology, which is essentially a constructivist philosophy, making it theoretically most similar to postmodern (Chapters 12–14) and systemic (Chapters 5–7) therapies (Gehart, 2012). Therapists working from these approaches may find the concepts naturally fit with their approach. These practices require therapists to have their own mindfulness practice and training before trying to teach clients to do the same.

A Brief History of Mindfulness in Mental Health

Mindfulness has an unusual history as a cognitive-behavioral approach: it was not developed in a researcher's lab or from a Western philosophical tradition. Instead, mindfulness comes from religious and spiritual traditions, making it a surprising favorite for cognitive-behavioral therapists, who ground themselves almost exclusively in Western scientific traditions. Most commonly associated with Buddhist forms of meditation, mindfulness is found in virtually all cultures and religious traditions, including Christian contemplative prayer (Keating, 2006), Jewish mysticism, and the Islamic-based Sufi tradition. Although it has religious roots, mindfulness entered mental health as a nonreligious "stress reduction" technique and was intentionally separated from religion and spiritual elements and adapted for use in behavioral health settings (Kabat-Zinn, 1990).

Over 30 years ago, Jon Kabat-Zinn (1990) began researching the *Mindfulness Based Stress Reduction* (MBSR) program at the University of Massachusetts, which has been highly influential in making mindfulness a mainstream practice in behavioral medicine. The MBSR program is an 8-week group curriculum that teaches participants how to practice mindful breathing, mindful yoga postures, and mindful daily activities. Participants are encouraged to practice daily at home for 20–45 minutes per day. MBSR shows great promise as an effective treatment for a wide range of physical and mental health disorders, including chronic pain, fibromyalgia, psoriasis, depression, anxiety, ADHD, eating disorders, substance abuse, compulsive behaviors, and personality disorders (Baer, 2003; Gehart, 2012). Closely related, Teasdale, Segal, and Williams (1995) have adapted the MBSR curriculum for depression relapse in their program, *Mindfulness Based Cognitive Therapy*. With 50% of "successfully" treated cases of depression ending in relapse within a year, MBCT therapists are using mindfulness to reduce the high relapse rate with promising findings.

In addition to including *Mindfulness-Based Stress Reduction* (Kabat-Zinn, 1990) and *Mindfulness-Based Cognitive Therapy* (Teasdale, Segal, and Williams, 1995), mindfulness has been integrated into two counseling approaches that have also shown great promise in treating a wide range of clinical conditions: *Dialectic Behavioral Therapy* (Linehan, 1993) and *Acceptance and Commitment Therapy* (Hayes et al., 1999, 2011).

Mindfulness Basics

The most common form of mindfulness involves observing the breath (or focusing on a repeated word, a *mantra*) while quieting the mind of inner chatter and thoughts (Kabat-Zinn, 1990). Focus is maintained on the breath, grounding the practitioner in the present moment *without judging* the experience as good or bad, preferred or not preferred. Usually within seconds, the mind loses focus and wanders off—thinking about the exercise, a fight that morning, to-do lists, past memories, future plans; feeling an emotion or itch; or hearing a noise in the room. At some point, the practitioner realizes that the mind has wandered off and then returns to the object of focus without berating the self for "failing" to focus but rather with compassion understanding that the loss of focus is part of the process—refraining from beating oneself up is usually the most difficult part. This process of focusing—losing focus—regaining focus—losing focus—regaining focus continues for an established period of time, usually 10–20 minutes.

Gehart and McCollum (2007, 2008; McCollum & Gehart, 2010; Gehart, 2012) have used mindfulness to help first-year therapists and therapists learn how to develop therapeutic presence with encouraging reports from trainees who say that they notice significant changes both in and out of the therapy session. Trainees report being better able to be emotionally present with clients, less anxious in session, and better able to respond in difficult moments. Most report that their overall level of stress noticeably decreases within two weeks of practicing 5 days per week for 2–10 minutes; they also report better relationships and a greater sense of inner peace. Not bad for a 10–15 minute per week investment. So you might want to give it a try.

Starting Your Personal Mindfulness Practice

1. *Find a Regular Time:* The most difficult part of doing mindfulness is finding time—2–10 minutes—several days a week. My colleague Eric and I have our students do five days per week. It is best to "attach" mindfulness practice to some part of your regular routine, such as before or after breakfast, working out, brushing your teeth, seeing clients, coming home, or going to bed (if you are not too tired).

(continued)

2. *Find a Partner or Group (Optional):* If possible, find a partner or meditation group with whom you can practice on a regular basis. The camaraderie will help keep you motivated.
3. *Find a Timer (Highly Advisable):* Using a timer helps structure the mindfulness session, and many find it helps them focus better because they don't wonder if time is up. Most mobile phones have alarms and timers that work well; you can also purchase meditation apps with Tibetan chimes for iPhones and other smartphones. Digital egg timers work well too—avoid the ones that tick.
4. *Sit Comfortably:* When you are ready, find a comfortable chair to sit in. Ideally, you should not rest your back against the chair, but rather sit toward the front so that your spine is erect. If this is too uncomfortable, sit normally with your spine straight—but not rigid.
5. *Breathe:* Set your timer for 2 minutes initially and the watch yourself breathe while quieting the thoughts and any other discourse in your mind.
   a. Don't try to change your breathing; just notice its qualities; not judging it as good or bad.
   b. Know that your mind will wander off numerous times—both to inner and outer distractions—each time it does gently notice it without judging—perhaps imagine it disappearing like a cloud drifting off, soap bubbles popping, or say "ah, that too"—and then gently return your focus to watching your breath.
   c. Accept your mind how it is each time you practice; some days it is easier to focus than others. The key is to practice acceptance of "what is," rather than fall into the common pattern of being frustrated with what is not happening.
   d. The goal is *not* to have extended periods without thinking but rather to practice nonjudgmental acceptance and cultivate a better sense of how the mind works.
6. *Notice:* When the bell rings, notice how you feel. The same, more relaxed, more stressed? Try only to notice without judging. You may or may not feel much difference; the most helpful effects are cumulative rather than immediate. If you happen to notice you wish to go on, go ahead and add a minute or two the next time you practice. Slowly, you will add minutes until you find a length of time that works well for you. Don't extend the time until you feel a desire to do so.
7. *Repeat:* Our students report the best outcomes with shorter regular practices rather than longer but infrequent practices. Thus, doing five 2-minute practices is likely to produce better outcomes than one 10-minute session each week.

Resources to Support Your Practice

- *Free Meditation Podcasts:* You can download free guided mindfulness meditations and other free resources to support your practice from my website (www.dianegehart.com) and from UCLA's Mindfulness Awareness Research Center (www.marc.ucla.edu).
- *Workbooks: The Mindfulness-Based Stress Reduction Workbook* (Stahl & Goldstein, 2010) and *Get Out of Your Mind and Into Your Life: The New Acceptance and Commitment Therapy* (Hayes & Smith, 2005) are excellent workbooks to teach your practical techniques.

Specific Mindfulness Approaches

Mindfulness-Based Stress Reduction and Mindfulness-Based Cognitive Therapy Not primarily about becoming a good meditation practitioner, these mindfulness-based group treatments are designed to help clients change *how they relate to their thoughts and internal dialogue.* Using a highly structured group process clients are introduced to mindfulness breathing (similar to the instructions above), as well as mindful yoga (stretching) positions, and mindful daily activities (e.g., washing dishes mindfully, walking, etc.; Kabat-Zinn, 1990). Depending on the group's focus, clients may be taught to apply mindfulness to physical conditions, difficult emotions, depressive thinking, etc. The group format is ideal for motivating clients to practice regularly at home and report back to the group on progress.

The eight groups in MBSR cover the following:

Session 1: Introduction to mindfulness: Foundations of mindfulness and body scan meditation.

Session 2: Patience: Working with perceptions and dealing with the "wandering mind."

Session 3: Nonstriving: Introduction to breathing meditation; mindful lying yoga; qualities of attention.

Session 4: Nonjudging: Responding vs. reacting; awareness in breath meditation; standing yoga; research on stress.

Session 5: Acknowledgment: Group check in on progress; sitting meditation.

Session 6: Let it be: Skillful communication; loving kindness meditation; walking meditation; day-long retreat.

Session 7: Everyday mindfulness: Mindful movement and everyday applications; practicing on one's own life.

Session 8: Practice never ends: Integrating with everyday life.

The eight groups of MBCT cover the following (Segal, William, & Teasdale, 2002):

Session 1: Automatic pilot: Introduce mindfulness; eating mindfulness exercise; mindfulness body scan.

Session 2: Dealing with barriers: Explore mental chatter using the body scan exercise.

Session 3: Mindfulness of the breath: Introduce mindfulness breath focus and 3-minute breathing space exercise.

Session 4: Staying present: Link mindfulness to automatic thoughts and depression.

Session 5: Allowing and letting be: Introduce acceptance and "allowing" things to just be.

Session 6: Thoughts are not facts: Reframe thoughts as "just thoughts" and not facts.

Session 7: How can I best take care of myself?: Introduce specific techniques for depressive thoughts.

Session 8: Using what has been learned to deal with future moods?: Motivate to continue practice.

Through these mindfulness exercises, clients learn to:

- Deliberately direct their attention and thereby better control their thoughts.
- Become curious, open, and accepting of their thoughts and feelings, even those that are unpleasant.
- Develop greater acceptance of self, other, and things as they are.
- Live in and experience themselves in the present moment.

Dialectical Behavior Therapy Originally developed for treating suicidal borderline clients, *dialectical behavior therapy* (DBT; Linehan, 1993) is named for the fundamental dialectical tension between change and acceptance: "The paradoxical notion here is that therapeutic change can only occur in the context of acceptance of what is; however, 'acceptance of what is' is itself change" (p. 99). In contrast to other cognitive-behavioral approaches, this approach is based on the premise that emotion *precedes* the development of thought and that strong emotions, traumatic experiences, and attachment wounds are the source of psychopathology. In a nutshell, the process of DBT helps clients be present with, tolerate, and accept strong emotions in order to transform them. In a sense, it is a client's desperate attempts to avoid painful emotions that are the root of the problem.

DBT therapists help clients "be with" difficult emotions by encouraging clients to manage dialectic tension, the tension between two polar opposites, such as both loving and hating someone. Rather retreat to either pole, the therapist encourages clients to experience how they can both simultaneously feel love and dislike by acknowledging the multiple levels of truth and reality in a given situation. These contradictory feelings and thoughts are first tolerated, then explored and eventually synthesized so that the reality of both extremes can be recognized. For example, an adult might come to acknowledge both ultimately loving a critical parent but at the same time hate how that parent speaks to her. As the client becomes able to accept having both loving and hateful feelings toward the parent, she will find herself less reactive and emotional about the situation. The process of DBT involves helping clients learn to increase balance in their lives by better managing the inherent dialect tensions in life:

- Being able to both seek to improve oneself as well as accept oneself.
- Being able to accept life as it is and also seek to solve problems.
- Taking care of one's own needs as well as those of others.
- Balancing independence and interdependence.

Acceptance and Commitment Therapy A behavioral approach that shares philosophical assumptions with collaborative and narrative approaches (see Chapter 13 and 14), acceptance and commitment therapy (ACT, pronounced "act" not A-C-T) is based on the postmodern premise that we construct our realities through language, which shapes our thoughts, feelings, and behaviors. ACT practitioners believe that human suffering is in large measure created and sustained through language; "It is not that people are thinking the wrong thing—the problem is thought itself and how the verbal community (contemporary culture) supports its excessive use as a mode of behavioral regulation" (Hayes et al., 1999, p. 49). Unlike traditional cognitive-behaviorists, ACT practitioners assert that attempts to control thoughts and feelings and avoid direct experience are the *problem*, not the solution. Instead, they advocate mindfulness-based *experiencing* to promote acceptance of the full range of human emotions: "In the ACT approach, a goal of healthy living is not so much to feel *good*, but rather to *feel* good. It is psychologically healthy to feel bad feelings as well as good feelings" (p. 77).

> The same acronym, ACT, is used to outline the process of counseling:
> A = Accept and embrace difficult thoughts and feelings
> C = Chose and commit to a life direction that reflects who the client truly is
> T = Take action steps toward this life direction.

The first phase in ACT is to accept and embrace the very thoughts and feelings clients have been trying to avoid via their symptoms: accepting loss, feeling fear, and acknowledging anger. They caution clients to not "buy into" their thoughts and challenge them to see the flimsy link between reasons (a.k.a., excuses) and causes of their behavior. At the same time, they help them develop a *willingness* to experience difficult thoughts and feelings with their *observing self*. Through this process of observation, clients are better able to identify their true values and selves, which helps them not only readily identify a life direction but commit to pursuing it. As you might imagine, this is not as simple as it sounds. In the action phase, most clients reexperience the resistance to experience and negative thoughts that brought them to counseling in the first place. However, with increased ability to accept these and a renewed commitment to pursue a meaningful life direction, the therapist can work with the client when obstacles arise in pursuing new action.

Mindfulness in Couple and Family Therapy Family therapists are just beginning to explore the potential of mindfulness in couple and family therapy both in group format and traditional couple/family therapy sessions (Carson et al., 2004; Gehart, 2012; Gehart & McCollum, 2007). Several studies provide support for the idea that cultivating mindfulness may be particularly helpful for couples. In a study by Wachs and Cordova (2008), mindfulness is positively correlated with marital adjustment. Similarly, Barnes, Brown, Krusemark, Campbell, and Rogge (2008) found that the personality trait of mindfulness predicted greater marital satisfaction, lower emotional stress after conflict, and better communication. Block-Lerner, Adair, Plumb, Rhatigan, and Orsillo (2008) found that mindfulness training increased empathetic responding in couples, and Carson, Carson, Gil, and Baucom (2008) found that couples in a mindfulness-based relationship enhancement group demonstrated greater relationship satisfaction and less relational distress.

Similarly, mindfulness has been used to enhance parent-child relationships as well as to help children with attention deficit, hyperactivity disorder (ADHD) and other conduct issues. Several mindfulness-based parenting programs have been developed and researched, including Mindful Parenting (Duncan et al., 2009), mindfulness-based parenting training (Dumas, 2005), and mindfulness-based childbirth and parenting (Duncan & Bardacke, 2010). Furthermore, mindfulness has been enthusiastically studied as potential nonmedication based treatment for ADHD that may actually "correct" brain functioning by increasing activity in the prefrontal cortex and decreasing activity of the limbic system (Zylowska, Smalley, & Schwartz, 2009). Finally, mindfulness has been used with adolescents to decrease aggressiveness and improve their ability to regulate emotion (Singh, Lancioni, Joy, Winton, Sabaawi, Wahler, & Singh, 2007).

Loving Kindness Meditation

When working with couples and families to improve their relationships, therapists often emphasize *loving kindness meditation* rather than mindfulness breath meditation (Carson et al., 2004; Gehart, 2012). Loving kindness meditation involves sending well wishes to various people, typically an acquaintance, a significant other, a person with whom one has a difficult relationship and one self. The therapist can guide couples in this practice in session and invite them to practice at home with a recording or using the basic formula below:

Loving Kindness Meditation

Acquaintance: Bring to mind an acquaintance.

May this person be happy and joyful.

May this person be free from suffering.

May this person experience radiant health.

May this person have ease of well-being.

May this person be deeply at peace.

May this person be at peace with people in his/her life.

Continue this with each of the following:

- A significant others (just one; you can do others later)
- A person with whom you have a difficult relationship
- Yourself
- All beings

—Based on Gehart, 2012, p. 169

Gottman Method Couples Therapy

In a Nutshell: The Least You Need to Know

Gottman (1999) developed his scientifically based marital therapy from observational and longitudinal research on communication differences between couples who stayed together and ones who divorced. In the therapy based on these findings, the therapist coaches couples to develop the interaction patterns that distinguish successful marriages from marriages that end in breakup. Gottman's model is one of the few therapy approaches that are grounded entirely in research results rather than theory. However, a "scientifically based" therapy differs from an empirically supported treatment (see Chapter 2) in that the former uses research to set therapeutic goals whereas the latter requires research on treatment outcomes.

Debunking Marital Myths
Myth 1: Communication Training Helps
Gottman's (1999) research debunks several myths, including the myth that improving communication helps couples stay together. His research indicates that better communication produces short-term gains but that training couples to talk using "I" statements and "nonblaming" statements does not significantly affect whether or not they stay together. Instead, he found that both happily and unhappily married couples engage in defensiveness, criticism, and stonewalling (three of the Four Horsemen; contempt, the fourth, is seen mostly in marriages heading for divorce; see following discussion of Four Horsemen). However, those who stay together maintain a ratio of 5:1 positive-to-negative interactions during conflict (20:1 during nonconflictual conversations). Thus, simply improving communication is not as important as increasing the ratio of positive to negative interactions during conflict.

Myth 2: Anger Is a Dangerous Emotion
Contrary to what many therapists and the public may assume, Gottman (1999) found that expressing anger did not predict divorce; however, contempt (feeling superior to one's partner) and defensiveness do. Furthermore, although anger was associated with lower marital satisfaction in the short term, it was associated with increased marital satisfaction over the long term.

Myth 3: Quid Pro Quo Error
Gottman (1999) also found that *quid pro quo* ("this for that"; see previous discussion) actually characterizes *unhappy* marriages. Thus, he argued that contingency contracting is *not* appropriate when treating couples.

The Big Picture: Overview of Treatment
Gottman (1999) uses a highly detailed assessment system with numerous written and oral assessment tools to assess couples and target areas of change. The intervention process involves extensive psychoeducation about what works and what does not, as well as structured exercises that sometimes include videotaping couple conversations, replaying them for analysis, and identifying where and how improvements can be made. Gottman also believes that couples therapy should be characterized as the following:

- *A Positive Affect Experience:* Therapy should primarily be a positive affect experience; it should be enjoyable for clients, and therapists should avoid criticizing or implying blame.
- *Primarily Dyadic:* Therapy should primarily be a dyadic experience between the couple rather than triadic with the therapist moderating all interactions.
- *Emotional Learning:* Learning is "state"-dependent, meaning that, in order to change an emotional state, couples must be in that emotional state and then work through it; thus couples have difficult conversations in session to learn how to handle them differently.
- *Easy:* Interventions should seem easy and nonthreatening.
- *Nonidealistic:* Therapists should not be idealistic about the potential for marital bliss and instead aim for realistic goals, such as reducing conflict.

Making Connection: The Therapeutic Relationship
Therapist as Coach
The therapist serves as a relationship coach, empowering couples to take ownership of their relationship (Gottman, 1999). The therapist does not soothe the couple during difficult conversations but rather coaches them on how to soothe themselves and each other.

The Viewing: Case Conceptualization and Assessment
Assessing Divorce Potential
After studying couples for over 30 years, Gottman (1999) can predict a couple's potential for divorce in the next 5 years with 97.5% accuracy with only five variables, an impressive achievement. Furthermore, his careful research has identified several key predictors of divorce, the most notorious of which are the Four Horsemen of the Apocalypse.

The Four Horsemen of the Apocalypse In Gottman's studies, the presence of the following four behaviors during a couple's argument predicted divorce with 85% accuracy.

1. *Criticism:* A statement that implies something is globally wrong with the partner (e.g., "always," "never," or a statement about personality). Women tend to criticize more than men.
2. *Defensiveness:* Used to ward off attack, defensiveness claims, "I'm innocent."
3. *Contempt:* The *single best predictor of divorce,* contempt is seeing oneself as superior to one's partner (e.g., "you are incapable of an intelligent thought"). Happy marriages had zero incidents of contempt.
4. *Stonewalling:* Stonewalling is when the listener withdraws from interaction, either physically or mentally. Men are more likely to stonewall than women.

5:1 Ratio Most couples criticize, defend, and stonewall to a certain extent. The difference between those who stay together and those who do not lies in the ratio of positive to negative interactions during conflict conversations. Stable couples have five times as many positive interactions as negative interactions *during conflict*; distressed couples may have a 1:1 ratio. Many couples find this research result very helpful in learning how to improve their marriage.

Negative Affect Reciprocity Negative affect reciprocity is the increased probability that one partner's emotions will be negative *immediately following* negativity in the other. Otherwise stated, "My negativity is *more predictable* after my partner has been negative than it ordinarily would be" (Gottman, 1999, p. 37). Negative affect reciprocity is the most consistent correlate of marital satisfaction and dissatisfaction, regardless of the culture studied, and is a far superior measure than the total amount of negative affect in the relationship.

Repair Attempts *Repair attempts* refer to when one partner tries to "make nice" and end the conflict, soothe the other, or soften the complaint. Because happy couples are more responsive to repair attempts, they need fewer of them. Distressed couples frequently reject repair attempts, resulting in a higher number of total attempts. When failed repair attempts are combined with the Four Horsemen, Gottman (1999) can predict with 97.5% accuracy whether a couple will divorce in the next five years.

Accepting Influence Marriages in which men are unwilling to accept influence from their wives (e.g., suggestions, requests) are 80% more likely to end in divorce.

Harsh Startup Harsh startup is raising an issue using negative affect in the first minute of a conversation. For 96% of couples, only the first minute of data is necessary to predict divorce or stability, and harsh startup is one of several key variables in predicting divorce in that first minute (Gottman, 1999). Relationships in which the woman uses harsh startup are more likely to end in divorce.

Distance and Isolation Cascade What if there are no horsemen in sight? Does that mean a couple is doing well? Not necessarily. If problems go unresolved, often couples become emotionally disengaged, starting the distance and isolation cascade (Gottman, 1999). These couples often say, "Everything is okay," but there is underlying tension and sadness. These couples are characterized by an absence of emotional expression, lack of friendship, unacknowledged tension, high levels of physiological arousal in conflict, and few efforts to soothe the other.

Typologies of Happy Marriages

Gottman (1999) has identified three different types of stable, happy marriages, which means that there is more than one way to get marriage right. All maintain the 5:1 ratio of positive-to-negative interactions but do so at different rates.

Volatile Couples: Volatile couples are more emotionally expressive, expressing more positive and negative emotions. Passionate fighting and passionate loving characterize their relationships.

Validating Couples: Validating couples have moderate emotional expression and strong marital friendships.

Conflict-Avoiding Couples: Conflict-avoiding couples have the least emotional expression, minimize problems, prefer to talk about the strengths of their marriage, and end conversations on a note of solidarity.

The Sound Relationship House Gottman (1999) maintains that marriages that work have two elements:

- An overall sense of positive affect
- An ability to reduce negative affect during conflict

He has designed a marriage therapy to increase these two qualities in ailing marriages in a model he terms *The Sound Marital House,* which has seven key aspects:

1. *Love Maps:* A cognitive understanding of who your partner is and what makes him/her happy is a basic component of marital friendship.
2. *Fondness and Admiration System:* This refers to the amount of respect and affection partners feel for each other and are willing to express.
3. *Turning Toward Versus Turning Away: The Emotional Bank Account:* This is a habitual turning toward and opening to each other emotionally in nonconflict interactions (e.g., wanting to share stories, spend time together).
4. *Positive Sentiment Override:* This refers to giving your partner "the benefit of the doubt," compared to negative sentiment override, in which even neutral comments are interpreted negatively.
5. *Problem Solving:* Another myth that Gottman's research has caused therapists to reconsider is that couples need to improve their problem-solving skills. That is only half the story, or more precisely 31%. Gottman's research indicates that both successful and unsuccessful couples argue about the same topics 69% of the time; he calls these the perpetual problems, which are due to inherent personality differences.
    - *Solving the Solvable:* Stable couples are able to successfully resolve solvable problems.
    - *Dialogue with Perpetual Problems:* Happy couples avoid gridlock and instead find a way to continue talking about their perpetual problems and core personality differences. Because it is impossible not to have perpetual problems, selecting a partner is really about choosing a particular set of perpetual problems.
    - *Physiological Soothing:* In successful marriages, partners are able to soothe themselves and their partner, enabling everyone to "calm down."

6. *Making Dreams Come True:* Couples avoid gridlock, especially around perpetual problems, by working together to make each partner's dreams come true.
7. *Creating Shared Meaning:* Happy couples develop a marital "culture" with rituals of connection and shared meanings, roles, and goals.

The Doing: Interventions
Session Format

Therapy begins with a highly structured assessment process. In the intervention phase, the typical session follows this format:

- *Catch-up:* The couple check in on marital events, homework, and major issues; they are directed to talk to one another, not report to the therapist.
- *Preintervention Marital Interaction: The Boxing Round:* The couple interact 6 to 10 minutes, usually by discussing a difficult topic.
- *Give an Intervention:* After the interaction, the therapist asks the couple for an intervention before suggesting one, based on the premise that people tend to accept their own ideas better.
- *The Spouses Make the Intervention Their Own:* The couple discuss their thoughts on how to improve their interactions, with the therapist facilitating and educating as necessary to help them master the process.
- *Got Resistance?* If there is resistance, the therapist needs to address the source of these concerns.
- *No Resistance?* Once the couple have a viable plan for altering their interaction, the therapist instructs them to engage in another 6-minute interaction to practice the suggested changes.
- *Homework:* Tasks are assigned based on the interactions in session.

Specific Interventions

Gottman uses highly detailed interventions that are outlined in *The Marriage Clinic* (Gottman, 1999). Some of the more notable interventions are as follows.

Love Maps Couples are encouraged to develop their knowledge of each other by answering the following questions:

- Who are your partner's friends?
- Who are your partner's potential friends?
- Who are the rivals, competitors, "enemies" in your partner's world?
- What are the recent important events (in your partner's life)?
- What are some important upcoming events?
- What are some current stresses in your partner's life?
- What are your partner's current worries?
- What are some of your partner's hopes and aspirations for self and others? (Gottman, 1999, p. 205)

Soften Startup Gottman teaches couples to use the following rules to help soften startup.

- *Be Concise:* Keep the initial statement brief and to the point.
- *Complain but Don't Blame:* Complain about a specific incident rather than blame or label.
- *Start With Something Positive:* Pose problems by starting with something positive.
- *Use "I" Instead of "You" Statements:* Start statements with "I" rather than "you" to avoid blaming and to increase personal responsibility.
- *Describe What Is Happening Rather Than Judge:* Keep statements behavioral rather than global.
- *Ask for What You Need:* Clearly describe the behavioral changes you desire.

- *Be Polite and Appreciative:* Express appreciation for what your partner does do, and be respectful.
- *Express Vulnerable Emotions:* When possible, describe more vulnerable than blaming emotions.

Dreams Within Conflict When working with the gridlock created by perpetual problems, the therapist asks each partner about the deeper meanings and dreams that are beneath his/her rigid stance in the gridlock. In session, partners are directed to ask each other the following questions in relation to a perpetual problem or gridlock issue:

- What do you believe about this issue?
- What do you feel about it? Tell me all of your feelings about it.
- What do you want to happen?
- What does this *mean* to you?
- How do you think your goals can be accomplished?
- What dreams or symbolic meanings (e.g., freedom, hope, caring) are behind your position on this issue? (Gottman, 1999, p. 248)

Negotiating Marital Power Gottman facilitates couple discussion of gender roles using an extensive checklist (Gottman, 1999, pp. 298–300). He does not advocate a particular division of labor but instead helps the couple arrive at their own definition of "fair and equitable." The purpose of this exercise is to increase respect for the roles and duties of each partner.

Research and the Evidence Base

Because CBFT's conceptual home is in experimental psychology, research is part of its culture; therefore, CBFT therapies are some of the best-researched approaches in family therapy. In fact, CBT is the most frequently listed evidence-based approach in the Substance Abuse and Mental Health Services Administration Registry of Evidenced-Based Practices (www.nrepp.samhsa.gov), high on the list for individual issues such as depression and anxiety.

Over the years, CBFT theorists have modified their approach based on research outcomes—sometimes dramatically—to incorporate more affective and relational components. Behavioral couples therapy is the premier example of this trend. Because of good short-term but poor long-term outcomes, Neil Jacobson reformulated his couples therapy to include more affective aspects; his new approach is called *integrative behavioral couples therapy* (Jacobson & Christensen, 1996). Similarly, because research indicates that a nonjudgmental therapeutic alliance is crucial, CBFT therapists have become increasingly attentive to this aspect of therapy (Beck, 2005).

Although CBFT has an extensive research history, therapists should not assume that it is therefore superior to other approaches in every case or in general. As discussed in Chapter 2, when confounding factors such as researcher allegiance are controlled for, *no therapy is consistently found to be superior to any other* (Sprenkle & Blow, 2004). Furthermore, for certain issues, such as adolescent conduct issues, systemic approaches have been found to be superior to CBT (see Chapter 5). Finally, many therapists have critiqued CBT on the basis that its claims to superiority have more to do with the volume of research by CBT proponents than any substantive advantage (Loewenthal & House, 2010). Along those lines Miller (2012) reports that a massive effort to transform all mental health in Sweden to CBT resulted in no effect on overall outcome of those diagnosed with depression and anxiety, with over 25% dropping out of treatment; in addition, a significant number of these clients who were not considered disabled prior to treatment *became* disabled, costing the government more money. The Swedish government soon ended its attempt to systematically institute CBT, encouraging new approaches. Thus, although CBT does have the most extensive evidence base in the field, it is by no means a panacea and has by no means provided a simple answer to the question of what works best in therapy.

Tapestry Weaving: Working with Diverse Populations

Ethnic, Racial, and Cultural Diversity

Because CBFT defines behavioral norms (e.g., each culture defines "rational" differently), therapists must carefully apply this approach with diverse populations to avoid conflicts in values and relational styles. CBFT's emphasis on the expert stance in relation to clients has strengths and weaknesses when working with diverse populations. Men and certain culture groups, such as Latinos, Asians, and Native Americans, often prefer active, directive therapy (Gehart & Lyle, 2001; Pedersen, Draguns, Lonner, & Trimble, 2002). However, hierarchical difference may cause a rebellious reaction in some clients (e.g., highly educated adults or teens) or an overly compliant and withholding response in others (e.g., women who have a habit of people pleasing; Gehart & Lyle, 2001).

By design, cognitive-behavioral therapies aim to help clients conform to dominant cultural values, because that is equated with "functional" within a given society. Therefore, therapists need to carefully evaluate treatment goals prior to intervention to ensure that they do not clash with religious, cultural, racial, socioeconomic, or other values and realities and to consider how the various cultures of which a client is a part may have conflicting values. Researchers are just now exploring the specific effects of ethnicity in CBT practice, identifying in which contexts ethnicity plays the greatest role. For example, initial studies have found no significant difference when using CBT for depression with European American, Latino, and Asian Americans (Marchand, Ng, Rohde, & Stice, 2010), yet another study found that ethnicity found a significant role in the strength of the therapeutic relationship over the course of therapy with perpetrators of domestic violence (Walling, Suvak, Howard, Taft, & Murphy, 2012); thus, ongoing research will further refine our understanding of how best to use CBT with diverse populations. Based on this emerging research and clinical expertise, CBT therapists have begun developing recommendations for specific ethnic groups.

Hispanic and Latinos CBFT is generally considered to be culturally consistent and appropriate for working with Hispanic and Latino clients (Organista & Muñoz, 1996). Therapists have developed several suggestions for successfully working with Hispanic and Latino clients using CBFT.

- *Addressing Language Needs:* Language issues are central when working with many immigrant Latino families, who often prefer to discuss private and emotional issues in their native language (most people prefer to discuss emotional issues in their native language, and Spanish speakers are particularly passionate about this). In addition, Spanish-speaking therapists need to consider how to translate not just words but concepts across cultures; for example, in one CBT program for Latinos the A-B-C method for disputing irrational beliefs was streamlined and translated as the *Si, Pero* (Yes, but ...) technique (Piedra & Byoun, 2012).
- *Increasing Experiential Components:* When modifying standard CBT or CBFT curricula, therapists often increase the applied and experiential components: rather than talk about it, do it (Piedra & Byoun, 2012).
- *Respecting Cultural Value of* Familismo: Family is highly valued, with individual interests often placed second to family needs; thus, CBFT therapists need to be careful to avoid labeling a Latino/a's choice to put family above personal needs as "irrational" (Duarté-Vélez, Bernal, & Bonilla, 2010; González-Prendes, Hindo, & Pardo, 2011).
- *Respecting Cultural Value of Personalism:* Latino culture values warm and trusting personal relationships, which therapists need to integrate into the therapeutic relationship (González-Prendes, Hindo, & Pardo, 2011). To create a context of *personalism*, therapists can begin their initial session in "small talk" to share their background in addition to learning about the client's (Organista & Muñoz, 1996).
- *Respecting the Cultural Value of* Respecto: Latino families are typically hierarchical with formal expectations of respect to parents and elders; in addition, clients may

prefer to be addressed with formal titles, Senor or Senora (González-Prendes, Hindo, & Pardo, 2011; Organista & Muñoz, 1996).

- *Respecting the Cultural Value of Machismo:* Often misunderstood in its negative extreme, within its native culture machismo refers to a man's sense of leadership, loyalty, and responsibility to provide and care for the family (Duarté-Vélez, Bernal, & Bonilla, 2010; González-Prendes, Hindo, & Pardo, 2011).
- *Working with Spirituality:* Many Hispanic/Latino clients have a strong sense of spirituality, most often Roman Catholic (Duarté-Vélez, Bernal, & Bonilla, 2010). Prayer, church, and the church community are resources that therapists can tap into; but therapists should also discuss the use of prayer to ensure that it facilitates active problem solving rather than reduce their sense of efficacy or responsibility (Organista & Muñoz, 1996).
- *Attitude to Gay/Lesbian Relationships:* Often due to their strong religious beliefs, Hispanic families may have a highly negative response to a child's coming out, often viewing same-sex attraction as a "sin," creating significant cultural dissonance and rejection for gay Latino youth (Duarté-Vélez, Bernal, & Bonilla, 2010).
- *Addressing Immigration, Migration, and Acculturation:* Many Hispanic clients have complex immigration and migration patterns, often moving back and forth between cultures multiple times and family members living in different countries at the same time. In addition, acculturation issues are often an issue between parents and children that CBFT therapists need to help them address (Piedra & Byoun, 2012).

African Americans Many also consider CBFT an appropriate choice for working with African Americans because it helps them to directly dispute problematic social beliefs and it focuses on present behaviors (Kelly, 2006; McNair, 1996). Recommendations for adaptation include:

- *Forming a Collaborative Relationship:* Clients should be invited to work with the therapist in setting goals, agreeing on time frames; this relationship should be used to foster a greater sense of self-efficacy in the client (Kelly, 2006; McNair, 1996).
- *Focusing on Behaviors and Skills.* Rather than analyzing the past, therapists should focus on problems that are experienced in the present and are of immediate importance and relevance to the client (McNair, 1996).
- *Empowerment:* CBFT therapists can help empower clients by helping them to develop coping and communication skills and to expand the support networks (Kelly, 2006). Furthermore, clients can be taught how to use thought records and other strategies to solve their own problems without the therapist.
- *Disputing Stereotypes and Expectations:* CBFT therapists can use the techniques of the approach to help African Americans identify and logically dispute many of the stereotypes and expectations they have internalized from the broader culture—such as having to be twice as good to be good enough—and to increase their personal sense of purpose and opportunities (McNair, 1996).
- *Addressing Discrimination:* Therapists can help African Americans mitigate the effects of discrimination by specifically identifying how they cope with discrimination and how it relates to how they perceive and relate to the world (McNair, 1996).
- *Functional Analysis:* Functional analysis that examines the symptom in behavioral terms can help to circumvent potential therapist bias and prejudice and can help clients do the same (Kelly, 2006).

Chinese Americans Therapists have also considered how best to use cognitive-behavioral approaches with Chinese Americans (Hwang, Wood, Lin, & Cheung, 2006). Hwang and colleagues (2006) recommend the following:

1. Explicitly educating clients about the process and goals of therapy to increase their understanding and acceptance.
2. Therapist learning more about the client's cultural background and its significance to the client.

3. Clearly establishing goals and measures of improvement early in treatment to reduce confusion about the process.
4. Focusing on psychoeducational aspects and reinforce efforts to learn to reduce stigma and empower clients.
5. Using *cultural bridging* to link Western psychological concepts to client's cultural beliefs and practices [e.g., discussing Qi (life energy) when discussing depression].
6. Presenting therapist as an expert often is helpful to Chinese American clients.
7. Explicitly discussing therapist-client relationship to clarify roles and set realistic expectations.
8. Attending to cultural differences in communication and deference to authority.
9. Spending extra time joining with client and learning about immigration history and family background.
10. Respecting family orientation of Chinese culture and involve family whenever possible.
11. Being sensitive to social stigma associated with mental illness in Chinese community and the desire to keep diagnosis private.
12. Being patient because Chinese may not be as comfortable discussing emotions.
13. Be aware of the push-pull between client's culture of origin and the culture of therapy.
14. Integrating Chinese cultural healing traditions, such as QiGong or acupuncture, into treatment plan.
15. Being aware of the ethnic differences in expressing distress, with Chinese notion of self involving a closer relationship between mind/body than in Western culture.

Sexual Identity Diversity

Several authors have considered how to adapt CBT and CBFT for working with gay, lesbian, bisexual, transgendered, and questioning (GLBTQ) clients (Safren & Rogers, 2001). Safren and Rogers (2001) note that it is common that therapists either over- or underestimate the impact that sexual orientation and difference has on a client's presenting problems, and therefore they recommend that therapists begin by examining their own beliefs and schemas related to sexual orientation and identity. Furthermore, Safren and Rogers (2001) encourage CBT therapists to examine the role that societal norms and stigma play in the development of client's beliefs and schemas.

When working with GLBTQ youth and adults, therapists need to assess additional areas of stress and functioning:

- Overt acts of harassment, abuse, and violence (Safren, Hollander, Hart, & Heimberg, 2001)
- Internalized homophobia (Safren et al., 2001)
- Existence of social support networks (Safren & Rogers, 2001)
- Development of identity as a sexual minority (Safren et al., 2001)
- Disclosure of sexual orientation to family, friends, others (Safren et al., 2001)
- Development of platonic and romantic relationships with other gay, lesbian, bisexual, or transgendered persons (Safren et al., 2001)
- Stress due to social stigma (Glassgold, 2009)
- Stress due to concealing stigma and distress (Glassgold, 2009)

Mylott (1994) has identified common irrational beliefs with which gay, lesbian, and bisexual adults frequently struggle; these beliefs are largely informed by societal norms and include:

- I need to be loved.
- I can't stand rejection.
- Because people will often accept or reject me on the basis of my physical attractiveness, age, socioeconomic status, masculinity or femininity, I have to use these same criteria and accept or reject myself.
- It will be awful if I grow old without a lover.

- I can't stand being alone, and since I can't stand being alone, it is better to be in an emotionally and even physically damaging relationship than to be alone.
- When gay people are the victims of homophobia, I have to get very angry and upset about it; I also have to become enraged at homophobic individuals, groups, and institutions.
- I can only accept my homosexuality if I know for certain that it is genetically determined, or that "God made me gay." Otherwise, I cannot accept myself.
- It's awful if people (family, friends, the church, etc.) don't accept my homosexuality.
- CBT therapists help clients to learn how to adapt more realistic beliefs about their life, sexuality, and others using techniques such as thought records or Socratic dialogue.

When working with families whose children are coming out, Willoughby and Doty (2010) adapt using Datillo's (2005) work to help the family adjust. In their case study, they began working with the parents for four sessions without the child present to identify schemas and automatic thoughts that were triggered by the child's coming out. In addition, the CBFT therapist invited the parents to explore expectations and attributions about being gay or lesbian from their parents and families of origin. In the second and later sessions, the therapist challenged parent's negative beliefs about sexual orientation. In the third and fourth session, the therapist made direct suggestions for alternative behaviors. Finally, in a fifth and sixth session, the therapist invited the child in to facilitate more effective communication with the parents.

Online Resources

Frank Dattilio:
www.dattilio.com

Gottman Method Couples Therapy:
www.gottman.com

Juvenile Justice Bulletin (Functional Family Therapy):
www.ncjrs.org/pdffiles1/ojjdp/184743.pdf

Mindfulness Awareness Research Center (UCLA):
www.marc.ucla.edu

Mindfulness Based Stress Reduction Clinic (Jon Kabat-Zinn):
www.umassmed.edu/cfm/mbsr/

Oregon Social Learning Center (Patterson and Forgatch Parenting Program):
www.oslc.org

Substance Abuse and Mental Health Services Administration Registry of Evidenced-Based Practices:
www.nrepp.samhsa.gov

References

*Asterisk indicates recommended introductory readings.

Baer, R. A. (2003). Mindfulness training as a clinical intervention: A conceptual and empirical review. *Clinical Psychology: Science and Practice, 10*(2), 125–143.

Barnes, S., Brown, K. W., Krusemark, E., Campbell, W. K., & Rogge, R. D. (2008). The role of mindfulness in romantic relationship satisfaction and responses to relationship stress. *Journal of Marital and Family Therapy, 33*, 482–500.

Baucom, D., Epstein, N., Sayers, S., & Sher, T. (1989). The role of cognitions in marital relationships: Definitional methodological, and conceptual issues. *Journal of Family Psychology, 10*, 72–88.

Beck, A. T. (1976). *Cognitive therapy and the emotional disorders.* New York: International Universities Press.

*Beck, A. T. (1988). *Love is never enough.* New York: Harper & Row.

*Beck, J. (2005). *Cognitive therapy for challenging problems: What to do when the basic don't work.* New York: Gilford.

Berg-Cross, L., Jennings, P., & Baruch, R. (1990). Cinematherapy: Theory and application. *Psychotherapy in Private Practice, 8*, 135–157.

Block-Lerner, J., Adair, C., Plumb, J. C., Rhatigan, D. L., & Orsillo, S. M. (2008). The case for mindfulness-based approaches in the cultivation of empathy: Does nonjudgmental, present-moment awareness increase capacity for perspective-taking and empathic concern? *Journal of Marital and Family Therapy, 33*, 501–516.

Carson, J. W., Carson, K. M., Gil, K. M., & Baucom, D. H. (2004). Mindfulness-based relationship enhancement. *Behavior Therapy, 35*(3), 471–494. doi:10.1016/S0005-7894(04)80028-5

Carson, J. W., Carson, K. M., Gil, K. M., & Baucom, D. H. (2008). Self expansion as a mediator of relationship improvements in a mindfulness intervention. *Journal of Marital and Family Therapy, 33*, 517–526.

*Dattilio, F. M. (2005). Restructuring family schemas: A cognitive-behavioral perspective. *Journal of Marital and Family Therapy, 31*, 15–30.

Duarté-Vélez, Y., Bernal, G., & Bonilla, K. (2010). Culturally adapted cognitive-behavioral therapy: Integrating sexual, spiritual, and family identities in an evidence-based treatment of a depressed Latino adolescent. *Journal of Clinical Psychology, 66*(8), 895–906. doi:10.1002/jclp.20710

Dumas, J. E. (2005). Mindfulness-based parent training: Strategies to lessen the grip of automaticity in families with disruptive children. *Journal of Clinical Child and Adolescent Psychology, 34*(4), 779–791. doi:10.1207/s15374424jccp3404_20

Duncan, L. G., & Bardacke, N. (2010). Mindfulness-based childbirth and parenting education: Promoting family mindfulness during the perinatal period. *Journal of Child and Family Studies, 19*(2), 190–202. doi:10.1007/s10826-009-9313-7

Duncan, L. G., Coatsworth, J., & Greenberg, M. T. (2009). A model of mindful parenting: Implications for parent–child relationships and prevention research. *Clinical Child and Family Psychology Review, 12*(3), 255–270. doi:10.1007/s10567-009-0046-3

Ellis, A. (1962). *Reason and emotion in psychotherapy*. New York: Lyle Stuart.

Ellis, A. (1978). Family therapy: A phenomenological and active-directive approach. *Journal of Marriage and Family Counseling, 4*, 43–50.

Ellis, A. (1994). Rational-emotive behavior marriage and family therapy. In A. M. Horne (Ed.), *Family counseling and therapy* (pp. 489–514). San Francisco, CA: Peacock.

Epstein, N. (1982). Cognitive therapy with couples. *American Journal of Family Therapy, 10*, 5–16.

Epstein, N., & Baucom, D. (2002). *Enhanced cognitive-behavioral therapy for couples: A contextual approach*. Washington, DC: American Psychological Association.

Epstein, N., Chen, F., & Beyder-Kamjou, I. (2005). Relationship standards and marital satisfaction in Chinese and American couples. *Journal of Marital and Family Therapy, 31*, 59–74.

Epstein, N., Schlesinger, S., & Dryden, W. (1988). *Cognitive-behavioral therapy with families*. New York: Brunner/Mazel.

*Falloon, I. R. H. (Ed.). (1988). *Handbook of behavioral family therapy*. New York: Guilford.

Falloon, I. R. H. (1991). Behavioral family therapy. In A. S. Gurman & D. P. Kniskern (Eds.), *Handbook of family therapy* (Vol. 2, pp. 65–95). Philadelphia: Brunner/Mazel.

Freeman, A., Epstein, N., & Simon, K. (1987). *Depression in the family*. New York: Routledge.

Gehart, D. (2012). *Mindfulness and acceptance in couple and family therapy*. New York: Springer.

Gehart, D. R., & Lyle, R. R. (2001). Client experience of gender in therapeutic relationships: An interpretive ethnography. *Family Process, 40*, 443–458.

*Gehart, D., & McCollum, E. (2007). Engaging suffering: Towards a mindful re-visioning of marriage and family therapy practice. *Journal of Marital and Family Therapy, 33*, 214–226.

Gehart, D., & McCollum, E. (2008). Teaching therapeutic presence: A mindfulness-based approach. In S. Hicks (Ed.), *Mindfulness and the healing relationship*. New York: Guilford.

Glassgold, J. M. (2009). The case of Felix: An example of gay-affirmative, cognitive-behavioral therapy. *Pragmatic Case Studies In Psychotherapy, 5*(4), 1–21.

González-Prendes, A., Hindo, C., & Pardo, Y. (2011). Cultural values integration in cognitive-behavioral therapy for a Latino with depression. *Clinical Case Studies, 10*(5), 376–394. doi:10.1177/1534650111427075

*Gottman, J. M. (1999). *The marriage clinic: A scientifically based marital therapy*. New York: Norton.

Gottman, J. M. (2002). *The seven principles for making marriage work*. New York: Three Rivers Press.

Gottman, J. M., & Gottman, J. S. (2008). *And baby makes three*. New York: Three Rivers Press.

Hayes, S. C., Strohsal, K. D., & Wilson, K. G. (1999). *Acceptance and commitment therapy: An experiential approach to behavior change*. New York: Guilford.

*Hayes, S. C., Strohsal, K. D., & Wilson, K. G. (2011). *Acceptance and commitment therapy: An experiential approach to behavior change* (2nd ed.). New York: Guilford.

Holtzworth-Munroe, A., & Jacobson, N. S. (1991). Behavioral marital therapy. In A. S. Gurman & D. P. Kniskern (Eds.), *Handbook of family therapy* (Vol. 2, pp. 96–133). Philadelphia, PA: Brunner/Mazel.

Hwang, W., Wood, J. J., Lin, K., & Cheung, F. (2006). Cognitive-behavioral therapy with Chinese Americans: Research, theory, and clinical practice. *Cognitive and Behavioral Practice, 13*(4), 293–303. doi:10.1016/j.cbpra.2006.04.010

Jacobson, N. S., & Addis, M. E. (1993). Research on couples and couples therapy: What do we know? Where are we going? *Journal of Consulting and Clinical Psychology, 57*, 5–10.

*Jacobson, N. S., & Christensen, A. (1996). *Integrative couple therapy*. New York: Norton.

Kabat-Zinn, J. (1990). *Full catastrophe living: Using the wisdom of your body and mind to face stress, pain, and illness*. New York: Delta.

Keating, T. (2006). *Open mind open heart: The contemplative dimension of the gospel*. New York: Continuum International Publishing Group.

Kelly, S. (2006). Cognitive-behavioral therapy with African Americans. In P. A. Hays & G. Y. Iwamasa (Eds.), *Culturally responsive cognitive-behavioral therapy: Assessment, practice, and supervision* (pp. 97–116). Washington, DC: American Psychological Association. doi:10.1037/11433-004

*Linehan, M. M. (1993). *Cognitive-behavioral treatment of borderline personality disorder*. New York: Guilford.

Loewenthal, D., & House, R. (2010). *Critically engaging CBT.* Berkshire, UK: Open University Press.

Marchand, E., Ng, J., Rohde, P., & Stice, E. (2010). Effects of an indicated cognitive-behavioral depression prevention program are similar for Asian American, Latino, and European American adolescents. *Behaviour Research and Therapy, 48*(8), 821–825. doi:10.1016/j.brat.2010.05.005

McCollum, E., & Gehart, D. (2010). Using mindfulness to teach therapeutic presence: A qualitative outcome study of a mindfulness-based curriculum for teaching therapeutic presence to master's level marriage and family therapy trainees. *Journal of Marital and Family Therapy, 36,* 347–360. doi: 10.1111/j.1752-0606.2010.00214.x

McNair, L. D. (1996). African American women and behavior therapy: Integrating theory, culture, and clinical practice. *Cognitive and Behavioral Practice, 3*(2), 337–349. doi:10.1016/S1077-7229(96)80022-8

Meichenbaum, D. (1997). The evolution of a cognitive-behavior therapist. In J. K. Zeig (Ed.), *The evolution of psychotherapy: Third conference* (pp. 95–106). New York: Brunner/Mazel.

Miller, S. (2012, May 12). Revolution in Swedish mental health practice: The cognitive behavioral therapy monopoly gives way. Retrieved from www.scottmiller.com/?q=node%F160.

Mylott, K. (1994). Twelve irrational ideas that drive gay men and women crazy. *Journal of Rational-Emotive & Cognitive Behavior Therapy, 12*(1), 61–71. doi:10.1007/BF02354490

Organista, K. C., & Muñoz, R. F. (1996). Cognitive behavioral therapy with Latinos. *Cognitive and Behavioral Practice, 3*(2), 255–270. doi:10.1016/S1077-7229(96)80017-4

Patterson, G., & Forgatch, M. (1987). *Parents and adolescents: Living together: Part 1: The Basics.* Eugene, OR: Castalia.

Pavlov, I. P. (1932). Neuroses in man and animals. *Journal of the American Medical Association, 99,* 1012–1013.

Pedersen, P. B., Draguns, J. G., Lonner, W. J., & Trimble, J. E. (Eds.). (2002). *Counseling across cultures* (5th ed.). Thousand Oaks, CA: Sage.

Piedra, L. M., & Byoun, S. (2012). Vida Alegre: Preliminary findings of a depression intervention for immigrant Latino mothers. *Research on Social Work Practice, 22*(2), 138–150. doi:10.1177/1049731511424168

Rogers, Carl. (1961). *On becoming a person: A therapist's view of psychotherap*y. London: Constable.

Safren, S. A., Hollander, G., Hart, T. A., & Heimberg, R. G. (2001). Cognitive-behavioral therapy with lesbian, gay, and bisexual youth. *Cognitive and Behavioral Practice, 8*(3), 215–223. doi:10.1016/S1077-7229(01)80056-0

Safren, S. A., & Rogers, T. (2001). Cognitive-behavioral therapy with gay, lesbian, and bisexual clients. *Journal of Clinical Psychology, 57*(5), 629–643. doi:10.1002/jclp.1033

Segal, Z. V., Williams, J. G., & Teasdale, J. D. (2002). Mindfulness-based cognitive therapy for depression: *A new approach to preventing relapse.* New York: Guilford.

Singh, N. N., Lancioni, G. E., Joy, S., Winton, A. W., Sabaawi, M., Wahler, R. G., & Singh, J. (2007). Adolescents with conduct disorder can be mindful of their aggressive behavior. *Journal of Emotional and Behavioral Disorders, 15*(1), 56–63. doi:10.1177/10634266070150010601

Skinner, B. F. (1953). *Science and human behavior.* New York: MacMillian.

Sprenkle, D. H., & Blow, A. J. (2004). Common factors and our sacred models. *Journal of Marital and Family Therapy, 30,* 113–129.

Teasdale, J. D., Segal, Z. V., & Williams, J. M. C. (1995). How does cognitive therapy prevent depressive relapse and why should attentional control (mindfulness) help? *Behaviour Research and Therapy, 33,* 25–39.

Wachs, K., & Cordova, J. V. (2008). Mindful relating: Exploring mindfulness and emotional repertoires in intimate relationships. *Journal of Marital and Family Therapy, 33,* 464–481.

Walling, S., Suvak, M. K., Howard, J. M., Taft, C. T., & Murphy, C. M. (2012). Race/ethnicity as a predictor of change in working alliance during cognitive behavioral therapy for intimate partner violence perpetrators. *Psychotherapy, 49*(2), 180–189. doi:10.1037/a0025751

Willougby, B.L.B., & Doty, N. D. (2010). Brief cognitive behavioral family therapy following a child's coming out: A case report. *Cognitive and Behavioral Practice, 17*(1), 37–44. doi:10.1016/j.cbpra.2009.04.006

Zylowska, L. L., Smalley, S. L., & Schwartz, J. M. (2009). Mindful awareness and ADHD. In F. Didonna (Ed.), *Clinical handbook of mindfulness* (pp. 319–338). New York: Springer. doi:10.1007/978-0-387-09593-6_18

Cognitive-Behavioral Case Study: Korean Family Conflict Over Grades

Sun-ja, 37 years old (AF37), Young-soo, 42 years old (AM42), and Mark, 15 years old (CM15) come to counseling because of the mother's concern about the father and son's constant quarrels. The couple immigrated to the United States from Korea during college and stayed to raise their son, wanting him to have the opportunities available to him in the United States. Since Mark began high school, he and his father have had frequent arguments over what Young-soo perceives as Mark's lack of motivation and school performance (3.5 GPA). The parents disagree about how to handle the situation: Young-soo believes they should set strict rules and punishment for missed chores and unsatisfactory grades and disobediences, whereas Sun-ja doesn't agree, believing the father has unreasonable expectations of the son. She reasons that they should be grateful that unlike other teenage boys Mark stays out of trouble. Sun-ja worries that continual pressure on Mark can backfire and spark further rebellion. In addition to working outside of the home as a dental hygienist, she maintains the household with little help from Mark or Young-soo, leaving her feeling unappreciated and too tired to be intimate with her husband. Young-soo feels pressured to support the family living in an upscale neighborhood for its good schools, putting in long hours at his engineering firm. Recent financial troubles at his work have added tension to the couple's relationship, requiring them to eliminate the minimal housekeeping help and tutoring they had, thus adding more responsibilities for Sun-ja at home.

Cognitive-Behavioral Case Conceptualization

- *Problem Definition*
    - AM and CM argue three to five times per week over the same subjects: school, chores, attitude.
    - CM is not maintaining a high school grade point average that will get him into a "good" university (3.5 GPA).
    - AF and AM do not agree on how to parent CM regarding grades and chores.
- *Baseline Assessment*

Problem Behavior	When?	How Long?	How Severe?	Events Before?	Events After?
Monday argument	After AM came home and discovered CM watching TV and not studying for test	10 minutes between AM and CM	Moderate; CM quickly left to go study.	AM had tough day at work	AM and AF argue for another 10 minutes without much resolution; CM stay in room until bedtime
Wednesday argument	After AM came home, discovered trash not taken out	3 minutes	Mild; CM left room to take trash out	AM had tough day at work; AF was busy with her housework	Whole family quiet and distance for rest of night
Friday argument	CM asks to go to party with friends	15-minute argument; AM very disparaging of CM, his friends, his future	Severe: AM was particularly harsh	AM had tough day at work	CM angrily shuts door; AF tries to mediate between CM and AM. CM stays in room; AF and AM silently watch TV.

- *Functional Analysis:*
    - *How does this specific problem handicap this person (and/or the family) in everyday life?*
      The family is in constant state of tension, leaving everyone emotionally exhausted and the CM closed off and feeling defensive.
    - *What would happen if the problem were reduced in frequency?*
      The family could interact more peacefully; family members would be more likely to cooperate and reach agreement on what are reasonable expectation from one another.
    - *What would this person (and his or her family) gain if the problem were resolved?*
      The family would gain emotional closeness; CM a sense of autonomy; AF a sense of peacefulness; AM feeling respected and that his sacrifices matter.
    - *Who (or what) reinforces the problem with attention, sympathy, and support?*
      AF and AM not working as a team is reinforcing the power struggle with CM. AF often offers sympathy to both AM and CM.

- *Under what circumstances is the specific problem reduced in intensity?*
  Fights better:
  —When CM brings home a high grade on a significant assignment.
  —When AM comes home and CM is studying.
  —When AM comes home and CM's chores are done.
- *Under what circumstances is the specific problem increased in intensity?*
  Fights are worse:
  —When CM brings home a particularly bad grade.
  —When CM asks for greater freedom for social activities.
  —When AM more stressed at work.
  —When AF less involved in managing CM's life.
- *What do family members currently do to cope with the problem?*
  They separate themselves into different rooms or pretend that nothing happens after a fight.
- *What are the assets and deficits of the family as a problem-solving unit?*
  Despite the tension in the family and individual's differences in defining the problem, the family's willingness to seek counseling shows that there is still an underlying sense of affection and connection. The family's biggest deficit is parental disagreement on rules and expectations and difficulty with problem solving.
- *What behaviors need to increase? Decrease?*
  AM and AF need to find more effective ways to jointly parent CM.
  AM and CM need to find more effective ways to problem solve and community.

- *A-B-C Theory: Identify Irrational Beliefs*
  - Identify A (Action/trigger event); C (Consequence: behavior, mood); and B (irrational belief that leads to C).
    A: CM's unsatisfactory school performance.
    B: AM believes CM is setting himself up for a difficult life and that CM is unappreciative of the sacrifices AM has made for him. AM believes that if AF would push CM harder, they would not have these problems.
    C: AM criticizes CM and AF.

- *Cognitive Schemas and Core Beliefs*
  - *Selective Abstraction:* AF and AM: We failed as a parent because our son's GPA is not high; AM ignores the fact that CM generally makes good decisions and stays out of trouble.
  - *Overgeneralization:* AF: Strict parenting results in defiance. AM: Mediocre grades in his first year of high school will keep him from going to a good college. As a minority in the United States, one has to be better than others to succeed in this world.
  - *Magnification and Minimization:* AM: If CM does not get into a great college, he will never be successful. There is too much working against him as a minority. AF: Setting limits now could make things far worse. CM: Does not see father's dedication as a provider as showing he cares.
  - *Personalization:* AM: Reasons that his wife must not love him, because she disagrees with him on how to parent. CM: Believes father does not love him because he is so harsh.
  - *Dichotomous Thinking:* CM: Thinks his parents think of him only as either good or bad if he gets straight As. He also thinks of his father as being "too Korean" and not understanding his life as an American teen ("We aren't in Korea anymore, Dad."). AM: Believes CM's grades are the only route to success.

- *Mislabeling:* AM: "CM is lazy if he is not getting perfect grades at school." CM: "My dad doesn't love me because he is gets angry at me."
- *Relational Cognition Patterns*
  - *Selective Perceptions:* AM: Focuses on CM's grades and ignores other aspects of his son's life. Focuses on the challenges of being a minority.
  - *Attributions:* AM: Believes CM's laziness is cause of mediocre grade. CM: Sees father as uncaring because he gets mad about grades and ignores all else; does not think his father understands him primarily because he is too Korean.
  - *Expectancies:* AM: Expects CM to continue to be a mediocre student and for AF to indulge him. AF: Expects that if AM continues to be harsh with CM things will get worse. CM: Expects his father to be unapproving of him not matter what he does.
  - *Assumptions:* AF: Does not believe AM cares about her struggles. CM: Does not think father is proud of him. AM: Assumes CM will not work harder unless forced to do so by the parents.
  - *Standards:* AM: Straight As grades are needed to ensure a successful future as a minority. CM: Expects to have all of his needs met by his parents. AF: Wants more assistance from husband, as is typical in the United States.

Cognitive-Behavioral Treatment Plan

CBT Initial Phase of Treatment with Family

Initial Phase Counseling Tasks

1. Develop working counseling relationship. *Diversity Note: Attend to possible cultural expectations to receive guidance from an expert; protect dignity and avoid embarrassment of any family member; respect family hierarchy.*
   a. Develop an *empathic, supportive* working relationship with all members of system in which therapist is able to effectively provide *education* from a position of *expertise.* Focus on strengths of family members.

2. Assess individual, systemic, and broader cultural dynamics. *Diversity Note: Consider impact of immigration, socioeconomic status, and Korean family norms. Assess each parent for level of acculturation and connection to native culture. Identify local cultural supports as well as connection to extended family.*
   a. *Baseline assessment* of fighting and *functional analysis* of the family's conflict.
   b. Identify each person's and jointly held *irrational beliefs, family schemas,* and *core beliefs* that are the source of relational conflict; examine cultural foundations for beliefs.

Initial Phase Client Goals

1. Increase parents' ability to effectively coparent CM to reduce conflicts between AM/AF and AM/CM.
   a. *Parent training* (possibly mindfulness-based if couple interested) to help parents jointly define rules and expectations for CM and decide how they will approach reinforcing these as a team.
   b. Identify *family schemas* that underlie each parent's position on how to manage CM, what defines "success," and how best to respond to CM.
   c. *Bibliotherapy* on parenting to provide parents with ideas for working together and to reframe issues, such as Dan Siegel's *The Whole-Brain Child.*

(continued)

CBT Working Phase of Treatment with Couple/Family

Working-Phase Counseling Task

1. Monitor quality of the working alliance. *Diversity Note: Monitor relationship; generally as an Asian family begins to feel safe in therapy, they will be less guarded and more open about intimate details.*
   a. Check in with client at regular intervals about satisfaction with therapy process and goals.

Working-Phase Client Goals

1. Increase *cooperative interactions* between CM and parents to reduce conflict.
   a. Set up *contingency contract* that clearly spells out privileges associated with CM's grades and behaviors; contract to be enforced by AF as well as AM. Contract to emphasize CM taking proactive steps toward grades and homework rather than rely on mother to oversee work.
   b. *Socratic dialogue* to help CM develop greater self-motivation to be successful.

2. Increase family's ability to *communicate* respectfully and effectively to reduce conflict.
   a. *Psychoeducation and communication training* to increase skills and knowledge with in-session role plays.
   b. *Homework tasks* that help family to practice new communication skills.

3. Increase couple's and family's ability to effectively *problem solve* to reduce conflict.
   a. *Problem-solving training* to help couple/family learn skills and strategies for effectively solving problems. Assist family with identifying and verbalizing emotional needs when problems arise.
   b. *Homework tasks* to practice problem solving with initially minor and then increasingly difficult issues. Demonstrate the process prior to assigning the client to do it on their own.

CBT Closing Phase of Treatment with Family

Closing-Phase Counseling Task

1. Develop aftercare plan and maintain gains. *Diversity Note: Connect with school, Korean church, or other cultural resources to help family feel supported.*
   a. Refer to *mindfulness-based couple/parenting group* or similar group to reduce potential for relapse.

Closing-Phase Client Goals

1. Increase couple's sense of emotional connection and enjoyable interactions to increase capacity for intimacy.
   a. *Bibliotherapy* to develop more realistic expectations of relationships, such as John Gottman's *7 Principles That Make a Marriage Work*.
   b. *Communication training* that emphasizes expression of acceptance and appreciation of one another.

2. Increase each family member's ability to make *realistic interpretations* of daily life and identify when thoughts are irrational to reduce conflict and increase sense of wellness.
   a. *Problem-solving training* to help client manage ongoing daily challenges.
   b. *Thought records* and *Socratic dialogue* with self to continue identifying and debunking unrealistic beliefs.

CHAPTER 12

Solution-Based Therapies

"We intend to influence clients' perceptions in the direction of solution through the questions we choose to ask and our careful use of solution language. Reflection upon these questions helps clients to consider their situations from new perspectives."

—O'Hanlon & Weiner-Davis, 1989, p. 80

Lay of the Land

The most well known and arguably the first strength-based therapies, solution-based therapies are positive, active approaches that help clients move toward desired outcomes. The following three main strands of practice share many more similarities than differences:

- *Solution-Focused Brief Therapy (SFBT):* Developed by Steve de Shazer (1985, 1988, 1994) and Insoo Berg at the Milwaukee Brief Family Therapy Center, this therapy focuses on the future with minimal discussion of the presenting problem or the past; interventions target small steps in the direction of the solution.
- *Solution-Oriented Therapy:* This therapy (O'Hanlon & Weiner-Davis, 1989) and the related approach, *possibility therapy* (O'Hanlon & Beadle, 1999), were developed by Bill O'Hanlon and colleagues to incorporate a similar future orientation while drawing more directly from the language techniques used in Ericksonian trance. Solution-oriented therapy also uses more interventions that draw from the past and present to identify potential solutions than does solution-focused brief therapy.
- *Solution-Oriented Ericksonian Hypnosis:* Both solution-focused and solution-oriented therapies were inspired by the work of Milton Erickson, whose strength-oriented trance work is one of the most popular modern approaches to hypnosis.

Because of their numerous commonalities, this chapter presents solution-focused and solution-oriented therapies together under the general term *solution-based therapies*; it then discusses Ericksonian hypnosis.

Solution-Based Therapies

In a Nutshell: The Least You Need to Know

Solution-based therapies are brief therapy approaches that grew out of the work of the Mental Research Institute in Palo Alto (MRI) and Milton Erickson's brief therapy and trance work (de Shazer, 1985, 1988, 1994; O'Hanlon & Weiner-Davis, 1989). The first and leading "strength-based" therapies, solution-based therapies are increasingly popular with clients, insurance companies, and county mental health agencies because they are efficient and respectful of clients. As the name suggests, solution-based therapists spend a minimum of time talking about problems and instead focus on moving clients toward enacting *solutions*. Rather than being "solution-givers," solution-based therapists work with the client to envision potential solutions based on the client's experience and values. Once the client has selected a desirable outcome, the therapist assists the client in identifying small, incremental steps toward realizing this goal. The therapist *does not* solve problems or offer solutions but instead collaborates with clients to develop aspirations and plans that they then translate into real-world action.

Common Solution-Based Therapy Myths

More so than others, solution-based therapists are haunted by myths and misconceptions about what actually happens in session. So let's straighten these out before we go any further.

Myth: Solution-Based Therapists Propose Solutions (Give Advice)

Closer to the Truth: Solution-based therapists do not suggest logical solutions to clients (O'Hanlon & Beadle, 1999). Instead, it is the client who identifies solutions with the help of the therapist, who identifies *exceptions* to the problem, descriptions of what is already working, and client resources to help the client envision potential solutions. Once a clear, behavioral goal is identified, the therapist works with the client to take small steps in this direction.

Myth: Solution-Based Therapists Never Talk About the Problem

Closer to the Truth: Solution-based therapists are not psychic and therefore, like all therapists, must spend some time talking about the problem. However, they spend *less* time talking about the problem than most other therapists, especially SFBT therapists (De Jong & Berg, 2002). Solution-based therapists typically follow their clients' lead in determining how much and how often they need to talk about the problem versus the solution. To add further myth-busting evidence, hallmark techniques, such as *exception questions,* require talking about the problem as part of identifying the solution.

Myth: Solution-Based Therapists Never Talk About the Past

Closer to the Truth: Again, solution-based therapists are not psychic, and they actually have numerous techniques that are grounded in talking about the past. However, when they talk about the past, they focus on strengths as well as the problem (Bertolino & O'Hanlon, 2002). Talking about the past is one of the most important means of identifying solutions: what has worked and what has not. The past is talked about in ways that facilitate the enactment of solutions.

Myth: Emotions Are Not Discussed

Closer to the Truth: Emotions cannot be avoided in any therapy. However, solution-based therapists do not view the expression of emotion as curative in and of itself, as is assumed in humanistic therapies (Lipchik, 2002). Instead, emotions are used as clues to what works and what does not and where clients want to go.

The Juice: Significant Contributions to the Field
If you remember one thing from this chapter, it should be this:

Assessing Client Strengths

Assessing client strengths is one of the key practices in solution-based therapies (Bertolino & O'Hanlon, 2002; DeJong & Berg, 2002; O'Hanlon & Weiner-Davis, 1989). Strengths include resources in a person's life (personally, relationally, financially, socially, or spiritually) and may include family support, positive relationships, and religious faith. Most therapists *underestimate* the difficulty of identifying strengths. In fact, identifying client strengths is often harder than diagnosing pathology because clients come in with a long list of problems they want fixed and are thus prepared to discuss pathology. Surprisingly, many clients, especially those with depressive or anxious tendencies (the majority of outpatient cases), have great difficulty identifying areas without problems in their lives. Similarly, many couples who have been distressed and arguing for long periods of time have difficulty identifying positive characteristics in their partners, happy times in the marriage, and, in the most desperate of cases, the reasons they got together in the first place. Therefore, therapists often have to ask more subtle questions and attend to vague clues in order to assess strengths well.

Solution-based therapists assess strengths in two ways: (a) by directly asking about strengths, hobbies, and areas of life that are going well, and (b) by listening carefully for *exceptions* to problems and for areas of unnoticed strength (Bertolino & O'Hanlon, 2002; De Jong & Berg, 2002). Furthermore, I have found that any strength in one context has the potential to be a liability in another context and that the inverse is also true: any weakness in one area is generally a *strength* in another area (see client strengths in Chapter 17). Therefore, if a client has difficulty identifying strengths and more readily discusses weaknesses and problems, a finely tuned solution-focused ear will be able to identify potential areas in which the "weakness" is a strength. For example, a person who is critical, anxious, or negative in a relational context typically excels at detailed or meticulous work and tasks. This insight can be useful in identifying ways for clients to move toward their goals. Solution-based therapists have been in the vanguard of a larger movement within mental health that emphasizes identifying and utilizing client strengths to promote better clinical outcomes (Bertolino & O'Hanlon, 2002). Increasingly, county mental health and insurance companies are requiring assessment of strengths as part of initial intake assessments. The key to successfully assessing strengths is having an unshakable belief that *all* clients have significant and meaningful strengths no matter how dire and severe their situations appear. It is helpful for therapists to remember that we see people typically in their worst moments; therefore, even if we are not seeing strengths in the present moment, they are undoubtedly there. Solution-based therapists maintain that all people have strengths and resources, and they make it their job to help identify and utilize those strengths toward achieving client goals. In the case study at the end of the chapter, the therapist is working with an interfaith/intercultural couple who is questioning whether they want to continue their relationship given Shirin's family's opposition. When assessing strengths, the therapist identifies not only the couple's individual and relational strengths but also about each partner's strong emotional bond with their families as well as a large number of relatives supportive of the relationship in Shirin's extended family as well as her rabbi. The therapist can then help the couple to draw upon these resources as they make the difficult decision of how to proceed with their relationship.

Rumor Has It: The People and Their Stories
Behind-the-Scenes Inspiration: Milton Erickson

Milton Erickson served as an inspiration to Bill O'Hanlon's solution-oriented therapy and Steve de Shazer's solution-focused brief therapy. Trained in medicine as a psychiatrist, Erickson was a master therapist who was well known for his brief, rapid, and creative interventions (Erickson & Keeney, 2006; Haley, 1993; O'Hanlon & Martin, 1992). Rather than follow a specific theory, Erickson relied on keen observation while listening with an open mind to each patient's unique story. He frequently employed light trance work to evoke patient strengths and latent abilities (Haley, 1993; O'Hanlon & Martin, 1992). At a time when most therapies focused on the past, Erickson directed his clients to focus on the present and future, often envisioning times without the problem. Although others have meticulously studied his work, there has been no consensus or singular definition of Ericksonian therapy. The influence of his trance work is clearly evident in interventions such as the *miracle question* and the *crystal ball technique*, which rely on a pseudo-orientation to time and the implicit assumption that change *will* occur.

Solution-Focused Brief Therapy: Milwaukee Brief Family Therapy Center

Steve de Shazer   Steve de Shazer was a deeply thoughtful leader in the field. After his early work at the MRI with giants such as Jay Haley, Paul Watzlawick, John Weakland, and Virginia Satir, the late Steve de Shazer developed with his wife Insoo Kim Berg solution-focused brief therapy: "An iconoclast and creative genius known for his minimalist philosophy and view of the process of change as an inevitable and dynamic part of everyday life, he was known for reversing the traditional psychotherapy interview process by asking clients to describe a detailed resolution of the problem that brought them into therapy, shifting the focus of treatment from problems to solutions" (Trepper, Dolan, McCollum, & Nelson, 2006, p. 133). De Shazer was a prolific writer (de Shazer, 1985, 1988, 1994; de Shazer & Dolan, 2007), laying much of the philosophical and theoretical foundations for solution-focused work. His early work was influenced by the trance work of Milton Erickson (de Shazer, 1985, 1988) and his later work by Ludwig Wittgenstein, who viewed language as inextricably woven into the fabric of life (de Shazer & Dolan, 2007). With Insoo Kim Berg, de Shazer founded the Solution-Focused Brief Therapy Association and the Milwaukee Brief Family Therapy Center, where they trained therapists until their deaths in 2005 (Steve) and 2007 (Insoo).

Insoo Kim Berg   Warm and exuberant, Insoo Kim Berg was an energetic developer and leading practitioner of SFBT, cofounding the Milwaukee Brief Family Therapy Center and the Solution-Focused Brief Therapy Association with her husband, Steve de Shazer (Dolan, 2007). She was an outstanding clinician who furthered the development of SFBT, developing solution-focused approaches to working with drinking issues (Berg & Miller, 1992), substance abuse (Berg & Reuss, 1997), family-based services (Berg, 1994), child protective services families (Berg & Kelly, 2000), children (Berg & Steiner, 2003), and personal coaching (Berg & Szabo, 2005).

Scott Miller   Originally trained at the Milwaukee Brief Family Therapy Center with de Shazer and Berg, Scott Miller, along with his colleagues Barry Duncan and Mark Hubble, has been a strong proponent of the *common factors* movement (see Chapter 2; Miller, Duncan, & Hubble, 1997) and client-centered, outcome-informed therapy (see Chapter 1).

Yvonne Dolan   Yvonne Dolan studied and worked with de Shazer and Berg at the Milwaukee Brief Family Therapy Center, specializing in sexual abuse and trauma treatment (Dolan, 1991, 2000).

Linda Metcalf   Linda Metcalf applies solution-focused therapy in school counseling contexts, including solution-focused school counseling (Metcalf, 2008), solution-focused children's groups (Metcalf, 2007), solution-focused parenting (Metcalf, 1998), and solution-focused teaching (Metcalf, 2003).

Solution-Oriented Therapy

Bill O'Hanlon   A former student of Milton Erickson, Bill O'Hanlon is an energetic and popular leader in solution-oriented, strength-based therapies, including solution-oriented therapy and possibility therapy (O'Hanlon & Beadle, 1999; O'Hanlon & Weiner-Davis, 1989).

A prolific and highly accessible writer and speaker, O'Hanlon emphasizes the significance of language, using subtle shifts in language to spark change. His work aims to transform client viewing as well as to solve the problem while attending to broader contextual issues that impact the client's situation (Bertolino & O'Hanlon, 2002). He has written extensively on numerous topics, including solution-oriented couples therapy (Hudson & O'Hanlon, 1991), solution-oriented approaches to treating sexual abuse (O'Hanlon & Bertolino, 2002), solution-oriented hypnosis (O'Hanlon & Martin, 1992), solution-oriented therapy with children and teens (Bertolino & O'Hanlon, 1998), and spirituality in therapy (O'Hanlon, 2006), as well as books for clients and popular audiences (O'Hanlon, 2000, 2005, 2006).

Michelle Weiner-Davis   Michelle Weiner-Davis developed a highly successful solution-oriented approach to working with divorce that she calls *divorce busting* (Weiner-Davis, 1992). She uses a brief, solution-oriented self-help approach for couples who want to prevent divorce.

Collaborative, Strength-Based Therapy
Matthew Selekman   Grounding his work in solution-oriented therapies and systems theory, Matthew Selekman has developed collaborative, strength-based therapies for working with children, adolescents, families, and self-harming adolescents (Selekman, 1997, 2005, 2006).

Ericksonian Hypnosis
Stephen Lankton   Having training with Erickson for several years, Stephen Lankton has written extensively about clinical applications of Ericksonian hypnosis and in many ways helped this approach to become a mainstream and widely recognized form of hypnosis (Lankton, 2008, 2009; Lankton & Erickson, 1994; Lankton & Lankton, 2008; Lankton & Matthews, 2010; Lankton, Gilligan, & Zeig, 1991; Lankton, Lankton, & Matthews, 1991). He has also been the editor-in-chief of the *American Journal of Clinical Hypnosis* since 2005.

Solution-Focused Associations
SFBT has been influential in the United States, Europe, Latin America, and Asia:

- *European Brief Therapy Association:* Founded in 1994, the European Brief Therapy Association serves as a network of brief therapy practitioners in Europe who have worked closely with de Shazer and Berg over the years. They sponsor research projects in brief therapy and meet annually, bringing together therapists from across Europe and the world.
- *Solution-Focused Brief Therapy Association (SFBTA):* In 2001, Steve de Shazer and Terry Trepper began organizing SFBTA to bring together North American SFBT practitioners and researchers. Under the dynamic leadership of Thorana Nelson, Eric McCollum, Terry Trepper, Yvonne Dolan, and others, SFBTA is an active and engaged community of practitioners, educators, and researchers who work to further develop and refine SFBT practices.

The Big Picture: Overview of Treatment
Small Steps to Enacting Solutions
In broad strokes, solution-based therapists help clients identify their preferred solution (by talking about the problem, exceptions, and desired outcomes) and work with clients to take small, active steps in this general direction each week (O'Hanlon & Weiner-Davis, 1989). In some cases, this is a very time-limited approach, often as few as 1 to 10 sessions; solution-focused therapists are advocates of the possibility of single-session therapy. In more complex cases, such as the treatment of sexual abuse or alcohol dependency, therapy may take years (O'Hanlon & Bertolino, 2002).

Making Connection: The Therapeutic Relationship

The Zen of Viewing: The Beginner's Mind

O'Hanlon and Weiner-Davis (1989) use the concept of a beginner's mind when forming a relationship, referring to the classic Zen saying "In the beginner's mind there are many possibilities; in the expert's mind there are few" (p. 8). Assuming a position of beginner's mind involves listening to each client's story as if you are listening for the first time, not filling in blanks with personal or professional knowledge. Most therapists underestimate how hard this is to do. When a client starts talking about "feeling depressed," most therapists believe they have useful diagnostic information, unthinkingly assuming that clients have read diagnostic manuals and use the term as a professional would. In contrast, solution-oriented therapists bring a beginner's mind to the conversation and are curious about *how this person experiences his/her unique depression*. If you get in the habit of asking, you will find that every depression is surprisingly "one of a kind." Thus, solution-oriented therapists make no assumptions when they are listening, asking to hear more about clients' unique experiences and understandings.

Echoing the Client's Key Words

Solution-based therapists carefully attend to *client word choice* and echo their key words whenever possible (De Jong & Berg, 2002). For example, rather than teaching clients to use psychiatric terms such as *depression* or *hallucinations* to describe their experience, the therapist prefers to use the clients' own language, such as "feeling blue" or "schizos." Using client language often makes the problem more solvable and engenders greater hope. For many, "ending the blahs" or "getting back to my old self" is a more attainable goal than treating a psychiatrically defined problem of "Major Depressive Disorder, Single Episode, Moderate." In the case study at the end of the chapter, Shirin frequently repeats that she fears her family will "cut her off" if she stays with Brian; the therapist uses this language and also identifies what this behaviorally and practically may translate to if it comes to pass, considering emotional, financial, relational, and other forms of cutoff, to help her think through her decision.

Carl Rogers With a Twist: Channeling Language

O'Hanlon and Beadle (1999) describe how solution-oriented therapists use reflection, including reflection of feelings, with clients to build rapport. This approach is similar to humanistic approaches, such as Carl Rogers's client-centered therapy, but with a twist: solution-oriented reflections *delimit* the difficult feeling, behavior, or thought by reflecting on a time, context, or relational limit. Such reflections generally take three forms:

1. *Past Tense Rather Than Chronic State or Characteristic:* Therapists reflect statements back to clients in the *past tense*, e.g., "You were feeling down yesterday."
2. *Partial Rather Than Global:* Therapists reflect global statements back to clients as *partial* statements, e.g., "Your partner sometimes/often (instead of always) does things that annoy you."
3. *Perception Rather Than Unchangeable Truth:* Therapists reflect a client's "truth" or "reality" claim as a *perception*; e.g., in response to "I'll never find anybody," they might say, "There does not seem to be anybody you are interested in right now."

For example, suppose a client is telling a story about how her boyfriend got angry at her for "no reason." A client-centered reflection would be something like "You aren't feeling understood" (present-focused statement about client's unexpressed emotion), whereas the solution-focus twist to delimit would be "You were not feeling understood *by your boyfriend last Saturday*." The solution-oriented twist emphasizes the limited time and relational context in order to (a) define the problem in more solvable ways and (b) create hope. O'Hanlon and Beadle (1999) refer to this process as "channeling language." This technique helps both the client and therapist transition to desired outcomes. The therapist in the case study at the end of the chapter uses channeling language to help the couple avoid totalizing language about each other and their families and to help them reduce their arguments and identify possibilities for viewing and relating to their families of origin.

Optimism and Hope

"Hope is like a road in the country; there was never a road, but when many people walk on it, the road comes into existence."—Lin Yutang

Optimism and hope are brilliantly palpable in solution-based therapies (Miller, Duncan, & Hubble, 1996, 1997). With all clients, solution-based therapists assume that change is inevitable and that improvement—in some form—is always possible (O'Hanlon & Weiner-Davis, 1989). Their optimism and hope do not stem from naïveté but rather from their ontology and epistemology: their theory of what it means to be human and how people learn. Because change is always happening—moods, relationships, emotions, and behaviors are in constant flux—change is inevitable (Walter & Peller, 1992). They have hope that the change will be positive because the client is in therapy to make an improvement and because over 90% of clients report positive outcomes in psychotherapy (Miller et al., 1997). Hope is cultivated early in therapy to develop motivation and momentum (Bertolino & O'Hanlon, 2002).

Assumption of Solution and Possibilities

A subtle but essential element of the therapeutic relationship, solution-focused view all clients as resilient and capable of enacting solutions (Trepper, McCollum, de Jong, Korman, Gingerich, & Franklin, 2012). Although philosophical at one level, this assumption is realized in their word choice, descriptions, and verb tense—"when you two are getting along better" or "as you recover from trauma." This belief that clients are fully capable of change tends to inspire clients to believe in themselves and is key to making the interventions associated with this approach work.

The Viewing: Case Conceptualization and Assessment
Strengths and Resources

Assessing strengths and resources is covered in Juice above and in Chapter 17.

Exceptions and "What Works"

Solution-based therapists listen for exceptions and examples of what works when clients are talking (de Shazer, 1985, 1988; O'Hanlon & Weiner-Davis, 1989). If you are listening closely, most clients spontaneously offer exceptions and examples of what works: "His ADHD is less of a problem in his math class" or "When his stepbrother helps him with homework, there does not seem to be much of a problem." These exceptions provide clues to what works and therefore what clients need to do more frequently.

Solution-based therapists identify exceptions and descriptions of what works in two ways: (a) indirectly, by listening for spontaneous descriptions, and (b) directly, by asking questions. They listen as carefully for exceptions as medical-model therapists listen for diagnostic symptoms. They then use these exceptions to help clients enact preferred solutions in their lives. They also ask exception questions to gather more information about what works.

Examples of Exception Questions

- Are there any times when the problem is less likely to occur or be less severe?
- Can you think of a time when you expected the problem to occur but it didn't?
- Are there any people who seem to make things easier?
- Are there places or times when the problem is not as bad?

The vast majority of clients can identify exceptions with questions such as these. The underlying assumption is that the problem varies in *intensity*; the times when the problem is *less severe* are considered exceptions and generally provide clues to what works (de Shazer, 1985; O'Hanlon & Weiner-Davis, 1989). Especially with diagnoses such as depression, which are experienced most of the time on most days, therapists need to focus on variations of intensity rather than absence of the symptom to identify exceptions. With couples, the therapist listens for times when the couple are not fighting, "get along," or get things done, which may be in relation to children, work, extended family, and so forth.

Client Motivation: Visitors, Complainants, and Customers

Steve de Shazer (1988) assessed client motivation for change using three categories: visitors, complainants, and customers.

- *Visitors* do not have a complaint, but generally others have a complaint against them. They are typically brought to therapy by an outside other, such as courts, parents, or spouse.
- *Complainants* identify a problem but expect therapy or some other person to be the primary source of change. They are there to have their problems fixed by an expert.
- *Customers* identify a problem and want to take action toward the solution.

Client Motivations

	Visitor	Complainant	Customer
Motivation	Low	Moderate to high	High
Source of Problem or Solution	Outside other (spouse, parent, court) thinks client has a problem.	Problem generally related to outside cause or person; expect therapist or another to be source of solution.	Self as part of problem and active agent in solution.
View of Who Needs to Change	Outside other needs to believe there is no problem.	Outside other needs to change and/or fix it.	Self needs to take action to fix things.
Building Therapeutic Alliance	Therapist identifies areas where client sees a problem and is willing to become a customer for change.	Therapist honors client's view of the situation while identifying specific instances where client can make a difference.	Therapist joins with client by complimenting readiness for change.
Focus of Interventions	Building alliance; understanding client perspective; framing outside request for change as "the problem."	Observation-oriented tasks (e.g., identifying exceptions over the week; Selekman, 1997).	Reframing; identifying what does not work; action-oriented tasks.
Readiness for Action	Not ready for making active changes in life until client believes there is some sort of problem and is motivated to change.	Not ready for action until open to the idea that client's actions can make a difference.	Ready to take action to make changes.

Assessing clients' motivation is helpful for knowing how to join with the client as well as knowing how to proceed. Many new therapists assume that all clients are customers for change: ready to take action to improve their situation simply because they showed up for a therapy session. However, this is often not the case. People come to therapy with mixed emotions and levels of motivation. Generally, most mandated clients are "visitors," and therapists need to find a way to connect with their agenda while still working with the referring party's agenda. This same dynamic is often the case with children, teens, and even one-half of a couple. With complainants, therapists need to either find ways that the client can contribute to making a difference or help shift the viewing of the problem to increase client willingness to take action.

Targeting Change: Goal Setting

Goal Language: Positive and Concrete

Solution-based therapists state their goals in positive, observable solution-based terms (De Jong & Berg, 2002; Bertolino & O'Hanlon, 2002). Positive goal descriptions emphasize what the client is *going to be doing* rather than focusing on symptom reduction, which is typical in the medical model and cognitive-behavioral therapies. Observable descriptions include clear, specific behavioral indicators of the desired change.

Examples of Observable and Nonobservable Goals

Positive, Observable Goals	Negative (Symptom-Reducing), Nonobservable Goals
Increase periods of enjoyable activity, social interaction, and hope for future	Reduce depression
Increase frequency of couple's emotionally intimate conversations	Reduce couple conflict
Increase cooperation and pro-social activities	Reduce defiance

You may have noticed that simply reading the left column generates more hope and provides greater direction for clinical change than the right column. Positive, observable goals provide a constant reminder to the therapist and client of the goal, and reinforce a solution-focused and solution-oriented perspective. Many solution-based techniques, such as scaling questions (discussed later in chapter), invite the client to measure goal progress weekly; thus carefully crafted goal language is particularly critical to success in solution-based therapies.

Solution-based goals should also have the following qualities (Bertolino & O'Hanlon, 2002; De Jong & Berg, 2002):

- *Meaningful to Client:* Goals must be personally important to the client.
- *Interactional:* Rather than reflect a general feeling (e.g., "feeling better"), the goals should describe how interactions with others will change.
- *Situational:* Goals are stated in situational terms (e.g., "improved mood at work") rather than global terms.
- *Small Steps:* Goals should be short-term and with identifiable small steps.
- *Clear Role for Client:* Goals should identify a clear role for the client rather than for others.
- *Realistic:* Goals need to be realistic for this client at this time.
- *Legal and Ethical:* Goals should be legal and adhere to client, therapist, and professional ethics.

Miracle and Solution-Generating Questions

Solution-based therapists use several related questions to help generate solutions and identify goals early in therapy: miracle questions (de Shazer, 1988), crystal ball questions (de Shazer, 1985), magic wand questions (Selekman, 1997), and time machine questions (Bertolino & O'Hanlon, 2002). When successfully delivered, these questions help clients envision a future without the problem, generating hope and motivation. When poorly delivered, they can be quite awkward.

> ### Examples of Solution-Generating Questions
>
> - *Miracle Questions:* "Imagine that you go home tonight and during the middle of the night a miracle happens: all the problems you came here to resolve are miraculously resolved. However, when you wake up, you have no idea a miracle has occurred. What are some of the first things you would notice that would be different? What are some of the first clues that a miracle has occurred?"
> - *Crystal Ball Questions:* "Imagine I had a crystal ball that allowed us to look into the future to a time when the problems you came here for are already resolved. I hold it up to you, and you look in. What do you see?"
> - *Magic Wand Questions:* "Imagine I had a magic wand (or imagine that this magic wand I have actually works), and after you leave I wave it and all of the problems you came here for are resolved overnight. You of course have no idea that your problems have been solved. When you wake up in the morning, what would be the first clues that something is different? What would you be doing differently? What would others be doing differently?"
> - *Time Machine Questions:* "Imagine I had a time machine that could propel you into the future to the point in time when the problems you came to see me for are totally resolved. Imagine you stepped in: Where do you end up? Who is with you? What is happening? How is your life different? How did your problems go away?"

Successfully delivering one of these solution-generating questions is far more difficult that it appears. As you can imagine, if done poorly these hit the floor like a lead balloon. To avoid such humiliation, de Shazer et al. (2007, pp. 42–43) describe seven steps for successfully delivering the miracle question.

1. *Obtain Client Agreement (Wait for Nod 1):* The first and most critical step is to prepare clients by changing their state of mind so that they are willing to engage in an atypical conversation. By asking, "Is it okay if I ask you a strange question?" or "Would you be willing to play along if I ask a somewhat odd question?," the counselor signals clients to change their frame of mind so that they are better able to enter a more fanciful, creative conversation. The counselor waits for the client to say, "yes," or nod in agreement.
2. *Custom Tailor Initial Setup (Wait for Nod 2):* After the client agrees to the odd question, begin delivering the question but customize it to include numerous little details from the client's everyday life to get them fully engaged in the story and enable them to better visualize the miracle. For example, you might say, "Let's imagine that after we are done talking here, you get in your car and drive home; you make dinner for the family like you usually do; you clean up dishes like you usually do; and check the kids' homework like you usually do." *Continue until the client starts to nod.*
3. *Setup for Miracle (Wait for Nod 3):* Once the client nods, continue with, "Then you get the kids to bed, maybe do some minor chores or watch television, and then you finally get to bed and fall asleep." Up to this point, you have only asked the client to imagine a regular day, but this is very important. The client needs to mentally leave the therapy office and vividly imagine being at home where the miracle is to occur. *Wait for the confirmatory nod or "yea" before moving on to the miracle.*
4. *Introduce the Miracle (Wait for Pause):* "Then, during the night ... while you are sleeping ... a miracle happens." Pause. Pause and wait for a reaction: a smile, lifted eyebrow, laugh, or questioning look. Insoo says she often looks intently at the client and smiles, but Steve warns that if the pause is too long the client is likely to respond with, "I don't believe in miracles"; so keep moving.
5. *Specifically and Clearly Define the Terms of the Miracle:* "And it's not just any miracle. This miracle makes the *problems that brought you here* today disappear ... just like that" (snapping your fingers at this point is optional but adds flair). The most common mistakes counselors make with the miracle question is to leave out the "problems that brought you here" part of the question. Without this "limit" on the

miracle, clients will spend a lot of time exploring vague and unrelated goals and the counselor will have to do a lot of repair work by later asking follow-up questions to get this information.

6. *Add the Mystery (Nod 4, Optional):* "But since the miracle happens while you are asleep, you won't know it has happened." If delivered well, most clients nod at this point, stare off into space, and begin to behave as if they were thinking about the proposition.
7. *Ask What Is Different:* "So, you wake up the morning after the miracle happens during the night. All the problems that brought you here are gone—poof—just like that. What is the very first thing you notice after you wake up? What are the little changes you start to notice that tell you the problem is gone?" At this point, many clients take a moment to think about their answer, often becoming quiet and still as their breath slows. The counselor needs to quietly and patiently *wait* for the response.

Once the client begins to describe what is different, the counselor helps focus the answer by asking about behavioral changes in the client and others. In typical solution-focused style, the counselor is not interested in what the client is *not* doing, but instead in what the client *is doing*. For example, if the client says, I will not be depressed any more, the counselor responds with, "What will you be doing instead?" Once the client identifies one new behavior, then the counselor asks for more: "What else would be different?" The counselor continues until several (three or more) concrete miracle behaviors are identified that can be useful for developing goals and clear direction for change.

Small Steps: Scaling Questions

Scaling Questions and the Miracle Scale If I were challenged to use only one technique and one technique only to work with clients from start to finish, scaling questions would have to top the list of possibilities, as they are arguably one of the most versatile and comprehensive interventions. Counselors can use scaling questions to (a) assess strengths and solutions, (b) set goals, (c) design homework tasks, (d) measure progress, *and* (e) manage crises with safety plans. They can be used in the first session and can be reused weekly until the last. Like many modern-day miracle, all-in-one products—such as shampoo-conditioners, moisturizer-sunscreen-foundations, and car wash-and-waxes—this technique is likely to be one of your go-to techniques for years to come.

As the name indicates, *scaling questions* involve asking clients to define their goals and rate their progress toward them using a scale, most often a 10-point scale but sometimes percentages or shorter, nonnumeric versions are used with children (Bertolino, 2010; de Shazer, 1994; O'Hanlon & Weiner-Davis, 1989; Selekman, 1997). de Shazer, Dolan, Korman, Trepper, McCollum, and Berg (2007) recommend having 0 represent "*when you decided to seek help*" rather than when things are "at their worst." When scaling 0 to mean "at their worst," this can refer to decades earlier or have different meanings to different people. Instead, scaling from "when you decided to seek help" allows for a clearer system for measuring progress from the beginning to end of counseling; they refer to these as *miracle scales* and use these scales to follow up with the miracle question (see Case Conceptualization below). Others use 0 or 1 to represent things at their *worst* (Bertolino, 2010).

Two Approaches to Scaling

Miracle Scale: Measures Progress From Beginning of Counseling

When you decided to seek help   Miracle Situation
0-------1-------2-------3-------4-------5-------6-------7-------8-------9-------10

Worst-to-Solution Scale: Measures Progress in General

Things at their Worst   Solution
0-------1-------2-------3-------4-------5-------6-------7-------8-------9-------10

(continued)

> Scaling questions can be used early in the counseling process to help identify meaningful long-term goals, much the way the miracle question is used (see below):
>
> *Scaling question for long-term goal setting (assessing solutions and setting goals)*
>
> If you were to describe your situation on a scale from 1 to 10 with 0 being where you were when you decided to seek help (or at their worst) and 10 being where you'd like them to be, where would you be today? What was happening and what were you doing at a 0? Can you describe to me what you would be doing (*not* what you would *not* be doing) if things were at a 9 or 10?

As the client responds, the counselor listens for clear, specific descriptions of what life would be like at a 10, helping the client to paint a clear, behavioral picture: what is the client doing, what are others doing, how does the day go? Once a clear description of the 10 scenario is developed, the counselor can then ask where things are today and obtain a behavioral description of the current situation. A concrete description of when things were at the point the client decided to seek help or at their worst (0 or 1) can also be useful assessment information. Once the big picture is assessed, the same scale can be used to identify *what works*.

> ### Scaling Question for Assessing What Works
>
> If 0 is where you were when you decided to seek help and 10 is where you would be if the problem you came here for was resolved, where are you today? [If above a 0] What are you doing or what is happening that tells us you are at a 3 and not 0? How did you get from a 0 to here?

This next set of questions helps identify what works, exceptions, and potential solutions that need to be assessed before more specific interventions can be developed. If the client gets stuck answering any of these questions, it can be helpful to ask what significant others might rate them on the scale. After exploring where a client is this week, where he/she was when deciding to seek counseling, and where he/she would like to be, then it is time to use scaling to identify the next step.

> ### Scaling Question for Designing Week-to-Week Interventions and Tasks
>
> On a scale from 0 to 10 with 10 being your desired goal, where are you this week? [Client responds and rates: e.g., 3]. If you are at a 3 this week, what things would need to be different in your life for you to come in and say you were at a 4, one step higher (or *half* a step higher if client tends toward pessimism or needs to keep goal smaller)?

This is not an easy question to answer. Often clients rush ahead and describe an 8, 9, or 10 when the next step is really a 4. In such cases, the counselor needs to help clients identify more realistic expectations. In other cases, clients say, "I don't know"; when this happens, counselors need to practice patience, silence, and encouragement to help clients identify mini-steps toward their goal, which is a critical part of the process. The client—not the counselor—is the only person who can answer this question in any meaningful way. Remember, the counselor is *not* a solution giver.

Once a clear description of the next step is developed, this information is used to identify specific, small tasks the client can take during the next week to move one step higher on the scale. It generally takes an entire session to fully flesh out concrete steps that are (a) realistic and meaningful, and (b) something that the client is motivated to try. For the intervention to work, the counselor and client need to develop micro-steps that take into account the client's motivation, willingness, schedule, variations in the schedule, reactions of others, etc. The counselor asks questions to identify and listens for potential barriers and pitfalls as well as helpful resources, working with the client to find ways to negotiate these. Together, the clients and counselors work together to develop specific homework tasks that clients believe will help move them toward their goals.

For example, if the client's initial response is that "At a 4, I would feel less anxious," the counselor needs to ask follow-up questions such as "How would you know you were less anxious?" "What would you be doing differently" "How would your days be different?" "What changes would other people notice in you?" These questions help the client identify one to two specific, small steps to be taken over the week such as "invite a friend over," "go to the mall alone," "watch a funny movie," etc. The steps need to be small enough that the client thinks the steps are easily attainable, especially early in the process.

> ### Scaling Question for Measuring Progress
> Last week you said you were at a 3; where would you say you are this week and why? [If things got better] What did *you do* that helped move you up the scale? [If things stayed the same or got worse] What happened that kept things the same (or made them worse)?

The counselor follows up the next week to see if the homework tasks helped move clients closer to their goals. If so, they explore what helped and how to do more of it. If clients don't report progress or report that things got worse [with or without doing the homework], the counselor does not despair, but simply goes back to more carefully assess whether (a) the solution has been meaningfully assessed and concretely identified, and (b) whether the task was small enough, concrete enough, and motivating enough. This scale can be used to measure progress from the beginning to end of the counseling process. In the case study at the end of the chapter, scaling questions are used extensively to help Shirin and Brian take small steps to resolving the dilemma of their inter-religious/intercultural relationship, of which Shirin's parents do not approve. The scaling questions allow them to make meaningful progress by identifying realistic, doable next steps rather than feeling stuck because they have not decided on whether to get married.

One Thing Different: Client-Generated Change

In the beginning, the key is to do one small thing different—call one friend—rather than visit with someone everyday (de Shazer, 1985, 1988; O'Hanlon, 2000; O'Hanlon & Weiner-Davis, 1989). Similarly, a goal such as "get up and exercise one day this week" is more likely to generate change and motivation for further change than a goal such as "get up and exercise everyday this week." In most cases, making this one small change starts a cascade of change events that are inspired from the client's *own* motivation rather than the prescription of the therapist (or even the cocreated solution with the therapist). Ideas generated in therapy are not viewed as the "best," "only," or "correct" solution but rather activities that will spark clients to identify what works for them.

The Doing: Interventions
Solution-Focused Tenants for Intervention

de Shazer et al. (2007, p. 2) identify basic tenants of solution-focused intervention that counselors can use to guide their work.

- *If It Isn't Broken, Don't Fix It:* Don't use therapeutic theory to determine areas of intervention.
- *If It Works, Do More of It:* Amplify and build upon things that are currently working.
- *If It's Not Working, Do Something Different:* Even if it's a good idea, if it's not working, find another solution.

- *Small Steps Can Lead to Big Changes:* Begin with small doable changes; these typically happen and quickly lead to more change.
- *The Solution Is Not Necessarily Related to the Problem:* Focus on moving forward, not understanding why there is a problem.
- *The Language for Solution Development Is Different From That Needed to Describe a Problem:* Problem-talk is negative and past-focused; solution-talk is hopeful, positive, and future-focused.
- *No Problem Happens All the Time; There Are Always Exceptions That Can Be Utilized:* Even the smallest exception is useful for identifying potential solutions.
- *The Future Is Both Created and Negotiable:* Clients have a significant role in designing their future.

Formula First Session Task

As the name implies, the formula first session task (de Shazer, 1985) is typically used in the first session with all clients, regardless of the issue, to increase client hope in the therapy process and motivation for change.

> #### Example of a Formula First Session Task
>
> *Formula First Session Task:* "Between now and the next time we meet, we [I] would like you to observe, so that you can describe to us [me] next time, what happens in your [pick one: family, life, marriage, relationship] that you want to continue to have happen" (de Shazer, 1985, p. 137).
>
> *Paraphrase With Introduction:* "As we are starting therapy, many things are going to change. However, I am sure that there are many things in your life and relationships that you do *not* want to have changed. Over the next week, I want (each of) you to generate a list of the things in your life and relationships that you *do not* want to have changed by therapy. Notice small things as well as big things that are working right now."

This directive stimulates clients to notice what is working, identifies their strengths and resources, and helps generate hope in their ability to change.

Scaling Questions for Weekly Task Assignments

Scaling questions are used for goal setting (see previous discussion) as well as for developing weekly tasks and homework (O'Hanlon & Weiner-Davis, 1989). Once a client has identified what behaviors would constitute moving up the scale, the therapist works with the client to identify specific activities and behaviors that will make the small changes needed to enact these behaviors. This intervention can be used weekly to develop homework assignments that will move clients step by step toward their goals. If clients do not follow through on the assigned tasks, therapists need to reassess by asking the following questions: (a) Am I expecting a complainant to be as motivated as a customer for change? (b) Was the task too big or can smaller steps be taken? (c) Are the right people involved? and (d) What actually motivates the client to take action and move toward change?

Asking Presuppositional Questions and Assuming a Future Solution

Presuppositional questions and talk that assumes future change help clients envision a future without the problem, generating hope and motivation (O'Hanlon & Beadle, 1999; O'Hanlon & Weiner-Davis, 1989). Solution-focused therapists assume change based on the observation that all things change: a client's situation cannot *not* change. Knowing that change is inevitable and that most clients benefit from therapy, therapists can be confident when they ask presuppositional questions such as the following:

- What will you be doing differently once we resolve these issues?
- Do you think there are other concerns you will want to address once we resolve these issues?
- When the problem is resolved, what is one of the first things you will do to celebrate?

Utilization

Drawing on the hypnotic work of Milton Erickson, de Shazer (1988) employed utilization techniques to help clients identify and enact solutions. *Utilization* refers to finding a way to use and leverage whatever the client presents as a strength, interest, proclivity, or habit to develop meaningful actions and plans that will lead in the direction of solutions. For example, if a client has difficulty making friends and close relations but has numerous pets, the therapist will utilize the client's interest in animals to develop more human connections, perhaps by having the client take a dog for a walk in public places, join a dog agility class, or volunteer at a pet shelter.

Coping Questions

Coping questions generate hope, agency, and motivation, especially when clients are feeling overwhelmed (De Jong & Berg, 2002; de Shazer & Dolan, 2007). They are used when the client is not reporting progress, describing an acute crisis, or otherwise feeling hopeless. Coping questions direct clients to identify how they have been coping with a current or past difficult situation.

Examples of Coping Questions

- "This sounds hard—how have you managed to cope with this to the degree that you are?" (de Shazer & Dolan, 2007, p. 10)
- "How have you managed to prevent it from getting worse?" (p. 10)

Compliments and Encouragement

Solution-based therapists use compliments and encouragement to motivate clients and highlight strengths. The key with compliments is to compliment *only* when clients are making steps toward *goals that they have set* or to compliment *specific strengths that relate to the problem*. This is so important I am going to say it again and put it in a special box so you don't forget.

Therapeutic Compliments

Rule for Making Compliments

Compliment only when clients are making steps toward goals they have set, or compliment specific strengths that relate to the problem; compliment their progress, not their personhood.

Examples

Therapeutic compliment:

- Wow! You made real progress toward your goal this week.

Even better therapeutic compliment:

- I am impressed; you not only followed through on the chart idea we developed last week but you came up with your own additional strategies—setting up a weekend outing with his friend—to improve your relationship with your son. (compliments specific behavior toward goal)

Not-so-good therapeutic compliments:

- *I really admire what you have done with your life.* (too personal and does not clearly relate to the problem; sounds like you are buttering up client)
- *You really are a great mom.* (nonspecific; evaluating mother skills globally; discouraging client from evaluating her own behavior and performance as a mother)

When therapists compliment the client on anything other than the client's goals and strengths, they are setting up a situation in which they are rendering judgment, albeit a positive one, on the client or his/her life. Compliments should not be used to be "nice" to clients (De Jong & Berg, 2002). They should only be used to reinforce progress toward goals that clients have set for themselves and stated in such a way that they encourage clients to validate themselves rather than rely on an outside authority figure to do so. When clients can set goals and make progress toward them, they develop a greater sense of *self-efficacy* ("I can do this"), which is a greater predictor of happiness than self-esteem (Seligman, 2004).

Interventions for Specific Problems
Couples Therapy and Divorce Busting

Solution-oriented couples therapy is popular because the emphasis on strength and hope is well suited for working with negative, interpersonal conflict, especially with couples who are in crisis or considering divorce (Hudson & O'Hanlon, 1991; Weiner-Davis, 1992). These solution-oriented couples therapies have several unique interventions.

Videotalk Videotalk is based on distinguishing between three levels of experience: facts, stories, and experience (Hudson & O'Hanlon, 1991). The *facts* are a behavioral description of what was done and said, what would be recorded on videotape during the couple's interaction. The *story* is the interpretation and meaning that a person associates with the behaviors and words. The *experience* is the internal thoughts and feelings each person had. When couples are having difficulty, therapists can help them sort through their differences by separating the facts from the story and the experience to increase each person's understanding of how he/she is interpreting the situation. Clients are encouraged to use videotalk (e.g., "She asked me three times" or "He went to his study after dinner without saying anything to me") instead of their default interpretations (e.g., "She was nagging" or "He is distant"). By using videotalk to separate the *behaviors* from the *interpretation of the behaviors*, couples become less defensive with one another and are able to engage in conversations in which they better understand each other and identify meaningful ways to reduce future conflict.

From Complaints to Requests Solution-oriented therapists encourage clients to ask for what they want rather than what they don't want or, in other words, to move from making *complaints* to making *requests* (O'Hanlon & Hudson, 1991; Weiner-Davis, 1992). For example, instead of saying, "You don't do anything romantic anymore" (a global, overgeneralizing complaint), the partner would learn to rephrase the complaint as a request: "I would really enjoy adding romance back into our relationship." These requests need to be behavioral and specific. Thus the person in this example needs to add a specific behavioral request: "I'd like to go away for a weekend; go to dinner and a movie; go to the beach for a sunset walk" and/or "It would be nice if we went back to giving each other a kiss before we left the house, leaving little love notes, or giving each other massages."

Therapy for Sexual Abuse and Trauma

Yvonne Dolan (1991) and Bill O'Hanlon and Bob Bertolino (2002) use solution-based therapies with child and adult survivors of childhood sexual abuse. Solution-based approaches stand out from traditional approaches to sexual abuse treatment in their optimistic and hopeful stance that emphasizes the resiliencies of survivors. Given that these clients have *survived* such a difficult trauma, solution-oriented therapists harness those strengths in new ways to help them resolve current issues. Some of the distinctive qualities of these approaches include the following.

Honoring the Agency of Survivors More so than traditional therapists, solution-based therapists honor the agency of survivors, allowing them to decide whether to tell their abuse stories and to determine the pacing of their treatment (Dolan, 1991; O'Hanlon & Bertolino, 2002). Although many therapists insist that survivors cannot heal without sharing the details of their abuse to their therapists, solution-based therapists would not readily agree. Instead, they work with clients to identify if, when, how, and to whom it is best to tell their stories. By fully honoring their agency, therapists create a relationship in which survivors regain full authority over the private aspects of their lives, reclaiming the autonomy that was lost through the abuse. Therapists who play a more directive role in working with a survivor may unintentionally replicate the abuse pattern by forcing clients to reveal parts of their sexual life in the name of treatment before they are ready, leaving them feeling violated and retraumatized.

The Recovery Scale: Focusing on Strengths and Abilities Dolan (1991) uses the *Solution-Focused Recovery Scale* to identify what areas of the client's life were not affected by the abuse, thus reducing the sense that the client's whole life and self have been affected. The success strategies in these areas are used to address areas that are affected by the abuse.

3-D Model: Dissociate, Disown, and Devalue O'Hanlon and Bertolino (2002) conceptualize the aftereffects of abuse and trauma using the 3-D model, which postulates that abuse leads people to dissociate, disown, and devalue aspects of the self, with the result that they develop symptoms that either inhibit experience (e.g., lack of sexual response, lack of memories, lack of anger) or that create intrusive experiences (e.g., flashbacks, sexual compulsions, or rage). The goal of therapy is to reconnect people with these disowned parts. O'Hanlon and Bertolino (2002) note that many of the symptoms related to sexual abuse are experienced as a sort of *negative trance*, a feeling that the experience is uncontrollable and involves only a part of the self. Solution-oriented therapists use *permissive, validating,* and *inclusive language* to encourage clients to revalue and include the devalued aspects of self that were disowned through the abuse.

Constructive Questions Dolan (1991) uses constructive questions to identify the specifics of clients' unique solutions:

- What will be the first (smallest) sign that things are getting better, that this (the sexual abuse) is having less of an impact on your current life?
- What will you be doing differently when this (sexual abuse trauma) is less of a current problem in your life?
- What will you be doing differently with your time?
- What will you be thinking about (doing) *instead* of thinking about the past?
- Are there times when the above is already happening to some (even a small) extent? What is different about those times? What is helpful about those differences?
- What differences will the above healing changes make when they have been present in your life over an extended period of time (days, weeks, months, years)?
- What do you think your significant other would say would be the first sign that things are getting better? What do you think your significant other will notice first?
- What do you think your (friends, boss, significant other, etc.) will notice about you as you heal even more?
- What differences will these healing changes you've identified make in future generations of your family? (pp. 37–38)

Videotalk (Action Terms) Because of the intense emotions that characterize abuse and trauma, survivors often have difficulty identifying the current effects of abuse in their present life. O'Hanlon and Bertolino (2002) use *videotalk* (previously discussed) with survivors to help identify specific actions and patterns of behavior that recreate the traumatic experience, including the sequence of events, antecedents, consequences, invariant actions, repetitive actions, and body responses. Once these recurrent patterns are identified, therapists help clients either change one part of the context to interrupt the cycle and create space for new responses or, if the client feels a certain degree of control over the symptoms, identify new, alternative solution-generating actions.

Putting It All Together

Case Conceptualization Template

- *Strengths and Resources*
  Identify individual and relational strengths in the following areas:
  - Personal: Areas of strength, resilience, and resources; current coping skills
  - Relational: Strengths of the relationship and family
  - Community: Friends, community support, job context, religious affiliations, etc.
  - Diversity: Support groups, communities, resources, etc.
- *Exceptions*
  - Identify as many times, places, relationships, context, etc. when the problem is less of a problem.
  - Identify what behaviors seem to make things a bit better.
- *Miracle Question*
  - Behaviorally describe what client would be doing (not *not* doing) differently if miracle occurred.
- *Client Motivation*
  - Customer for change
  - Complainant
  - Visitor

Treatment Plan Template for Individual

Solution-Based Initial Phase of Treatment with Individual

Initial Phase Counseling Tasks

1. Develop working counseling relationship. *Diversity Note: [Describe how you will adjust to respect cultured, gendered, and other styles of relationship building and emotional expression.]*
   a. Develop *collaborative* relationship and inspire *hope* and *optimism*, using "beginner's mind" and listening for *strengths*.

2. Assess individual, systemic, and broader cultural dynamics. *Diversity Note: [Describe how you will adjust assessment based on cultural, socioeconomic, sexual orientation, gender, and other relevant norms.]*
   a. Use *miracle question* to concretely and behaviorally define solutions in positive terms.
   b. Identify *exceptions* to the effects of abuse in all areas of life, times when trauma symptoms are less severe, areas of life not affected by the trauma, *strengths and resources,* supportive relationships, and client level of *motivation*.
   c. Use the *recovery scale* to identify areas of strengths and resources.

Initial Phase Client Goals

1. Increase sense of *control and safety when remembering abuse* to reduce flashbacks, dissociation, and hypervigilance.
   a. *Formula first session task* to identify existing areas of life that are working and can be used when feeling overwhelmed.
   b. *Scaling for safety* to develop a safety plan for when emotions and memories become overwhelming.
   c. *Coping questions* to explore how client has been coping and how to expand these abilities.

(continued)

Solution-Based Working Phase of Treatment with Individual

Working-Phase Counseling Task

1. Monitor quality of the working alliance. *Diversity Note: [Describe how you will attend to client response to interventions that indicate therapist using expressions of emotion that are not consistent with client's cultural background.]*
   a. *Assessment Intervention: Session rating scale* to assess client experience of the relationship.

Working-Phase Client Goals

1. Increase *client's sense of safety* with friends, family, and at home to reduce hypervigilance and dissociation.
   a. *Scaling questions* to identify small steps to increase sense of trust and openness in safe relationships.
   b. *Constructive questions* to help clients identify the specific behaviors that will characterize more trusting relationships.

2. Increase client's sense of *agency and survivorship* and decrease self-blame related to abuse to reduce sense of victimization, hopelessness, and stigma.
   a. *Permissive, validating, and inclusive language* to help clients reconnect with disowned and undervalued parts of themselves.
   b. *Channeling language* to help clients separate past from the present experience and to delimit the effects of the abuse.
   c. *Exception and coping questions* to enable client to restory abuse in such a way that it includes attempts to avoid the abuser, stop the abuse, ask for help and/or cope to increase sense of agency and decrease self-blame.

3. Increase client's sense of *positive connection with her body and sexuality* to reduce physical symptoms of trauma, sexual issues, body image issues, and/or eating issues.
   a. *Exception questions* to help client identify positive associations with body and sexuality.
   b. *Presuppositional questions* that assume the client will return to a health relationship with body and identify strategies for doing so.

Solution-Based Closing Phase of Treatment with Individual

Closing-Phase Counseling Task

1. Develop aftercare plan and maintain gains. *Diversity Note: [Describe how you will access resources in the communities of which they are a part to support them after ending therapy.]*
   a. *Coping questions* that identify how they *will* cope with future problems and setbacks.

Closing-Phase Client Goals

1. Increase client's *sense of agency* and engagement in [specify enjoyable activities] to reduce hopelessness and depressed mood.
   a. *Scaling questions* to identify small, doable steps to reclaiming agency in own life.
   b. *Therapeutic compliments* of client's attempts to take steps toward agency.

2. Increase client's ability to engage in *satisfying intimate relationships* to reduce isolation, hypervigilance, and extreme sense of vulnerability.
   a. *Scaling questions* to identify small steps to forming and/or improving intimate relationships.
   b. *Videotalk* for helping client identify patterns of interpreting partner's behavior based on trauma rather than present.

Treatment Plan Template for Couple/Family

Solution-Based Initial Phase of Treatment with Couple/Family

Initial Phase Counseling Tasks

1. Develop working counseling relationship. *Diversity Note:* [Describe how you will adjust to respect cultured, gendered, and other styles of relationship building and emotional expression.]
   a. Develop *collaborative* relationship with all members; inspire *hope* and *optimism*, using *"beginner's mind"* and listening for *strengths*.

2. Assess individual, systemic, and broader cultural dynamics. *Diversity Note:* [Describe how you will adjust assessment based on cultural, socioeconomic, sexual orientation, gender, and other relevant norms.]
   a. Use *miracle question* with each person to concretely and behaviorally define solutions in positive terms.
   b. Identify *exceptions* to conflict, times of connection, areas of life not affected by the conflict, *strengths and resources*, supportive relationships, and client level of *motivation*.

Initial Phase Client Goals

1. Increase engagement in *shared, enjoyable activities* to reduce conflict.
   a. *Formula first session task* to identify existing areas of relationship that are working.
   b. *Exception questions* to identify areas of relationship that are more satisfying and identify between session tasks for increasing or expanding these activities.

Solution-Based Working Phase of Treatment with Couple/Family

Working-Phase Counseling Task

1. Monitor quality of the working alliance. *Diversity Note:* [Describe how you will attend to client response to interventions that indicate therapist using expressions of emotion that are not consistent with client's cultural background.]
   a. *Assessment Intervention: Session rating scale* to assess clients' experience of the relationship.

Working-Phase Client Goals

1. Increase *satisfying communication* between couple/family to reduce conflict.
   a. *Videotalk* to separate behavior from interpretations and help each person better understand the other.
   b. Suggest clients move from *making complaints to making positive, behavioral requests* of each other.

2. Increase [specify positive couple/family interactions and relational patterns] to reduce conflict.
   a. *Scaling questions* to identify small steps *(one thing different)* with instructions for each member.
   b. *Channeling language* to reduce sense that other members are "always" a certain way.
   c. *Therapeutic compliments* to encourage small steps toward desired solutions.

(continued)

3. [For couples] Increase *mutually enjoyable physical and sexual intimacy* to reduce conflict and increase intimacy.
    a. *Exception questions* to help clients' areas of intimate life that are or have been satisfying for both.
    b. *Scaling questions* to identify small, concrete steps each could take to make desired changes.

Solution-Based Closing Phase of Treatment with Couple/Family

Closing-Phase Counseling Task

1. Develop aftercare plan and maintain gains. *Diversity Note: [Describe how you will access resources in the communities of which they are a part to support them after ending therapy.]*
    a. *Coping questions* that identify how they *will* cope with future problems and setbacks.

Closing-Phase Client Goals

1. Increase *sense of shared identity and connection* to reduce conflict and increase sense of intimacy.
    a. *Scaling questions* to identify small, doable steps to reclaiming agency in own life.
    b. *Compliments* of client's attempts to take steps toward agency.

Solution-Oriented Ericksonian Hypnosis

Solution-oriented hypnosis, also known as Ericksonian hypnosis or naturalistic trance, is a unique form of hypnosis that aims to evoke client strengths and resources to resolve client problems (Erickson & Keeney, 2006; Lankton & Lankton, 2008 O'Hanlon & Martin, 1992). As the name suggests, solution-oriented Ericksonian hypnosis grew out of the work of Milton Erickson, *preceding* solution-focused and solution-oriented therapies. Many solution-focused premises and techniques directly evolved from Erickson's approach to hypnosis and therapy. Stephen Lankton (Lankton & Lankton, 2008), Bill O'Hanlon (O'Hanlon & Martin, 1992), and Jay Haley (1993) have been some of the leaders in preserving and continuing Erickson's work.

Difference From Traditional Hypnosis

Solution-oriented hypnosis is different from traditional hypnosis in two significant ways (O'Hanlon & Martin, 1992):

- *Permissive Rather Than Hierarchical:* Traditional hypnosis is hierarchical; the therapist runs the show ("You 'vil become very sleepy"), whereas solution-oriented hypnosis is permissive: "You may find yourself wanting to close your eyes, or you may prefer to keep them open."
- *Evokes Client's Natural Resources:* In traditional hypnosis, the therapist effectively "reprograms" the client once he/she is in a trance; in solution-oriented therapy, the therapist tries to "evoke" or stimulate the client's natural resources for healing, serving more as a midwife who allows the natural process to occur.

The Big Picture: Overview of Treatment

Ericksonian therapy invites the client to go into a trance in order to evoke resources and strengths that the client already has: this may be done with or without a clear induction into a hypnotic trance state. Erickson was known for using stories, analogies, and directives to activate latent abilities that will enable clients to resolve their problems (O'Hanlon & Martin, 1992). His work was arguably the first *brief therapy*, developed during a time when psychiatry and psychotherapy were dominated by psychodynamic ideas about problem development and problem resolution (Erickson & Keeney, 2006; O'Hanlon & Martin, 1992). Many of his students have gone on to develop his work, including Richard Bandler, John Grinder, Jay Haley, Bill O'Hanlon, Ernest Rossi, and Jeffrey Zeig, and the Milton H. Erickson Foundation continues training in his approach.

The Doing: Interventions

Permission

When inviting clients to enter a trance state, Ericksonian therapists give their clients *permission* to think, experience, and feel whatever they are experiencing without any pressure to *do* something (O'Hanlon & Martin, 1992). Clients are given explicit permission to have doubts, allow their mind to chatter, lose focus, entertain distracting thoughts, and accept other thoughts or feelings.

Presuppositions

Ericksonian therapists who are inviting clients into a trance state *presuppose* that clients will enter a trance state by delivering certain questions and comments to the client (O'Hanlon & Martin, 1992); for example, "Have you ever been in a trance before today?" "You can choose to keep your eyes open or closed when you go into a trance." "Don't go into a trance too quickly" (p. 18).

Splitting

The therapist may split two things that the client may habitually consider as one thing, such as the conscious and unconscious mind or the left brain and right brain (O'Hanlon & Martin, 1992). Generally, problem-saturated thoughts and feelings are attributed to the conscious mind and positive thoughts to the unconscious to allow the person to trust the trance process. For example, the therapist may attribute doubt or apprehension to the conscious mind and trust to the unconscious mind: "Your conscious mind may doubt that trance is even possible, but your unconscious mind knows exactly what to do."

Class of Problems Versus Class of Solutions

To the untrained eye, Erickson often seemed to discuss issues and topics that were totally unrelated to the problem. For example, when working with a child with enuresis, he might spend the entire session chatting about seemingly unrelated topics, such as baseball and digestion, without ever discussing the presenting problem (O'Hanlon & Martin, 1992). Erickson would assess the class of problem (e.g., lack of muscle control) and then identify the class of solution that would most likely solve the problem (e.g., muscle control; O'Hanlon & Martin, 1992). Similarly, if a person's depressed mood was characterized by pessimism, he would find areas where the client had hope and optimism; if the client's depression was related to a recent failure, he would find ways to evoke a sense of success, whether actual or potential. Thus, for Erickson, the focus was on evoking the class of solution—*in any other area of the person's life*—rather than trying to solve the literal problem. He would focus on evoking the solution through both dialogue and trance.

The practice of identifying the class of solution is a particularly useful concept when assessing strengths because it highlights that all strengths are not created equal. Erickson draws our attention to the fact that it is most important *to identify client strengths that are in the same class of solutions that relate to the presenting problem.* Thus, if a child is having difficulty following parental requests at home, therapists should carefully listen for other situations in which the client is able to follow rules, whether playing on a soccer team, paying attention in class, or playing board games with a friend.

Research and the Evidence Base

Solution-based therapies have a steadily growing foundation of empirical support. Two recent meta-analyses of well-controlled, clinical trial studies (Stams, Dekovic, Buist, & De Vries, 2006; Kim, 2008) found that the solution-focused brief therapy has modest effect sizes and typically has equivalent outcomes to other approaches but typically does so in less time and therefore at less cost (Gingerich, Kim, Stams, & MacDonald, 2012). As further evidence of its growing evidence base, the Office of Juvenile Justice and Delinquency Prevention has recognized it as a promising practice, setting the stage for recognition as a evidence-based practice (Kim, Smock, Trepper, McCollum, & Franklin, 2010).

The majority of research has been conducted on solution-focused brief therapy, which was first described in the literature in 1986, had its first controlled study in 1993, and has had a total of 48 published studies and two meta-analytic reviews (Gingerich et al., 2012). The quality of studies is steadily improving, including randomized samples, treatment fidelity measures, and standardized outcome measures. Researchers have studied the effectiveness of solution-focused brief therapy with a wide range of clinical populations and problems, including domestic violence offenders, couples, schizophrenics, child abuse, troubled youth, school settings, parenting, alcohol, foster care (Franklin, Trepper, Gingerich, & McCollum, 2012). Research on the process of solution-focused brief therapy suggests that presuppositional questions, the miracle question, the first-session task, scaling questions, and solution talk have been found to accomplish their intended therapeutic effect and that the model in general engenders hope and optimism in clients (McKeel, 2012).

Currently, quantitative research in solution-focused brief therapy has focused on refining and improving clinical trial studies through the expansion of a standardized treatment manual (Trepper et al., 2012) and development of improved adherence and fidelity measures (Lehman & Patton, 2012). In addition, process research is being done on in-session therapeutic alliance and client progress as both an outcome measure and collaborative intervention (Duncan, Miller, & Sparks, 2004; Gillaspy & Murphy, 2012). Finally, of great current in psychotherapy research is the examination of the mechanisms of change which underlie an evidenced-based clinical approach such as solution-focused brief therapy. Laboratory experiments have provided evidence for some of the major underlying theoretical assumptions of solution-focused brief therapy; for example, that collaboration and coconstruction led significantly better outcomes in the laboratory (Bavelas, 2012). Also, microanalysis research has shown that solution-focused therapists are overwhelmingly positive in their formulations and questions as compared with client-centered and cognitive-behavioral therapists, whose formulations and questions were primarily negative (Tomor & Bavelas, 2007; Korman et al., 2012; Smock et al., 2012). It was also shown that this "positive talk" led to more positive talk, and negative talk led to more negative talk (Smock et al., 2012). Thus "… a therapist's use of positive content seems to contribute to the coconstruction of an overall positive session, whereas negative content would do the reverse" (Bavelas, 2012; p. 159). This area of research may someday refine understanding of the exact mechanisms of change of solution focused brief therapy and how this approach differs from other clinical models.

Tapestry Weaving: Working with Diverse Populations

Ethnic, Racial, and Cultural Diversity

Because it does not use a theory of health to predefine client goals (O'Hanlon & Weiner-Davis, 1989), solution-focused therapy can be adapted to a wide range of populations and value systems. This therapy is widely used with diverse populations in North America, South America, Europe, the Middle East, Asia, and Australia (Gingerich & Patterson, 2007). It has also been studied with a range of clients, including immigrants, African Americans, Hispanics, Saudi Arabians, Chinese, and Koreans; and in a wide range of contexts, such as schools, prisons, hospitals, businesses, and colleges (Gingerich & Patterson, 2007). When working with diverse clients, solution-focused therapists access their unique emotional, cognitive, and social resources, which often relate to issues of diversity.

Corcoran (2000) identifies several reasons why solution-focused therapy is generally a good fit for diverse populations:

- *Behavior Considered in Context:* The solution-based viewing of behaviors in context allows for a more fair understanding of problem behaviors of marginalized populations.
- *Client-Generated, Behavioral Goals*: The goal-setting process of solution-based therapies is a good fit for many ethnic groups because the goals are set in client language, they are concrete and behavioral, and they are generally short term.
- *Behavioral Rather Than Emotional Focus:* The focus on behavior rather than emotion is more comfortable and value consistent for many ethnic minority groups.
- *Future Orientation:* The focus on solving problems in the future rather than understanding the past makes sense to many ethnic minorities.

Asian Americans

Several therapists have explored using solution-based therapies with Asians and Asian Americans. Hsu and Wang (2011) describe solution-focused as being highly compatible for Asian clients because of its positive reframing (avoids losing face), relation-based perspective, and its focus on pragmatic solutions. Considerations for working with Asian clients include:

- *Filial Piety:* Several therapists have used solution-focused therapy to help clients navigate issues of filial piety, a common Asian value in which elders are obeyed and honored (Lee & Mjelde-Mossey, 2004; Hsu & Wang, 2011). Solution-focused approaches help reframe each generation's perspective in positive terms, so the good intentions of each are better understood.
- *Form and Content:* Many Asian societies value the *form*—how something is done—as much as the *content*—*what* is done (Berg & Jaya, 1993). This can translate to family members being more concerned about how someone approached a problem as much as what was done to solve it.
- *Practical Solutions:* Asian cultures generally discourage dramatic displays of emotion, and this often translates to a preference for therapeutic approaches that focus on pragmatic solutions rather than exploration of feelings and causes (Berg & Jaya, 1993).
- *Saving Face:* Solution-focused therapy is unique in that its focus on "what to do from here" rather than identify root causes more easily allows for Asian clients to save face. Asian cultures are *shame*—not guilt—based, and thus positive reframes and compliments work particularly well with these populations (Berg & Jaya, 1993).
- *Brief Treatment:* As therapy is usually a last resort for Asian families who prefer privacy, the brief approach of solution-focused therapy is an excellent fit (Berg & Jaya, 1993).

Sexual Identity Diversity

Because it does not have a predefined theory of health and focuses on client-defined goals, solution-based therapies can be appropriate when working with gay, lesbian, bisexual, or transgendered clients. However, to be sensitive to these issues, therapists should inquire with "beginner's mind" about their possible role in the client's presenting problem rather than assume that because it is not mentioned that it is not part of the problem or solution. Little has specifically been written on solution-based therapies and these populations. However, this approach has been used to facilitate adjustment to one's spouse coming out. Treyger, Ehlers, Zajicek, and Trepper (2008) found solution-based therapy appropriate for a case in which a woman's spouse announced he was gay. The approach was used to help the client identify her own solutions to the situation in a supportive, nonpathologizing context. They used Buxton's (2004) seven stage model for partner adjustment to news that their spouse is homosexual or bisexual; the stages include: disorientation/disbelief, facing and acknowledging reality, accepting, letting go, healing, reconfiguring and refocusing, and transforming. The client in Treyger et al. (2008) reported that what helped the most from the approach was talking without being judged, not being given advice (which her family and friends offered plenty of), positive feedback and compliments, and scaling questions to concretely measure progress.

Online Resources

Milton H. Erickson Foundation:
www.erickson-foundation.org

Solution-Focused Brief Therapy Association:
www.sfbta.org

Solution-Oriented, Possibility Therapy:
www.billohanlon.com

Divorce Busting:
www.divorcebusting.com

European Brief Therapy Association:
www.ebta.nu

References

*Asterisk indicates recommended introductory books.

Bavelas, J. B. (2012). Connecting the lab to the therapy room: Microanalysis, co-construction, and Solution-Focused Brief Therapy. In Cynthia Franklin, Terry S. Trepper, Wallace J. Gingerich, & Eric E. McCollum (Eds.), Solution-focused brief therapy: A handbook of evidence-based practice (pp. 144–164). London: Oxford University Press.

Berg, I. K. (1994). Family based services: A solution-focused approach. New York: Norton.

Berg, I. K., & Jaya, A. (1993). Different and same: Family therapy with Asian-American families. Journal of Marital and Family Therapy, 19, 31–38.

Berg, I. K., & Kelly, S. (2000). Building solutions in child protective services. New York: Norton.

Berg, I. K., & Miller, S. (1992). Working with the problem drinker: A solution-focused approach. New York: Norton.

Berg, I. K., & Reuss, N. H. (1997). Solutions step by step: A substance abuse treatment manual. New York: Norton.

Berg, I. K., & Steiner, T. (2003). Children's solution work. New York: Norton.

Berg, I. K., & Szabo, P. (2005). Brief coaching for last solutions. New York: Norton.

Bertolino, B. (2010). Strengths-based engagement and practice: Creating effective helping relationships. Boston, MA: Pearson.

Bertolino, B., & O'Hanlon, B. (1998). Therapy with troubled teenagers: Rewriting young lives in progress. New York: Wiley.

*Bertolino, B., & O'Hanlon, B. (2002). Collaborative, competency-based counseling and therapy. New York: Allyn & Bacon.

Buxton, A. P. (2004). Paths and pitfalls: How heterosexual spouses cope when their husbands or wives come out. Journal of Couple and Relationship Therapy, 3, 95–109.

Corcoran, J. (2000). Solution-focused family therapy with ethnic minority clients. Crisis Intervention & Time-Limited Treatment, 6(1), 5–12.

*De Jong, P., & Berg, I. K. (2002). Interviewing for solutions (2nd ed.). New York: Brooks/Cole.

*de Shazer, S. (1985). Keys to solution in brief therapy. New York: Norton.

*de Shazer, S. (1988). Clues: Investigating solutions in brief therapy. New York: Norton.

de Shazer, S. (1994). Words were originally magic. New York: Norton.

*de Shazer, S., & Dolan, Y. (with Korman, H., Trepper, T., McCollum, & Berg, I. K.). (2007). More than miracles: The state of the art of solution-focused brief therapy. New York: Haworth.

*Dolan, Y. (1991). Resolving sexual abuse: Solution-focused therapy and Ericksonian hypnosis for survivors. New York: Norton.

Dolan, Y. (2000). One small step: Moving beyond trauma and therapy into a life of joy. New York: Excel Press.

Dolan, Y. (2007). Tribute to Insoo Kim Berg. Journal of Marital and Family Therapy, 33, 129–131.

Duncan, B., Miller, S. D., & Sparks, J. A. (2004). The heroic client: A revolutionary way to improve effectiveness through client-directed, outcome-informed therapy. New York: Jossey-Bass.

Erickson, B. A., & Keeney, B. (Eds.). (2006). Milton Erickson, M.D.: An American healer. Sedona, AZ: Leete Island Books.

Franklin, C., Trepper, T. S., Gingerich, W. J., & McCollum, E. E. (Eds.) (2012). Solution-focused brief therapy: A handbook of evidence-based practice. New York: Oxford University Press.

Gillaspy, A., & Murphy, J. J. (2012). Incorporating outcome and session rating scales in solution-focused brief therapy. In C. Franklin, T. S. Trepper, W. J. Gingerich, & E. E. McCollum (Eds.), Solution-focused brief therapy: A handbook of evidence-based practice (pp. 73–94). New York: Oxford University Press.

Gingerich, W. J., Kim, J. S., Stams, G. J. J. M., & MacDonald, A. J. (2012). Solution-focused brief therapy outcome research. In C. Franklin, T. S. Trepper, W. J. Gingerich, & E. E. McCollum (Eds.), *Solution-focused brief therapy: A handbook of evidence-based practice* (pp. 95–111). New York: Oxford University Press.

Gingerich, W. J., & Patterson, L. (2007). The 2007 SFBT effectiveness project. Retrieved March 20, 2008, from http://gingerich.net/SFBT/2007_review.htm.

Haley, J. (1993). *Uncommon therapy: The psychiatric techniques of Milton H. Erikson, M.D.* New York: Norton.

Hsu, W., & Wang, C. C. (2011). Integrating Asian clients' filial piety beliefs into solution-focused brief therapy. *International Journal for the Advancement of Counseling, 33*(4), 322–334. doi:10.1007/s10447-011-9133-5

Hudson, P. O., & O'Hanlon, W. H. (1991). *Rewriting love stories: Brief marital therapy*. New York: Norton.

Kim, J. S. (2008). Examining the effectiveness of solution-focused brief therapy: A meta-analysis. *Research on Social Work Practice, 18*(2), 107–116. doi:10.1177/1049731507307807

Kim, J. S., Smock, S., Trepper, T. S, McCollum, E. E., & Franklin, C. (2010). Is solution-focused brief therapy evidence based? *Families in Society, 91*, 300–306. DOI: 10.1606/1044-3894.4009.

Korman, H., Bavelas, J. B., & De Jong. P. (2012). Microanalysis of formulations, Part II: Comparing solution-focused brief therapy, cognitive-behavioral therapy, and motivational interviewing. Manuscript submitted for publication.

Lankton, S. (2008). An Ericksonian approach to clinical hypnosis. In M. R. Nash, A. J. Barnier (Eds.), *The Oxford handbook of hypnosis: Theory, research, and practice* (pp. 467–485). New York: Oxford University Press.

Lankton, S. R. (2009). *Tools of intention: Strategies that inspire change*. New York: Crown House.

Lankton, S. R., & Erickson, K. K. (1994). *The essence of a single-session success*. Philadelphia, PA: Brunner/Mazel.

Lankton, S. R., Gilligan, S. G., & Zeig, J. K. (1991). *Views on Ericksonian brief therapy, process and action*. Philadelphia: Brunner/Mazel.

Lankton, S. R., Lankton, C. H. (2008). *Answer within: A clinical framework for Ericksonian hypnosis*. New York: Crown House.

Lankton, S. R., Lankton, C. H., & Matthews, W. J. (1991). Ericksonian family therapy. In A. S. Gurman & D. P. Kniskern (Eds.), *Handbook of family therapy*, Vol. 2 (pp. 239–283). Philadelphia, PA: Brunner/Mazel.

Lankton, S. R., & Matthews, W. J. (2010). An Ericksonian model of clinical hypnosis. In S. Lynn, J. W. Rhue, & I. Kirsch (Eds.), *Handbook of clinical hypnosis* (2nd ed.) (pp. 209–237). Washington, DC: American Psychological Association.

Lee, M., & Mjelde-Mossey, L. (2004). Cultural dissonance among generations: A solution-focused approach with East Asian elders and their families. *Journal of Marital and Family Therapy, 30*(4), 497–513.

Lehmann, P., & Patton, J. D. (2012). The development of a solution-focused fidelity instrument: A pilot study. In C. Franklin, T. S. Trepper, W. J. Gingerich, & E. E. McCollum (Eds.), *Solution-focused brief therapy: A handbook of evidence-based practice* (pp. 39–54). New York: Oxford University Press.

Lipchik, E. (2002). *Beyond technique in solution-focused therapy: Working with emotions and the therapeutic relationship*. New York: Guilford.

McKeel, J. (2012). What works in solution-focused brief therapy: A review of change process research. In C. Franklin, T. S. Trepper, W. J. Gingerich, & E. E. McCollum (Eds.), *Solution-focused brief therapy: A handbook of evidence-based practice* (pp. 130–143). New York: Oxford University Press.

Metcalf, L. (1998). *Parenting towards solutions*. Paramus, NJ: Prentice Hall.

Metcalf, L. (2003). *Teaching towards solutions* (2nd ed.). Wales, UK: Crown House Publishing.

Metcalf, L. (2007). *Solution-focused group therapy*. New York: Free Press.

Metcalf, L. (2008). *Counseling towards solutions: A practical solution-focused program for working with students, teachers, and parents* (2nd ed.). New York: Jossey-Bass.

Miller, S. D., Duncan, B. L., & Hubble, M. (Eds.). (1996). *Handbook of solution-focused brief therapy*. San Francisco, CA: Jossey-Bass.

Miller, S. D., Duncan, B. L., & Hubble, M. A. (1997). *Escape from Babel: Towards a unifying language for psychotherapy practice*. New York: Norton.

O'Hanlon, B. (2000). *Do one thing different: Ten simple ways to change your life*. New York: Harper.

O'Hanlon, B. (2005). *Thriving through crisis: Turn tragedy and trauma into growth and change*. New York: Penguin/Perigee.

O'Hanlon, B. (2006). *Pathways to spirituality: Connection, wholeness, and possibility for therapist and client*. New York: Norton Professional.

*O'Hanlon, B., & Beadle, S. (1999). *A guide to possibilityland: Possibility therapy methods*. Omaha, NE: Possibility Press.

O'Hanlon, B., & Bertolino, B. (2002). *Even from a broken web: Brief and respectful solution-oriented therapy for resolving sexual abuse*. New York: Norton.

O'Hanlon, W. H., & Martin, M. (1992). *Solution-oriented hypnosis: An Ericksonian approach*. New York: Norton.

*O'Hanlon, W. H., & Weiner-Davis, M. (1989). *In search of solutions: A new direction in psychotherapy*. New York: Norton.

*Selekman, M. D. (1997). *Solution-focused therapy with children: Harnessing family strengths for systemic change*. New York: Guilford.

Selekman, M. (2005). *Pathways to change: Brief therapy with difficult adolescents*. New York: Guilford.

*Selekman, M. (2006). *Working with self-harming adolescents: A collaborative, strength-oriented therapy approach*. New York: Norton.

Seligman, M. (2004). *Authentic happiness*. New York: Free Press.

Smock, S., Froerer, A., & Bavelas, J. B. (2012). Microanalysis of positive and negative content in solution-focused brief therapy and cognitive-behavioral therapy expert sessions. Manuscript submitted for publication.

Stams, G., Deković, M., Buist, K., & de Vries, L. (2006). Effectiviteit van oplossingsgerichte korte therapie: Een meta-analyse [Efficacy of solution-focused brief therapy: A meta-analysis]. *Gedragstherapie, 39*(2), 81–94.

Tomori, C., & Bavelas, J. B. (2007). Using microanalysis of communication to compare solution-focused and client-centered therapies. *Journal of Family Psychotherapy, 18*, 25–43.

Trepper, T. S., Dolan, Y., McCollum, E. E., & Nelson, T. (2006). Steve de Shazer and the future of Solution-Focused Therapy. *Journal of Marital and Family Therapy, 32*, 133–140.

Trepper, T. S., McCollum, E. E. de Jong, P., Korman, H., Gingerich, W. J., & Franklin, C. (2012). Solution-focused brief therapy treatment manual. In C. Franklin, T. S. Trepper, W. J. Gingerich, & E. E. McCollum (Eds.) *Solution-focused brief therapy: A handbook of evidence-based practice* (pp. 20–38). New York: Oxford University Press.

Treyger, S., Ehlers, N., Zajicek, L., & Trepper, T. (2008). Helping spouses cope with partners coming out: A solution-focused approach. *American Journal of Family Therapy, 36*(1), 30–47. doi:10.1080/01926180601057549

Walter, J. L., & Peller, J. E. (1992). *Becoming solution-focused in brief therapy*. New York: Brunner/Mazel.

Weiner-Davis, M. (1992). *Divorce busting*. New York: Summit Books.

Solution-Based Therapy Case Study: Interfaith/Intercultural Couple Considers Marriage

Shirin (AF), a 28-year-old from a wealthy Persian Jewish family, and Brian (AM), a 32-year-old from an American Catholic family of moderate means, are seeking treatment because of tension and conflict within their relationship and talks about possible separation. Shirin and Brian have been dating for three years; they kept their relationship hidden from Shirin's family for over two years because it is forbidden to date outside of the Jewish faith. Shirin revealed to her family nine months ago that she had been dating Brian for the past three years and that they were discussing marriage in the near future. Shirin's family disapproves of the relationship and threatened to cut off their financial and emotional support if she decided to move forward with the marriage unless Brian would convert. Brian has strong ties to his religion and has told Shirin from the beginning that he would not convert but would be willing to observe the Jewish traditions if Shirin was willing to observe his Catholic traditions. They made an agreement that they would have an interfaith marriage if they decided to get married and have children; recently Shirin has been having doubts about interfaith marriage because of the pressure from her family. Shirin has been feeling depressed since she revealed the relationship to her family; she feels like she has little power to move forward in the relationship and to make a decision based on her own needs and wants. Brian has been having difficulty understanding Shirin's feelings of helplessness and has resentment and anger toward her family. Brian has a close relationship to his family and has difficulty understanding how Shirin's family could threaten to cut her out of their lives. Brian's parents are strict Catholics and had reservations about Shirin initially but have fully accepted her into their lives.

Over the past nine months Shirin and Brian have had continuous fights about the influence of Shirin's family on their lives. Brian believes that they can make a life on their own and that the next chapter in life is about "creating our own family." He also believes that Shirin's family will eventually come around and accept the decision she makes. Shirin doesn't have the same views as Brian; she believes that their next chapter should include her family involvement and cannot envision her life any other way. Shirin has disengaged both emotionally and physically from Brain since her family threatened to cut her out of their lives. They described once having a relationship that was full of laughter, excitement, and unconditional love, and they now describe their relationship as one filled with tension, anger, and resentment. They have been talking about ending the relationship but neither one is willing to make the move. Brian stated that he is willing to do whatever it takes to make the relationship work; Shirin is more conflicted but is interested in getting professional assistance.

Solution-Based Case Conceptualization

- *Strengths and Resources*
  Identify Individual and Relational Strengths in the Following Areas:
  Personal: AF has strong family values and belief system. AF was confident enough to tell her family about the relationship although she knew that they would

disapprove and possibly cut her out of their lives. Although she has been feeling depressed she has been able to function in other areas of her life, such as graduate school and work. AM has strong family values and belief system. AM is strong-willed and has been standing up for himself against AF's family. AM is also strongly invested in the relationship and is willing to do whatever it takes to help make the relationship work once again.

Relational: Both AF and AM are willing to go to therapy to help strengthen their relationship. There is loyalty and commitment to the relationship; they are making an attempt to make it work despite rejection from family. AF and AM report having had a strong emotional connection with one another and a high level of trust and compassion for one another. AF and AM have strong relationships with their family of origin, and each have close relationships with their siblings; there is honor and respect for their parents.

Community: Friends, community support, job context, religious affiliations, etc. AF has a large support group of friends who help her cope with her struggles. AF also has support from extended family (a number of close cousins) that support the relationship and any decision she chooses to make. AF is part of a reformed synagogue and finds a sense of calmness and ease when attending services. She has been able to speak to her rabbi about her situation and has received support and advice with her concerns about interfaith marriage. AM has a large social group and a tight knit group of friends that he receives support from. The church AM is affiliated with is accepting and supports the relationship.

Diversity: AF is somewhat involved with the Jewish community; she attends High Holiday services and has a supportive relationship with the rabbi. Support groups, communities, resources, etc. AM attends mass infrequently, but reports having an active spiritual life in terms of prayer and faith.

- *Exceptions*
- *When Problem Is Less of a Problem:*
    - AF feels that the problem is less of a problem when she is willing to let go of her parent's beliefs and be fully present with AM.
    - Times they do not feel the problem as much is when they go back to doing the things they once enjoyed, such as the beach, cooking, hiking, and going to concerts.
    - AF feels a sense of hope when she sees her situation with AM as less of a problem and when she sees extended family that are in support of their relationship.
    - AF is more open minded to interfaith marriage after speaking to her rabbi, therefore causing fewer fights about their future together.
    - AM explains that when they are out with friends they are able to reconnect and bring humor back into their relationship which is the main reason that they were initially attracted to each other.
    - AM states that when he supports AF and her feelings of depression instead of criticizing her helps with bringing them close to one another.
- *What Makes Things Better:*
    - When AM doesn't criticize or judge AF's family.
    - When AM listens to AF's personal struggles without passing judgment.
    - When AF sits and listens to AM without getting up to leave the conversation prematurely.
    - When AM uses less sarcasm and is more accepting of AF's circumstances.
    - When AF shows even a little affection.
- *Miracle Question*
    - AF's miracle involves having a good relationship with AM and moving forward in their relationship together while maintaining a relationship with her family. AF would like to unite the two worlds she kept separate for so long and have support from both AM and family. AF would be able to support and be there for both AM and her family emotionally. AF would also be more affectionate

and understanding of AM's feelings. AF would have emotional closeness with AM and stand up for what it is that she wants and believes in.
  - AM's miracle involves having a relationship with AF that they once had, a relationship filled with laughter, compassion, intimacy, and love. AM would support AF fully and understand her needs. He would be accepting of her thoughts, feelings, and emotions in regards to their relationship and her family.
- *Client Motivation*
  - *Customer for Change:* AM would like to do whatever it takes to fix the problems and sees areas he needs to work on in order for the relationship to work. He understands that he needs to be more conscious of AF's situation, feelings and, cultural background, and also be less defensive and critical.
  - *Complainant:* AF feels that the problem is due to her parents' disapproval and AM's lack of support and understanding. After answering the miracle question, AF understands that she needs to make changes as well, therefore becoming more of a customer for change. She identifies that she needs to be more supportive, understanding, and be psychically and emotionally available.

Solution-Based Treatment Plan

Solution-Based Initial Phase of Treatment with Family

Initial Phase Counseling Tasks

1. Develop working counseling relationship. *Diversity Note: Ensure AF feels that her ethnicity and religion are being respected in the counseling relationship. Assess AF's level of acculturation and adapt relationship accordingly, considering the importance of family loyalty, respecting elders, and gender norms. Be careful to not subtly impose Western values of independence when discussing her fears of losing family approval by verbally acknowledging the reality of her situation. Ensure AM feels that his religious views are being respected. Address socioeconomic differences between the two families.*
    a. Develop *collaborative* relationship with all members; inspire *hope* and *optimism*, using *"beginner's mind"* and listening for *strengths*.

2. Assess individual, systemic, and broader cultural dynamics. *Diversity Note: Adjust assessment of family norms to account for Persian, Jewish, and Catholic family dynamics, especially collectivist versus individualist cultural norms. Assess level of acculturation of AF and her family. Carefully assess AF and AM's understanding of their cultural differences.*
    a. Use *miracle question* with each person to concretely and behaviorally define solutions in positive terms.
    b. Identify *exceptions* to conflict, times of connection, areas of life not affected by the conflict, *strengths and resources*, supportive relationships, and client level of *motivation*.

Initial Phase Client Goals

1. Increase engagement in *shared, enjoyable activities* to reduce conflict and increase connectedness.
    a. *Formula first session task* to identify existing areas of relationship that are working.
    b. *Exception questions* to identify areas of relationship that are more satisfying and identify between session tasks for increasing or expanding these activities.

(continued)

Solution-Based Working Phase of Treatment with Family

Working-Phase Counseling Task

1. Monitor quality of the working alliance. *Diversity Note: Ensure both partners respond in such a way that evidences a sense of safety and trust with therapist and that neither feels the therapist is taking sides or that therapist is imposing cultural norms.*
   a. *Session rating scale* to assess clients' experience of the relationship. Adjust alliance as AF becomes more of a customer of change.

Working-Phase Client Goals

1. Increase certainty of each partner's *position regarding commitment to relationship* and desired future to reduce conflict.
   a. *Scaling questions* to scale level of commitment and certainty of commitment.
   b. *Action plan based on scaling question* to help AF in particular pursue actions that will help her to make a decision about what she wants to do with the relationship, which may include sessions with her family, dialogues with herrabbi and extended family, considering any financial implications of decision, and discussing risks she is willing and not willing to take.

2. Increase *satisfying communication* between couple to reduce conflict.
   a. *Videotalk* to separate behavior from interpretations and help each person better understand the other, such as AM better understanding AF's need to distance and AF understanding how her distancing affects AM.
   b. Suggest AF and AM move from *making complaints to making positive, behavioral requests* of each other.

3. Increase each partner's ability to *supportively respond* to issues related to family of origin and cultural/religious differences to reduce conflict.
   a. *Scaling questions* to identify small steps *(one thing different)* with instructions for each partner.
   b. *Channeling language* to reduce sense that the others are "always" a certain way and to identify alternative possibilities when tension arises.
   c. *Therapeutic compliments* to encourage small steps toward desired solutions.

Solution-Based Closing Phase of Treatment with Family

Closing-Phase Counseling Task

1. Develop aftercare plan and maintain gains. *Diversity Note: If couple decides to pursue marriage, identify potential sources of support in from both AF and AM's religious communities, including meetings with rabbis and priests in their community that support interfaith marriages. Support groups for interfaith and interethnic couples.*
   a. *Coping questions* that identify how they will cope with future problems and setbacks. Identify how AF and AM will cope with possible cutoff from AF's family.

Closing-Phase Client Goals

1. Increase *sense of shared identity and connection* as an interfaith/intercultural marriage to reduce conflict and increase sense of intimacy.
   a. *Scaling questions* to identify small, doable steps to building a shared identity that both partners believe in wholeheartedly.
   b. Reflect on *miracle question* from beginning of treatment using *miracle scales* and discuss where they are in relation to that miracle and possible next steps.

2. Increase each partner's *sense of agency* as an adult child in their families of origin to reduce conflict with families and increase sense of personal wellness.
   a. *Videotalk* to identify how each wants to relate to own parents and partner's parents as an adult child and identify corresponding action steps.
   b. *Coping questions* to explore how each will cope with their parents not relating to them as hoped.

CHAPTER 13

Narrative Therapy

"Many people who seek therapy believe that the problems of their lives are a reflection of their own identity, or the identity of others, or a reflection of the identity of their relationships. ... Externalizing conversations can provide an antidote to these internal understandings by objectifying the problem. They employ practices of objectification of the problem against cultural practices of objectification of people. This makes it possible for people to experience an identity that is separate from the problem; the problem becomes the problem, not the person."

—White, 2007, p. 9

Lay of the Land

The most recently developed family therapies are called *postmodern therapies*, which can be broadly divided into two streams of practice:

- *Narrative therapy* developed by Michael White and David Epston (1990) in Australia and New Zealand
- *Collaborative therapy* (Chapter 14) developed by Harlene Anderson and Harry Goolishian in Texas (1988, 1992; Anderson, 1997; Goolishian & Anderson, 1987) and by Tom Andersen in Norway (1991, 1992)

These two approaches share many of the social constructionist premises described in Chapter 4, each being an approach to coconstructing new meanings with clients. Like solution-based therapists (Chapter 12), postmodern therapists optimistically focus on client strengths and abilities. Despite many similarities, however, collaborative and narrative therapies differ in significant ways, most notably in their philosophical foundations, the therapist's stance, the role of interventions, and the emphasis on political issues. Broadly speaking, narrative therapists have well-defined sets of questions and strategies for helping clients enact preferred narratives, whereas collaborative therapists avoid standardized techniques, instead using postmodern and social constructionist assumptions to facilitate a unique relational and dialogical process. The following table summarizes the difference between these two therapies.

Collaborative and Narrative Therapies

	Collaborative Therapy	Narrative Therapy
Primary Philosophical Foundations	Postmodernism; social constructionism; hermeneutics (study of interpretation)	Foucault's philosophical writings; critical theory; social constructionism
Therapeutic Relationship	Therapist more facilitative; facilitates a dialogical process	Therapist more active: "coeditor," "coauthor"
Therapeutic Process	No interventions; therapist focus is facilitating particular processes	Structured interventions
Politics and Social Justice	Political issues raised tentatively for client consideration	Social justice issues regularly included in therapy conversations

Narrative Therapy

In a Nutshell: The Least You Need to Know
Developed by Michael White and David Epston in Australia and New Zealand, narrative therapy is based on the premise that we "story" and create the meaning of life events using available *dominant discourses*—broad societal stories, sociocultural practices, assumptions, and expectations about how we should live. People experience "problems" when their personal life does not fit with these dominant societal discourses and expectations. The process of narrative therapy involves *separating the person from the problem,* critically examining the assumptions that inform how the person evaluates himself/herself and his/her life. Through this process, clients identify alternative ways to view, act, and interact in daily life. Narrative therapists assume that all people are resourceful and have strengths, and they do not see "people" as having problems but rather see problems as being imposed upon people by unhelpful or harmful societal cultural practices.

The Juice: Significant Contributions to the Field
If you remember one thing from this chapter, it should be this:

Understanding Oppression: Dominant Versus Local Discourses
Narrative therapy is one of the few psychotherapeutic theories that integrates societal and cultural issues into its core conceptualization of how problems are formed and resolved. Narrative therapists maintain that problems do not exist separately from their sociocultural contexts, which are broadly constituted in what philosopher Michel Foucault called *dominant discourses* (Foucault, 1972, 1979, 1980; White, 1995; White & Epston, 1990). Dominant discourses are culturally generated stories about how life should go that are used to coordinate social behavior, such as how married people should act, what happiness looks like, and how to be successful. These dominant discourses organize social groups at all levels: large cultural groups down to individual couples and families. They are described as dominant because they are so foundational to how we behave and evaluate our lives that we are rarely conscious of their impact or origins.

Foucault contrasts dominant discourses with *local discourses,* which occur in our heads, our closer relationships, and marginalized (not mainstream) communities. Local discourses have different "goods" and "shoulds" than dominant discourses. A classic example is that women value relationships whereas men value outcome in typical work environments. Both discourses have a value they are working toward; however, men's discourse is generally privileged over women's and thus is considered a dominant discourse, with women's discourse being local. Narrative therapists closely attend to the fluid interactions of local and dominant discourses and how these different stories of what is "good" and valued collide in our web of social relationships, creating problems and difficulties. By attending to this level of social interaction, narrative therapists help clients become aware of how these different discourses are impacting their lives; this awareness increases clients' sense of agency in their struggles, allowing them to find ways to more successfully resolve their issues. In the case study at the end of this chapter, Darren and Ernesto, a multiethnic, gay couple, struggle with several dominant discourses, such as being gay, Nigerian/Mexican, and HIV+, which affect both their individual identity narratives as well as their couple identity narratives.

Rumor Has It: The People and Their Stories

Michael White A pioneer in narrative therapy and the first to write about the process of *externalizing* problems, Michael White was based at the Dulwich Centre in Adelaide, Australia, which provides training and publishes books and newsletters on narrative therapy. Along with David Epston, he wrote the first book on narrative therapy, *Narrative Means to Therapeutic Ends* (White & Epston, 1990). His last publication, *Maps of Narrative Practice* (White, 2007), describes his later work before his death in 2008.

David Epston From Auckland, New Zealand, David Epston worked closely with Michael White in developing the foundational framework for narrative therapy. His work emphasized creating unique sources of support for clients, such as writing letters to clients to solidify the emerging narratives and developing communities of concern or *leagues* (see discussion of leagues under Interventions) in which clients provide support to each other.

Jill Freedman and Gene Combs Based in the United States, husband-and-wife team Jill Freedman and Gene Combs (1996) developed the narrative approach emphasizing social construction of realities and further developed the narrative metaphor for conceptualizing therapeutic intervention. They are the codirectors of the Evanston Family Therapy Center in Illinois.

Gerald Monk and John Winslade After beginning their work in New Zealand, Gerald Monk and John Winslade now work in the United States and have developed narrative approaches for schools in counseling, multicultural counseling, mediation, and consultation (Monk, Winslade, Crocket, & Epston, 1997; Monk, Winslade, & Sinclair, 2008; Winslade & Monk, 2000, 2007, 2008).

The Big Picture: Overview of Treatment
Treatment Phases

The process of narrative therapy involves helping clients find new ways to view, interact with, and respond to problems in their lives by redefining the role of those problems (White, 2007). From a narrative perspective, persons are not the problem; problems are the problem. Although there is variety among practitioners, narrative therapy broadly involves the following phases (Freedman & Combs, 1996; White & Epston, 1990):

- *Meeting the Person:* Getting to know people as *separate* from their problems by learning about the hobbies, values, and everyday aspects of their lives
- *Listening:* Listening for the effects of dominant discourses and identifying times without the problems
- *Separating Persons From Problems:* Externalizing and separating people from their problems to create space for new identities and for life stories to emerge
- *Enacting Preferred Narratives:* Identifying new ways to relate to problems that reduce their negative effects on the lives of all involved
- *Solidifying:* Strengthening preferred stories and identities by having them witnessed by significant others in a person's life

Use of Thickening Descriptions

The narrative therapy process is a thickening and enriching of the person's identity and life accounts rather than a "story-ectomy." Instead of replacing a problem story with a problem-free one, narrative therapists *add* new strands of identity to the problem-saturated descriptions with which clients enter therapy. In any given day, an infinite number of events can be storied into our accounts of the day and who we are. When people begin to experience problems, they tend to notice only those events that fit with the problem narrative. For example, if they are feeling hopeless, they tend to notice when things do not go their way during the day and do not give much weight to the good things that happened. Similarly, when couples start a period of fighting, they start to notice only what the other person is doing that confirms their position in the fight and ignore and/or forget other events. In narrative therapy, the therapist helps clients create more balanced, rich, and appreciative descriptions of events that will enable them to build more successful and enjoyable lives.

Making Connection: The Therapeutic Relationship
Meeting the Person Apart from the Problem

Narrative therapists generally begin their first session with clients by meeting clients "apart from the problem," that is, as everyday people (Freedman & Combs, 1996). Therapists ask questions such as the following to familiarize themselves with clients' everyday lives:

Questions for Meeting the Person (Not the Problem)

- What do you do for fun? Do you have hobbies?
- What do you like about living here? What don't you like?
- Tell me about your friends and family.
- What is important to you in life?
- What is a typical weekday like? Weekend?

The answers to these questions enable narrative therapists to know and view their clients in much the same way that clients view themselves, as everyday people. In the case study at the end of the chapter, the therapist emphasizes meeting Darren and Ernest apart from the problem because of their sense of inadequacy and defeat in relation to many of the problems in their lives.

Separating People from Problems: The Problem Is the Problem

In narrative therapy, the motto is: "The problem is the problem. The person is not the problem" (Winslade & Monk, 1999, p. 2). Once therapists have come to know the client apart from the problem and have a clear sense of who the client is as a person, they begin to "meet" the problem in much the same way, keeping their identities separate. The problem—whether depression, anxiety, marital conflict, ADHD, defiance, loneliness, or a breakup—is viewed as a separate entity or situation that is *not* inherent to the person of the client. Therapists maintain a polite, social, "getting to know you" attitude.

Questions for "Meeting" the Problem

- When did the problem first enter your life?
- What was going on with you then?
- What were your first impressions of the problem? How have they changed?
- How has your relationship with the problem evolved over time?
- Who else has been affected by the problem?

Narrative therapists can take an adversarial stance toward the problem (wanting to outwit, outsmart, or evict it; White, 2007) or a more compassionate stance (wanting to understand its message and concerns; Gehart & McCollum, 2007).

Optimism and Hope

Because narrative therapists view problems as problems and people as people, they have a deep, abiding optimism and hope for their clients (Monk et al., 1997; Winslade & Monk, 1999). Their hope and optimism are not sugar-coated, naïve wishes but instead are derived from their understanding of how problems are formed—through language, relationship, and social discourse—having confidence that their approach can make a difference. Furthermore, by separating people from problems, they quickly connect with the "best" in the client, which reinforces a sense of hope and optimism. In the case study at the end of the chapter, the therapist's sense of hope is particularly important for this couple to have the courage to envision and take action toward enacting their preferred narratives, which includes living without shame as a committed gay couple.

Therapist as Coauthor and Coeditor

The role of the therapist is often described as a *coauthor* or *coeditor* to emphasize that the therapist and client engage in a joint process of constructing meaning (Freedman & Combs, 1996; Monk et al., 1997; White, 1995; White & Epston, 1990). Rather than attempting to offer a "better story," the therapist works alongside the client to generate a more useful narrative. Although the degree and quality of input vary greatly, narrative therapists tend to focus on the sociopolitical aspects of a client's life. Some narrative therapists maintain that therapists should take a stance on broader sociocultural issues of injustice with all clients (Zimmerman & Dickerson, 1996), but not all narrative therapists share this agenda (Monk & Gehart, 2003).

Therapist as Investigative Reporter

In his later works, White (2007) describes his relationship to problems as that of an *investigative reporter*:

> The form of inquiry that is employed during externalizing conversations can be likened to investigative reporting. The primary goal of investigative reporting is to develop an exposé on the corruption associated with abuse of power and privilege. Although investigative reporters are not politically neutral, the activities of their inquiry do not take them into the domains of problem-solving, of enacting reform, or of engaging indirect power struggles ... their actions usually reflect a relatively "cool" engagement. (pp. 27–28)

Thus, rather than rushing in to fix problems, the therapist uses a calm but inquisitive stance to explore the origins of problems and thus to inspire clients to develop a better understanding of their larger contexts.

The Viewing: Case Conceptualization and Assessment

Problem-Saturated Stories

As clients are talking, narrative therapists listen for the problem-saturated story (Freedman & Combs, 1996; White & Epston, 1990), the story in which the "problem" plays the leading role and the client plays a secondary role, generally that of victim. The therapist attends to how the problem affects the client at an *individual level* (health, emotions, thoughts, beliefs, identity, relationship with the divine) and at a *relational level* (with significant other, parents, friends, coworkers, teachers), as well as how it affects each of these significant others at a personal level. While listening to a client's problem-saturated story, the therapist listens closely for alternative endings and subplots in which the problem is less of a problem and the person is an effective agent; these are referred to as *unique outcomes*.

Unique Outcomes and Sparkling Events

Unique outcomes (White & Epston, 1990) or *sparkling events* (Freedman & Combs, 1996) are stories or subplots in which the problem-saturated story does not play out in its typical way: the child cheerfully complies with a parent's request; a couple are able to stop a potential argument from erupting with a soft touch; a teenager decides to call a friend rather than allow herself to cut. These stories often go unnoticed because they have no dramatic ending or particularly notable outcome that warrants attention, and therefore they are not "storied" in clients' or others' minds. These unique outcomes are used to help clients create the lives they prefer and to develop a more full and accurate account of their own and others' identities. When working with clients such as Darren and Ernesto in the case study at the end of the chapter, unique outcomes are particularly essential because they report a sense of hopelessness due to the many seemingly "permanent" and "unchangeable" aspects of the problems in their lives, such as religious families who disapprove of their sexuality, HIV+ diagnosis, and disapproval from their ethnic communities.

Dominant Cultural and Gender Discourses (see also Juice)

As already discussed, narrative therapists listen for dominant cultural and gender themes that have informed the development and perception of a problem (Monk et al., 1997; White & Epston, 1990). The purpose of all discourses is to identify the set of "goods" and "values" that organize social interaction in a particular culture. All cultures are essentially a set of dominant discourses: social rules and values that make it possible for a group of people to meaningfully interact (see Chapter 4).

Dominant discourses are the societal stories of how life "should" happen; for example, to be a happy and good person, you should get married, get a stable, high-paying job, have kids, get a nice car, buy a house, and volunteer at your child's school. Whether you comply with this vision of happiness, rebel against it, or are not even in the game because of social or physical limitations, problems can arise in relation to it. In working with clients, narrative therapists listen closely for the dominant discourses that are most directly informing the perception of a problem. In response, they inquire about *local* or *alternative discourses*.

Local and Alternative Discourses: Attending to Client Language and Meaning

Local and alternative discourses are those that do not conform to the dominant discourse (White & Epston, 1990): couples who choose not to have children, same-sex relationships, immigrant families wanting to preserve their roots, speaking English as a second language, teen subculture in any society, and so forth. The local discourses offer a different set of "goods," "shoulds," and ethical "values" than what is portrayed in the dominant discourse. For example, teens have created a subculture with different beauty standards, sexual norms, vocabulary, and friendship rules than are found in adult culture. The teen culture represents an alternative discourse that therapists can tap into to understand the teen's worldview and values, as well as to explore with the teen how this alternative discourse can successfully coexist with the dominant discourse. Thus the local discourse provides a resource for generating new ways of viewing the self and for talking and interacting with others around the problem.

Targeting Change: Goal Setting
Preferred Realities and Identities

As a postmodern approach, narrative therapy does not include a set of predefined goals that can be used with all clients. Instead, goal setting in narrative therapy is unique to each client. In the broadest sense, the goal of narrative therapy is to help clients *enact their preferred realities and identities* (Freedman & Combs, 1996). In most cases, enacting preferred narratives involves increasing clients' sense of *agency*, the sense that they influence the direction of their lives. When identifying preferred realities, therapists work with clients to develop thoughtfully reflected goals that consider local knowledges rather than simply adopting the values of the dominant culture. Clients often redefine their preferred reality to incorporate these local knowledges and to lessen the influence of dominant discourses. For example, a couple may come in wanting things to go back to the way they were while dating, but as they move through the therapeutic process, they realize that they want and need something different than what they had before because they are entering a new chapter in their lives as individuals and as a couple.

Thus the key is defining the "preferred" reality and identity thoughtfully and with intention after considering the impact of dominant and local discourses as well as the meanings and impact of the proposed preferred reality. This process is often a gradual shift from "make this problem go away" to "I want to create something beautiful/meaningful/great with my/our life(ves)." The therapist allows the client to take the lead in defining the preferred realities and acts as a coeditor to help the client reflect on where the idea came from and the effects it will have on the client's life.

Middle-Phase Goals

Middle-phase goals target immediate symptoms and the presenting problem; the following are some examples:

- "Increase sense of agency in problem-resolution conversations with spouse."
- "Increase opportunities to interact with friends using 'confident, social' self."
- "Reduce number of times mother and father allow Anger to take over in response to child's defiance."
- "Increase instances of defiance in response to anorexia's directions to not eat."

Late-Phase Goals

Late-phase goals target personal identity, relational identity, and the expanded community:

- *Personal Identity:* "Solidify a sense of personal identity that derives self-worth from meaningful activities, relationship, and values rather than body size."
- *Relational Identity:* "Develop a family identity narrative that allows for greater expression of differences while maintaining family's sense of closeness and loyalty."
- *Expanded Community:* "Expand preferred 'outgoing' identity to social relationships and contexts."

The Doing: Interventions

Externalizing: Separating the Problem from the Person

The signature technique of narrative therapy, externalizing involves conceptually and linguistically separating the person from the problem (Freedman & Combs, 1996; White & Epston, 1990). To be successful, externalization requires a sincere belief that people are separate from their problems; thus the *attitude* of externalization is key to its effectiveness (Freedman & Combs, 1996). More than a single-session intervention, externalization is an organic and evolving process of shifting clients' perception of their relationship to the problem: from "having" it to seeing it as outside the self. Therapists can externalize by naming the problem as an external other or by changing a descriptive adjective into a noun (e.g., from a client being depressed to having a relationship with Depression, or changing from being a conflictual couple to having a relationship with Conflict). At other times, clients respond better by talking about "sides" of themselves or a relationship: "the little girl in me who is afraid" or "the competitive side of our relationship."

For externalization to work, it cannot be forced onto the client but rather needs to emerge from the dialogue or be introduced as a possibility for how to think about the situation. In most cases, techniques such as mapping the influence of persons and problem (see next section) invite a natural, comfortable process for externalizing the problem. Alternatively, therapists can ask clients if they want to refer to the problem as something separate from themselves when the conversation allows. If clients already have a name for the problem and conceptualize the problem as a sort of external entity or very discrete part of themselves, therapists need only build on the externalization process they have started.

Relative Influence Questioning: Mapping Influence of the Problem and Persons

Relative influence questioning was the first detailed method for externalization (White & Epston, 1990). Used early in therapy, it serves simultaneously as an assessment and an intervention and is composed of two parts: (a) mapping the influence of the problem and (b) mapping the influence of persons.

Mapping the Influence of the Problem When mapping the influence of the problem, therapists inquire about how the problem has affected the lives of the client and significant others, often *expanding* the reach of the problem beyond how the client generally thinks of it; thus it is critical that this is followed up by *mapping the influence of person* questions to ensure that the client does not feel worse afterwards.

Questions for Mapping the Influence of the Problem

How has the problem affected:

- Clients at a physical, emotional, and psychological level?
- Clients' identity stories and what they tell themselves about their worth and who they are?
- Clients' closest relationships: partner, children, parents?
- Other relationships in clients' lives: friendships, social groups, work or school colleagues, etc.?
- The health, identity, emotions, and other relationships of significant people in clients' lives (e.g., how parents may pull away from friends because they are embarrassed about a child's problem)?

Mapping the Influence of Persons Mapping the influence of persons begins the externalization process more explicitly. This phase of questioning, which should immediately follow mapping the influence of problems, involves identifying how the person has affected the life of the problem, reversing the logic of the first series of questions.

Questions for Mapping the Influence of Persons

When have the persons involved:

- Kept the problem from affecting their mood or how they value themselves as people?
- Kept the problem from allowing themselves to enjoy special and/or casual relationships in their lives?
- Kept the problem from interrupting their work or school lives?
- Been able to keep the problem from taking over when it was starting?

White and Epston (1990) report that externalization has the following beneficial effects:

- Decreases unproductive conflict and blame between family members
- Undermines sense of failure in relation to the problem by highlighting times the persons have had influence over it
- Invites people to unite in a struggle against the problem and reduce its influence
- Identifies new opportunities for reducing the influence of the problem
- Encourages a lighter, less stressed approach to interacting with the problem
- Increases interactive dialogue rather than repetitive monologue about the problem

Mapping influence questions are used extensively in the case study at the end of the chapter because they not only help separate person from problems but help create a sense of agency, which both men are seeking.

Externalizing Conversations: The Statement of Position Map

White (2007) describes his more recently developed process for facilitating externalizing conversations as "the statement of position map." This map includes four categories of inquiry, which are used multiple times throughout a session and across sessions to shift the client's relationship with the problem and open new possibilities for action.

Inquiry Category 1: Negotiating an Experience-Near Definition White begins by defining the problem using the client's language (experience-near language) rather than in professional or global terms (e.g., a "diagnosis"). Thus, "feeling blue" is preferred to "depressed."

Inquiry Category 2: Mapping the Effects As in White's early work (White & Epston, 1990), mapping the effects of problems involves identifying how the problem has affected the various domains of the client's life: home, work, school, and social contexts; relationships with family, friends, and himself/herself; and the client's identity and future possibilities.

Inquiry Category 3: Evaluating the Effects After identifying the effects of the problem, the therapist asks the client to evaluate these effects (White, 2007, p. 44):

- Are these activities okay with you?
- How do you feel about these developments?
- Where do you stand on these outcomes?
- Is the development positive or negative—or both, or neither, or something in between?

Inquiry Category 4: Justifying the Evaluation In the final phase, the therapist asks about how and why clients have evaluated the situation the way they have (White, 2007, p. 48):

- Why is or isn't this okay for you?
- Why do you feel this way about this development?
- Why are you taking this stand or position on this development?

These "why" questions must be offered in a spirit of allowing clients to give voice to what is important to them rather than creating a sense of moral judgment. They should open up conversations about what motivates clients and how they want to shape their identities and futures. In the case study at the end of the chapter, statement of position mapping is used frequently with Darren and Ernesto to help them consciously define themselves and their relationship in terms of their sexuality, substance use, and anxieties.

Externalizing Metaphors

When externalizing using these four categories, White (2007, p. 32) employs various metaphors for relating to problems:

- Walking out on the problem
- Going on strike against the problem
- Defying the problem's requirements
- Disempowering the problem
- Educating the problem
- Escaping the problem
- Recovering or reclaiming territory from the problem
- Refusing invitations from the problem
- Disproving the problem's claims
- Resigning from the problem's service
- Stealing their lives from the problem
- Taming the problem
- Harnessing the problem
- Undermining the problem

Avoiding Totalizing and Dualistic Thinking

White (2007) avoids totalizing descriptions of the problem—the problem being all bad—because such descriptions promote dualistic, either/or thinking, which can be invalidating to the client and/or obscure the problem's broader context.

Externalizing Questions

Narrative therapists use externalizing questions to help clients build different relationships with their problems (Freedman & Combs, 1996). In most cases, these questions transform adjectives (e.g., *depressed, anxious, angry*) to nouns (e.g., *Depression, Anxiety, Anger*; capitalization is used to emphasize that the Problem is viewed as a separate entity). Externalizing questions *presume* that the person's are separate from the Problem and that they have a two-way relationship with the Problem: it affects them, and they affect it.

To experience the liberating effects of externalizing, Freedman and Combs (1996, pp. 49–50) have developed the following two sets of questions: one representing conventional therapeutic questions and the other externalizing questions. To do this exercise, choose a quality or trait that you or others find problematic, usually an adjective; substitute this for X in the questions below. Then find a noun form of that trait; substitute this for Y in the questions below. For example: X = depressed/Y = Depression; X = critical/Y = Criticism; X = angry/Y = Anger.

Conventional Versus Externalizing Questions

Conventional Questions (insert problem description as *adjective* for X)	*Externalizing Questions* (insert problem description as *noun* for Y)
When did you first become X?	What made you vulnerable to the Y so that it was able to dominate your life?
What are you most X about?	In what contexts is the Y most likely to take over?
What kinds of things happen that typically lead to your being X?	What kinds of things happen that typically lead to the Y taking over?
When you are X, what do you do that you wouldn't do if you weren't X?	What has the Y gotten you to do that is against your better judgment?
Which of your current difficulties come from being X?	What effects does the Y have on your life and relationship?
What are the consequences for your life and relationships of being X?	How has the Y led you into the difficulties you are now experiencing?
How is your self-image different when you are X?	Does the Y blind you from noticing your resources, or can you see them through it?
If by some miracle you woke up some morning and were not X anymore, how, specifically, would your life be different?	Have there been times when you have been able to get the best of the Y? Times when the Y could have taken over but you kept it out of the picture?

Problem Deconstruction: Deconstructive Listening and Questions

Drawing from the philosophical work of Jacques Derrida, narrative therapists use deconstructive listening and questions to help clients trace the effects of dominant discourses and to empower clients to make more conscious choices about which discourses they allow to affect their life (Freedman & Combs, 1996). In deconstructive listening, the therapist listens for "gaps" in clients' understanding and asks them to fill in the details or has them explain the ambiguities in their stories. For example, if a client reports feeling rejected because friends did not call when they said they would, the therapist listens for the meanings that led to the sense of feeling "rejected."

Deconstructive questions help clients to further "unpack" their stories to see how they have been constructed, identifying the influence of dominant and local discourses. Typically used in externalizing conversations, these questions target problematic beliefs, practices, feelings, and attitudes by asking clients to identify the following:

- *History:* The *history of their relationship* with the problematic belief, practice, feeling, or attitude: "When and where did you first encounter the problem?"
- *Context:* The *contextual influences* on the problematic belief, practice, feeling, or attitude: "When is it most likely to be present?"
- *Effects:* The *effects or results* of the problematic belief, practice, feeling, or attitude: "What effects has this had on you and your relationship?"
- *Interrelationships:* The *interrelationship with other* beliefs, practices, feelings, or attitudes: "Are there other problems that feed this problem?"
- *Strategies: Tactics and strategies* used by the problem belief, practice, feeling, or attitude: "How does it go about influencing you?"

Mapping in Landscapes of Action and Identity or Consciousness

Based on the narrative theory of Jerome Bruner (1986), mapping the problem in the landscapes of action and identity (White, 2007) or consciousness (Freedman & Combs, 1996) is a specific technique for harnessing unique outcomes to promote desired change. Mapping in the landscapes of action and identity generally involves the following steps:

1. *Identify a Unique Outcome:* The therapist listens for and asks about times when the problem could have been a problem but was not.
2. *Ensure That the Unique Outcome Is Preferred:* Rather than assume, the therapist asks clients about whether the unique outcome is a preferred outcome: "Is this something you want to do or have happen more often?"
3. *Map in Landscape of Action:* First, the therapist begins by mapping the unique outcome in the landscape of action, identifying what actions were taken by whom in which order. The therapist does this by asking about specific details: "What did you do first? How did the other person respond? What did you do next?" The therapist carefully plots the events until there is a step-by-step picture of the actions of the client and involved others, gathering details about the following:
   - Critical events
   - Circumstances surrounding events
   - Sequence of events
   - Timing of events
   - Overall plot
4. *Map in the Landscape of Identity or Consciousness:* After obtaining a clear picture of what happened during the unique outcome, the therapist begins to map in the landscape of identity. This phase of mapping *thickens the plot* associated with the successful outcome, thus directly strengthening the connection of the preferred outcome with the client's personal identity. Mapping in the landscape of identity focuses on the psychological and relational implications of the unique outcomes. The following sample questions cover various areas of impact:
   - "What do you believe this says about you as a person? About your relationship?"
   - "What were your intentions behind these actions?"
   - "What do you value most about your actions here?"
   - "What, if anything, did you learn or realize from this?"
   - "Does this change how you see life, God, your purpose, or your life goals?"
   - "Does this affect how you see the problem?"

Intentional Versus Internal State Questions

White (2007) privileges intentional state questions (questions about a person's intentions in a given situation: "What were your intentions?") over internal state questions (questions about how a person was feeling or thinking: "What were you feeling?") because intentional state questions promote a sense of *personal agency,* whereas internal state questions can have the effect of diminishing one's sense of agency, increasing one's sense of isolation, and discouraging diversity.

Scaffolding Conversations

Drawing on Vygotsky's concept of *zones of proximal development,* White (2007) uses scaffolding conversations to move clients from that which is familiar to that which is novel. Vygotsky was a developmental psychologist who emphasized that because learning is relational, adults should structure children's learning in ways that help them interact with new information. The zone of proximal development is the distance between what the child can do independently and what the child can do in collaboration with others. *Scaffolding* is a term White developed with clients to describe five incremental movements across this zone of learning:

- *Low-Level Distancing Tasks:* These tasks *characterize a unique outcome.* Because they are at a low-level distance from the client (very close to what is familiar to him/her), they encourage the client to attribute meanings to events that have previously gone unnoticed: e.g., "Are there times when you don't seem to get into an argument even though there is tension?"
- *Medium-Level Distancing Tasks:* These tasks allow *unique outcomes to be taken into a chain of association.* They introduce greater "newness," encouraging more comparisons and contrasts with other unique outcomes: e.g., "How was last night's 'effective problem-solving conversation' similar or different from the one you described last week?"
- *Medium-High-Level Distancing Tasks:* These tasks *reflect on a chain of associations.* They encourage clients to reflect on, evaluate, and learn from the differences and similarities with other tasks: e.g., "Looking back over these examples of effective problem solving, is there anything that stands out as useful in preventing arguments?"
- *High-Level Distancing Tasks:* These tasks promote *abstract learning and realizations.* They require clients to assume a high level of distance from their immediate experience, promoting increased abstract conceptualization of life and identity: e.g., "What do these effective problem-solving conversations say about you as a person and about your relationship?"
- *Very-High-Level Distancing Tasks:* These are *plans for action.* They promote a high-level distancing from immediate experience to enable clients to identify ways of enacting their newly developed concepts about life and identity: e.g., "Do you have some ideas of how you want to translate these ideas into future action?"

Over the course of a conversation, therapists move back and forth between various levels of distancing tasks, progressively moving to higher levels of action planning.

Permission Questions

Narrative therapists use *permission questions* to emphasize the democratic nature of the therapeutic relationship and to encourage clients to maintain a clear, strong sense of agency when talking with the therapist. Quite simply, permission questions are questions therapists use to ask permission to ask a question. This goes against the prevailing assumption that therapists can ask any question they want to gather information they purportedly need to help the client. Socially, therapists are exempt from the prevailing social norms of polite conversation topics and are free to bring up taboo subjects such as sex, past abuse, relationship problems, death, fears, and weaknesses. Many clients feel compelled to answer these questions even if they are not comfortable doing so. Narrative therapists are sensitive to the power dynamic related to taboo and difficult subjects and therefore *ask for the client's permission* before asking questions that are generally taboo or that the therapist anticipates will make a particular client feel uncomfortable. For example, they might say, "Would it be okay if I ask you some questions about your sex life?"

In addition, permission questions are used throughout the interview regarding *what* is being discussed and *how* to ensure that the conversation is meaningful and comfortable for the client. For example, often when starting a session the therapist may briefly outline his/her ideas for how to use the time, asking for client input and permission to continue with a particular topic or line of questioning. Similarly, when therapists find themselves asking one person more questions than the others in a family session, they pause to ask permission to continue to ensure that everyone is okay with what is going on.

Situating Comments

Like permission questions, *situating comments* are used to maintain a more democratic therapeutic relationship and to reinforce client agency by ensuring that comments from the therapist are not taken as a "higher" or "more valid" truth than the client's (Zimmerman & Dickerson, 1996). Drawing on the distinction between dominant and local discourses, narrative therapists are keenly aware that any comment made by the therapist is often considered more valid than anything the client might say. Thus, therapists *situate* their comments by revealing the source of their perspective, emphasizing that it is only one perspective among others. When the source and context of a therapist comment are revealed, a client is less likely to overprivilege the comment.

Examples of Situating Comments

Therapist Comment Without Situating	Situating Therapist Comment
I am noticing that you tend to ...	Having grown up on a farm, my attention is of course drawn to ...
Research indicates that ...	There is one therapist who has developed a theory (or done a study) that suggests. ... Does this sound like something that would be true for you?
I suggest that you ...	Since you are asking me for a suggestion, I can only tell you what I think as someone who believes action is more productive than talk ...

Narrative Reflecting Team Practices

On the basis of Tom Andersen's collaborative practice of using reflecting teams (see earlier discussion), narrative therapists have developed a similar practice that supports their work. Using Andersen's format, Freedman and Combs (1996) assign the team *three primary tasks*:

1. Develop a thorough understanding by closely attending to details of the story.
2. Listen for differences and events that do not fit the dominant problem-saturated narrative.
3. Notice beliefs, ideas, or contexts that support the problem-saturated descriptions.

In addition, they propose the following *guidelines* for the team:

1. During the reflecting process, the reflecting team members participate in a back-and-forth conversation rather than in a monologue.
2. Team members should not talk to each other while observing the interview.
3. Comments should be offered in a tentative manner (e.g., "perhaps," "could," "might").
4. Comments are based on what actually occurs in the room (e.g., "At one point mom got very quiet; I was wondering what was going on for her at that moment").
5. When appropriate, comments are situated in the speaker's personal experience (e.g., "Having been a teacher, I may have been the only one who focused on this ... ").
6. All family members should be responded to in some way.
7. Reflections should be kept short.

Re-Membering Conversations

White (2007) uses *re-membering conversations* to develop a multivoiced sense of personal identity that enables clients to make sense of their lives in a more coherent and orderly way. In these conversations, clients develop the sense of an identity that is grounded in *associations of life* rather than in a singular, core self. The associations of life include a "membership" of significant people and identities from the client's past, present, and projected future. In these conversations, clients are encouraged to identify who's a member, assess the influence of each member, and decide whether their membership should be upgraded, downgraded, or canceled (e.g., canceling the membership of a high school bully whose taunting still haunts the client). The process of re-membering includes the following components:

- Identifying the other person's contribution to the client's life
- Articulating how the other person may have viewed the client's identity
- Considering how the client may have affected the other person's life
- Specifying the implications for the client's identity (e.g., "I am a person who values justice")

In the case study at the end of the chapter, the therapist uses re-membering conversations to help Darren and Ernesto more consciously decide how and to what extent they will let the various persons in their past and current life define who they are and their ultimate value as a person.

Leagues

To solidify a new narrative and new identities, narrative therapists have created leagues (or clubs, associations, teams), membership in which signifies an accomplishment in a particular area. In most cases, leagues are virtual communities of concern (e.g., a Temper Tamer's Club to which a child is given a membership certificate), although some meet face to face or interact via the Internet (Anti-Anorexia/Anti-Bulimia League; see www.narrativetherapy.com).

Definitional Ceremony

Generally used toward the end of therapy to solidify the emerging preferred narrative and identity, definitional ceremonies involve inviting significant others to *witness* the emerging story. This ceremony has three phases:

1. *The First Telling:* The client tells his/her life story, highlighting the emerging identity stories as the invited witnesses listen.
2. *Retelling:* The witnesses take turns retelling the story from their perspectives; they are prepared for the process by being asked to refrain from offering advice, making judgments, or theorizing and are asked to *situate* their comments.
3. *Retelling of the Retelling:* The client then retells the story incorporating aspects of the witnesses' stories.

Letters and Certificates

Narrative letters are used to develop and solidify preferred narratives and identities (White & Epston, 1990). Therapists can write letters detailing a client's emerging story after a session in lieu of doing case notes (unless they work in a practice environment that requires a specific format). Narrative letters use the same techniques used in session to reinforce the emerging narrative; they perform the following functions:

- *Emphasize Client Agency:* Letters highlight clients' agency in their lives, including small steps in becoming proactive.
- *Take Observer Position:* The therapist clearly takes the role of *observing* the changes the client is making, citing specific, concrete examples whenever possible.
- *Highlight Temporality:* The time dimension is used to plot the emerging story: where clients began, where they are now, and where they are likely to go.

- *Encourage Polysemy:* Rather than propose singular interpretations, multiple meanings are entertained and encouraged.

Letters can be used early in therapy to engage clients, during therapy to reinforce the emerging narrative and reinforce new preferred behaviors, or at the end of therapy to consolidate gains by narrating the change process.

Sample Letter White and Epston (1990, pp. 109–110) offer numerous sample letters, such as the following:

> Dear Rick and Harriet,
> I'm sure that you are familiar with the fact that the best ideas have the habit of presenting themselves after the event. So it will come as no surprise to you that I often think of the most important questions after the end of an interview. ...
> Anyway, I thought I would share a couple of important questions that came to me after you left our last meeting:
> Rick, how did you decline Helen's [the daughter] invitation to you to do the reasoning for her? And how do you think this could have the effect of inviting her to reason with herself? Do you think this could help her to become more responsible?
> Harriet, how did you decline Helen's invitations to you to be dependable for her? And how do you think this could have the effect of inviting her to depend upon herself more? Do you think that this could have the effect of helping her to take better care of her life?
> What does this decreased vulnerability to Helen's invitations to have her life for her [e.g., take responsibility for her life by making decisions, solving problems, and handling consequences] reflect in you both as people?
> By the way, what ideas occurred to you after our last meeting? M.W.

Certificates Certificates are often used with children to recognize the changes they have made and to reinforce their new "reputation" as a "temper tamer," "cooperative child," and so forth.

Certified Temper Tamer

This is to certify that

_____

has proven himself as a skilled Tamer of Tempers,

having gone two months without temper problems at home or school

using the following taming techniques that he developed for himself:

1. Asking for help when confused

2. Taking three deep breaths when the scent of Temper appears

3. Using soft words to talk about anger and frustration

Date of award: _____

Witnessed by: _____ (therapist)

_____ (parents)

_____ (teacher)

Interventions for Specific Problems

Children

Numerous narrative therapists have developed interventions for working with children (Freeman, Epston, & Lobovits, 1997; Smith & Nylund, 2000; Vetere & Dowling, 2005; White & Morgan, 2006). The externalization process seems to come more naturally to children, perhaps because it is reflected so often in cartoons and children's literature (e.g., the devil and angel on a cartoon character's shoulders). Externalization adapts well to play and art therapies: externalized problems (e.g., Temper, Sadness, Anger) can be portrayed in art media (drawings, clay, paintings) or acted out with puppets and dolls. In addition to externalized problems, children enjoy drawing or acting out unique outcomes and preferred narratives; this process often accelerates their adaptation of new behaviors.

Domestic Violence

Narrative therapists have developed unique and promising alternatives to the standard treatment for those who batter (Augusta-Scott & Dankwort, 2002; Jenkins, 1990). Unlike the feminist-based Duluth model, which traces the cause of violence to men's attempts to gain power and control (Pence & Paymar, 1993; see Chapter 16), narrative approaches work from within clients' lived reality, which usually includes the experience of helplessness and powerlessness that they say leads them to try to regain control through violence (Augusta-Scott & Dankwort, 2002).

Jenkins (1990) warns therapists against accepting responsibility for the violence, which therapists inadvertently do when they challenge the man's explanations, give advice on how to stop abusive behavior, offer strong arguments against violence, or try to break down his denial, all of which are common therapist responses to violence. Instead Jenkins uses a nine-step model that requires the *client* to take full responsibility for the violence and for ending it. Throughout this process the therapist is supportive without condoning violence or attacking it; instead the focus is on facilitating the process in the nine-step model.

Jenkins's Nine-Step Model for Working with Men Who Batter

1. Invite the man to address his violence.
2. Invite the man to argue for a nonviolent relationship.
3. Invite the man to examine his misguided efforts to contribute to the relationship.
4. Invite the man to identify time trends in the relationship.
5. Invite the man to externalize restraints (*Note:* He avoids externalizing anger and violence to prevent possible minimizing of responsibility).
6. Deliver irresistible invitations to challenge restraints.
7. Invite the man to consider his readiness to take new action.
8. Facilitate the planning of new action.
9. Facilitate the discovery of new action (p. 63).

Throughout the process, the therapist identifies the dominant discourses, particularly patriarchal discourses, that have contributed to the violence; these are deconstructed and externalized to help the client develop more effective ways of relating. The narrative approach is careful not to replicate the abusive pattern of harshness and criticism in the therapeutic relationship and instead models the respect, tolerance, and boundaries that clients are aspiring to enact.

Putting It All Together

Case Conceptualization Template

- *Meeting Person Apart from the Problem:* Describe who the person/people are apart from the problem: strengths, hobbies, interests, career, etc.

- *Problem-Saturated Narrative:* Describe how the problem is affecting the persons involved at (a) a personal level (emotional, behavioral, identity narrative, etc.), (b) relational level (conflicts or distance in significant relationships), and (c) in broader life circumstances (work, school, etc.).
- *Unique Outcomes/Sparkling Events*: Describe:
  - Times, contexts, relationships, etc. when the problem is less of a problem or not a problem.
  - The effect of persons on the problem: What things do people do that make the problem less of a problem?
- *Dominant Discourses*
  - *Cultural, Ethnic, SES, Religious, etc.*: How do key cultural discourses inform what is perceived as a problem and the possible solutions?
  - *Gender, Sexual Orientation, etc.*: How do the gender/sexual discourses inform what is perceived as a problem and the possible solutions?
  - *Contextual, Family, Community, School, and Other Social Discourses*: How do other important discourses inform what is perceived as a problem and the possible solutions?
- *Local/Alternative Narratives*
  - *Identity/Self-Narratives:* How has the problem shaped each family member's identity?
  - *Local or Preferred Discourses:* What is the client's preferred identity narrative and/or narrative about the problem? Are there local (alternative) discourses about the problem that are preferred?

Treatment Plan Template for Individual

Narrative Initial Phase of Treatment with Individual

Initial Phase Counseling Tasks

1. Develop working counseling relationship. *Diversity Note: [Describe how you will adjust to respect cultured, gendered, and other styles of relationship building and emotional expression.]*
   a. *Meet person apart from problem*, inquiring about identity outside of problem.
   b. Engage client from *hopeful, optimistic* position and as *coauthor/investigative reporter.*

2. Assess individual, systemic, and broader cultural dynamics. *Diversity Note: [Describe how you will adjust assessment based on cultural, socioeconomic, sexual orientation, gender, and other relevant norms.]*
   a. *Relative influence questioning* to map the effects of Depression (Anxiety) and the effects of persons on Depression; identify *unique outcomes.*
   b. Identify the *dominant discourses* that support Depression (Anxiety) and the *local, alternative discourses* that may be a resource in changing relationship to Depression (Anxiety).

Initial Phase Client Goals

1. Increase *client's influence over Depression* (Anxiety or crisis symptom) to reduce severity of depressed mood and anxiety.
   a. *Externalizing* Depression to identify current areas of person's influence and possibilities for expanding this influence.
   b. *Map unique outcomes in landscapes of action and consciousness* to identify new possibilities for reducing influence of Depression on person.

(continued)

Narrative Working Phase of Treatment with Individual

Working-Phase Counseling Task

1. Monitor quality of the working alliance. *Diversity Note: [Describe how you will attend to client response to interventions that indicate therapist using expressions of emotion that are not consistent with client's cultural background.]*
    a. *Permission questions and situating comments* to increase client sense of safety and agency within relationship.
    b. Administer *session rating scale*.

Working-Phase Client Goals

1. Decrease *influence of Depression (Anxiety)* and *dominant discourses* to reduce depressed mood/anxiety.
    a. *Externalizing questions* to separate client from problem.
    b. *Mapping Depression vs. unique outcomes in landscapes of action and consciousness* to separate client from Depression and strengthen alternative identity.
2. Increase *sense of agency* by developing *new relationship* with Depression (Anxiety) to reduce depressed mood/anxiety.
    a. *Externalizing metaphors* to define relationship with Depression (Anxiety).
    b. *Statement of position map* to evaluate effects of Depression and inspire new relationship to Depression.
3. Increase actions that support *preferred identity* to reduce depressed mood/anxiety.
    a. *Scaffolding conversations* to move client from contemplation to action.
    b. *Statement of position map* to evaluate effects of actions and identify where and how to make adjustments.

Narrative Closing Phase of Treatment with Individual

Closing-Phase Counseling Task

1. Develop aftercare plan and maintain gains. *Diversity Note: [Describe how you will access resources in the communities of which they are a part to support them after ending therapy.]*
    a. Map potential setbacks in *landscapes of action and consciousness* and identify alternative behaviors.
    b. *Letter from therapist* and/or from client preferred to self to future self-experiencing setback to solidify narrative of overcoming Depression.

Closing-Phase Client Goals

1. Increase and expand influence of *preferred identity* in [specify work/school, relational, and other significant areas of life] to reduce depressed mood and increase sense of wellness.
    a. *Scaffolding conversations* to move client from contemplation to action.
    b. *Statement of position map* to evaluate effects of actions and identify where and how to make adjustments.
2. Increase the number of relationships that support client in *preferred identity* to reduce depression and increase sense of wellness.
    a. *Definitional ceremony* to expand network of friends and family who support client's preferred identity.
    b. *Therapeutic letter* from therapist to document journey from being overwhelmed by Depression (Anxiety) to client being able to manage/subdue (or preferred metaphor) it.
    c. *Narrative reflecting team* to support client in newly enacted identity.

Treatment Plan Template for Couple/Family

Narrative Initial Phase of Treatment with Couple/Family

Initial Phase Counseling Tasks

1. Develop working counseling relationship. *Diversity Note: [Describe how you will adjust to respect cultured, gendered, and other styles of relationship building and emotional expression.]*
    a. *Meet persons apart from problem*, inquiring about identities outside of problem.
    b. Engage couple/family from *hopeful, optimistic* position and as *coauthor/ investigative reporter.*

2. Assess individual, systemic, and broader cultural dynamics. *Diversity Note: [Describe how you will adjust assessment based on cultural, socioeconomic, sexual orientation, gender, and other relevant norms.]*
    a. *Relative influence questioning* to map the effects of Conflict and the effects of persons on Conflict; identify *unique outcomes.*
    b. Identify the *dominant discourses* that support Conflict and the *local, alternative discourses* that may be a resource in changing relationship to Conflict.
    c. If possible, *externalize a common enemy* that couple/family can unite against.

Initial Phase Client Goals

1. Increase *couple/family's influence over the frequency and severity of [Externalized Problem: Anger, Stress, Conflict, etc.]* to reduce conflict.
    a. *Externalizing* [Problem] to identify current areas of each person's influence and possibilities for expanding this influence.
    b. *Map unique outcomes in landscapes of action and consciousness* to identify new possibilities for reducing influence of [Externalized Problem].

Narrative Working Phase of Treatment with Couple/Family

Working-Phase Counseling Task

1. Monitor quality of the working alliance. *Diversity Note: [Describe how you will attend to client response to interventions that indicate therapist using expressions of emotion that are not consistent with client's cultural background.]*
    a. *Permission questions and situating comments* to increase clients' sense of safety and agency within relationship.
    b. Administer *session rating scale.*

Working-Phase Client Goals

1. Decrease *influence of [Externalized Problem]* and *dominant discourses* to reduce conflict and hopelessness about the relationships.
    a. *Externalizing questions* to separate clients from problem.
    b. *Mapping Conflict vs. unique outcomes in landscapes of action and consciousness* to separate clients from [Externalized Problem] and strengthen alternative identities.

2. Increase *each person's sense of agency* by developing *new relationship* with [Externalized Problem] to reduce conflict and increase hope.
    a. *Externalizing metaphors* to define relationship with [Externalized Problem].
    b. *Statement of position map* to evaluate effects of [Externalized Problem] and inspire new relationship to it.

3. Increase actions that support *preferred relational identities* to reduce conflict.
    a. *Scaffolding conversations* to move clients from contemplation to action.
    b. *Statement of position map* to evaluate effects of actions and identify where and how to make adjustments.

(continued)

Narrative Closing Phase of Treatment with Couple/Family

Closing-Phase Counseling Task

1. Develop aftercare plan and maintain gains. *Diversity Note: [Describe how you will access resources in the communities of which they are a part to support them after ending therapy.]*
   a. *Map potential setbacks in landscapes of action and consciousness* and identify alternative behaviors; address potential blame.
   b. *Letter from therapist* and/or from client preferred to self to future self-experiencing setback to solidify narrative of overcoming [Externalized Problem].

Closing-Phase Client Goals

1. Increase and thicken *narrative of preferred couple/family identity* to reduce conflict and increase relational satisfaction.
   a. *Scaffolding conversations* to move clients from contemplation to action.
   b. *Map unique outcomes in landscape of action and consciousness* to solidify new narrative.

2. Increase external relationships that support couple/family's *preferred identity* to reduce conflict and increase sense of cohesion.
   a. *Definitional ceremony* to expand network of friends, family, and/or school personnel who support client's preferred identity.
   b. *Therapeutic letter* from therapist to document journey from fighting while under the influence of [Externalized Problem] to uniting against its effects and redefining their relationship/family.
   c. *Narrative reflecting team* to support couple/family in newly enacted identity.

Research and the Evidence Base

Consistent with their philosophical underpinnings, postmodern therapists, including narrative, have conducted more qualitative than quantitative investigations about their approaches to therapy (Anderson, 1997; Gehart et al., 2007). However, outcome research on narrative therapy has increased in recent years. For example, one study from Australia examined the effectiveness of narrative therapy for treating major depressive disorder and found its effects were comparable to other approaches, with 74% of clients achieving reliable improvement (Vromans & Schweitzer, 2011). In another Australian study, women with eating disorder and depressive issues engaged in a 10-week narrative therapy group that resulted in reduced self-criticism and changes in daily practices/activities. Process research on narrative therapy indicates that unique outcomes that specifically enable clients to reconceptualize their problems and foster new experiences are correlated with positive outcomes (Matos, Santos, Gonçalves, & Martins, 2009).

Tapestry Weaving: Working with Diverse Populations

Ethnic, Racial, and Cultural Diversity

If you have not already noticed, more than any other approach covered in this book, narrative therapy integrates consideration of cultural issues at the most fundamental level of their method. Considerations of how society and its norms affect individuals are the guiding premises in postmodern philosophical literature, making these therapies particularly suitable for clients from marginalized groups (for an in-depth discussion, see Monk et al., 2008). Unlike most mental health therapies, narrative therapy places societal issues of oppression at the heart of its therapeutic interventions, and many narrative therapists are active agents of social justice (Zimmerman & Dickerson, 1996). Narrative therapy has international roots and is practiced in numerous countries around the world.

Applications with Native American, First Nations, and Aboriginals

Narrative therapy approaches have been widely used and researched with native cultures in Canada, Australia, New Zealand, and the United States. In one study, therapists who worked with Native Americans in Wyoming identified communication and therapy practices that seemed best suited for this culture, all of which were descriptive of narrative therapy practices (Lee, 1997). Some of the observations from these therapists include:

- *Subtle Eye Contact:* Native Americans generally do not maintain sustained eye contact, and therapists should follow the client's lead.
- *Active Listening:* Therapists should use nonverbal and verbal feedback to signal that the client's meanings are understood.
- *Subtle Emotional Expression:* Native Americans are not prone to strong, demonstrative expressions of emotions, and therapists must respect this by not pushing hard for emotional expression.
- *Spirituality:* Spirituality is generally highly valued by Native Americans, and therapists need to respectfully address and find ways to integrate these practices into therapeutic work.
- *Self-in-Relation:* Native Americans are likely to construct their identity within relationships, seeing much of their identity as tied to family or the tribe.
- *Home Visits:* The therapist making home visits is generally appreciated by Native Americans, helpful in terms of comfort, familiarity, and time flexibility.
- *Gentle, Reflective Stance:* Therapists described a therapeutic stance that was calm, reflective, nonconfrontational, and nonconflictual.
- *Art, Storytelling, and Metaphor:* The use of art, drawing, storytelling, and metaphor were found to be particularly helpful interventions.

In another study set in Manitoba, Canada, native healers and their clients were interviewed for what they believed to be helpful in changing emotions, cognitions, and behaviors (McCabe, 2007). This study included many of the above practices in addition to:

- *Ceremonies and Rituals:* Ceremonies and rituals are powerful allies in the healing process and are used throughout.
- *Spiritual Guidance:* Native Americans with traditional spiritual beliefs generally believe in a Creator spirit that provides guidance in the healing process.
- *Self-Acceptance:* For many, themes of self-acceptance of one's identity, especially within the larger context of marginalization.
- *Lessons of Daily Living:* Using opportunities in daily living to apply learning.
- *Empathy:* Expressions of empathy from the healer help in the healing process.
- *Role Modeling:* The healer serves as a role model for the client in terms of lifestyle, behaviors, and attitudes.

Narrative therapists readily use native rituals, ceremonies, spiritual beliefs, and cultural stories to create therapeutic contexts that are respectful and responsive to the needs of native and aboriginal clients, often inviting tribal elders, leaders, and healers into the broader therapeutic dialogue (Carey & Russell, 2011). In particular, separating the person from the problem and meeting the person apart from the problem can be particularly useful. In addition, therapists can use situating and permission comments to increase the client's sense of voice and autonomy in the therapeutic relationship and to help create a more respectful healing context.

Hispanic Youth

Narrative therapy has also been used with Hispanic children and adolescents with depression, high-risk behaviors, and/or substance abuse (Malgady & Costantino, 2010). Some of their unique applications with this population include:

- *Cuento Therapy:* Puerto Rican folktales, *cuentos*, were used to convey themes and morals to provide models for adaptive responses to problems, such as acting out,

self-esteem, and anxiety. They adapted the tales to incorporate traditional Puerto Rican values as well as Anglo cultural values.
- *Hero/Heroine Therapy:* As many children were in single-parent households, the therapists had children identify male and female Puerto Rican heroes and heroines to help bridge bicultural, intergenerational, and identify conflicts commonly experienced in adolescences.
- *Temas Storytelling Therapy:* Therapists selected pictures from the Thematic Apperception Tests that represented Hispanic cultural elements: traditional food, games, family scenes, and neighborhoods. Then the group of children was asked to develop a story with the cards. Then each group member was invited to share personal experiences that relate to the story the group created. The therapist reinforced adaptive, preferred narratives and helped find alternatives to maladaptive responses. Finally, the group did a dramatization of the story in class to practice preferred behaviors and responses.

Sexual Identity Diversity

Postmodern therapies, such as narrative, are often the theory of choice for working with gay, lesbian, bisexual, transgendered, and questioning (GLBTQ) clients because they help to deconstruct the heterosexist discourses that often are the source of greatest suffering in GLBTQ clients. In addition, many GLBTQ clients benefit from therapeutic conversations that help them to construct positive labels and identity narratives for themselves and their relationships (Perez, 1996). Furthermore, it is important for therapists to identify how much of the presenting problem is due directly to a client's sexual identity versus more general life or relationship circumstances (Aducci & Baptist, 2011).

Narrative therapists have developed a specific approach that can be used with GLBTQ clients struggling with dissonance between heterosexual dominant discourses and their own lived experience. Yarhouse (2008) describes *narrative sexual identity therapy* as a middle ground to *gay-affirmative therapy* on one hand, which affirms the inherent goodness of identifying as GLBT, and *reorientation therapy* on the other, a highly controversial approach (so much so it is under legislative scrutiny in California at the moment) designed to help clients change their sexual orientation to heterosexual. In contrast, narrative sexual identity therapy, which is based on narrative therapy principles, helps clients seek to live their lives in *congruence* between their personal beliefs and behaviors, a process that focuses on deconstructing dominant discourses that constrain and confuse clients' sexual identity. Designed to help clients struggling with their sexual identity, this approach is particularly relevant for clients from religious and cultural backgrounds that have strong anti-gay messages.

Drawing from standard narrative therapy practices, the therapist helps clients identify the problem narrative and then identify potential counter-narrative that is more congruent with the client's lived experience and values. In this approach, the client determines what values and intentions they want to define their lives. The approach has six general steps or phases, which are of course fluid but can nonetheless be helpful to therapists in conceptualizing treatment.

1. *Client Presents Sexual Identity Concern:* Initially, a client presents with concerns about sexual identity. These concerns may be related to external (e.g., religious, cultural, or family) or internal voices (e.g., personal values) that believe their lived experience of sexual attraction is somehow wrong, immoral, or otherwise problematic.
2. *Map Dominant Narratives:* The next step is to explore the dominant discourse, its origin, and specific nuanced meanings. Questions therapists can ask to map the dominant discourse include:
    - What were some of the messages you received growing up about identifying as gay? (Yarhouse, 2008, p. 205)
    - How was the message communicated to you that feelings toward same-sex meant you were gay? (Yarhouse, 2008, p. 205)
    - How did you respond to these views of same-sex attraction?

3. *Identify Preferred Narratives:* In the next phase, clients reflect on their own personal lived experience of same-sex attraction. Most often the metaphor used to describe this is *"discovery"*: discovering a pre-exiting fact about their true selves. In other cases, the metaphor is more one of "integration": integrating their experience of same-sex attraction with their other aspects of their identity. The metaphor of integration is more common when a person chooses not to assume a gay identity but instead chooses to only integrate certain elements.
4. *Recognize Exceptions/Emerging Counter-Narrative:* Next, the therapist helps clients identify exceptions to the dominant discourses to create space for a counter-narrative to emerge. Questions therapists can use to facilitate this process include:
   - In what ways are you understanding your sexual identity differently than when you first thought of yourself? (Yarhouse, 2008, p. 206)
   - Have you had experiences that call into question the meaning of same-sex attraction that you learned when you were young?
   - In what ways have your personal experiences of same-sex attraction contrasted to messages you have heard about what gay people are like?
5. *Highlight Identity-Congruent Attributes, Activities, and Resources:* As the counter-narrative emerges, the therapist listens for and highlights personal attributes, activities, and resources that are congruent with the client's preferred identity.
   - In what ways would you like to challenge some of the messages you received about your sexual identity? (Yarhouse, 2008, p. 207)
   - In what ways are you already living in accordance—even in small measure—with your preferred values and identity?
   - What type of relationships, activities, or communities might further provide support for your preferred identity and values?
6. *Resolution/Congruence:* Finally, clients are able to live according to their preferred sexual identity, one that is congruent with their lived experience, spiritual beliefs, and cultural values. Questions therapists can use to facilitate this process include:
   - In the course of the next few months or so, what do you see as the relationship between your sexual identity and your religious or spiritual identity? (Yarhouse, 2008, p. 208)
   - Can you share a little of the way you would like to describe your sexual identity in the year to come? (Yarhouse, 2008, p. 208)

Online Resources

Anti-Anorexia/Anti-Bulimia League:
www.narrativeapproaches.com/antianorexia%20folder/anti_anorexia_index.htm

Dulwich Centre (Michael White's Narrative Therapy):
www.dulwichcenter.com

Evanston Family Therapy Center (Freedman and Combs' Narrative Therapy):
www.narrativetherapychicago.com

Narrative Approaches:
www.narrativeapproaches.com

Yaletown Family Therapy (Narrative Therapy, Canada):
www.yaletownfamilytherapy.com

References

*Asterisk indicates recommended introductory readings.

Aducci, C. J., & Baptist, J. A. (2011). A collaborative-affirmative approach to supervisory practice. *Journal of Feminist Family Therapy: An International Forum, 23*(2), 88–102. doi:10.1080/08952833.2011.574536

*Andersen, T. (1991). *The reflecting team: Dialogues and dialogues about the dialogues.* New York: Norton.

Andersen, T. (1992). Relationship, language and pre-understanding in the reflecting process. *Australian and New Zealand Journal of Family Therapy, 13*(2), 87–91.

*Anderson, H. (1997). *Conversations, language, and possibilities: A postmodern approach to therapy.* New York: Basic Books.

Anderson, H., & Goolishian, H. (1988). Human systems as linguistic systems: Preliminary and evolving ideas about the implications for clinical theory. *Family Process, 27,* 157–163.

*Anderson, H., & Goolishian, H. (1992). The client is the expert: A not-knowing approach to therapy. In S. McNamee & K. J. Gergen (Eds.), *Therapy as social construction* (pp. 25–39). Newbury Park, CA: Sage.

Augusta-Scott, T., & Dankwort, J. (2002). Partner abuse group intervention: Lessons from education and narrative therapy approaches. *Journal of Interpersonal Violence, 17,* 783–805.

Bruner, J. (1986). *Actual minds, possible worlds.* Cambridge, MA: Harvard University Press.

Carey, M., & Russell, S. (2011). Pedagogy shaped by culture: Teaching narrative approaches to Australian Aboriginal health workers. *Journal of Systemic Therapies, 30*(3), 26–41. doi:10.1521/jsyt.2011.30.3.26

Foucault, M. (1972). *The archeology of knowledge* (A. Sheridan-Smith, trans.). New York: Harper & Row.

Foucault, M. (1979). *Discipline and punish: The birth of the prison.* Middlesex: Peregrine Books.

Foucault, M. (1980). *Power/knowledge: Selected interviews and other writings.* New York: Pantheon Books.

*Freedman, J., & Combs, G. (1996). *Narrative therapy: The social construction of preferred realities.* New York: Norton.

*Freeman, J., Epston, D., & Lobovits, D. (1997). *Playful approaches to serious problems.* New York: Norton.

Gehart, D. R., & McCollum, E. E. (2007). Engaging suffering: Towards a mindful re-visioning of family therapy practice. *Journal of Marital and Family Therapy, 33*(2), 214–226. doi:10.1111/j.1752-0606.2007.00017.x

Goolishian, H., & Anderson, H. (1987). Language systems and therapy: An evolving idea. *Psychotherapy, 24*(3S), 529–538.

Jenkins, A. (1990). *Invitations to responsibility: The therapeutic engagement of men who are violent and abusive.* Adelaide, Australia: Dulwich Centre Publications.

Lee, S. (1997). Communication styles of Wind River Native American clients and the therapeutic approaches of their clinicians. *Smith College Studies in Social Work, 68*(1), 57–81. doi:10.1080/00377319709517516

Malgady, R. G., Costantino, G. (2010). Treating Hispanic children and adolescents using narrative therapy. In J. R. Weisz & A. E. Kazdin (Eds.), *Evidence-based psychotherapies for children and adolescents* (2nd ed.) (pp. 391–400). New York: Guilford Press.

Matos, M., Santos, A., Gonçalves, M., & Martins, C. (2009). Innovative moments and change in narrative therapy. *Psychotherapy Research, 19*(1), 68–80. doi:10.1080/10503300802430657

McCabe, G. H. (2007). The healing path: A culture and community-derived indigenous therapy model. *Psychotherapy: Theory, Research, Practice, Training, 44*(2), 148–160. doi:10.1037/0033-3204.44.2.148

Monk, G., & Gehart, D. R. (2003). Conversational partner or socio-political activist: Distinguishing the position of the therapist in collaborative and narrative therapies. *Family Process, 42,* 19–30.

Monk, G., Winslade, J., Crocket, K., & Epston, D. (1997). *Narrative therapy in practice: The archaeology of hope.* San Francisco, CA: Jossey-Bass.

Monk, G., Winslade, J., & Sinclair, S. (2008). *New horizons in multicultural counseling.* Thousand Oaks, CA: Sage.

Pence, E., & Paymar, M. (1993). *Education groups for men who batter: The Duluth Model.* New York: Springer.

Perez, P. J. (1996). Tailoring a collaborative, constructionist approach for the treatment of same-sex couples. *The Family Journal, 4*(1), 73–81. doi:10.1177/1066480796041016

Smith, C., & Nylund, D. (2000). *Narrative therapy with children and adolescents.* New York: Guilford.

Vetere, A., & Dowling, E. (2005). *Narrative therapies with children and their families: A practitioner's guide to concepts and approaches.* New York: Routledge.

Vromans, L. P., & Schweitzer, R. D. (2011). Narrative therapy for adults with major depressive disorder: Improved symptom and interpersonal outcomes. *Psychotherapy Research, 21*(1), 4–15. doi:10.1080/10503301003591792

White, M. (1995). *Re-authoring lives: Interviews and essays.* Adelaide, Australia: Dulwich Centre Publications.

*White, M. (2007). *Maps of narrative practice.* New York: Norton.

*White, M., & Epston, D. (1990). *Narrative means to therapeutic ends.* New York: Norton.

White, M., & Morgan, A. (2006). *Narrative therapy with children and their families.* Adelaide, Australia: Dulwich Centre Publications.

Winslade, J., & Monk, G. (1999) *Narrative counseling in schools: Powerful and brief.* Thousand Oaks, CA: Corwin Press.

Winslade, J., & Monk, G. (2000). *Narrative mediation.* San Francisco, CA: Jossey-Bass.

Winslade, J., & Monk, G. (2007). *Narrative counseling in schools: Powerful and brief* (2nd ed.). Thousand Oaks, CA: Corwin Press.

Winslade, J., & Monk, G. (2008). *Practicing narrative mediation: Loosening the grip of conflict.* San Francisco, CA: Jossey-Bass.

Yarhouse, M. A. (2008). Narrative sexual identity therapy. *American Journal of Family Therapy, 36*(3), 196–210. doi:10.1080/01926180701236498

Zimmerman, J. L., & Dickerson, V. C. (1996). *If problems talked: Narrative therapy in action.* New York: Guilford.

Narrative Case Study: Multiethnic Gay Couple

Darren (AM40) is a 40-year-old, gay, HIV–, second-generation Nigerian American who presents as disconnected from self and others. He feels isolated, dejected, and lonely, even while in his current relationship. He has been in a relationship with a 32-year-old, gay, HIV+, Mexican American, Ernesto (AM32) for the last year, and they state that they each have difficulty trusting each other and repeatedly question the other about where he has been, whom he has been with, and if he has a sexual or flirtatious relationship with that person. AM40 works as a server in a restaurant and distrusts his manager and his coworkers, believing their motives to be racist. AM40 admits to having a lot of anxiety, a depressed mood, and low self-esteem. AM40 experienced a physically abusive childhood from his father and an emotionally abusive childhood from both parents. He was parentified by his mother at a young age and given the responsibility to help take care of his brothers and sisters. His father was an alcoholic that would "disappear—sometimes for days at a time."

AM40 endured homophobia in his family. His father warned him "not to be a sissy" from a young age and would be physically abusive "to toughen him up." His aunt, whom he was very close with, distanced herself from AM40 and told other family members that she "is very disappointed in him" because he is gay. He has internalized homophobia and finds being gay in conflict with values taught to him by his Christian family—his mother, in particular. He feels that he "has let her down."

AM32, a store manager, admits to having a "fear of commitment" and "intimacy issues." He is a habitual marijuana smoker who drinks "sometimes excessively on the weekends" and feels that he might "be partying too much." AM32 is only "out" to his sister and a few close friends. He worries about the rest of his family of origin discovering that he is gay. AM32 was in one relationship prior to being with AM40, and was lied to by his ex-boyfriend concerning his HIV status. AM32 feels betrayed because he trusted a man that infected him with HIV.

AM40 and AM32 recently moved in together but have been frequently arguing and finding it difficult to trust one another. They have come to counseling to try and work through these issues together and salvage their relationship.

Narrative Case Conceptualization
Meeting the People Apart from Problems: Strengths and Interests

AM40 Values:
Honesty
Caring for others
Responsibility
Views home as sanctuary
Spirituality

AM32 Values:
Family
Finding humor in things
Gratitude
Kindness
Spirituality

AM40 Likes:
Reading
Traveling
Art

AM32 Likes:
Dancing
Hiking
Cooking

AM40: Sexual Identity Development
Out to family

AM32: Sexual Identity Development
Positively identifies with couple relationship

Positively identifies with relationship
Accepting of sexual orientation

Taking responsibility for HIV+ status

Problem-Saturated Narratives

Both AM40 and AM32 view life and others as dangerous and untrustworthy, which currently color the stories they tell themselves about the other and the relationship. After experiencing a childhood fraught with abuse, AM1 views the world as a dangerous place in which he must be hypervigilant against trouble lurking around every corner. As his father would "disappear for days at a time" during drinking binges, he finds it difficult to trust that AM32 (or anyone else) will consistently be there for him, especially since AM32 uses alcohol and other substances. AM32 carries resentment for being lied to by his ex-boyfriend and infected with HIV. He finds it difficult to trust others as a result and uses marijuana, alcohol, and occasionally other substances as a maladaptive coping mechanism to manage his affect and avoid feeling difficult feelings. These fears and distrust creep into the relationship and spark arguments over fidelity and sincerity.

In addition, AM40 and AM32 each carry internalized homophobic shame. The fact that they are gay is in direct conflict with values instilled in them by their families and religions of origin. AM40 has been rejected by close family members, and AM32 lives in fear of rejection by his family if he were to allow himself to be honest with them about his lifestyle. Although they have each other for support, they did not experience support from their families of origin and acceptance for being who they are. On one level, they each carry a worldview of being somewhat unlovable as a result, which makes them suspicious of the other's true intent.

Unique Outcomes/Sparkling Events
- Both describe good times in the relationship when they feel safe and loved.
- Both state that they have "no friends" but then talked about recent night out with one.
- Both report several positive and supportive relationships at work.
- Each has members in their family of origin that support and love them, ones that know they are gay.
- Each has taken the risk to trust the other in this relationship.
- AM32 sometimes uses more preferred coping mechanisms, such as hiking or cooking.

Dominant Discourses
Cultural, Ethnic, Religious Both men come from religious—Christian (AM40) and Catholic (AM32)—and familial backgrounds that define being gay as a sin. So, although these religious traditions instilled good humanitarian values, they also created homophobic shame. These religious discourses often make them feel guilty, dirty, without purpose, and lacking integrity. Furthermore, AM40 reports feeling marginalized as second-generation Nigerian American in some communities, as gay in others, or as both. He does not feel a shared connection with most African Americans and maintains few connections with the local Nigerian community. AM32 also reports feeling marginalized as Latino gay man and carries secret shame of being HIV+. Both feel judged for being in an interracial, gay relationship from several communities and their families of origin.

Gender/Sexual Orientation AM40 describes greater acceptance of his sexual orientation, but still experiences significant alienation and sense of inferiority related to his sexual orientation. AM32 struggles more with his sexual orientation, primarily because he feels he will lose the close connection he now enjoys with his family. Neither is particularly well connected with the local gay community.

Other Social Discourses Both men work in service industries and often report feeling victimized and poorly treated in their roles at work.

Local Discourse and Preferred Narrative
Identity/Self-Narratives Both men report that they have to a certain degree accepted that they are gay and that they want a committed relationship, even if they have to keep it secret from others at the moment. Even though they sometimes feel ashamed, a more significant part of each of them wants to pursue a fulfilling intimate relationship with the other. They have considered attending a gay-inclusive church, HIV+ support groups, and a gay 12-step program.

Local Discourses Both men report knowing gay men in committed, happy relationships, and they use these as their role models and inspirations.

Narrative Treatment Plan

Initial Phase Counseling Tasks

1. Develop working therapeutic relationship. *Diversity Note: Allow clients to share their narrative of what it means to be gay and from Nigerian/Mexican culture.*
   a. Meet the people apart from the problems by identifying strengths and interests.
   b. Meet couple apart from the problem by noting couple strengths.
   c. Engage couple from hopeful/optimistic position as coauthor, being openly supportive of gay relationship.

2. Assess individual, systemic, and broader cultural dynamics. *Diversity Note: Attend to effects of sexual orientation and cultural backgrounds on the life of the problem.*
   a. Identify key dominant discourses that affect the life of the problem.
      - Identify impact of heterosexist society.
      - Religious discourses from families of origin.
   b. Identify local and preferred discourses.
      - Sexual identity as gay men in committed relationship.
      - Integrating values with lifestyle.
        - Support system in gay community.
        - Gay-inclusive church.
   c. Externalize common enemy: Anxiety aroused by
      - Fear of intimacy.
      - Lack of trust.
      - Internalized homophobic shame.
   d. Relative influence questioning: Map effects of anxiety/conflict on people and effect of people on anxiety/conflict.

Initial Phase Couple Goals

1. Increase couple's influence over substance use and abuse history to reduce conflict and increase capacity for intimacy.
   a. *Externalize* substances and abuse to identify strategies for increasing influence and to increase sense of agency and choice related to substance use and past abuse.
   b. *Map problem-saturated, unique, and then preferred outcomes in landscapes of action and consciousness* for both substance use and past abuse.

Working-Phase Counseling Tasks

1. Monitor quality of the working alliance. *Diversity Note: Use warmth and humor that help create an especially safe and accepting environment.*
   a. *Respect clients' reality* of gay couple in heterosexist society and experience of marginalization and homophobia.
   b. *Permission questions and situating comments* to increase client's sense of safety, trust, and agency within relationship.
   c. Administer *session rating scale* to help ensure couple feels understood and that therapy process is addressing their concerns.

Working-Phase Couple Goals

1. Decrease influence of homophobic and HIV shame stemming from childhood as well as adult life events to increase ability to trust and experience intimacy.
   a. *Externalizing questions* to identify the effects of homophobic and HIV shame as well as unique outcomes.
   b. *Map unique outcomes in landscapes of action and consciousness* to strengthen preferred identities.

(continued)

2. Increase couple's sense of agency by enabling them to develop a new relationship with their anxieties and fears to reduce anxiety, substance use, depressive symptoms, and conflict at work.
   a. *Scaffolding conversations* that move them from contemplation to action.
   b. *Statement of position map* to evaluate effects of anxiety and inspire new relationship to it.

3. Increase actions that support preferred identities, to include greater acceptance of own sexual orientation and ethnicity, to reduce conflict and increase sense of safety in relationship.
   a. *Connect with persons and communities* who would support their sexual orientation and relationship, such as gay-friendly church and the local gay/lesbian community center.
   b. Utilize *statement of position map* to help restory individual and relational identities.

Closing-Phase Counseling Tasks

1. Develop plan for aftercare and to maintain gains. *Diversity Note:* Connect with local gay community resources and supportive family members; explore potentials for support in subsets of Nigerian and Mexican American communities.
   a. *Map potential setbacks in landscapes of action and consciousness,* including ongoing discrimination and conflict as each contends with coming out to families of origin.
   b. *Letters from therapist* to solidify narrative new identities, document couple's journey, cope with family-of-origin position, and prepare them for future challenges.

Closing-Phase Client Goals

1. Increase and thicken narrative of preferred couple identity—committed gay couple—to increase sense of intimacy.
   a. *Map preferred narrative in land of action and consciousness* to identify new behaviors and restory new identity.
   b. *Scaffolding questions* to move couple from contemplation to action.

2. Increase external and family relationships that support couple's preferred identity to increase sense of wellness and acceptance.
   a. Discuss pros/cons of coming out to more members of family and possibilities for successfully doing so.
   b. *Map preferred relational narrative* for family members of key significant, such as AM40's mother and AM32's sister, to identify ways to improve these relationships and potentially relationship with whole family.
   c. *Definitional ceremony* to expand network of friends, family, and/or work colleagues who support client's preferred identity.
   d. *Narrative reflecting team,* ideally with other gay individuals/couple, to support couple/family in newly enacted identity.

CHAPTER 14

Collaborative Therapy and Reflecting Teams

"This kind of listening, hearing, and responding requires that a therapist enter the therapy domain with a genuine posture and manner characterized by an openness to the other person's ideological base—his or her reality, beliefs, and experiences. This listening posture and manner involve showing respect for, having humility toward, and believing that what a client has to say is worth hearing. ... This is best accomplished by actively interacting with and responding to what a client says by asking questions, making comments, extending ideas, wondering, and sharing private thoughts aloud. Being interested in this way helps a therapist to clarify and prevent misunderstanding of the said and learn more about the unsaid."

—Anderson, 1997, p. 153

Collaborative Therapy and Reflecting Teams

In a Nutshell: The Least You Need to Know

Putting postmodern, social constructionist principles into action, collaborative therapy is a two-way dialogical process in which therapists and clients coexplore and cocreate new and more useful understandings related to client problems and agency. Avoiding scripted techniques, therapists focus on the *process* of therapy, on *how* the client's concerns are explored and exchanged. They listen for how clients interpret the events of their lives and then ask questions and make comments to better understand how the client's story "hangs together." These questions and comments naturally emerge from conversation as the therapist strives to understand the values and internal logic of the client's perspective: to understand the client *from within the client's worldview*. As this process unfolds, the client is naturally invited to share in the therapist's curiosity, joining the therapist in a mutual or shared inquiry—a *mutual puzzling* about how things came to be and how things might best move forward. As therapist and client engage in this *shared inquiry*, asking questions and tentatively sharing their perspectives, alternative views and future options emerge on the client's situation. This process provides an opportunity for clients to see their situation differently, allowing them to make new interpretations and develop fresh ideas. Therapists do not try to control or direct the content of this meaning-making process; instead, they honor the client's ability to determine what to do with these new ideas (i.e., they honor the client's *agency*).

I am guessing this process still sounds vague, so perhaps it is best to offer an example. If a client says she is feeling "depressed," rather than hearing concrete, diagnostic information, collaborative therapists are profoundly aware of how little they know about *this* client's unique experience of depression, thus becoming sincerely curious about how the client came to this understanding of her experience. With no predetermined set of questions, the therapist asks questions that emerge from a genuine desire to better understand, such as: Does she cry often about something? About nothing? Has life gone to gray and nothing seems interesting anymore? Is her heart broken? Does she feel like a failure? There are as many unique depression stories as there are people who say they are depressed. As the therapist explores the client's view, the client is invited to join in the curiosity about her depression. Each new understanding informs alternative actions, thoughts, and feelings, thus shifting experience on multiple levels until the client has found a way to manage or resolve her initial concern.

The Juice: Significant Contributions to the Field
If you remember one thing from this chapter, it should be this:

Not Knowing and Knowing With
Perhaps one of the most frequently misunderstood concepts in collaborative therapy (Anderson, 2005), the idea of "not knowing" was first introduced by Goolishian and Anderson in 1988. At first blush, the not-knowing stance sounds contradictory: how can a paid professional like a therapist "not know"? Isn't that what they are paid for? What do you do with all that you have learned in graduate school? *Not knowing* refers to how therapists think about what they think they know and the intent with which they introduce this knowing (expertise, truths, etc.) to the client. Obviously, collaborative therapists are avoiding a particular type of knowing that Anderson calls "pre-knowing" (Anderson, 1997, 2007). In common English it's called *assuming*: believing that you can fill in the gaps or that you have enough information without sufficient evidence. Drawing from a postmodern social constructionist epistemology, collaborative therapists maintain that clients with apparently similar experiences, such as "psychosis," "mania," or "sexual abuse," have unique understandings of their situations (Anderson, 1997). Each client's understanding has evolved through conversations with significant others, acquaintances, professionals, and strangers, as well as through the larger societal discourse and stories in the media and literature. Therapists choose to *know with* and *alongside* clients as they engage in a process of better understanding clients' lives (Anderson, 1993, 2007). They view the client's knowledge as equally valid with their own.

This not-knowing, not-assuming stance requires the therapist to ask what on the surface appear as obvious or trivial questions: "You say you are sad about the loss of your mother. Can you tell me what aspects of her loss touch you most deeply?" or "Tell me how you experience that sadness in your daily life." When clients begin to explore the ideas, experiences, and influences that led to the perception of a problem, they often hear themselves saying things they have never told anyone before. Hearing these thoughts aloud for the first time inevitably shifts their perspective of the situation, sometimes subtly and sometimes dramatically. These new perspectives inform new action and identities related to the problem (e.g., from viewing her mother as an entirely separate person, the client may shift to seeing that she is part of how her mother lives on). In the case study at the end of the chapter, the therapist works with a Pakistani/Indian couple dealing with the wife's recent revelation of childhood sexual abuse; the husband and wife have very different understandings of what this means, with one drawing on a more traditional view and the other a more modern perspective.

Rumor Has It: The People and Their Stories

Harlene Anderson and Harry Goolishian Harlene Anderson and Harry Goolishian developed collaborative therapy with their colleagues at the University of Texas Medical Branch in Galveston and later established the *Houston Galveston Institute* (Anderson, 1997; Anderson, 2005, 2007; Goolishian & Anderson, 1987). Their collaborative approach has roots in the early model developed by the Galveston group called "multiple impact therapy," a multidisciplinary approach to working with hospitalized adolescents, their families, and the broader social system. Their interest in hermeneutics, social construction, postmodern assumptions, and related social and natural science theories was initially fueled by their curiosity with the ideas of the Mental Research Institute (MRI), but in their work at Galveston they began to listen differently to what clients were saying rather than trying to learn clients' language to use it as a strategic tool. They noticed that it was not the family, but rather each *member* of the family, that seemed to have his/her own language, using words and phrases with unique meanings.

These interests naturally led to postmodern ideas, social construction theory, and then to the work of theorists such as Ludvig Wittgenstein, Mikhail Bakhtin, Ken Gergen, and John Shotter (see Chapter 4). As a result, Anderson and Goolishian began to conceptualize their work from a postmodern perspective, focusing on the construction of meaning in relationships. They also had a mutually influencing relationship with Tom Andersen, and over the years their therapy became known as collaborative language systems (Anderson, 1997) and, more recently, collaborative therapy (Anderson & Gehart, 2007).

After Harry's death in 1991, Harlene and her colleagues at the Houston Galveston Institute continued developing this internationally practiced approach. Ken Gergen (see Chapter 4), Harlene Anderson, Sheila McNamee, and others joined to form the *Taos Institute,* an organization of collaborative practitioners working in the fields of education, business, consultation, therapy, medicine, and other disciplines. Having found that the assumptions on which collaborative therapy is based have applications beyond therapy systems, Harlene currently refers to her work as "collaborative practices."

Tom Andersen No relation other than a close friend to Harlene (note the "e" versus the "o" in Anders*e*n), Tom Andersen was a Norwegian psychiatrist who is best remembered for his gentle demeanor, respect for client privacy, and elegant therapeutic conceptions. Having originally studied with the Milan team using one-way mirrors, Tom transformed the systemic practice of the observation team using postmodern sensibilities that reduced the team-client hierarchy and made the process dialogical rather than strategic. His descriptions of *inner and outer dialogues* as well as *appropriately unusual comments* provide collaborative therapists with practical concepts that can be used to facilitate therapeutic conversations without the use of technique.

Lynn Hoffman Known for her keen theoretical insights and broad vision, Lynn Hoffman has worked closely with many of family therapy's most influential thinkers, including Virginia Satir, Jay Haley, Paul Watzlawick, Salvador Minuchin, Dick Auserwald, Gianfranco Cecchin, Luigi Boscolo, Tom Andersen, Harlene Anderson, and Peggy Penn. Her first book, *Foundations of Family Therapy* (Hoffman, 1981), provides one of the most comprehensive overviews of systemic family therapy available. She began learning about family therapy at the MRI, where she served as an editor for Satir's books. She was so inspired by these ideas that she went on to pursue a career as a social worker, training in systemic family therapies. She befriended the Milan team, and along with Peggy Penn helped further their later development of the model (Boscolo, Cecchin, Hoffman, & Penn, 1987). In her later years, Hoffman became increasingly attracted to postmodern, collaborative approaches (Hoffman, 1990, 1993, 2001). She has detailed her remarkable journey in *Family Therapy: An Intimate History* (Hoffman, 2001), a favorite with my students who want to learn about the theories of family therapy yet prefer a little more "juice" and excitement than is offered in a textbook such as this. Hoffman is currently exploring the notion of rhizome theory in human systems.

Peggy Penn A former training director of the Ackerman Institute and a published poet (Penn, 2002), Peggy Penn has developed unique approaches to using writing in collaborative therapy (Penn, 2001; Penn & Frankfurt, 1994; Penn & Sheinberg, 1991). Like Hoffman, Penn began her training in systemic therapies, most notably the Milan approach (Boscolo et al., 1987), but her work has since evolved into a more postmodern approach. She uses various forms of writing in therapy to help clients access multiple voices and perspectives.

Jaakko Seikkula Psychologist Jaakko Seikkula and his colleagues (Haarakangas, Seikkula, Alakare, & Aaltonen, 2007) developed and researched the *open dialogue* approach to working with patients with psychotic symptoms in the Lapland region of Finland. As a result of 20 years of work, their hospital no longer has chronic cases of psychosis, and patients with psychotic symptoms need fewer medications and return to work more often. Jaakko's research provides some of the best empirical evidence for postmodern therapies (see Clinical Spotlight later in chapter).

Houston Galveston Institute Originally founded by Harlene Anderson, Harry Goolishian, and their colleagues, the Houston Galveston Institute continues to be the premier training center for collaborative therapy, providing services to local child protection agencies, schools, and trauma survivors. Sue Levin currently serves as the executive director, and her research focuses on women who have been abused by their partners (Levin, 2007). Saliha Bava serves as the associate director of the institute, and her current work focuses on trauma (Bava, Levin, & Tinaz, 2002), qualitative research (Gehart, Tarragona, & Bava, 2007), and transformative performance.

Grupo Campos Elísios: Collaborative Therapy Training Center in Mexico City Located in Mexico City and working closely with the Houston Galveston Institute, the bilingual faculty at Grupo Campos Elísios offer training in collaborative therapy and provide therapy and consultation services to local families, schools, and hospitals; the faculty and cofounders include Sylvia London, Margarita Tarragona, Irma Rodriguez-Jazcilevich, and Elena Fernandez.

Klaus Deissler: The Marburg Institute Klaus Deissler and his colleagues at the Marburg Institute in Marburg, Germany, have developed a four-year postgraduate training program in collaborative therapy and collaborative business consultation, working closely with European businesses, school districts, and psychiatric hospitals.

The Big Picture: Overview of Treatment

Collaborative therapists do not have set stages of therapy or an outline for how to conduct a session. Instead, they use a single guiding principle: facilitate *collaborative relationships and generative, two-way dialogical conversations*, regardless of the topic and the participants. In short, they "keep the dialogue going." The key to facilitating dialogue is avoiding monologues.

Avoiding Monologues and the Therapeutic Impasse

Harry Goolishian often said that it is easier to identify what *not* to do as a therapist than what to do. Extending this logic, collaborative therapy is often easier to understand by identifying what is *not* a collaborative conversation, namely, a *monologue* (Anderson, 1997, 2007). A monologue can be a conversation with others or a silent conversation with oneself or an imagined other. In a spoken monological conversation between two people, each person is trying to sell his/her idea to the other person: a duel of realities. In such conversations, participants listen only to, or long enough, to plan their next defense—they are not trying to understand the other out of genuine curiosity or attempting to develop new understandings. In silent conversations, monologues occur when the same description, opinion, or thought consistently occupies one's thoughts, leaving no room for new ones or curiosity and being closed to other thoughts.

In therapy, monological conversations lead to a *therapeutic impasse*, at which point the therapeutic discussion no longer generates useful meanings or understandings. For most, it is easy to identify monological conversations because tension develops and the conversational task becomes trying to convince the other of a particular point. Therapists may also begin to have pejorative descriptions of clients, such as "resistant." When this happens—whether between therapist and client or between any two people in the room—the therapist's job is to gently shift the conversation back to a dialogical exchange of ideas. Therapists can achieve this by shifting back into a curious stance—asking to better understand the client's perspective or inquiring if there is a particular point that the client thinks the therapist is not fully understanding. However, to re-engage others in dialogue, the therapist must also be in an internal dialogical mode. In the simplest terms, a collaborative therapist's primary job is to ensure that the conversations in the room—whether between members of the client system or between the therapist and the client—do not become dueling monologues. As long as conversations are dialogical, change and transformation are inevitable. In the case study at the end of the chapter in which the couple has very different views on how to interpret the wife's revelation of childhood sexual abuse, the therapist will be particularly vigilant for monologues, which are likely when parties have diametrically opposed views.

Making Connection: The Therapeutic Relationship
Philosophic Stance
Collaborative therapists conceptualize the therapist's position as a philosophical stance, a particular *way of being in relation with others*. This stance informs how therapists speak, think about, act with, and respond in session, focusing their attention on the *person* of the client and shifting attention away from roles and functions. The philosophical stance essentially encompasses a sincere embodiment of the postmodern, social constructionist ideas that inform the collaborative approach, such as viewing the client as expert and valuing the transformative process of dialogue.

Conversational Partners: "Withness"
The therapeutic relationship in collaborative therapy is best described as a conversational partnership (Anderson, 1997), a process of being *with* the client. In this way of relating, sometimes referred to as "withness" (Hoffman, 2007), the conversational partners "touch" and move one another through their mutual understandings. Withness also involves a willingness to go along for the roller coaster ride (Anderson, 1993)—the ups and downs—of the client's transformational process, regardless of how uncomfortable, unpredictable, or scary it may be. It is a commitment to walk alongside the client, no matter where the journey leads.

Curiosity: The Art of Not Knowing
A hallmark of the collaborative therapeutic stance (Anderson, 1995, 1997), curiosity refers to the therapist's sincere interest in clients' unique life experiences and the meanings that are generated from these experiences. This curiosity is fueled by a *social constructionist epistemology* (assumptions about knowledge and how we know what we know; see Chapter 4), which posits that each person constructs a unique reality from the webs of relationships and conversations in which he/she is engaged. Thus no two people experience marriage, parenting, depression, psychosis, or anxiety the same.

Client and Therapist Expertise
In 1992, Anderson and Goolishian radically proposed, "The client is the expert." Although sometimes misunderstood to mean that the therapist has no opinion and no role in the therapeutic process, the concept of "client as expert" means that the therapist's attention is focused on sincerely valuing clients' thoughts, ideas, and opinions. Therapists ultimately have very limited information about the fullness and complexity of clients' lives; they can never acquire the complete history and "insider" perspective that clients themselves have (Anderson, 1997). Thus the concept of "client as expert" is more about respect for the client than a description of how the therapeutic process is conducted.

During the therapy session, however, therapists have a different expertise because they are responsible for ensuring that an effective and respectful dialogical conversation is conducted. They rely on the generative quality of the conversation to support client transformation rather than dictate the content, direction, or outcome of the conversation.

In broad strokes (which are always inaccurate), it may be helpful in the beginning to think of the client as holding more expertise in the area of *content* (what needs to be talked about) and the therapist as holding more expertise in the area of *process* (how things are talked about); however, in this collaborative process, both therapist and client have input on both content and process. If a collaborative therapist believes the client is not addressing an important area of content, the therapist will nonhierarchically raise the issue: "I know you prefer not to talk about the past, but I wonder if it might not be worthwhile to spend a little time exploring how your childhood abuse affects your marriage today." Such a comment is offered in such a way that the client feels truly free to say yes or no, and the therapist honors the client's wishes.

Conversely, the therapist is also open to client feedback about the therapeutic process, allowing clients' input on which processes work best for them, including who is in the room, the pacing, the types of homework or suggestions, the types of questions, and so forth. The therapist does not necessarily take the client's request as a dictate for how to do therapy, but thoughtfully considers the request and the need that underlies it and works to find the best possible ways to address it. This back-and-forth exchange is a sincere partnership in which the therapist works side by side with the client to find useful ways of talking. Anderson (1997) talks about this continual openness to client feedback as "research as part of everyday practice." The therapist uses the feedback to fine-tune the therapy process, lessening the opportunity for therapeutic impasse and ensuring that therapy is tailored to each client's unique needs. With the couple in the case study at the end of the chapter, the therapist will need to work closely with each client to honor their respective expertise and values related to negotiating their conflict related to the wife's childhood sexual abuse.

Everyday, Ordinary Language: A Democratic Relationship

Collaborative therapists listen, hear, and speak in a natural, down-to-earth way that is more congruent with the client's language and more democratic than hierarchical (Andersen, 1991; Anderson, 2007). Although they are responsible for facilitating a dialogical process that clients find useful, they do not approach the task from a position of leadership or expertise. Instead, they assume a more humble position, using everyday language, a relaxed style, and a willingness to learn that invites clients to join them in exploring how best to proceed.

Inner and Outer Talk

Tom Andersen conceptualized conversations as involving both inner and outer talk (Andersen, 2007). In a conversation, we most quickly recognize the *outer talk,* the verbally spoken conversation between the therapy participants. Andersen also recognized that there were other dialogues going on, namely, *inner talk,* the thoughts and conversations each person has within while participating in a conversation. Thus, if a therapist is working with one client, at least three conversations are simultaneously occurring: (a) the client's inner dialogue, (b) the therapist's inner dialogue, and (c) the outer spoken dialogue. The therapist needs to allow space and time for each one of these conversations.

As Andersen (2007) pointed out, when clients are speaking, they are speaking not only to the therapist but, more importantly, *to themselves.* Often in therapy, clients are saying something aloud for the first time, and they may need time to reflect on the weight or unexpected content of what they hear themselves saying to the therapist. Andersen strongly admonished therapists to not pressure clients to share their inner dialogue, as is common in more content-based therapies. Thus, if a client does not want to speak about her sexual abuse or a difficult relationship, the therapist does not force the issue but instead leaves an open invitation for the client to speak about it when ready. Unlike most therapists, Andersen was a champion for client privacy and autonomy even in session, a reflection of his abiding faith that clients have the ability to navigate their lives in a way that works best for them.

In addition to tracking the outer dialogue with the client, Andersen encouraged therapists to track their own inner dialogues: their thoughts, feelings, and reactions to the client and the outer dialogue. The therapist's inner dialogue provides many forms of information that can facilitate the therapeutic relationship: the therapist's reaction to the client may provide information about how others are relating to the client, or it may indicate that the therapist is reacting to the client based on personal history or issues rather than professional knowledge. The therapist's inner dialogue might also include insights, ideas, or metaphors that could further the outer dialogue (Anderson, 1997). When the therapist's inner dialogue is distracting from the outer conversation—as in Anderson's notion of silent monologue—the therapist is encouraged to bring up the issue with the client if doing so furthers the dialogue in useful ways (Anderson, 1997). For example, if a client continually minimizes the role of alcohol in his stories of one-night stands and yet in each incident the therapist notices there is a clear link, the therapist can *gently* put forth this observation, while verbally and nonverbally giving permission for the client to maintain his/her opinion without feeling that the relationship is threatened (e.g., "I know from past conversations you don't think there is a link here, but I want to say that I keep seeing a link between your nights out partying and getting into these relationships you later regret. If you do not see alcohol as the main cause, is there a minor role it might be playing?"). The key is to offer the perspective in such a way that invites curiosity rather than defensiveness.

The Viewing: Case Conceptualization and Assessment

Case conceptualization in collaborative therapy involves asking two key questions:

- *Who* is talking about the problem?
- *How* does each understand the problem?

Therapists answer the first by assessing who is in the *problem-organizing system,* or who is in conversation with whom, about what. The second question is approached using the therapist's *philosophical stance* to understand the client's worldview.

Who's Talking? Problem-Organizing, Problem-Dissolving Systems

Anderson and Goolishian (1988, 1992) initially conceptualized therapeutic systems as *linguistic systems* that organize around the identification of a problem: therapists and clients come together because someone has identified a problem, issue, or concern; the word *problem* may not always be explicitly used by the client. They referred to these systems as problem-organizing, problem-dissolving systems. They are "problem-organizing" because they only come into being after someone has identified a problem. They are "problem-dissolving" in that they dissolve when the participants—therapists, clients, and interested third parties—no longer have a problem to discuss. Additionally, *dissolving* refers to the idea that the problem often is not "solved" in the traditional sense of finding a solution. Instead, the participants' understandings evolve through dialogue, allowing for new thoughts, feelings, and actions. In the end, the client may not feel that the problem was solved as much as it dissolved. For example, if a client initially reports feeling stressed because of a recent breakup, the problem is not solved, but rather the client comes to interpret the situation differently and therefore acts and feels differently.

Aware that all persons talking about the problem are part of the problem-organizing, problem-dissolving system, collaborative therapists ask the following questions:

Questions About the Problem

- Who is talking about the problem in session and outside of session?
- How does each define it?
- What does each think should be done about it?

As the understanding of the problem shifts and evolves through dialogue, the therapist continually assesses who is involved in talking about it outside of session and continually inquires about the multiple perspectives about the problem, encouraging all perspectives to be heard without trying to reconcile them or identify the "truth." Clients and therapists are most likely to generate new and more useful perspectives when they allow multiple, contradictory perspectives to constantly linger in the air.

Philosophical Stance: Social Constructionist Viewing

As mentioned, collaborative therapists' primary tool in therapy is not a technique or intervention but a system of viewing, their philosophical stance (Anderson, 1997). Collaborative therapists work from a social constructionist, postmodern perspective, which maintains that our realities are constructed in language and through relationships. Rather than seeing identities and meanings as fixed, social constructionism describes how we engage in a constant process of revising and reinterpreting our personal identities and social realities by the way we tell ourselves what it means to be "a good person," "happy," "successful," "cared for," "living a meaningful life," "respected," and so forth. These stories are shaped by conversations with friends, news stories, fiction pieces, and any exchange of ideas, whether in person or through media. Rather than being bent on showing how clients are "incorrect" or "off," the therapist is curious about clients and focuses on *how clients construct meaning about the events in their lives.*

Assessing the Client's Worldview

This curiosity means that collaborative therapists focus on better understanding clients' worldview, their system for interpreting life events. They do not look for "errors" or even "the source of the problem" but rather approach clients with a gentle, nonjudging curiosity, much like a child exploring a tide pool for the first time, careful not to crush the intriguing creatures in this fascinating new world (Anderson, 1997; Hoffman, 2007). The therapist is looking for the internal logic that makes the client's world, hopes, problems, and symptoms make sense. For example, if a woman is feeling that her marriage is failing, how did she first get this idea? How did she respond? How did she make sense of her partner's changing behaviors and her own? What does she fear it says about her as a person? What does she think happened, and what does she see as the options from here? Why did her marriage work up until now, and what would it take to get it back to where it was or better? Such questions would not be in the therapist's toolbox but rather would be responses that remain congruent with the conversation at any point. Thus "assessment" in collaborative therapy is a continuous "coassessment" that occurs through conversation. In the case study at the end of the chapter, the therapist spends a lot of time exploring each partner's worldview, considering not just personal views, but cultured and gendered worldviews, both from their cultures of origin and Americanized perspectives.

Targeting Change: Goal Setting

Self-Agency

Like other postmodern approaches, collaborative therapists do not have a predefined model of health toward which they steer all clients in cookie-cutter fashion. Instead, the overall goal is to increase clients' sense of *agency* in their lives: the sense that they are competent and able to take meaningful action. Anderson (1997) believes that agency is inherent in everyone and can only be *self-accessed,* not given by someone else, as is implied in the concept of client "empowerment"; instead collaborative therapists see their role as participating in a process that maximizes the opportunities for agency to emerge in clients.

Transformation

Rather than conceptualizing the output of therapy as change, collaborative therapists conceptualize the process as transformation, emphasizing that some "original" aspects remain while other aspects are added or diminished. In therapy, clients' narrative of self-identity, who they tell themselves they are, is transformed through the dialogical process, opening new possibilities for meaning, relating to others, and future action. This transformational process is not controlled or directed by the therapist but emerges from within clients as they listen to themselves, the therapist, and others share their ideas, thoughts, and hopes.

The process of transformation through dialogue is inherently and inescapably *mutual*. When therapists participate in dialogical conversations, they risk being changed themselves because the same dialogical process that allows clients to change creates a context in which therapists are also transformed (Anderson, 1997). Although this transformation may not be as dramatic or immediately evident as the client's transformation, the worldview of therapists inevitably evolves and shifts as they learn from their clients about other ways to make sense of and engage life.

Setting Collaborative Goals

As the name implies, therapeutic goals are constructed collaboratively with clients using their everyday language rather than professional terms. In collaborative therapy, goals continually evolve as meanings and understandings change. The evolution of goals may be gradual—from arguing less to having more positive conversations—or dramatic—from focusing on school performance to focusing on emotionally connecting with one's parent. Therapists do not have a set of predefined goals they use with all clients. Instead, goals are negotiated with each client individually.

Examples of Middle-Phase Goals That Address Presenting Problems
- Reduce arguments between couple by increasing the number of conversations where they "get" each other.
- Increase periods of "harmony" between the children.
- Increase the times when the child can do his homework without being monitored.
- Expand social network by reconnecting with old friends and family.

Examples of Late-Phase Goals That Target Agency and Identity Narratives
- Increase sense of agency and assertiveness when relating to colleagues at work.
- Increase the mother's sense of agency and ability to prioritize where her time and energy go.
- Develop a family identity narrative that retains a strong sense of connection while allowing for individuality and differences of opinion.
- Develop a sense of identity that honors the difficulties of the past without living in the shadow of the past.

The Doing: Interventions and Ways of Promoting Change
Conversational Questions: Understanding from Within the Dialogue

Conversational questions are questions that come naturally from within the dialogue rather than from professional theory (Anderson, 1997). They are not canned or pre-planned but instead follow logically from what the client is saying and are generated from the therapist's curiosity and desire to understand more. For example, if a client describes her frustration with her husband's lack of help around the house, the therapist asks questions that logically flow from the conversation in the moment, such as "What chores would you like help with? Has it always been this way?" rather than therapeutic or theoretically informed questions such as the miracle question in solution-based therapy (Chapter 12), externalizing questions in narrative therapy (see following text), or systemic interaction questions in a systems approach (Chapter 5).

Using the client's preferred words and expressions, therapists ask conversational questions, which help both the therapist and the client to better understand the client's situation. In research on the therapy process, clients reported that questions asked out of genuine curiosity are received quite differently than "conditional" or "loaded" questions, which are driven by a professional agenda to assess or intervene (Anderson, 1997).

Making "Appropriately Unusual" Comments

One of the most elegant and practically useful therapeutic concepts, *appropriately unusual comments* enable therapists to offer clients reflections that make a difference. On the basis of his work with reflecting teams, Tom Andersen (1991, 1992, 1995) recommends that therapists avoid comments and questions that are "too usual" or "too unusual." Comments that are *too usual* essentially reflect the client's worldview, offering no possibility for generating new understanding or change: agreeing to or reflecting back the client's current perspective is not likely to promote change. Alternatively, comments that are *too unusual* are too different to be useful in developing new meanings. Some clients give immediate signals that a comment is too unusual by becoming "resistant," re-explaining themselves, or rejecting the comment or suggestion. Other clients give little indication in session that the comment is too unusual but afterward do not follow up on the comment and may even lose faith in the therapist and the therapy process.

Appropriately unusual comments are comments that clearly fit within the client's worldview while inviting curiosity and perhaps offering a new perspective that is easily digestible. For example, if a client comes in feeling overwhelmed with a new job that is more multifaceted than the previous job, an appropriately unusual response from the therapist might be: "It sounds like your new job may require skills in multitasking and prioritizing that weren't necessary in your old job," which speaks to the client's current experience while offering a slightly different viewpoint. Such comments capture the client's attention because they are familiar enough to be safe and viable yet different enough to offer a fresh perspective (Anderson, 1997).

Listening for the Pause When clients hear an appropriately unusual comment, suggestion, or question, they almost always have to pause and take time to integrate the new perspective with their current perspective: in these moments it is most important for the therapist to allow the client time for inner dialogue. Sometimes a client says, "I have to think about that" or "I never thought of it that way." A client's initial response may be "I don't know," but after taking a few moments to reflect on the new idea, the client usually begins to generate a response that reflects thoughts and ideas the client never had before.

How Far to Go? How unusual is appropriately unusual? The trick here is that each client needs a different level of unusualness; alternatively stated, each client finds a different level of difference useful for generating new ideas. I often find that when I am first working with a teen, mandated client, or someone unsure of therapy, appropriately unusual comments cannot include significant differences from their current worldview until they have developed greater trust in me. Additionally, the more emotionally distraught clients are, the less useful they find highly unusual comments. Other clients require and prefer that the therapist deliver comments that are quite different from their own, often in a very direct manner that verges on being socially impolite. I have had clients, particularly men, say to me, "Just tell me where you think I got it wrong" or "Just tell it to me straight—don't sugarcoat it—I hate when therapists do that." Thus, "appropriately unusual" depends on the client's preferred style of communication and the quality of the therapeutic relationship. Collaborative therapists fine-tune their communication skills to deliver a range of appropriately unusual comments and carefully observe client responses to assess whether or not the comments are useful. When working with a multiethnic couple on a highly sensitive matter, such as the couple in the case study at the end of the chapter, the therapist would be cautious and conservative when introducing "appropriately unusual" ideas to the conversation.

Mutual Puzzling Questions and Process: "Kicking Around" New Meanings

As already mentioned, the process by which collaborative therapists invite their clients to join them in becoming curious about clients' lives is referred to as "mutual puzzling" (Anderson, 1997). Anderson suggests that the therapist's curiosity becomes contagious, and clients are naturally invited into it. What begins, therefore, as the therapist's one-way inquiry shifts to a joint one. When clients join the therapist in the meaning-making process, their rate of talking may slow down, there may be more pauses in the conversation, and there is an inquisitive yet hopeful air to the conversation. Often the shift in clients is visible: their body posture softens, the head may tilt to the side, and they move more slowly or more quickly (Andersen, 2007). Mutual puzzling can occur only when therapists are successful in creating a two-way dialogical conversation in which both parties are able to sincerely take in and reflect on each other's contributions.

For example, if a client lives in daily fear of having another psychotic episode after not having one for over 10 years and says that it is her illness that keeps her from moving forward in life, the mutual-puzzling process may be sparked by a question such as: "That's interesting. You say you haven't had an episode in 10 years, so hallucinations don't seem to be plaguing you these days. But it does sound like the *worry about* hallucinations is the problem at this point. Do you think of this as part of the original problem or is it a new problem that only developed after the first was resolved?" In this case, a new distinction is highlighted and the client is invited to "kick it around" and see what, if any, new ideas emerge and to follow where they lead. The therapist does not politely insist that worrying is the new problem but rather listens for how the client made sense of the comment and continues to follow the client's thinking, kicking around the next idea that evolves from the conversation. The therapist is always most curious about how the client is making sense of what is being discussed.

Being Public: Sharing One's Inner Dialogue

In "being public," therapists share their inner dialogue for two potential reasons: (a) to respect clients by honestly sharing their thoughts about significant issues affecting treatment, and (b) to prevent monological conversation by offering their private thoughts to the dialogue (Anderson, 1997, 2005). When therapists make their perspectives publicly known, they do so tentatively and, even when discussing professional knowledge, are careful not to overshadow the client's perspective (Anderson, 2007). When therapists are open with their silent thoughts, this helps prevent them from slipping into a monological view of the client and creates a situation where something different may be created for the therapist as well.

Being public generally occurs in two situations: (a) in communications about professional information with clients or outside agencies or professionals (e.g., courts, psychiatrists), and (b) when the therapist has significant differences from the client in values, goals, and purposes.

Being Public with Professional Communication Whenever collaborative therapists handle professional matters, such as when making a diagnosis, speaking with a social worker, or filing a report with the court, they "make public" their thoughts, rationales, and intentions by discussing them *directly with clients*. Openly discussing what the therapist will reveal in an upcoming conversation with another professional and/or recapping what happened in the last conversation goes against traditional procedures, in which communications between professionals were kept confidential from the client, ostensibly because it could do "harm" to the client to know what professionals were actually thinking. The apparent "harm," however, was usually that clients would be angry.

In dramatic contrast, collaborative therapists have been pioneers in lifting the veil on dialogues between professionals, engaging in honest, direct conversation with clients about the contents of these conversations. Such conversations are not always easy, such as when a therapist has to tell a client that she cannot recommend unification through child protective services until x, y, and z happen (typically spelled out by the social worker or court). In the past, the parent learned this in court or from a social worker; in collaborative therapy, the therapist has an upfront conversation from the beginning, clearly laying out the types of behaviors that need to be seen for the desired recommendations, and then wholeheartedly and enthusiastically working with clients to reach this goal.

Most clients greatly respect the therapist's honesty and integrity and respond with increased motivation to make needed changes. They fully understand when they are not given the report they hoped for because the therapist and client have been discussing progress—or lack thereof—consistently along the way. When working with court-mandated clients, collaborative therapists St. George and Wulff (1998) have clients help write the first drafts of letters to the courts about their progress, including clinical recommendations, and then use the letters to discuss their progress and goals.

Similarly, when working with a client who self-injures, the therapist may "make public" concerns about safety, especially if the client is not highly motivated to stop cutting. When the client is invited into a discussion to address the therapist's concern about the client's safety—without having rigid requirements for the client—the client and therapist can work together to develop a plan that is meaningful to the client while also addressing the therapist's safety concerns.

Being Public with Significant Differences in Values and Goals The other situation in which collaborative therapists make public their voice is when there are significant differences in values or goals that make it hard for the therapist to move forward as an active participant in the conversation. For example, I recently had a teenager who discussed his plans to meet someone who had challenged him to a fight at a park and who had said, "Don't bring weapons or friends." The teen believed that if he didn't show up, more guys at school would gang up on him and that could lead to more events such as this. Although I saw his point, I also saw that he was at risk for seriously being hurt, a concern I decided to make "public." I invited him to explore my concerns: the guy might come with friends or weapons, there might be legal ramifications, and so forth. I offered my list of dangers from a place of serious concern without demanding a particular course of action on his part. Instead, I asked him how he would manage the dangers I saw; by the end of the conversation we arrived at a place where my concerns and his fears were addressed and we both felt good about his chosen course of action, namely, to avoid the park that day and to try to find out about this person's social network.

Accessing Multiple Voices in Writing

Peggy Penn and her associates (Penn, 2001; Penn & Frankfurt, 1994; Penn & Sheinberg, 1991) access multiple, alternative voices using various forms of writing (e.g., letters, poems, journals) to generate alternative perspectives and make room for silenced inner voices or the voices of significant persons not currently in the therapeutic dialogue. Penn and Frankfurt (1994) have found that "writing slows down our perceptions and reactions, making room for their thickening, their gradual layering" (p. 229). They have also found that the performative aspect, the reading aloud of letters to witnesses (the therapist, family, and others), makes things happen. Penn's writing has a different intent than writing in experiential therapies, which is meant to express repressed emotions, bring resolution to a past situation, or achieve a similar clinical aim. Instead, Penn's writing invites different voices into the conversation to generate alternative possibilities for understanding. Furthermore, writing promotes agency: "to write *gives us agency: we are not acted on by a situation, we are acting!*" (Penn, 2001, p. 49; emphasis in original). Clients may be asked to write the following:

- Letters to themselves from aspects of themselves and/or from newly emerging, future, or past selves
- Letters to themselves from significant others from the present, past, or future
- Letters to and from significant others (alive or dead) speaking from a voice or perspective that was formerly kept private
- Letters or journal entries to speak from parts of the self that are typically not expressed and/or are emerging in therapy
- Letters to the world or general audience
- Multivoiced biographies that describe one's life from various perspectives
- Poems that express inner voices and perspectives that are not readily articulated in other ways

Reflecting Teams and the Reflecting Process

Tom Andersen trained at the Milan Institute, where a small team of therapists would observe the therapist talking with families behind a one-way mirror, the preferred method for interviewing in early family therapy. Influenced by postmodern thinking as well as a gut feeling of discomfort because of the distance (Andersen, 1995), Andersen and his colleagues wanted to make the process more democratic and developed the idea of having the families listen to the team's conversation behind the mirror: thus, the reflecting team practice began. With the earliest reflecting teams, the family and team would literally switch rooms if sound could only be heard in one room, or they would turn off the lights in the family's room and turn on the lights in the team's room, reversing the one-way mirror. In later years, the team was invited to sit in the *same room but separate from* the family and therapist having a conversation. Over the years, the practice has developed into more of a general *process* of reflecting that is used for talking with clients, with or without a team.

The idea behind a collaborative reflecting team is to develop diverse strands of conversation so that the client can choose that which resonates and that which does not. This is in contrast to the private team conversations, which are synthesized by the team and in which the team chooses what is important for the client to hear. Collaborative reflecting teams avoid coming to agreement on any one description of what is going on with the client, allowing for *multiple, contradictory perspectives* to linger and promoting the development of new meanings and perspectives. Teams avoid comments that evaluate or judge the client in any way, positively or negatively. Instead, they focus on offering what is called *reflections*, observations, questions, or comments that are clearly owned by the person making them (e.g., "As I listened, I was wondering …").

General Guidelines for Reflecting Teams

Andersen (1991, 1995) provides the following guidelines for teams:

- *Only Use with the Client's Permission:* The therapist should obtain the client's permission to use a team *before* the session starts. When the therapist has a strong rapport with the client and confidently explains how the reflecting process works, most clients enthusiastically agree.
- *Give the Client Permission to Listen or Not to Listen:* Andersen gives the clients *explicit* permission to listen or not listen. I find it helpful to tell clients that they will probably hear some comments that resonate deeply and others that fit less well with their experience, and I recommend they focus on the comments that "strike a chord."
- *Comment on What Is Seen or Heard, Not What Is Observed:* Team members should comment on a specific event or statement in the conversation and then "wonder" or be "curious" about it. The wondering or curiosity statement should be *appropriately unusual* to help generate new perspectives.
- *Talk From a Questioning, Speculative, and Tentative Perspective:* Team members avoid offering opinions or interpretations and instead use "wondering" questions ("I am wondering if …") or offer a tentative perspective ("I am aware I don't know enough to know the whole story, but it seems like there might be …"). If a team member offers a strong opinion, another team member may ask, "What did you see or hear in the conversation that made you think that?" to open the conversation up and invite multiple perspectives.
- *Comment on All That You Hear but Not All That You See:* If the family members try to cover something up, allow them the *right to not talk about* all that they think and feel. Andersen warned: "Don't confuse therapy with confession." Unlike in psychodynamic and humanistic traditions, Andersen explicitly stated that if a client wants to hide an emotion or not say something, the client should be free to do so. He was a rare advocate for client privacy in therapy, believing that clients will share when they are ready. If a therapist notices a client getting agitated or holding back tears, he does not comment on it, allowing the client to speak about these emotions when he or she is ready to do so.
- *Separate the Team and the Family:* The team and family can be in the same room but should not talk to each other. Andersen believed that an important psychological space is created by the physical space between the team and client and by the two not talking directly; later research studies supported his view (Sells, Smith, Coe, & Yoshioka, 1994). This space invites all participants to focus on their inner dialogue, stimulating new thoughts and ideas more readily.
- *Listen for What Is Appropriately Unusual: Avoid what is too usual or too unusual.* To identify useful reflections, Andersen asked himself: "Is what is going on now appropriately unusual or is it too unusual?" (Andersen, 1995, p. 21).

- Ask: *"How Would You Like to Use This Session Today?"* This question, although likely to be asked at the beginning of any session, is critical when a team is involved. If the client is nervous about using a reflecting team, the therapist can also add, "Are there particular topics you want to avoid with the team here?"

Related Reflecting Processes

Over time, the concept of the reflecting team has developed into a number of reflecting processes:

- *Multiple Reflectors:* A team of two to four therapists observe the therapist-client conversation, sitting either in a different room using a one-way mirror (or camera) or in a separate space in the same room.
- *Single Reflector:* If only one colleague is available, the therapist may turn and have a reflecting conversation with this one reflector while the client listens.
- *No Outside Reflector When Working With a Family:* When there is no outside colleague available, the therapist may choose to speak with a single family member while other family members listen.
- *No Outside Reflector When Working With an Individual:* When the therapist is working with an individual client, a reflective process can be created by talking about issues from the perspective of someone who is not present (e.g., a parent, friend, spouse, or famous person of personal significance).
- *With Young Children:* When working with children, reflections can include play media. A single therapist working with an individual child can create reflecting teams using puppets or other such media (Gehart, 2007a).

"As If" Reflecting

Developed by Anderson (1997), the "as if" reflecting process involves having the team members or other witnesses to the conversation speak or reflect "as if" they are some of the people in the problem-organized system (i.e., the people talking about the problem), which includes the client, family members, friends, bosses, teachers, school personnel, medical professionals, probation officers, and so forth. This process can be used with clients or with supervisees staffing a case.

Putting It All Together

Case Conceptualization Template

- *Who's Talking About the Problem: Multiple Definitions of the Problem*
  - Client's definition of the problem
  - Extended family's definition of the problem
  - Broader system definition (school, work, friends, etc.)
- *Construction of the Problem:* Describe client's significant meanings and constructions related to the problem (e.g., constructions of love, depression, duties of family members).
- *Client's Worldview:* Describe the client's worldview, inner dialogue, and/or narrative of the problem and/or their life/relational circumstances related to it.
- *Dominant Discourses*
  - *Cultural, Ethnic, SES, Religious, etc.:* How do key cultural discourses inform what is perceived as a problem and the possible solutions?
  - *Gender, Sexual Orientation, etc.:* How do the gender/sexual discourses inform what is perceived as a problem and the possible solutions?
  - *Contextual, Family, Community, School, and Other Social Discourses:* How do other important discourses inform what is perceived as a problem and the possible solutions?

- *Local/Alternative Narratives*
  - *Identity/Self-Narratives:* How has the problem shaped each family member's identity?
  - *Local or Preferred Discourses:* What is the client's preferred identity narrative and/or narrative about the problem? Are there local (alternative) discourses about the problem that are preferred?

Treatment Plan Template for Individual

Collaborative Initial Phase of Treatment with Individual

Initial Phase Counseling Tasks

1. Develop working counseling relationship. *Diversity Note: [Describe how you will adjust to respect cultured, gendered, and other styles of relationship building and emotional expression.]*
   a. Develop a *collaborative partnership* with client that *honors client's expertise*.
   b. Use *everyday, ordinary language* that is comfortable for client.

2. Assess individual, systemic, and broader cultural dynamics. *Diversity Note: [Describe how you will adjust assessment based on cultural, socioeconomic, sexual orientation, gender, and other relevant norms.]*
   a. Identify *who is talking about the problem* and inquire about how each describes and constructs it.
   b. Assess *client's worldview* on issues related directly to the problem and other areas of significance/interest, listening for the personal meanings and *interpretations* that they use to construct their lived reality.

Initial Phase Client Goals

1. Increase *client's sense of agency and possibilities* for managing [specify crisis symptom or other issue prioritized by client] to reduce depressed mood/anxiety [or specific crisis behavior].
   a. *Conversational questions* to transform meanings associated with crisis behaviors and to identify realistic alternative responses.
   b. *Mutual puzzling* about where is the best place to start and how best to approach the identified issue.

Collaborative Working Phase of Treatment with Individual

Working-Phase Counseling Task

1. Monitor quality of the working alliance. *Diversity Note: [Describe how you will attend to client response to interventions that indicate therapist using expressions of emotion that are not consistent with client's cultural background.]*
   a. Monitor sessions for client or therapist *monologues* or other breakdowns in dialogical exchange.
   b. *Being public* if therapist perceives problems in the therapeutic relationship.
   c. Administer *session rating scale*.

Working-Phase Client Goals

1. Increase *fluidity of the description/construction of the problem* to identify new possibilities for action to reduce hopelessness.
   a. *Appropriately unusual comments* to generate new meaning.
   b. *Reflecting team/process* to generate new descriptions and understandings.
   c. Suggest various *written options* to access multiple voices and perspectives.

(continued)

2. Increase client *possibilities for responding* to [specify factor client identifies as related to symptoms] to reduce depression/anxiety.
    a. *Mutual puzzling* to consider alternative responses that will have more desired outcomes.
    b. *Conversational and not-knowing questions* to expand meaning and possibilities.
3. Increase *effectiveness of client response* to [specify factor client identifies as related to symptoms] to reduce depression/anxiety.
    a. *Appropriately unusual questions* to generate new possibilities for responding.
    b. *Reflecting team/processes* to expand view of possibilities.

Collaborative Closing Phase of Treatment with Individual

Closing-Phase Counseling Task

1. Develop aftercare plan and maintain gains. *Diversity Note: [Describe how you will access resources in the communities of which they are a part to support them after ending therapy.]*
    a. Have *client identify* potential setbacks or future problems and discuss possibilities for managing.

Closing-Phase Client Goals

1. Increase sense of *personal agency* in [life area related to depression/anxiety] to reduce depression and increase sense of wellness.
    a. *Not-knowing questions* to explore client construction of meaning and possibilities.
    b. *Writing* to access multiple voices and perspectives.
2. Increase the number of relationships in which the client has a *sense of being heard* and has *generative conversations* to reduce depression and increase sense of wellness.
    a. *Collaborative questions* to explore relational network and possibilities for supportive relationships.
    b. *Mutual puzzling* to identify how best to build relational support network.

Treatment Plan Template for Couple/Family

Collaborative Initial Phase of Treatment with Couple/Family

Initial Phase Counseling Tasks

1. Develop working counseling relationship. *Diversity Note: [Describe how you will adjust to respect cultured, gendered, and other styles of relationship building and emotional expression.]*
    a. Develop a *collaborative partnership* with each member of the system that *honors the expertise of each*.
    b. Use *everyday, ordinary language* that is comfortable for client.
2. Assess individual, systemic, and broader cultural dynamics. *Diversity Note: [Describe how you will adjust assessment based on cultural, socioeconomic, sexual orientation, gender, and other relevant norms.]*
    a. Identify *who is talking about the problem* (both in and outside of session) and inquire about how each describes and constructs it; if two have similar definitions, look for nuanced differences.

(continued)

b. Assess each *client's worldview* on issues related directly to the problem and other areas of significance/interest, listening for the *personal meanings* and *interpretations* that they use to construct their lived realities.

Initial Phase Client Goals

1. Increase *fluidity of the description/construction of the problem* to identify new possibilities for action to reduce conflict.
    a. *Conversational questions* to explore each person's meanings and interpretations.
    b. *Appropriately unusual comments* to generate new meaning.
    c. *Reflecting team/process* to generate new descriptions and understandings.

Collaborative Working Phase of Treatment with Couple/Family

Working-Phase Counseling Task

1. Monitor quality of the working alliance. *Diversity Note: [Describe how you will attend to client response to interventions that indicate therapist using expressions of emotion that are not consistent with client's cultural background.]*
    a. Monitor sessions for client or therapist *monologues* or other breakdowns in dialogical exchange.
    b. *Being public* if therapist perceives problems in the therapeutic relationships.
    c. Administer *session rating scale*.

Working-Phase Client Goals

1. Increase couple/family's ability to engage in *productive dialogue* to handle problems in daily living to reduce conflict.
    a. *Conversational questions* to explore relational dynamics that characterize conflict interactions and explore alternative possibilities.
    b. *Mutual puzzling* for how couple/family can interact in such a way that each person's perspective is included/considered.

2. Increase couple/family's ability to honor the *multiple realities* in the relationship to reduce conflict.
    a. *Conversational and not-knowing questions* to expand meaning and possibilities.
    b. *Reflecting team/processes* to expand views of situation.

Collaborative Closing Phase of Treatment with Couple/Family

Closing-Phase Counseling Task

1. Develop aftercare plan and maintain gains. *Diversity Note: [Describe how you will access resources in the communities of which they are a part to support them after ending therapy.]*
    a. Have *client identify* potential setbacks or future problems and discuss possibilities for managing.

Closing-Phase Client Goals

1. Increase cohesiveness of *shared relational narrative* to reduce conflict and increase relational satisfaction.
    a. *Not-knowing questions* to explore client construction of meaning and possibilities.
    b. *Writing* to access multiple voices and perspectives.

2. Increase each person's sense of *personal agency* as it relates to sustaining relationship to reduce conflict and increase sense of wellness.
    a. *Collaborative questions* to help each person story his/her contribution to improvements.
    b. *Mutual puzzling* to identify how best to build on new sense of agency.

Clinical Spotlight: Open Dialogue, an Evidence-Based Approach to Psychosis

Using the collaborative approach described by Anderson, Goolishian, and Andersen, Jaakko Seikkula (2002) and his colleagues (Haarakangas et al., 2007) in Finland developed the open dialogue approach in their work with psychosis and other severe disorders. They report impressive outcomes in their 20 years of research, including 83% of first-episode psychosis patients returning to work and 77% with no remaining psychotic symptoms after two years of treatment. In comparison with standard treatment, the patients in the open dialogue treatment had more family meetings, fewer days of inpatient care, reduced use of medication, and a greater reduction in psychotic symptoms.

This approach uses collaborative dialogue and reflecting practices, as well as the following:

- *Immediate Intervention:* Within 24 hours of the initial call, the person who has had a psychotic break, the significant people in his/her life, and a treatment team of several professionals (e.g., for psychosis the team often includes a psychiatrist, psychotherapist, and nurse) meet to discuss the situation using collaborative dialogue.
- *Social Network and Support Systems:* Significant persons in the client's life and other support systems are invited to participate in all phases of the process.
- *Flexibility and Mobility:* Treatment is uniquely adapted to clients and their situations, with the treatment team sometimes meeting in their homes and sometimes in a treatment setting, depending on what is most useful.
- *Teamwork and Responsibility:* The treatment team is built on the basis of client needs; all team members are responsible for the quality of the process.
- *Psychological Continuity:* The team members remain consistent throughout treatment regardless of the stage of treatment.
- *Tolerance of Uncertainty:* Rather than employ set protocols, the team allows time to see how each situation will evolve and what treatment will be needed.
- *Dialogue:* The focus of each meeting is to establish an open dialogue that facilitates new meanings and possibilities. This process requires establishing a sense of safety for all participants to say what needs to be said.

Outside the Therapy Room

Because collaborative therapy is more a way of talking and being in the world, the collaborative conversational process has been applied to numerous other contexts, including education, research, and business consultation.

Education and Pedagogy

Collaborative practices have been used as a way to conceptualize educational pedagogy in K-12 and college settings (Anderson, 1997; Gehart, 2007b; London & Rodriguez-Jazcilevich, 2007; McNamee, 2007). Using social constructionist epistemology, educators use relational, collaborative practices to engage student curiosity and agency in the learning process, which is seen as a *community learning process* rather than an individual process. Students are invited to participate in designing the learning experiences, engage with multiple perspectives, and contribute to the learning of all class members.

Research

The same collaborative process used in therapy settings is used in research contexts to access client voices (Gehart, Tarragona, & Bava, 2007). Typically used in qualitative, interview studies, this form of research inquiry views data as coconstructed with participants, meaning that the participants play an active role not only in identifying what is important for the researchers to know about their experience but also in providing feedback on the final presentation of the results to ensure that they fairly represent the participants' intentions and meanings. Like the therapist, the researcher approaches clients from a not-knowing stance of curiosity, wanting to learn more about the participants' experiences rather than testing a preconceived hypothesis.

Business Consultation

Collaborative conversational and reflecting practices have also been used for consulting with businesses and other large systems (Anderson, 1997; Deissler, 2007). The consultant approaches the system from a curious position, taking time to learn from those within it what they view as working and not working, what they value most, and what they would like to see happen. Much as in the therapy process, the consultant facilitates two-way dialogues in which members are able to hear and say things they were not able to before. Various reflecting processes are used to create forums for new dialogue and understanding.

Research and the Evidence Base

Consistent with their philosophical underpinnings, postmodern therapists, including collaborative, have conducted more qualitative than quantitative investigations about their approaches to therapy (Anderson, 1997; Gehart et al., 2007). Qualitative research on postmodern therapies has focused on clients' lived experience of therapy and its effects on their lives, emphasizing the clients' experience over researcher-defined measures of successful therapy (Andersen, 1997; Gehart & Lyle, 1999; Levitt & Rennie, 2004; London, Ruiz, Gargollo, & M.C., 1998). A notable exception, Finnish psychiatrist Jaakko Seikkula (2002) and his team (Haarakangas et al., 2007) have used qualitative and quantitative methods to study their open dialogue approach to working with psychosis and other severe diagnoses for over 20 years, generating significant evidence for their model's effectiveness (see Clinical Spotlight discussion). They report that they have nearly eradicated chronic cases of psychosis, with most clients returning to normal functioning within two years; their work is an exemplary, evidence-based approach for recovery-oriented approaches to treating severe mental health issues.

The assumptions and premises of postmodern therapy are also receiving support from an unexpected arena: psychiatry, more specifically interpersonal neurobiology (Beaudoin & Zimmerman, 2011; Gehart, 2012; Siegel, 2012). Like postmodern therapists, Siegel proposes that storytelling is how humans make sense of life:

> We are storytelling creatures, and stories are the social glue that binds us to one another. Understanding the structure and function of narrative is therapy a part of understanding what it means to be human. (pp. 31–32)

Siegel describes how in childhood, humans are born into story, take on cultural meaning, and then in adolescence and adulthood continue to reshape these meanings, thus evolving human culture. These narratives constrain or expand a person's options for responding to life. Siegel suggests that *reflection*—a detached examination of these narratives—enable clients to liberate themselves from these limiting stories of who they are and what their life is; this same process is emphasized in postmodern therapies.

In describing how people change, Siegel (2012) emphasizes the importance of *bottom-up processing*, which is compared *top-down process*. Bottom-up processing refers to processing experience using the bottom three layers of the prefrontal cortex to generate new understandings, categories, and stories for what is happening. In contrast, top-down processing involves using the top three layers of the prefrontal cortex to categorize lived experience with pre-existing categories. Both forms of processing are important, but most people who feel "stuck" related to a problem are stuck in top-down processing, unable to change how they see their situation or possibilities for changing it. The *not-knowing* position of postmodern therapists is an excellent approach for facilitating bottom-up processing: in fact, the entire collaborative therapy process can be viewed as an approach for expanding bottom-up processing (Gehart, 2012).

Additionally, Siegel (2012) describes trauma resolution as a process of creating an integrative narrative, similar to thickening the plot in narrative therapy. Traumatic experiences overwhelm the hippocampus and explicit (narrative) memory not encoded with implicit memory, as happens in nontraumatic situations. Thus, traumatic memories are encoded primarily as implicit memories, which do not have a sensation of being from the past and are often experienced as fragmented auditory, visual, and sensual experiences that are occurring in the here and now (i.e., flashbacks). These fragmented memories create chaotic and rigid patterns that impair the brain's ability to enter into health-integrated neural states. Siegel suggests that the treatment from trauma essentially involves pulling the implicit memories together into a coherent narrative, which is most easily achieved by entering into "interpersonal attunement" with a therapist. He proposes that when two minds become entrained, in sync, the client can "borrow" the therapist's neural integrated state to help cope with the traumatic memories long enough to put them into a coherent narrative. Through this process of recreating an integrated explicit narrative of the trauma the client is then able to recall the trauma while remaining in an integrated neural state (i.e., not have symptoms when recalling the trauma).

Finally, Beaudoin and Zimmerman (2011) note that the postmodern techniques that help clients "give voice" to their lived experience and put into words things that have never been said before help move clients from their anxiety-focused limbic system to the more calm and reason-focused prefrontal cortex. By labeling and giving voice to their stress-based experiences, clients are able to reduce the firing of the limbic system, allowing the client to make more conscious choices and responses. Furthermore, therapeutic conversations about preferred realities strengthen neural connections that support clients' preferred identities and associated behaviors. In particular, *affect-infused descriptions* of preferred realities help to solidify new identity narratives, which are predominantly associated with the right brain hemisphere; such affect-infused descriptions include not only emotion but also sensory experiences, such as imagery, scent, and tactile experiences.

Tapestry Weaving: Working with Diverse Populations

Ethnic, Racial, and Cultural Diversity

Collaborative therapy is widely used in Europe, Latin America, and Asia. The collaborative, not-knowing approach makes it an excellent approach for working with diverse clients because the therapist avoids making assumptions and strives to work from within the client's lived reality. In addition, by using the *client's* definition of the problem—rather than an external theory of health that would inherently include cultural bias—this approach can support the local knowledges of marginalized clients. This focus on local knowledges ensures that the client's cultural values and beliefs are a central part of the therapy process. In addition, the approach is highly adaptable and flexible, allowing it to adapt to cultural and institutional needs, including using family members in reflecting teams in Cuban hospitals (Deissler, 2007), involving prison wardens in sessions with prisoners in Sweden (Wagner, 2007), or exploring sociocultural aspects of eating disorders in Mexico (Fernández, Cortés, & Tarragona, 2007).

Intercultural Couples

Collaborative therapy has been identified as a particularly appropriate approach for working with intercultural couples (Biever, Bobele, & North, 1998). Collaborative therapy is based on the premise that there are multiple realities in any given situation and that these conflicting realities should be respected without attempting to force a single reality or interpretation. Instead, the dialogical process enables intercultural couples to coconstruct new meanings that may incorporate, recontextualize, and/or transcend their earlier understandings through a highly collaborative process of mutual inquiry and exploration. Specifically, Biever, Bobele, and North (1988) suggest the following:

- *Collaborative, Curious Stance:* Therapists use a curious stance to understand how each partner understands the problem and whether each sees cultural as related to it. In addition, therapists can explore in what ways they see cultural differences have helped and hindered the relationship.

- *Openness and the Generation of New Meanings:* Collaborative therapy can be used to explore alternative descriptions and explanations of the problem. These multiple, contradictory meanings enable the couple to coconstruct more useful understandings of the problem that lead to more desirable outcomes.
- *Tentativeness and Not Understanding Too Quickly:* The therapist uses tentative language—such as "wondering if," "kind of" and "maybe"—to keep meanings fluid and to encourage clients to generate their own interpretations and meaning. With intercultural couples, this may take the form of asking, "I am wondering if there are differences in how your families would see this situation?"
- *Exploration of Client's Ideas About the Problem:* Client understandings of the problem are highly valued because they are often key to promoting desired outcomes, not because they are necessarily more "correct" or "true." Collaborative therapists are cautious about how the questions they ask may infer that they believe a certain cause to client problems, such as asking too much about childhood experiences. Instead, they prefer to focus on client explanations.
- *View of Cultural Differences as One Explanation of Conflicts:* Therapists may tentatively use their knowledge of cultural differences to understand the couple's patterns, but they do not assume culture necessarily plays a role in their problems. In addition, they are careful to not make cultural stereotypes, aware that each person has a unique interpretation and experience of their cultural background and norms based on a myriad of factors, including age, gender, immigration status, family of origin, etc.
- *Encouraging Both/and Stance:* Collaborative therapists can use a both/and stance to extract themselves from couple arguments and explore how each partner's position came to be and when each position is useful or makes sense.
- *Search for Liberating Traditions Within Each Culture:* Therapists can help couples identify strengths and values within each culture that help support the couple in recontextualizing the problem or identifying new possibilities for relating.
- *View Impasses as Attempts to Impose Beliefs/Values on the Other:* When a couple's dialogue becomes stuck, it is often because one or both are trying to impose their belief or position on the other. Therapists may want to ask themselves: Who is trying to convince whom of what? What is hoped for? What is feared?

Sexual Identity Diversity

Collaborative therapy is often the theory of choice for working with gay, lesbian, bisexual, transgendered, and questioning (GLBTQ) clients because they help to deconstruct the heterosexist discourses that often are the source of greatest suffering in GLBTQ clients (Aducci & Baptist, 2011; Perez, 1996). When working with GLBTQ clients, many therapists recommend an advocacy stance with expertise in the subject matter rather than a strict not-knowing stance (Aducci & Baptist, 2011; Perez, 1996).

Same-Sex Couples

Perez (1996) suggests using collaborative therapy for working with same-sex couples. Same-sex couples tend to have the same types of concerns as heterosexual couples, such as problems with intimacy and communication, with the exception of two key issues: (a) stage differences in coming out and sexual identity, and (b) less social support and greater stress due to marginalization. Related to these two issues, same-sex couples might also struggle to negotiate their identity as a couple, in terms of both positively identifying as being a couple and presenting themselves as a couple to family, friends, and coworkers. He provides the following suggestions for using collaborative therapy with same-sex couples.

- *Creating a Safe Therapeutic Environment:* The not-knowing stance of collaborative therapists can be ideal for working with same-sex couples, being genuinely curious about the couple's experience and the meanings they make from it. However, Perez's primary caution when using collaborative therapy with same-sex couples is that by assuming a not-knowing stance, the therapist may de-emphasize the need for therapist knowledge of and sensitivity to same-sex couples issues.

- *Educating Clients About Community Resources:* As collaborative therapists view identity as relationally constructed through dialogue, therapists can help clients identify relationships and community resources that can help them restory their individual and couple identity in positive ways.
- *Benefits of Self-Labeling:* Because positively self-labeling oneself as gay or lesbian is correlated with better mental health outcomes, collaborative therapists can use mutual inquiry to help same-sex couples deconstruct negative labels and coconstruct more positive self and couple identity labels.
- *Use of Client Language:* Collaborative therapists typically work within and respect client's language systems. With same-sex couples, it is important to attend to homophobic and derogatory language they may use toward themselves. Collaborative therapists can use conversational questions to explore where the language comes from and if more positive language would be helpful to the couple.
- *Therapist Expertise with Sexual Identity Issues:* Although collaborative therapists generally do not assume an expertise position, they still need to develop competency and expertise in the types of concerns faced by same-sex couples. Therapists should be knowledgeable about a range of issues, including:
  - Contemporary gay and lesbian culture, including local, rural, and urban
  - Understanding the processes of these communities
  - Awareness of current political issues related to same-sex couples
  - Familiarity with community resources
  - Understanding the risks, benefits, and losses associated with coming out
  - Knowing the impact of social pressures on the relationship
  - Being aware of gay and lesbian literature
  - Being supportive of gay and lesbian relationships

Online Resources

Harlene Anderson:
www.harleneanderson.org

Grupo Campos Elíseos (Collaborative Therapy, Mexico City):
www.grupocamposeliseos.com

Houston Galveston Institute (Collaborative Therapy):
www.talkhgi.com

Marburg Institute (Collaborative Therapy, Germany):
www.mics.de

Taos Institute (Collaborative Practices in Therapy, Consultation, Education, Business):
www.taosinstitute.net

References

*Asterisk indicates recommended introductory readings.

Aducci, C. J., & Baptist, J. A. (2011). A collaborative-affirmative approach to supervisory practice. *Journal of Feminist Family Therapy: An International Forum, 23*(2), 88–102. doi:10.1080/08952833.2011.574536

*Andersen, T. (1991). *The reflecting team: Dialogues and dialogues about the dialogues.* New York: Norton.

Andersen, T. (1992). Relationship, language and preunderstanding in the reflecting process. *Australian and New Zealand Journal of Family Therapy, 13*(2), 87–91.

Andersen, T. (1995). Reflecting processes; acts of informing and forming: You can borrow my eyes, but you must not take them away from me! In S. Friedman (Ed.), *The reflecting team in action: Collaborative practice in family therapy* (pp. 11–37). New York: Guilford.

Andersen, T. (1997). Researching client-therapist relationships: A collaborative study for informing therapy. *Journal of Systemic Therapies, 16*(2), 125–133.

*Andersen, T. (2007). Human participating: Human "being" is the step for human "becoming" in the next step. In H. Anderson & D. Gehart (Eds.), *Collaborative therapy: Relationships and conversations that make a difference* (pp. 81–97). New York: Brunner/Routledge.

Anderson, H. (1993). On a roller coaster: A collaborative language systems approach to therapy. In S. Friedman (Ed.), *The new language of change* (pp. 323–344). New York: Guilford.

Anderson, H. (1995). Collaborative language systems: Toward a postmodern therapy. In R. Mikesell, D. D. Lusterman, & S. McDaniel (Eds.), *Family psychology and systems therapy* (pp. 27–44). Washington, DC: American Psychological Association.

*Anderson, H. (1997). *Conversations, language, and possibilities: A postmodern approach to therapy.* New York: Basic Books.

Anderson, H. (2005). Myths about "not knowing." *Family Process, 44*, 497–504.

Anderson, H. (2007). Historical influences. In H. Anderson & D. Gehart (Eds.), *Collaborative therapy: Relationships and conversations that make a difference* (pp. 21–31). New York: Brunner/Routledge.

*Anderson, H., & Gehart, D. (2007). *Collaborative therapy: Relationships and conversations that make a difference.* New York: Brunner/Routledge.

Anderson, H., & Goolishian, H. (1988). Human systems as linguistic systems: Preliminary and evolving ideas about the implications for clinical theory. *Family Process, 27*, 157–163.

*Anderson, H., & Goolishian, H. (1992). The client is the expert: A not-knowing approach to therapy. In S. McNamee & K. J. Gergen (Eds.), *Therapy as social construction* (pp. 25–39). Newbury Park, CA: Sage.

Bava, S., Levin, S., & Tinaz, D. (2002). A polyvocal response to trauma in a postmodern learning community. *Journal of Systematic Therapies, 21*(2), 104–113.

Beaudoin, M. N., & Zimmerman, J. (2011). Narrative therapy and interpersonal neurobiology: Revisiting classic practices, developing new emphases. *Journal of Systemic Therapies, 30*, 1–13.

Biever, J. L., Bobele, M., & North, M. (1998). Therapy with intercultural couples: A postmodern approach. *Counseling Psychology Quarterly, 11*(2), 181–188. doi:10.1080/09515079808254053

Boscolo, L., Cecchin, G., Hoffman, L., & Penn, P. (1987). *Milan systemic family therapy.* New York: Basic Books.

Deissler, K. (2007). Dialogues in a psychiatric service in Cuba. In H. Anderson & D. Gehart (Eds.), *Collaborative therapy: Relationships and conversations that make a difference* (pp. 291–309). New York: Brunner/Routledge.

Fernández, E., Cortés, A., & Tarragona, M. (2007). You make the path as you walk: Working collaboratively with people with eating disorders. In H. Anderson & D. Gehart (Eds.), *Collaborative therapy: Relationships and conversations that make a difference* (pp. 129–147). New York: Routledge/Taylor & Francis Group.

Gehart, D. (2007a). Creating space for children's voices: A collaborative and playful approach to working with children and families. In H. Anderson & D. Gehart (Eds.), *Collaborative therapy: Relationships and conversations that make a difference* (pp. 183–197). New York: Brunner/Routledge.

Gehart, D. (2007b). Process-as-content: Teaching postmodern therapy in a university context. *Journal of Systemic Therapies, 18*, 39–56.

Gehart, D. (2012). *Mindfulness and acceptance in couple and family therapy.* New York: Springer.

Gehart, D. R., & Lyle, R. R. (1999). Client and therapist perspectives of change in collaborative language systems: An interpretive ethnography. *Journal of Systemic Therapy, 18*(4), 78–97.

Gehart, D., Tarragona, M., & Bava, S. (2007). A collaborative approach to inquiry. In H. Anderson & D. Gehart (Eds.), *Collaborative therapy: Relationships and conversations that make a difference* (pp. 367–390). New York: Brunner/Routledge.

Goolishian, H., & Anderson, H. (1987) Language systems and therapy: An evolving idea. *Psychotherapy, 24*(3S), 529–538.

Haarakangas, K., Seikkula, J., Alakare, B., & Aaltonen, J. (2007). Open dialogue: An approach to psychotherapeutic treatment of psychosis in Northern Finland. In H. Anderson & D. Gehart (Eds.), *Collaborative therapy: Relationships and conversations that make a difference* (pp. 221–233). New York: Brunner/Routledge.

*Hoffman, L. (1981). *Foundations of family therapy: A conceptual framework for systems change.* New York: Basic Books.

Hoffman, L. (1990). Constructing realities: An art of lenses. *Family Process, 29*, 1–12.

Hoffman, L. (1993). *Exchanging voices: A collaborative approach to family therapy.* London: Karnac Books.

*Hoffman, L. (2001). *Family therapy: An intimate history.* New York: Norton.

Hoffman, L. (2007). The art of "withness": A bright new edge. In H. Anderson & D. Gehart (Eds.), *Collaborative therapy: Relationships and conversations that make a difference* (pp. 63–79). New York: Brunner/Routledge.

Levin, S. (2007). Hearing the unheard: Advice to professionals from women who have been battered. In H. Anderson & D. Gehart (Eds.), *Collaborative therapy: Relationships and conversations that make a difference* (pp. 109–128). New York: Brunner/Routledge.

Levitt, H. M., & Rennie, D. L. (2004). Narrative activity: Clients' and therapists' intentions in the process of narration. In L. E. Angus & J. McLeod (Eds.), *The handbook of narrative and psychotherapy: Practice, theory, and research* (pp. 299–313). Thousand Oaks, CA: Sage. doi:10.4135/9781412973496.d23

London, S., & Rodriguez-Jazcilevich, I. (2007). The development of a collaborative learning and therapy community in an educational setting: From alienation to invitation. In H. Anderson & D. Gehart (Eds.), *Collaborative therapy: Relationships and conversations that make a difference* (pp. 235–250). New York: Brunner/Routledge.

London, S., Ruiz, G., Gargollo, M., & M.C. (1998). Clients' voices: A collection of clients' accounts. *Journal of Systemic Therapies, 17*(4), 61–71.

McNamee, S. (2007). Relational practices in education: Teaching as conversation. In H. Anderson & D. Gehart (Eds.), *Collaborative therapy: Relationships and conversations that make a difference* (pp. 313–336). New York: Brunner/Routledge.

Penn, P. (2001). Chronic illness: Trauma, language, and writing: Breaking the silence. *Family Process, 40*, 33–52.

Penn, P. (2002). *So close.* Fort Lee, NJ: Cavankerry.

Penn, P., & Frankfurt, M. (1994). Creating a participant text: Writing, multiple voices, narrative multiplicity. *Family Process, 33*, 217–231.

Penn, P., & Sheinberg, M. (1991). Stories and conversations. *Journal of Systemic Therapies, 10*(3–4), 30–37.

Perez, P. J. (1996). Tailoring a collaborative, constructionist approach for the treatment of same-sex couples. *The Family Journal, 4*(1), 73–81. doi:10.1177/1066480796041016

Seikkula, J. (2002). Open dialogues with good and poor outcomes for psychotic crises: Examples from families with violence. *Journal of Marital and Family Therapy, 28*(3), 263–274.

Sells, S., Smith, T., Coe, M., & Yoshioka, M. (1994). An ethnography of couple and therapist experiences in reflecting team practice. *Journal of Marital and Family Therapy, 20,* 247–266.

Siegel, D. (2012). *Pocket guide to interpersonal neurobiology.* New York: Norton.

St. George, S., & Wulff, D. (1998). Integrating the client's voice within case reports. *Journal of Systemic Therapies, 17*(4), 3–13.

Wagner, J. (2007). Trialogues: A means to answerability and dialogue in a prison setting. In H. Anderson & D. Gehart (Eds.), *Collaborative therapy: Relationships and conversations that make a difference* (pp. 203–220). New York: Routledge/Taylor & Francis Group.

Collaborative Case Study: Pakistani/Indian Couple with Sexual Abuse History

Naseer (AM), a 36-year-old Pakistani engineer, is married to 34-year-old Indian woman named Shiva (AF), who is also an engineer. They have a daughter, Saman 15 (CF15) and a 7-year-old son, Atta (CM7). The couple has been experiencing intense marital problems after Shiva revealed her experience of sexual assault by an uncle at age 12. Previously, emotionally connected, Naseer has become cold and distant toward Shiva, making her feel victimized all over again. She had hoped that her opening up would help him understand her difficulties with certain aspects of their intimate relationship. Both husband and wife work long hours, but since the incident Naseer now works even longer than before and often comes home late at night. He used to spend time with the kids, especially helping them with schoolwork but now rarely does that. Shiva is very frustrated and says that she can't trust him anymore. When they are together, they fight constantly. Their 15-year-old daughter is frustrated with her parent's constant quarrels and prefers to spend most of her time at a friend's home; the mother then argues with her for being out too much. Saman has shared her frustrations about the parents' dilemma to her aunt, Naseer's sister. Their 7-year-old son has been experiencing bedwetting recently and complains about going to school almost every morning. Naseer's sister, who was a college roommate of Shiva, maintains good standing with her sister-in-law; she believes it was wrong of Shiva not to have shared this with her husband before marriage, but she also feels that she would have done the same and encourages the couple to get professional help before making decisions about the future of their family. Shiva's mother, who recently moved back home to India to be close to her dying mother, blames her daughter's marital problems on Shiva's naiveté and for bringing out such a dark past; this conversation has only added more tension to the family's situation.

Postmodern Case Conceptualization

- *Who Is Talking About the Problem? Multiple Definition of the Problem:*
  - Client's definition of the problem: "problem organizing"
    - —AF thinks the husband should be empathic and is being unfair; she also fears he doesn't love her anymore.
    - —AM believes the wife should have shared this before marriage and feels lied to.
    - —CF15 believes AF is too controlling and caused the problems.
    - —CM7 sees that AM works too much, that AF is sad, and that his parents fight too much.
  - Extended family definition of the problem: AF's mother says AF shouldn't have shared her "dark secret" with the husband and now needs to ask for forgiveness. AM's sister believes AF should not have kept/revealed the secret, but now thinks the couple needs to work past their issues of shame and trust.
  - Broader system definition (school, work, friend, etc.)
    CM's school has noted his regression and difficulties at school.

- *Construction of the Problem*
  - AF feels that it is evident from AM's recent behavior that he doesn't love her anymore because of what happened to her: that she is tainted and broken. AF believes that if AM truly cared for her, he would understand why she had not shared this with him earlier. AF had hoped that after many years of harmonious marriage and being acculturated, AM would have developed enough open-mindedness and connection with AF to look past their cultural norms.
  - AM feels betrayed by AF and lied to, believing she should have revealed the secret earlier. He also dismisses her concerns that he does not love her, stating that working long hours should be enough for any woman to see that her husband cares for the family.
  - CM (7) thinks parent's conflict may have something to do with his problem bed-wetting.
  - CF (15) believes parent's marital issues have made them more unreasonable and not attentive to the children's needs.
- *Client's Worldview*
  - AF believes AM is "educated" and should have a more "modern" understanding of childhood sexual abuse.
  - AM feels betrayed and that AF lied to him when she said she was a virgin when they got married. He says the circumstances do not justify the lie, especially about such an important issue.
- *Dominant Discourses*
  - *Cultural/Ethnic:* AM expectations of his family roles and his emotional reservation about his wife's abuse is in line with Pakistani cultural norms in which men tend to be emotionally reserved and do not always engage in discussions with women about emotionally charged issues, particularly about highly stigmatized matters such as sexual abuse of their spouse. He believes, "A man's family role is to maintain family's financial health and women's role is to attend to domestic responsibilities and attend to her children's need"; this is a common dialogue considering his background. Although AF's Indian culture shares similar values, AF identifies also with more American values, especially in terms of gender relations.
  - *Gender/Sexuality Discourses:* AM's cultural background values sexual purity in marriage, and he now feels shamed and betrayed by his wife's revelation; AF's mother and AM' sister in large measure agree. AF knows this but also believes that a more modern view of childhood sexual abuse should be considered.
  - *Other Discourses:* AF brings in a significant income to the family and believes that because of this "modern" woman's role, she should not be seen as a piece of property that must be "perfect" to be loved.
- *Local/Alternative Narratives*
  - *Identify/Self-Narratives:*
    —AF has mixed feelings. On one hand, she feels more broken and defiled than ever; on the other, she feels AM is acting unfairly, by blaming the victim and failing to view the situation with modern sensibilities.
    —AM feels his wife's lies early in their relationship were the real betrayal. He now questions what else she may have told him that was a lie. Although part of him knows it is not rational, he feels that somehow he got "defective merchandise."
    —CF feels invisible to her parents now and for that she retreats to her friend's home.
    —CM explains that he would much rather stay at home with his mother than go to school, and that he is much too tired to wake up in time in the middle of the night to take care of his business.
  - *Local or Preferred Discourse*
    —AF wants to feel that AM loves her and has empathy for her pain.
    —AM wants to feel that he can trust AF again. He does want to stay with her.
    —CF and CM just want their parents to "go back to normal again."

Postmodern Treatment Plan

Collaborative Initial Phase of Treatment with Couple

Initial Phase Counseling Tasks

1. Develop working counseling relationship. *Diversity Note: Attend to cultural and cultured gender norms, including respecting privacy, attitudes toward sexuality, and recognizing gender roles; avoid assuming preference to be called by first name or surname.*
    a. Develop a *collaborative partnership* with each partner that *honors the expertise of each*. Maintain two-way dialogue, by avoiding monologues, monitoring for tension buildups and for when conversation becomes attempt to convince the other.
    b. Use *everyday, ordinary language* that is comfortable for both and responsive to cultural values.

2. Assess individual, systemic, and broader cultural dynamics. *Diversity Note: Consider cultural norms of each, especially gender roles, as well as acculturation levels of each. Also, consider role of extended family system.*
    a. Identify *who is talking about the problem* (both in and outside of session) and inquire about how each describes and constructs it. Be certain to demonstrate sincere curiosity and interest in each of the family member's stories and perspectives shared. Look for internal logic that defines hopes, problems, and symptoms.
    b. Assess each *client's worldview* on issues related directly to the problem and other areas of significance/interest, listening for the *personal meanings* and *interpretations* that they use to construct their lived realities. Ask questions to clarify beliefs and traditions. For instance, how did AF make sense of changes in AM's behavior, or what does she think happened, or what are options from here? Or as why did the marriage worked prior to AF's confession and what it will take to get that back?

Initial Phase Client Goals

1. Increase *fluidity of the description/construction of the problem* to identify new possibilities for each partner to view the situation and each other's behavior to reduce conflict.
    a. *Conversational questions* to explore each partner's meanings and interpretations related to the past abuse, current revelation, each person's response, and cultural associations.
    b. *Appropriately unusual comments* to generate new meaning related to the past abuse, cultural meanings, and current behaviors of each.

Collaborative Working Phase of Treatment with Couple

Working-Phase Counseling Task

1. Monitor quality of the working alliance. *Diversity Note: Monitor to ensure clients feel safe, typically demonstrated with increased openness with private information. Adjust expressions of empathy and emotional expression as is comfortable with couple.*
    a. Monitor sessions for client or therapist *monologues* or other breakdowns in dialogical exchange. Ask questions for clarification and engage others to break down monologue.
    b. *Being public* if therapist perceives problems in the therapeutic relationships.

(continued)

Working-Phase Client Goals

1. Increase couple's ability to engage in *productive dialogue* to address past abuse and current reactions to reduce conflict.
   a. *Conversational questions* to explore relational dynamics that characterize conflict interactions and explore alternative possibilities. Use client's preferred words and expression how to engage by genuinely asking about meaning and reasoning for the action.
   b. *Mutual puzzling* for how couple can interact in such a way that each person's perspective is included/considered. Demonstrate meaning-making process through mutual puzzling, allowing the couple to generate their own new interpretations together.
   c. *Monitor children's behavior* to see if reduction in couple conflict helps improve children's symptoms.

2. Increase couple's ability to honor the *multiple realities* in the relationship to reduce conflict.
   a. *Conversational and not-knowing questions* to expand meaning and possibilities; explore cultures of origin and American cultural views of the situation without privileging one over the other.
   b. Use *appropriately unusual comments* to invite curiosity and new perspective.

Collaborative Closing Phase of Treatment with Couple

Closing-Phase Counseling Task

1. Develop aftercare plan and maintain gains. *Diversity Note: Identify potential cultural or religious resources for the couple in moving forward.*
   a. Have *client identify* potential setbacks or future problems and discuss possibilities for managing.

Closing-Phase Client Goals

1. Increase cohesiveness of *shared couple narrative* to reduce conflict and increase relational satisfaction.
   a. *Not-knowing questions* to explore how couple will redefine their relationship as they move forward.
   b. *Writing* to access multiple voices and perspectives. Introduce AM and AF to explore multiple voices in writing, to generate alternative possibilities, and to promote agency. For instance: Write on aspect of themselves from newly emerging, future, or past selves.

2. Increase cohesiveness of *family identity narrative* to increase each member's sense of being supported and loved.
   a. *Collaborative questions* to help family identify what they would like to improve and ideas for doing so.
   b. *Mutual puzzling* to identify how best to support each member of the family and to develop a vision of their shared family identity.

CHAPTER 15

Evidence-Based Treatments in Couple and Family Therapy: Emotionally Focused Therapy and Functional Family Therapy

"Behavior change is a therapeutic activity that requires great creativity on the therapist's part—it involves taking very general principles and matching them to the particular family."

—Sexton, 2011, p. 5/13

Lay of the Land

Systemic couple and family therapies enjoy a robust and growing evidence base that includes several highly regarded evidence-based treatments (which you may remember from Chapter 2 are manualized treatments for specific populations). Most schools of family therapy have at least one associated evidence-based treatment, with systemic and structural approaches the most prominent. In this chapter, I review two of the most widely used evidence-based treatment in family therapy: *emotionally family therapy* and *functional family therapy*. The primary reason for selecting these two theories for this chapter (because there are many to choose from) is that they have therapist-friendly literature that clearly describes the approach in ways that can be used in private practice or agency settings, unlike some approaches that are designed primarily for larger agencies or group formats. But before I go into detail about the theories, perhaps it is best if I bust some myths about evidence-based treatments before we start.

Myths About Evidence-Based Treatments

Myth: In the age of evidence, we don't need theory anymore.

Closer to the Truth: Evidence-based treatment generally requires more stringent adherence to a single theory or careful integration of a handful of traditional therapy theories. Working from an evidence-based approach often requires *greater* competency in *more* theories.

Myth: MFT theory and research is separable.

Closer to the Truth: Family therapy theory and research have been and will always be two sides of the same coin. MFT theories were originally developed through careful observational research at places such as the Mental Research Institute in Palo Alto. Virtually all research in the field is grounded in some form of theory. The two work synergistically together; any separation of the two decontexualizes the other (decontexualization considered high crime in MFT; if you didn't get that joke, please review Chapter 4).

Myth: Evidence-based treatments are rigid and robotic.

Closer to the Truth: One of the most common fears of those new to EBTs is that they are rigid, robbing the therapist of creativity, freedom, and self-expression (Sexton & van Dam, 2010). Just like any other theory in this book, evidence-based treatments have elements that are structured and areas that are flexible (that's what makes any theory meaningful). The reputation for rigidity is more applicable for how they may be "forced" on employees in an agency: but in that case, it is the agency, not the theory that is rigid. In some areas, EBTs are far *more flexible* than traditional approaches. For example, in EFT they are far more flexible than other MFT approaches in how the negative interaction cycle is traced and assessed; approaches such as strategic, CBFT, SBT, or narrative have far more restrictive approaches to tracing interaction cycles.

Myth: I must give up my personal theory of choice to use an evidence-based treatment.

Closer to the Truth: I have a secret to share after years of working with therapists: you really cannot subtract from a person's knowledge base, only add to it. That means, in evidence-based treatment therapists bring with them their personal style and philosophy that the evidence-based treatment *adds to it,* a specific approach for handling particular types of problems. Each therapist still has to integrate it into his/her style and make it a natural fit for the approach to work.

Myth: Researchers agree on what constitutes research.

Closer to the Truth: As discussed in Chapter 2, there are at least two distinct camps of researchers: those who value specific evidence-based treatments and those who focus on common factors. The EBT camp insists clinical trials are the truly "meaningful" studies that should be the primary focus of the field. In contrast, common factors supporters emphasize the research that suggests that the similarities across approaches more than the differences account for outcome. As the academic debate wages on, we are likely to discover the answer is a both/and.

Myth: Cognitive-behavioral therapy is the only evidence-based treatment and it is an evidence-based treatment for every disorder.

Closer to the Truth: Due to its long research history and generally easy to operationalize treatment protocols, CBT enjoys a robust evidence base, with CBT earning recognition for being an evidence based treatment for a wide range of clinical concerns. That said, systemic family therapies are a close second in terms of strength of evidence base, often in areas that CBT has not been particularly effective, such as conduct disorder and couple distress.

Myth: "Standard practice" = up-to-date practice.

Closer to the Truth: Experts estimate that standard practice generally lags about 17–20 years behind the state-of-the-art/science (Institute of Medicine, 2001; Slomski, 2010). That means, that if you are practicing according to standard practice in the field, you are probably about 18.5 (don't you love research stats) years behind the leading edge. This fact has made evidence-based treatments of great interest to those outside the field. Would you want your most precious relationships in the hands of someone who was 20 years behind the current knowledge base?

Emotionally Focused Therapy

In a Nutshell: The Least You Need to Know

Emotionally focused therapy (EFT) is one of the most thoroughly researched approaches in the field and is an empirically validated treatment for treating couples and families

(Johnson, 2004). It more than qualifies as an evidence-based treatment; it is an efficacious treatment and one that has been clinical tested in independent trials (see Chapter 2 for a painfully detailed definition; Furrow & Bradley, 2011; Lebow, Chambers, Christensen, & Johnson, 2012). Sue Johnson and Les Greenberg (Johnson & Greenberg, 1985, 1994) developed the model by integrating individual and systemic theories, and by careful observation of what works in effective couples therapy. EFT uses an integration of (a) attachment theory, (b) experiential theory (specifically, Carl Rogers's person-centered therapy), and (c) systems theory (specifically, systemic-structural therapies; see Chapters 5 and 6; Johnson, 2004). Emotionally focused therapists focus on the emotional system, which includes both interpersonal (i.e., couple and family) and intrapsychic (i.e., individual) systemic processes. In the first stage of therapy, the therapist identifies the couple's behavioral interactional patterns (just like most other systemic therapists) but also identifies the perceptions, emotions, and underlying attachment needs (need for safe connection) that fuel the problematic behavioral interactions. In the middle stage of therapy, the therapist helps couples fundamentally to restructure their interactions so that each partner is able to experience a sense of safety and connection with the other (i.e., gets attachment needs met). In the last phase of therapy, the therapist helps the couple to solidify the new patterns. The approach is generally brief (approximately 8–12 sessions) with 70%–73% of couples resolving issues and recovering from distress and 86% showing significant improvement within twelve sessions with a formally trained EFT therapist (Johnson, Hunsley, Greenberg, & Schindler, 1999). Although most of the research has been conducted with couples, the same principles and techniques can also be used for family therapy (Johnson, 2004).

The Juice: Significant Contributions to the Field

If you remember one thing from this chapter, it should be this:

Attachment and Adult Love

Sue Johnson's (2004, 2008) emotionally focused couples therapy is based on a new paradigm for understanding adult love relationships. The premise is basically this: *that humans have the need for secure attachment relationships across the lifespan not just in infancy and early childhood.* This claim was not widely accepted a few decades ago, but is now enjoying a strong evidence base, which is likely to radically reshape the direction of couple and family therapy in the 21st century (Furrow & Bradley, 2011). This, dear reader, is big news. So, you may want to pull out a highlighter for this section.

Bowlby's (1988) attachment research with human and primate infants established the physiological, emotional, and survival need for a safe, nurturing relationship between caregiver and child and has long been accepted in the human sciences. However, it was commonly believed that this need dissipated with time, and that adults could theoretically be psychological and physically healthy without such a secure relationship. Current medical research not only supports the claim that secure attachment affects physical health, but that specific correlations have been identified: avoidant attachment styles are correlated with pain-related complaints, anxious attachment styles with cardiovascular conditions, yet secure attachment was not correlated with any specific health condition (McWilliams, & Bailey, 2010). Insecure attachment has long been correlated with higher rates of psychopathology (Mason, Platts, & Tyson, 2005). Additional research has shown that securely attached adults are better able to withstand physical pain than those with insecure attachment styles (Meredith, Strong, & Feeney, 2006). Neurological research suggests that humans need secure relationships in order to emotionally self-regulate (Siegel, 2010).

Sue Johnson's work has been on the vanguard of the application of adult attachment theory to therapy, offering the most comprehensive approach for helping couples to develop secure attachments in their intimate relationships. The theory of adult attachment helps explain much of the puzzling behavior around intimate couple and family relationships: how come humans so often treat poorly those they claim to love the most? The problem is humans are trapped in a paradox when tension arises in attachment relationships: the other person is both the primary source of comfort and safety and—during conflict—the most threatening danger (Johnson, 2008). This paradox goes a long way in explaining much of the extreme behavior therapists often see in couple and family relationships.

Using Bowlby's (1988) theory of attachment to conceptualize adult love, Johnson (2004) identifies 10 tenets of this theory, which I paraphrase below (pp. 25–32).

1. *Attachment Is an Innate Motivating Force:* The desire to be connected to others is an intrinsic physiological need of all humans. This implies that a therapist's work is not done with any client until they have at least one secure relationship in their life.
2. *Secure Dependence Complements Autonomy:* Neither complete independence nor overdependence is possible, only effective or ineffective dependency. Thus, in therapy, the therapist's goal is to help clients develop an effective dependency (or interdependency if you prefer) with significant persons in their lives.
3. *Attachment Offers an Essential Safe Haven:* Secure attachment provides a buffer against the stresses of life and measurably reduces their psychological and physiological effects.
4. *Attachment Offers a Secure Base:* Just like young children, adults also need a secure base to enable them to feel free to experience exploration, innovation, and openness, which are all correlated with positive psychological health and growth.
5. *Emotional Accessibility and Responsiveness Build Bonds:* Secure attachment is established by being emotionally accessible and responsive, and thus, this is the focus of the working phase of EFT.
6. *Fear and Uncertainty Activate Attachment Needs:* When threatened, a person experiences an unusually strong emotional need for comfort and connection; it is this need that fuels the destructive patterns too often observed in couples and families.
7. *The Process of Separation Distress Is Predictable:* If attachment needs are not met, the person experiences predictable responses of anger, clinging, depression, and despair. *This distress is experienced as a primal survival need, thus helping to explain—but NOT justify—the often extreme and cruel acts committed in the name of love.*
8. *There Are a Finite Number of Insecure Attachment Styles:* When a secure relationship no longer feels secure, a person will use one of these three typical patterns to defend themselves against the trauma of having a secure relationship threatened.
   - *Anxious and Hyperactivated:* When needs are not met, the person becomes anxious, relentlessly pursues connection, and may become clingy, aggressive, blaming, and/or critical.
   - *Avoidance:* When needs are not met, the person suppresses attachment needs and instead emotionally and physically withdraws, often focusing on unrelated tasks or other distractions.
   - *Combination Anxious and Avoidant:* In this style, the person pursues closeness and then avoids it once offered.
9. *Attachment Involves Working Models of Self and Other:* People use the quality of attachments to define themselves and others as lovable, worthy, and competent and develop internal models about what can be expected and how to engage in attachment relationships.
10. *Isolation and Loss Are Inherently Traumatizing:* Isolation and loss of connection are inherently traumatic experiences that can trigger a panicked survival responses. Solitary confinement is a universal form of torture.

Rumor Has It: The People and Their Stories

Susan Johnson With Les Greenberg, Sue Johnson began developing emotionally focused therapy in the 1980s. They developed this approach by refining their methods based on the outcomes of their research and their observations of what worked and what did not work. Sue Johnson has continued research and development of the model, particularly as it applies to couples and families, and also teaches internationally throughout the world. She has applied emotionally focused therapy to the treatment of a host of issues, including couples therapy with trauma survivors (Johnson, 2000), depression, and couples with chronically ill children (Johnson, 2004). She has also created, with her colleagues, a detailed workbook for people learning EFT (Johnson, Bradley, Furrow, Lee, Palmer, Tilley, & Woolley, 2005) as well as a self-help book for couples (Johnson, 2008).

Les Greenberg With Sue Johnson, Les Greenberg codeveloped EFT while serving as a professor at York University in Toronto and the director of the York University Psychotherapy Research Clinic. He teaches internationally and continues to research and refine a version of the model he calls emotion-focused therapy. He has focused primarily on its application to individuals, although he has recently written a book called *Emotion-Focused Couple Therapy: The Dynamics of Emotion, Love, and Power* (Greenberg & Goldman, 2008).

The Big Picture: Overview of Treatment

The EFT therapy process is clearly structured (this is where the manualized treatment part comes in) in *three stages with nine steps* that describe the progression of therapy (Johnson, 2004). But before you have pie-in-the-sky fantasies of a step-by-step recipe for success, let me bring you down to earth and say that these steps are not perfectly linear and most clients cycle back and forth. Alas, not even in evidence-based treatment does therapy go perfectly according to plan. Thus, the steps are mainly instructive for the therapist to help guide the process of therapy and track progress and setbacks as well as have a very clear treatment plan (note to self: therefore, these are some of the easiest treatment plans to write).

Stage 1: De-escalation of Negative Cycles

Step 1: Create an alliance and delineate conflict in the attachment struggle.

Step 2: Identify the negative interaction cycle.

Step 3: Access unacknowledged emotions and underlying interactional positions.

Step 4: Reframe the problem in terms of the negative cycle and attachment needs, with the cycle being the common enemy.

Stage 2: Change Interactional Patterns and Creating Engagement

Step 5: Promote identification of disowned attachment needs and aspects of self, integrating these into relational interactions.

Step 6: Promote acceptance of the partner's experience along with new interaction sequences.

Step 7: Encourage direct expression of needs and wants while strengthening emotional engagement and attachment bonds.

Note: Steps 5–7 are generally first done focusing on the withdrawing partner, and then the same steps are repeated focusing on the pursuing partner.

Stage 3: Consolidation and Integration

Step 8: Facilitate new solutions to old problems.

Step 9: Consolidate new positions and new cycles of attachment.

Just in case your eyes glazed over reading that list, let me break it down a bit. The first stage of therapy is about two things: (a) establishing a strong therapeutic relationship needed for the exploration of difficult emotions, and (b) developing a clear case conceptualization using attachment theory. In the final step of this phase, the therapist will help the couple both reframe (see Chapters 5 and 6) and externalize (see Chapter 13) their negative interaction cycle in such a way that the couple/family conflict can be seen as failed attempts to connect (rather than the believing that their partner has become a horrible, terrible person, which is where many start the conversation).

The second stage of therapy is where the therapy gets intense and where EFT therapists earn their keep. In this stage, the therapist starts with the most emotionally *withdrawn* partner to help this person identify, own, and directly express to the other partner his/her unmet attachment needs. This strategy is arguably her second most brilliant contribution—second of course to her theory of adult attachment described in The Juice above. This move is radical because the withdrawer is generally the partner with the fewest complaints, so most therapists end up focusing on the *pursuer*, the one who is pursuing connection, which often takes the form of complaining (hint for case conceptualization: the pursuer is generally the partner who makes the call to therapist unless the withdrawer was just threatened with divorce). However, once the couple is de-escalated (stage 1), Johnson starts with the withdrawer because once the withdrawer reconnects it becomes easier for the pursuer to vulnerably ask for needs to be met rather than criticize because he/she is getting the connection he/she has been pursuing in some form. So, in this phase the therapist first works with the withdrawer to identify and express unmet attachment needs, and then does the same with the pursuer. In both cases, the therapist helps the partner learn how to respond effectively to bids for connection. Interestingly, it is often more difficult to get the complaining pursuer to soften because there is a painful history of reaching out and being rejected by the withdrawer (Furrow, Ruderman, & Woolley, 2011).

The third phase is a more hopeful phase as couples learn how to work through old issues using their new safely connected relational patterns (generally with far better results). Since this is real life, not a fairytale, there are typically quite a few bumps along the way as couples learn to relate to each other differently. EFT therapists are not surprised when old patterns reemerge; the therapist helps the couple to identify what is going on and help them reconnect and create a sense of safety.

In addition to the stages and steps, Johnson (2004) identifies three primary therapeutic tasks:

Task 1: Creating and maintaining alliance

Task 2: Assessing and formulating emotion (most critically attachment emotions)

Task 3: Restructuring interactions

If you were conscious while reading any of the previous 10 chapters, the first therapeutic task probably should look familiar (see, you are almost an expert). But the second is unique to EFT in that the couple or family's interaction cycle is conceptualized in terms of attachment needs. Finally, the last task (which is where most of the action happens) involves helping couples learning to change—primarily through experiential means—how they relate with one another so that they can reliably create a strong sense of safety and love.

Making Connection: The Therapeutic Relationship
Empathetic Attunement

As the name suggests, *emotionally* focused therapy focuses on emotions, and thus the therapist's ability to deeply empathize and attune is essential—without it, the therapy will not work. In EFT, the therapist attunes to each partner's emotions with the intent to make contact with the emotional worlds of each (Johnson, 2004). Empathetic attunement requires listening to clients, connecting what they say with the therapist's personal experience, and then staying within the client's subjective perspective. Such attunement typically happens more at the nonverbal level, both in perceiving the client's emotional state and in reflecting it back through nonverbal communication (e.g., by softening the voice, nodding the head). The key here is for *the client* to feel deeply heard and understood—by whatever means the therapist does it—and to experience a sense of emotional safety in the therapy room. The therapist's empathic attunement with each partner serves as a role model for them to relate to one another.

Expression of Empathy: RISSSC

If you get a chance to watch Sue Johnson doing therapy, you will note that she talks a bit different than many others. At critical points, she will slow down significantly, repeat herself, and clearly highlight words or images. As you might imagine, this is done on purpose. Johnson (2004) summarizes her unique approach to empathy in the following techniques (RISSSC) to express understanding of the client's affective reality:

- *Repeat:* Repeat key words and phrases the client says.
- *Images:* Use images to capture emotion in a way that abstract words cannot.

- *Simple:* Use simple words and phrases.
- *Slow:* Maintain a slow pace that enables emotional experience to unfold.
- *Soft:* Use a soft voice to soothe and encourage deeper experiencing and risk taking.
- *Clients' Words:* Adopt clients' words and phrases in a validating way.

Examples of RISSC from the EFT training manual (Johnson et al., 2005) include:

- "Livid and disappointed, yes? So how did that go between you?" (p. 166)
- "Sounds like that was a pivotal time for you ... that you needed him." (p. 167)
- "All your life it's been like that. Never getting the message that you're precious and you count, wondering if there's something wrong with you." (p. 168)

Genuineness

Genuineness requires that the therapist be real and emotionally present with clients without being impulsive or constantly disclosing personal information (Johnson, 2004). Therapists are humble and able to admit mistakes and misunderstandings. As a result, clients experience the relationship between them and the therapist as an authentic human encounter.

Acceptance

Therapists maintain a nonjudgmental stance that is grounded in a positive view of human nature and acceptance of human struggles and a humble recognition of how hard it can be to create secure relationships with those we love (Johnson, 2004). Acceptance involves honoring and prizing clients *as they are* and acknowledging the fullness of their humanity. If you are nodding your head yes at this point, I should perhaps point out that this is far more challenging in practice than it sounds in theory (just remember, you heard it here first).

Self-Disclosure

Used infrequently, self-disclosure can build rapport or intensify validation of client responses (Johnson, 2004). However, the therapist keeps self-disclosure to a minimum to maintain a focus on the emotional process of the couple.

Continuous Monitoring of the Alliance

The therapist continually monitors the therapeutic alliance with each partner to ensure that a strong affective connection and sense of safety continue through all stages of therapy (Johnson et al., 2005). When expressing empathy in couples therapy, it is particularly easy for the alliance with one partner to weaken as the therapist expresses an understanding of the other partner, because for a moment the therapist may appear to be taking sides in an attempt to understand. Thus therapists need to constantly balance their focus between the two partners. Therapists monitor the alliance both verbally (directly asking about their experience of therapy) and nonverbally (identifying signs of defensiveness, checking out, etc.).

Joining the System

The therapist must build an alliance not only with each individual but also with the relationship as a system, accepting it as it is, just as each individual is accepted. Joining the system involves not only identifying relational patterns (e.g., nag/withdraw, pursue/distance, criticize/defend) but also reflecting these back to the couple so that they can better understand their relationship.

The Therapist's Role

According to Johnson (2004), the therapist is the following:

- A process consultant who helps the couple reprocess their emotional experiences
- A choreographer who helps the couple restructure their relationship dance
- A collaborator who follows and leads the therapeutic alliance

The therapist is *not* the following:

- A coach teaching communication skills
- A "wise creator of insight" into the past and/or future
- A strategist employing paradox or problem prescription

The Viewing: Case Conceptualization and Assessment
Intrapsychic and Interpersonal Issues

Much like other experiential family therapists (Chapters 8 and 9), the EFT therapist attends to both intrapsychic and interpersonal issues.

- *Intrapsychic:* How individuals process their experiences, particularly their key attachment-oriented emotional responses
- *Interpersonal:* How partners organize their interactions into patterns and cycles

Negative Interaction Cycle: Pursue/Withdraw

Similar to other systemic therapists (see Chapter 5 for a refresher), one of the therapist's first tasks is to identify the couple or family's negative interaction cycle, which is typically conceptualized in EFT as a pursue/withdraw pattern (Johnson et al., 2005). The *pursuer* is protesting the separation and distance they are experiencing in the relationship; this is indicative of an anxious attachment style. In contrast, the *withdrawer* creates distance to protect him or herself from the perceived lack of safety in the relationship, often in the form of criticism or rejection, and is typical of an avoidant attachment style. Of course, the more the pursuer tries to connect (often through nagging, criticizing, and demanding closeness), the more the withdrawer feels a need to distance in order to create safety. Pursuers typically express underlying emotions, such as feeling hurt, alone, and unwanted; whereas withdrawers typically report feeling rejected, inadequate, or judged.

Couples typically present with one of four basic pursue/withdraw patterns (Johnson et al., 2005):

- *Pursue/Withdraw:* The most common cycle, pursue/withdraw cycle involves a readily identifiable partner who pursues connection and one who withdraws. In most cases, the female is the pursuer, and the male is the withdrawer. In some cases, the distancing male pursues for sex but otherwise distances from connection; this is still seen as the typical pursue/withdraw pattern.
- *Withdraw/Withdraw:* Some couples present with what appears to be two withdrawers; however, in these cases there is typically a *burned-out pursuer,* a pursuer who has given up trying to connect because of repeated failures to connect (Furrow, Ruderman, & Woolley, 2011). Generally, the prognosis for these couples is less optimistic because typically more damage has been done to the relationship. These couples often describe feelings of numb distancing and refusals to engage.
- *Attack/Attack:* This pattern typically involves a withdrawer who sometimes turns, erupts into anger and fights when provoked by his/her partner's pursuit (often in the form of criticism). However, after the fight, the withdrawer typically reverts back to the withdrawn position.
- *Complex Cycles:* Trauma survivor couples often present with a complex multimove cycle that involves both high levels of anxiety and avoidance.

In session, the pursue-withdraw pattern is initially traced in the first phase of therapy by asking couples to describe the behaviors and secondary emotions that characterize a typical negative interaction cycle. The therapist slows down the process, inquiring about each partner's inner emotional experience in the cycle, and connecting each partner's description to the context of the other's experience. In this process, the therapist tries to move the couple to identifying and acknowledging more primary attachment emotions.

Primary and Secondary Emotions

How does an emotionally focused therapist know which emotions to focus on? Is any emotion a client expresses worthy of focus? The answers to these questions are multilayered. First, emotionally focused therapists distinguish between primary emotions and secondary emotions:

- *Primary Emotions:* Primary emotions are the initial reactions to a given situation; these typically represent attachment fears and needs (usually softer, more vulnerable emotions, such as feeling abandoned, alone, helpless, unloved, unwanted, inadequate, etc.)
- *Secondary Emotions:* Secondary emotions are emotions *about the primary emotions*, not the actual situation. For example, if the given situation is a critical remark, the primary emotion may be feeling inadequate, but the secondary emotion is likely to be anger about feeling inadequate. Secondary emotions often take the form of anger, frustration, or withdrawal; these emotions often allow the person to avoid the sense of vulnerability associated with primary emotions. They must be explored so that the primary emotion underneath can be identified.

In the early phases of therapy, the therapist focuses on the salient secondary emotions because these are typically what couples initially present and are the only emotions of which many are conscious. As therapy progresses, the therapist begins to raise each partner's awareness of the primary emotion underlying the secondary emotion, such as the hurt that is beneath the anger of having one's partner no longer expressing sexual interest. By the working stage of therapy, the focus is primarily on the primary emotions.

Attachment History

EFT therapists assess clients' attachment history, both with their primary caregivers and in adolescent and adult intimate relationships. Therapists do this by asking questions about the qualities of these relationships. To assess the quality of attachment bonds, EFT therapists use the acronym A.R.E.: Are you there for me? Accessible, Responsive, and Engaged (Johnson, 2008).

Questions for Assessing Attachment History

- Describe your parents' relationship? Were they close? Did they regularly express affection for one another? Did they fight and if so how? Were they generally able to resolve their conflicts in a way that restored harmony and safety in the family?
- When you were young, to whom did you turn to for comfort and nurturing? Was this source of comfort reliable?
- Describe your relationship with your parents and significant childhood caregivers? How did you know you were loved? Did you feel safe, physically and emotionally? How was conflict or disappointment handled? How was it resolved? Was there abuse or trauma?
- Describe significant love relationships in your adulthood. Did you feel safe, cared for, and nurtured? Were there significant betrayals or other trauma?
- Describe the early phases of your relationship when you felt more connected to your partner. How accessible were each of you? How did you show responsiveness? How did you engage one another? (Furrow, Ruderman, & Woolley, 2011)

Attachment Injury

In addition to an attachment history, the therapist also assesses for *attachment injuries* that may have occurred in the couple's relationship. An attachment injury is a specific type of betrayal, abandonment, or violation of trust in a couple's relationship (Furrow, Ruderman, & Woolley, 2011). Attachment injury occurs when one partner is in a moment of high need and vulnerability (e.g., pregnancy, loss, crisis, affairs, etc.) and the other partner fails to offer the needed support and nurturance. This injury fundamentally redefines the relationship for the injured party as an unsafe relationship. The person may still remain in the relationship but the quality of their participation is different in order to protect him/herself.

Therapists must identify and directly address attachment injuries for the EFT process to work. Thus, the therapist needs to assess for attachment injuries and repair them (using the steps 5–7). Identifying attachment injuries can sometimes be done directly, especially in the individual assessment sessions (see Initial Assessment Sessions): "Is there any event in your relationship that was so painful, possibly felt traumatic, that you feel has left you feeling fundamentally unsafe in your relationship?" But most often you will blindly stumble upon these because therapy gets stuck: you hit an impasse. And often, when you probe deeper, you learn about the attachment injury, which may have even been discussed earlier but in a way that glossed over its significance.

Initial Assessment Sessions

EFT therapists have a somewhat structured series of initial sessions to determine if the couple is appropriate for EFT (Furrow, Ruderman, & Woolley, 2011). Typically the assessment sessions are scheduled as follows: a joint couple session, followed by individual sessions with each partner, and then a return to regular conjoint sessions. In the conjoint session, the therapist assesses for:

- Perceptions of problems and strengths
- Negative and positive interaction cycles
- Relationship history and key events
- Brief attachment history of each
- Observed interactions in session
- Check for violence, abuse, or other contraindications
- Assess for prognostic indicators: degree of reactivity, strength of attachment, and openness (response to therapist)

In the individual sessions, the therapist assesses for each partner's:

- Commitment to the relationship
- Sense of physical and emotional safety
- Potential, current, and past affairs
- Trauma history
- Detailed attachment history

Contraindications to EFT

Not all couples or families are appropriate for EFT, and the initial case conceptualization process involves identifying whether or not this is an appropriate approach. Common contraindications for EFT (reasons for not using EFT) include (Johnson et al., 2005):

- *Different Agendas for the Relationship and Therapy:* Such as one wanting to marry and the other to remain single or one partner having an outside romantic relationship.
- *Separating Couples:* When it is clear one partner has emotionally left the relationship.
- *Abusive Relationships:* Couples in which there is physical, sexual, and/or emotional abuse are not appropriate for EFT. More subtle forms of abuse are often identified by one partner being afraid of the other.
- *Untreated Addiction:* Alcohol and substance abuse do not preclude EFT, but it must be determined if both parties can participate in therapy while sober. If one or both parties have substance abuse issues, an EFT therapist refers them out for substance abuse treatment while focusing the EFT process on how substances affect the interaction cycle.

Targeting Change: Goal Setting

The overarching goals of EFT are straightforward and focused, providing couple therapists with consistent direction during what is often a tumultuous adventure (Furrow, Ruderman, & Woolley, 2011):

- Creating secure attachment for both partners
- Developing new interaction patterns that nurture and support each partner
- Increasing direct expression of emotions, especially those related to attachment needs

The Doing: Interventions
Interventions by Stage of Therapy
EFT therapists have several interventions they use, with certain ones being more common during particular phases of therapy. They tend to be divided as follows:

Early Phase Interventions for De-escalating and Identifying the Cycle
- Validation
- Reflecting emotions
- Tracking the cycle (see Case Conceptualization)
- Evocative responding
- Empathic conjecture

Working-Phase Interventions for Restructuring Interactions
- Evocative responding
- Empathic conjecture
- Heightening
- Reframing
- Restructuring Interactions (enactments)

Closing-Phase Interventions
- Validation
- Evocative responding
- Reframing
- Restructuring interactions (enactments)

Validation
Rather than implying some kind of god-like approval, therapists use *validation* to communicate to clients that their emotional experiences are understandable and understood by the therapist, rather than their emotions are "correct" and "only to be expected" (such implications can weaken alliance with the other partner). Therapists use validation to convey that each partner is entitled to his or her experience and emotional responses, helping to articulate the underlying logic and emotions of each person's behaviors: "At first you feel very sad that he does not seem to want to spend time with you, but after a while you become angry and then try to force him to spend time with you."

Reflecting Primary and Secondary Emotions
The EFT therapist attends to poignant emotions and *reflects* back to the client a deep understanding and acceptance of these emotions (Johnson, 2004; Johnson et al., 2005). The goal of reflection is to help clients more fully experience their emotions, both primary (attachment-based) and secondary. Reflecting secondary emotions involves helping clients identify what they are feeling in the negative interaction cycle: "So when you shut down and pull away, it's when you see her disappointment and anger. It is too much to handle." As therapy progresses, reflections should highlight primary emotions related to unmet attachment needs: "What I hear you saying is not only that you feel like you need to hide when she gets angry but that you begin to fear she does not love or respect you anymore."

Tracking Interaction Patterns and Cycles
Using similar methods to other systemic therapists, therapists track patterns and cycles of interactions and then reflect on these patterns to help couples better understand the nature of their relationship (Johnson, 2004). For example, they look for common pursue/withdraw and blame/defend patterns and attend to the unique sequences of each couple's interactions. In addition, as therapy progresses, therapists track positive interaction cycles in a similar way, helping to reinforce the new interaction cycle.

Evocative Responding: Reflections and Questions
Whether phrased as reflections or questions, therapists use evocative responses to bypass superficial issues and identify unexpressed emotions and needs (Johnson, 2004). Because

evocative responses are based on conjecture, the therapist offers these *tentatively*, allowing the client to correct or rephrase: "I think that part of the reason you may be pulling away is that you want closeness and connection so badly; if you were to be rejected, you would be devastated."

Empathetic Conjecture and Interpretation

At times the EFT therapist will offer an *empathetic conjecture* or interpretation, typically addressing defensive strategies, attachment longings, and attachment fears (Johnson, 2004). This process is similar to evocative responding in that the therapist aims to deepen emotional experiencing, but empathic conjecture and interpretation more critically involves generating new meanings, not just touching on the emotion just below the surface. Often the therapist is making a link between the secondary and primary emotions: "So, on the surface you look angry and sure of yourself when you are arguing with him. But inside, something very different seems like it might be going on. Inside, it seems you feel more like a sad and lonely little girl who isn't too sure of herself ... sure if she is good enough or worthy of his love. Is that what is going on here?" In the later phases of therapy, conjecture is used to *seed attachment* by highlighting the desire for attachment that is blocked by fear or anger.

Heightening

Heightening involves using repetition, metaphors, images, and enactments to "heighten" key emotions and interactions that play a crucial role in maintaining the couple's negative cycle (Johnson, 2004). For example, the therapist can heighten by repeating a high-impact phrase (e.g., "feeling betrayed"), using nonverbal gestures like leaning forward or lowering the voice, using images and metaphors, or directing clients to enact responses. The therapist may also ask the client to repeat poignant moments: "Can you say that again—that you need her? Can you look at her and say that again?" Let's be honest: out-of-context these can seem phony and fake. But in the right context and at the right moment, these can be exactly what a person needs to explore complex and often painful feelings that arise in couples therapy.

Attending to Nonverbal Communication

Nonverbal communications often provide clues to primary emotions. EFT therapists closely monitor for nonverbal signs of unexpressed primary emotions. These may include:

- Changes in tone of voice
- Bodily reactions
- Looks and glances
- Giggles, jokes, and laughter
- Sighs
- Silence
- Deep breath
- Tight lips (Furrow, Ruderman, & Woolley, 2011)

Reframing in Context of Cycle and Attachment Needs

EFT therapists regularly use two types of reframing of each partner's behavior: in the context of the negative cycle and in the context of attachment needs. Consistent with Johnson's systemic perspective, problems are reframed in the broader context of the relationship to emphasize to help the couple see the cycle as the common enemy and to help each see how they contribute to the negative cycle (Johnson, 2004). For example, one partner's anger and the other's silence can be reframed to see how each response serves as a protective mechanism but that each contributes to maintaining the negative cycle. To increase solidarity, the negative interaction cycle is always framed as the couple's common enemy.

Similarly, reframing is used to help move clients from secondary to primary emotions. For example, "your stone-cold silence and distance is all that is showing on the outside. But inside, you are panicking; you're running scared. Because it seems that you have failed her again, and you just can't bear the thought. You can't bear the thought of having failed again."

Enactments, Restructuring, and Choreography

As in structural therapy, couples are asked to enact their present positions so that the therapist can help them more fully experience their underlying emotions (Johnson, 2004). The therapist then redirects or "choreographs" the couple's interaction to be less constricting and more accepting, perhaps by asking a partner to share a new affective insight directly with the other or by physically repositioning the couple to increase emotional intensity. In the later stages of therapy, the therapist usually uses enactments to choreograph partners' requests of each other and to create new positive responses that will lead to new bonding experiences that redefine the relationship as safe and secure.

Common Prompts for Choreographing New Interactions
- "What just happened here? Let's go back and see what was going inside for each of you?"
- "Can you turn to her (him) and tell him that right now?"
- "Can you turn to him (her) and ask for what you are saying you need right now?"

Softening Emotions

A hallmark EFT technique, softening of emotions is used to create emotional bonding, change interactional positions, and redefine the relationship as safe and connected. A softening occurs when a previously blaming, critical partner asks, from a position of emotional vulnerability, a newly accessible partner to meet his/her attachment needs and longings (Bradley & Furrow, 2004; Johnson, 2004). The more critical partner softens his/her stance and words, allowing the more vulnerable or anxious partner to reduce emotional reactivity and defensiveness. Therapists can facilitate softening by encouraging partners to express their underlying attachment-based fears, including hurt and disappointment, when discussing conflict areas. Typically, this is done by underscoring the attachment issues, such as fear of rejection, in a particular area of tension. For example, if the wife is complaining that the husband does not spend enough time with her and the children, emotionally focused therapists help her to articulate the fear of abandonment and/or feelings of rejection that underlie the complaint, thus revealing the wife's vulnerability and fears rather than her anger and frustration. When she expresses these softer emotions of vulnerability and asks her husband directly for comfort and connection, generally a more productive and healing conversation occurs that creates new bonding events. Bradley and Furrow (2004) have developed an extensive mini-theory about blamer softening to help guide therapists step by step through this challenging process.

Turning the New Emotional Experience into a New Response

After helping one partner explore an emotional experience, the therapist uses this experience to allow the other partner to respond in new ways, creating a positive interaction cycle in which each partner is better able to understand himself/herself and the other: "What is happening for you when you hear him say that his avoidance is not motivated by disinterest but by fear of rejection?" (Johnson, 2004).

Putting It All Together

Case Conceptualization Template
- *Negative Interaction Cycle:* Describe the cycle and who does what using one of the four types below:
  - Pursue/withdraw
  - Withdraw/withdraw
  - Attack/attack
  - Complex cycles

- *Primary and Secondary Emotions:* Describe each person's primary and secondary emotions related to the negative interaction cycle.
- *Attachment History:* Describe key events in each person's attachment history, including current attachment patterns.
- *Attachment Injuries:* Describe any known attachment injuries in the current relationship.
- *Contraindications to EFT:*
  - Different agendas for the relationship and therapy
  - Separating couples
  - Abusive relationships
  - Untreated addiction

For couples experiencing relational distress and/or conflict, EFT therapists might use the following treatment plan to help them conceptualize and guide their treatment.

Treatment Plan Template for Couple

Initial Phase of Treatment: EFT for Couples

Initial Phase Counseling Tasks

1. Create a working alliance with both partners. *Diversity Note: Adjust expressions of empathy to respect cultured, gendered, and other styles of emotional expression.* Relationship building approach/intervention:
   a. Use *empathic attunement, RISSSC, and genuineness* to develop a safe emotional context for the therapy process.

2. Assess individual, systemic, and broader cultural dynamics. *Diversity Note: Adjust assessment based on cultural, socioeconomic, sexual orientation, gender, and other relevant norms.*
   Assessment strategies:
   a. Identify the *negative interaction cycle*, including *pursuer/distance roles*.
   b. Identify *secondary and primary (attachment) emotions* that characterize cycle.
   c. Assess *attachment history*, including *attachment injuries*, as well as history of trauma.

3. Define and obtain client agreement on treatment goals. *Diversity Note: Modify goals to correspond with values from the client's cultural, religious, and other value systems.*
   a. *Reframe problem* in terms of *negative interaction cycle*, with the cycle being the common enemy.

4. Identify needed referrals, crisis issues, collateral contacts, and other client needs.
   a. Assess for *appropriateness of EFT for couple*, ruling out substance abuse, trauma, violence, conflicting agendas, or other contraindications.

Initial Phase Client Goals

1. Increase couple's awareness of *negative interaction cycle* and the *primary emotions* that fuel it to reduce conflict and hopelessness.
   a. *Validation, reflecting emotions, evocative responding, and empathic conjecture* to identify secondary and primary emotions.
   b. *Tracking the negative interaction cycle,* first with secondary emotions and later identifying primary emotions.
   c. *Reframing* in the context of the negative interaction cycle and attachment needs.

(continued)

Working Phase of Treatment: EFT for Couples

Working-Phase Counseling Task

1. Monitor quality of the working alliance with both partners. *Diversity Note: Attend to family's response to interventions that indicate therapist's approach not connecting with family's culturally informed meaning systems.*
   a. *Assessment intervention*: Attending to *nonverbal communication* as well as client's sense of *emotional safety* in the session.

2. Monitor client progress. *Diversity Note: Attend to cultural, gender, social class, and other diversity elements when assessing progress.*
   a. *Assessment Intervention*: Track couple's progress through the *EFT stages* in weekly progress notes.

Working-Phase Client Goals

1. *Increase engagement* and *emotional expression* of withdrawn partner to reduce conflict/avoidance.
   a. Use *empathy, validation, and conjecture* to facilitate identification and expression of *attachment needs*.
   b. *Enactments* to allow for direct communication of needs, the *acceptance* by partner, and *new interaction sequences*.

2. *Decrease criticism* from pursuing partner and increase pursuer's expression of *attachment emotions* to reduce conflict.
   a. *Heighten* pursuer's primary emotions to facilitate *softening* of blaming position.
   b. Use *enactments* to promote *acceptance* by partner and to facilitate *new interaction sequences*.

3. Increase the ability of both partners to respond to the other in ways that *create a sense of relational safety and bonding* even in moments of tension to reduce conflict, depressed mood, and/or anxiety.
   a. *Tracking the interaction cycle* and *empathetic conjecture* to help each partner see how his/her response affects his/her partner.
   b. *Enactments* that help partners to more directly express primary emotional needs as well as respond in supportive ways when the other reaches out.

Closing Phase of Treatment: EFT for Couples

Closing-Phase Counseling Task

1. Develop aftercare plan and maintain gains. *Diversity Note: Access resources in the communities of which they are a part to support them after ending therapy.*
   a. *Track positive as well as negative interaction cycles* to help couple prepare for potential setbacks.

Closing-Phase Client Goals

1. Increase couple's ability to *respond effectively* to new stressors to reduce conflict and hopelessness.
   a. *Track positive interaction cycles* to reinforce positive changes.
   b. *Reframe* both positive and negative interaction cycles in terms of *attachment needs*.

2. Increase couple's ability to consistently respond to one another in ways that *solidify a secure bond* to reduce conflict, depression, and anxiety.
   a. Enactments to facilitate *direct expression of emotional needs*.
   b. Facilitate turning the *new emotional experience* into a new response.

EFT can also be used to help families dealing with conflict. The treatment plan follows.

Treatment Plan Template for Family

Initial Phase of Treatment: EFT for Families

Initial Phase Counseling Tasks

1. Create a working alliance with all family members. *Diversity Note: Adjust expressions of empathy to respect cultured, gendered, age, and other styles of emotional expression; if young children are involved, use play and art to connect.*
   Relationship building approach/intervention:
   a. Use *empathic attunement, RISSSC, and genuineness* to develop a safe emotional context for the therapy process.

2. Assess individual, systemic, and broader cultural dynamics. *Diversity Note: Adjust assessment based on cultural, socioeconomic, sexual orientation, gender, and other relevant norms.*
   Assessment strategies:
   a. Identify the *negative interaction cycle*, including alliances, boundaries, dominance, and conflict resolution strategies.
   b. Identify *secondary and primary (attachment) emotions* for each member.
   c. Assess *attachment history* for all members of family, including *attachment injuries*, as well as history of *trauma*.

Initial Phase Client Goals

1. Increase awareness of *negative interaction cycle* and the *primary emotions* that fuel it to reduce conflict and hopelessness.
   a. *Validation, reflecting emotions, evocative responding, and empathic conjecture* to identify secondary and primary emotions.
   b. *Tracking the negative interaction cycle*, first with secondary emotions and later identifying primary emotions.
   c. *Reframing* in the context of the negative interaction cycle and attachment needs.

Working Phase of Treatment: EFT for Families

Working-Phase Counseling Task

1. Monitor quality of the working alliance with family members. *Diversity Note: Attend to family responses to interventions that indicate therapist using expressions of emotion that are not consistent with family's cultural backgrounds.*
   a. Assessment Intervention: Attending to *nonverbal communication* as well as family members' sense of *emotional safety* in the session.

Working-Phase Client Goals

1. Increase *engagement* and *emotional expression* of withdrawn family member(s) to reduce conflict/avoidance (may be done with family subsystems).
   a. Use *empathy, validation, and conjecture* to facilitate identification and expression of *attachment needs*.
   b. *Enactments* to allow for direct communication of needs, the *acceptance* by partner, and *new interaction sequences.*

2. *Decrease criticism* from pursuing family members(s) and increase pursuer's expression of *attachment emotions* to reduce conflict (may be done with family subsystems).
   a. *Heighten* pursuer's primary emotions to facilitate *softening* of blaming position.
   b. Use *enactments* to promote *acceptance* by partner and to facilitate *new interaction sequences.*

(continued)

3. Increase the ability of family members to respond to the other in ways that *create a sense of relational safety and bonding* even in moments of conflict to reduce conflict, depressed mood, and/or anxiety.
   a. *Tracking the interaction cycle* and *empathetic conjecture* to help each family member see how his/her response affects others.
   b. *Enactments* that help family members more directly express primary emotional needs as well as respond in supportive ways when others reach out.

Closing Phase of Treatment: EFT for Families

Closing-Phase Counseling Task

1. Develop aftercare plan and maintain gains. *Diversity Note: Access resources in the communities of which they are a part to support them after ending therapy.*
   a. *Track positive as well as negative interaction cycles* to help family prepare for potential setbacks.

Closing-Phase Client Goals

1. Increase family's ability to *respond effectively* to new stressors to reduce conflict and hopelessness.
   a. *Track positive interaction cycles* to reinforce positive changes.
   b. *Reframe* both positive and negative interaction cycles in terms of *attachment needs*.

2. Increase family's ability to consistently respond to one another in ways that *solidify a secure bond* to reduce conflict, depression, and anxiety.
   a. Enactments to facilitate *direct expression of emotional needs*.
   b. Facilitate turning the *new emotional experience* into a new response.

Tapestry Weaving: Diversity Considerations

Ethnic, Racial, and Cultural Diversity

EFT has been studied on a range of ethnically diverse couples (Liu & Wittenborn, 2011) and has been specifically adapted for work with Latino clients (Parra-Cardona, Córdova, Holtrop, Escobar-Chew, & Horsford, 2009). Additionally, attachment has been studied cross-culturally since 1967, and the general consensus is that attachment needs are normative and universal across theory; thus, this theory has elements that should be applicable with diverse clients. Liu and Wittenborn (2011) highlight that EFT therapists need to adapt and tailor their interventions to work effectively with diverse clients. Many of the signature nonverbal interventions, such as heightening emotion by leaning forward, may be interpreted as inappropriate or invasive depending on the client's cultural background. Furthermore, cultural norms generally have highly specific rules for emotional expression, which is a key focus of EFT, and thus EFT therapists need to understand these norms to be effective. When working with diverse clients, EFT therapists need to identify the specific cultural meanings and functions associated with the expression of emotion and attachment behaviors as well as recognize the socially constructed meaning of emotion. Finally, EFT therapists need to be mindful of these cultural meanings when they choose the words and metaphors to reflect feeling to ensure that they clearly convey their intended meaning with diverse clients.

Sexual Identity Diversity

Working with gay, lesbian, bisexual, or transsexual (GLBT) couples using EFT requires attention to their unique stresses and circumstances that often make forming a secure attachment more difficult for these couples (Zuccarini & Karos, 2011). Zuccarini and Karos (2011) describe how EFT can be adapted to be sensitive to the needs of GLBT couples. Although the general process remains the same, there are many predictable issues to which therapists must attend. In the first stage, therapists need to consider the impact on sexual identity-related stress on the formation of the couple and their generally complex interaction cycle: "negative emotional experiences and attachment related to the GL [gay-lesbian] identity block the acceptance of sexual orientation and seriously undermine any kind of safe emotional engagement in GL relationships" (p. 320). In addition, the coming out process as well as the effects of living in a heterosexist society frequently involve significant trauma and betrayals in significant relationships; the effect of these traumatic betrayals frequently affect a person's ability to develop secure relationships. Furthermore, when assessing the couple's negative interaction cycle, the therapist needs to consider the stresses of living in a heterosexist context and how rejection in the relationship is compounded by ongoing rejection in society.

Stage 2 involves the standard process of withdrawer re-engagement and pursuer softening, but with GLBT couples therapists must explore the multiple and complex traumas they have experienced due to their sexual orientation that make it particularly difficult to risk being vulnerable enough to form a secure attachment bond. Exploring these past traumas can help both partners feel compassion for one another and safe enough to trust the other. In stage 3, therapists should help each partner to address lingering identity issues and address the chronic stress related to being a sexual minority.

Evidence Base

Researched for over 25 years, EFT is currently one of two empirically validated couples therapy (the other being integrative behavioral couples therapy; Lebow, Chambers, Christensen, & Johnson, 2012). EFT has an overall 70%–73% recovery rate in 10 to 12 sessions, with 90% of all couples showing significant improvement (Johnson, Hunsley, Greenberg, & Schindler, 1999). However, often a course of treatment of more than 12 sessions is needed to alleviate couple distress (Johnson & Wittenborn, 2012). Follow-up studies show that the results tend to be stable, even with difficult to treat couples, such as trauma survivors and parents of chronically ill children (Clotheir, Manion, Gordon-Walker, & Johnson, 2002). In addition, three studies have examined EFT with sexual abuse survivors and their partners, with clinically significant improvement in martial adjustment and trauma symptoms in at least half of the participants (Johnson & Wittenborn, 2012).

Researchers have also studied the active ingredients of the change process in EFT: (a) the depth of emotion in key sessions in stage 2, (b) facilitating new interactions in which couples are able to express their attachment needs and be responsive to the other, and (c) the quality of the therapeutic relationship, especially the perceived relevance of the focus and tasks of therapy by clients (Johnson & Wittenborn, 2012; Lebow et al., 2012). In addition, researchers have identified factors that moderate the impact of EFT. Surprisingly, couple's initial level of distress does not seem to matter as much as their level of engagement in session or the male partner's level of trust in heterosexual couples. One study also considered treating attachment injuries within the relationship using EFT, with nearly two-thirds of couples resolving such injuries with a brief intervention; these results were stable at a three-year follow-up (Johnson & Wittenborn, 2012). To meet the highest level of validation, EFT has also been studied by independent researchers, those not affiliated with the originator of the theory; two such studies have been conducted and have shown that even novice therapists with limited supervision can learn to conduct successful EFT sessions (Johnson & Wittenborn, 2012).

Functional Family Therapy

In a Nutshell: The Least You Need to Know

An empirically validated family therapy treatment for working with conduct disorder and delinquency, functional family therapy (FFT) has been studied for over 40 years (Alexander & Parsons, 1982; Alexander & Sexton, 2002; Sexton & Alexander, 2000;

Sexton, 2011). The approach integrates cognitive theory, systems theory, and learning theory using a combination of strategic, cognitive, and behavioral interventions. In FFT, all behavior is viewed as *adaptive* to serve a particular *function* in the system. Behaviors are viewed as attempts to achieve two basic functions:

- *Relational Connection:* The relative balance of closeness and independence
- *Relational Hierarchy:* Defining who has influence and control

The therapist's primary task is to identify the *function* of the problem behaviors—how the behaviors maintain connection and defines hierarchy—and then find more effective behaviors that achieve the same basic function (i.e., sense of connection, influence, independence, etc.). Interventions aim to achieve the desired goal or function without the negative consequences that brought the family to therapy. The approach offers therapists a coherent approach for assessing and effectively intervening with families who have children with significant behavioral problems.

The Juice: Significant Contributions to the Field

If you remember one thing from this chapter, it should be this:

Multisystemic and Family Focus

Similar to other evidence-based treatments for troubled youth, FFT uses a *multisystemic focus,* meaning that the therapy process addresses individual, family, peer, and community system dynamics (Alexander & Parsons, 1982; Sexton, 2011). Furthermore, the evidence is quite clear that the preferred unit of treatment to affect these multiple systems is the *family,* not the individual youth. Research has identified a strong family bond to be one of the most critical protective factors that keep youth out of trouble. Furthermore, current research indicates that group treatment of troubled youth actually *augments antisocial behavior* rather than reduce it (Lebow, 2006). If you check with your local juvenile justice courts, you may find that teen anger management or adolescent substance abuse classes are mandated for many youth in trouble with the law: it is likely these groups are making the problem worse rather than better. Findings like this underscore the importance of research to help therapists identify when commonsense solutions don't work.

In FFT, the therapist works with the youth's school, probation officer, peers, community, and extended family as well as immediate family to effect change. This may involve direct interventions with persons in these systems or more indirectly affecting the system by helping the youth and family interact with them differently. In the case study at the end of this chapter, the therapist helps a young teen mother and her mother to identify resources in their community to help her in her sobriety so that she may keep her 2-year-old daughter. The therapist helps the family identify resources at the adult school, extended family, and church to help and encourages the daughter to distance herself from her substance-abusing friends for the sake of her daughter.

Rumor Has It: People and Places

James Alexander James Alexander developed FFT with Bruce Parsons in the 1960s while working at the University of Utah (Sexton, 2011). Alexander had a strong background in systemic theory, and he used it to build his theory, which was originally developed to help youth in the juvenile justice system.

Bruce Parsons Bruce Parsons worked with James Alexander starting as a graduate student to develop FFT in the 1960s.

Thomas Sexton Thomas Sexton (2011), one of the recent model developers of FFT, authored the most recent Blueprint manual, developed the newest FFT model description, and wrote the first FFT book in 35 years that describes how the approach can be used in standard outpatient clinical settings.

The Big Picture: Overview of Treatment
Early Phase: Engagement and Motivation

In the first phase, the therapist aims (a) to develop a connection with all members of the family and (b) to assess the *function* of the problem behaviors (Alexander & Sexton, 2002; Sexton, 2011). During this phase the therapist works to reduce anger, blame, and hopelessness. Therapists create a context conducive to change by using cognitive techniques to reduce parents' tendencies to blame the problem on negative child characteristics (e.g., laziness or irresponsibility) and to replace these characterizations with descriptions that do not impute negative motives (e.g., experimenting with freedom, exploring identity).

Middle Phase: Behavioral Change

In the middle phase, the therapist aims to modify cognitive sets, attitudes, expectations, labels, and beliefs so that family members see how their actions are interrelated (Alexander & Sexton, 2002). Therapists specifically target parenting skills, negativity, and blaming and intervene by making comments about the impact of a behavior on others; describing the interrelation of feelings, thoughts, and behavior; offering interpretations; stopping negative interactions; relabeling behaviors in nonblaming terms; discussing the implications of symptom removal; changing the context of a symptom; and shifting the focus from one person or problem to another.

Once the therapist has changed the family's cognitive set, the therapist focuses on building interpersonal and practical skills, such as *parent training, problem solving, conflict resolution,* and *communication skills.* Parent training is emphasized with younger children and follows traditional behavioral parenting interventions using operant conditioning principles. Problem solving and conflict resolution are favored when helping parents and older adolescents address their conflict. Communication skills training is based on traditional behavioral techniques that encourage brevity, directness, and active listening.

Late Phase: Generalization

The focus during this phase is to generalize change to the larger social systems in which the family interacts (Alexander & Sexton, 2002). The therapist now works within the family to help reduce relapse and outside the family to develop positive relations with community systems, such as mental health and juvenile justice authorities, and to develop a strong social network.

Making Connection: The Therapeutic Relationship
Alliance: Between Family Members and with Therapist

In FFT, the therapist develops an *alliance* with the family and also helps to develop a sense of alliance between family members (Alexander & Sexton, 2002; Sexton, 2011). This alliance is considered a personal connection that includes feeling understood and having trust in the other. What makes the alliance specifically therapeutic is the additional agreement on the goals and tasks of the therapy process. In FFT, a typically family approach, alliance means all members of the family feel safe, heard, and are in agreement about the direction of therapy. The family feels that they are "on the same page" with each other as well as the therapist.

Motivation and Engagement

If the therapist is successful in creating a sense of alliance within the family and between the therapist and family, this typically results in a subtle yet essential element for therapeutic success: *motivation* (Alexander & Sexton, 2002; Sexton, 2011). Otherwise stated, the hoped for outcome of a strong therapeutic alliance is client motivation to take the action necessary for change. Although some family members may have some motivation for change upon entering therapy, often the unspoken—or sometimes said aloud—hope is to have the therapist do something to fix another person in the family. Thus, through a strong alliance, the therapist helps inspire all family members to see how they can be a part of the solution and to be willing to do so.

Engagement involves having all members of the family actively participating in session. Therapists facilitate engagement by using humor, demonstrating respect, sincerely attempting to understand, and bringing therapeutic presence to the room. Moreover, therapists encourage family engagement by bringing a nonblaming and strength-based perspective to the conversation, one that allows each person to feel valued and respected.

Mandated and Reluctant Clients

Because FFT targets delinquent youth, many of the families that come are mandated by an external third party, such as a court or school, or contain at least one member reluctant to be in therapy (sometimes the parent, sometimes the child, sometimes both; Alexander & Sexton, 2002). In addition, most of them have had many painful experiences as well as difficulties and unfair treatment in school, justice, and other systems. Thus, FFT therapists *expect* their clients to come feeling wary, hopeless, blaming, resistant, negative, or otherwise not in a good place (Sexton, 2011).

The FFT therapist's systemic perspective is often experienced as a refreshing reprieve from the family's experiences with other professionals and systems. This systemic perspective refrains from blaming any member of the family and instead encourages them to see the bigger picture and how all members—as well as external systems—contribute to and sustain the problem. This new perspective inspires hope that things can be different and that the family can take action to make meaningful change.

Spirit of Respect and Collaboration

A subtle but particularly important element of FFT is the attitude that the therapist brings to working with youth and families that typically have not been treated well by the system (Sexton, 2011). Often, long before seeing a therapist, delinquent youth have been labeled as failures and outcasts, and their parents are often seen in a similar light. Thus, therapists need to be particularly aware of engaging these families from a place of respect, valuing their experience and creating space for them to share their side of the story. This requires patience and openness to learning from the client as well as a willingness to engage in a sincere collaborative partnership in the therapeutic process.

Credible Helper

FFT therapists are aware of the importance of having credibility in the eyes of the family. Credibility is not established by promises of future gains but rather by what the therapist says and does in the therapy room, starting with the first session. Therapists need to be able to demonstrate that they understand the family's situation and have effective ways of intervening and assisting the family. This relates closely to the common factor of establishing hope.

The Viewing: Case Conceptualization and Assessment
Relational Functions: The Glue

Like other systemic therapists, when assessing families, FFT identify the *relational functions* of the problem behavior (Sexton, 2011; Sexton & Alexander, 2000). In general, FFT therapists focus on two essential relational functions that behaviors can have: (a) relational connection, and (b) relational hierarchy.

Relational Connection Relationship patterns help families balance a sense of interdependence (connectedness) and independence (autonomy). Generally, there are three ways families can balance these, with none being necessarily better than the other, each family's preference highly influenced by cultural norms:

- *High Independence:* Families that value high independence support autonomy and independence but may also risk distance and disengagement.
- *High Interdependence:* Families that value high interdependence may enjoy closeness and connection but also risk enmeshment and dependency.
- *Midpointing:* These families strive for a balance between these two.

Relational Hierarchy Relational hierarchy describes relational control and influence in the relationship. There are three general patterns that families can fall into:

- *Parent-Up/Adolescent-Down:* The parent-child relationship can have a traditional hierarchy in which the parent has more power and the child less; of course, this can range from a small power difference to an extreme power difference.
- *Adolescent-Up/Parent-Down:* In some families the power hierarchy is reversed and the adolescent has more influence over outcomes than the parent; this is rarely an appropriate arrangement.
- *Symmetrical:* Some families have a strong democratic structure, in which both have similar levels of power.

Changing the Expression, Not the Function So, if you have been reading along thinking that the therapist's job is to help families get into certain of the above categories and out of others, you are in for a surprise. FFT therapists are *not* trying to help all families have the same style or to achieve some "optimal" form of functioning: that would be disrespectful to cultural and individual family differences and needs. Instead, the goal is to help families find better *expressions* of the same function. For example, if parents are using verbally and physically abusive methods to maintain hierarchy, the functional family therapist will help these parents learn new ways to maintain hierarchy that are not abusive. Similarly, functional family therapists would help parents who overfunction for their children to find ways to express their connection and affection that are contingent on the child's appropriate behavior. In the case study at the end of the chapter, the therapist helps the single-parent Mexican American family find more functional expressions of their desire for high interdependence and also helps the family address the adolescent-up/parent-down hierarchical arrangement by encouraging the 19-year-old daughter to take more responsibility for herself and her daughter.

Relational Function Assessment

So, the question is this: what function does the symptom serve?

- To create independence or interdependence?
- To establish hierarchy or distribute power?

The next question is this:

- How can the family achieve a similar function with more effective relational interactions?

Answer these questions and you have a plan for success.

Risk and Protective Factors

FFT therapists are quick to identify the various known risk and protective factors for troubled youth. Some of these include (Sexton & Alexander, 2000; Sexton, 2011):

Individual Youth and Parent Risk Factors
- History of violence or victimization
- History of early aggressive behavior and general poor behavior control
- Substance, alcohol, and/or tobacco use/abuse
- Diagnosis of an emotional or psychological concern, including ADHD, or other deficits in social, cognitive, or information processing
- Low IQ
- Antisocial beliefs or attitudes

Family Risk Factors
- Lack of mutual attachment and nurturing by parents
- Ineffective parenting
- Chaotic home environment
- Lack of a significant relationship with a caring adult
- Caregiver who abuses drugs, commits crimes, or is diagnosed with a mental disorder

Peer/School Risk Factors
- Associates with other troubled youth, including gang involvement
- Frequent social rejection by peers
- Lack of involvement in conventional activities
- Poor academic performance; little commitment to school

Community Risk Factors
- Diminished economic opportunities; high concentration of poor residents
- High level of transience and low levels of community participation
- High levels of family disruption

Protective Factors
- Strong bond between children and family
- Parental involvement in child's life
- Supportive parenting that meets financial, emotional, cognitive, and social needs of child
- Clear limits and consistent enforcement of discipline

Multisystemic Assessment

In functional family therapy, troubled youth are never assessed apart from the multiple systems they inhabit; thus, it is an *ecosystemic* approach (Alexander & Parsons, 1982; Sexton, 2011; Sexton & Alexander, 2000). FFT therapists view people as made up of internal systems (physiological, cognitive, emotional, behavioral, etc.) that are in constant interaction with multiple other external systems, such the family, neighborhood, school, peers, employment, human service agencies, cultural groups, region, etc. A client's behavior is assessed in terms of the *function* it plays in each of these systems. For example, a youth's or parent's poor choices are not seen as isolated, individual problems but rather as having particular meaning and effects in the multiple systems of which they are a part. Similarly, the interconnection of these systems is also the source of resilience and support, both potential and actualized. Therapists attend to both the untapped resources, potential resources, as well as negative influences in the multiple systems and uses these to inform the direction of therapy. Identifying the effects of multiple systems helps therapists know where and with whom to intervene.

Community and Culture

When assessing families, FFT therapists carefully attend to the family's cultural and community contexts (Sexton & Alexander, 2000). Informed by ethnic and religious norms, "Cultural expectations contribute to patterns of interaction within the family, the ways it expresses emotion and organizes around roles, how the roles generally look and feel, and parenting style" (Sexton, 2011, pp. 2–13). Community contexts refer to the family's local community, which is influenced by the cultures that compose them but are a unique expression of them, often combined with regional and other social influences.

When working with delinquent youth, culture and community are of particular importance. Although ethnic minorities comprise approximately one-third of the youth population, they account for two-thirds of the population in juvenile detention and correction centers (Sexton, 2011). Furthermore, youth of color are arrested more often, spend more time in detention, and tend to be given longer sentences. Thus, minority and white delinquent youth come to therapy with very different contexts, and the therapy process needs to be responsive to these different experiences.

> ### Cultural and Community Questions to Consider
>
> - *Cultural Background:* What is the family's cultural background(s)? In which ways do they identify with this background? In which ways do they not?
> - *Culture and Family Norms:* What are the norms for family structure, hierarchy, role, and emotional expression in the family's culture? In which ways does the family embrace these? In which ways do they not? How does this compare with norms from their local and regional community?
> - *Local Community:* How does the family fit within their local community? Are they connected to meaningful and supportive groups within the community? How are their problems viewed within this community? Are there support persons that can be engaged?
> - *Socioeconomic Context:* How does the family's socioeconomic status affect their role in the community and contribute to the problems they experience?
> - *Adapting Therapy:* How does the therapy process need to be adapted to be respectful of the family's cultural and community contexts, values, and norms? What type of therapeutic relationship, goals, assessments, and interventions would be most useful?

Strengths and Resiliency

FFT therapists strive to reach a healthy balance between seeing both client problems (glass half empty) and strengths (glass half full; Sexton, 2011). When working with troubled youth, this balance can be especially challenging because their problems are often in the realm of criminal and harmful to others, not just themselves. Thus, it is easy to fall into the trap of either labeling the teen as "bad" or "antisocial" or naively viewing it all as a "big mistake." Instead, the therapist is challenged to find a much more uncomfortable position of acknowledging both the bad and the good, even if the good is difficult to see at first.

Commonly Assessed Strengths in Troubled Youth and Their Parents
- A bond of love between one or more family members
- Extended family members or community members who care and are reliable
- At least one pro-social friend
- A meaningful hobby, interest, or ability
- Passing grades
- Holding a job
- A history of doing well socially, in school, or with the family

Targeting Change: Goal Setting

Initial Phase: Engagement
Goals:

1. Reduce within family risk factors.
2. Reduce blame and negativity in the family.
3. Increase family alliance and family-focused view of problem.

Working Phase: Behavior Change
1. Increase behavioral competencies (e.g., parenting, communication, problem solving) that fit for the family.
2. Match these competences to family's relational function.

Closing Phase: Generalization
Goals:

1. Increase within context protective factors.
2. Generalize.
3. Support and maintain gains.

The Doing: Interventions
Developing a Family-Focused Problem Description
In the initial sessions, FFT therapists help families move from a blame-focused definition of the problem to developing a family-focused definition. This family-focused definition helps build a sense of understanding, alliance, and motivation. The therapist begins by asking each family member to describe the problem, its causes, and how it affects him/her (Sexton, 2011). Based on this, the therapist can identify blaming, problem attributions, and emotions as well as develop a sense of relational patterns and family structure.

Next, the therapist uses reframing to help family members see how their seemingly individual behaviors are part of a larger family interaction pattern: for example, it may be that a teen stays out late to avoid getting pulled into his parents' arguments.

Identifying the Problem Sequence
Similar to strategic and behavioral family therapies (Chapters 6 and 11), FFT therapists identify the relational sequence around the presenting problem. This boils down to:

- What behaviors (from all family members) came before the problem?
- What behaviors (from all family members) come after the problem?

Therapists can often assess this indirectly by observing interactions in session as well as from the family's description of the presenting problem. Additionally, therapists sometimes directly ask about the problem sequence to more carefully assess it.

Relational Reframing
In the initial phases of FFT, therapists help families to change the interpretations and meanings about the problem, a process known as *cognitive restructuring*. Therapists help families move from blaming another member of the family for the problems to help them see how the problem is relational: everyone plays a part in maintaining the problem behaviors (Sexton, 2011). Thus, problems are *reframed relationally* to reduce malicious and negative attributions and increase understanding and hope.

FFT therapists use a three-phase process for relational reframing (Alexander & Sexton, 2002; Sexton, 2011):

1. *Acknowledgment*: Acknowledge each person's initial position, views, understandings, and feelings. Acknowledgement statements—such as "You got really angry"—show the therapist understands and supports the importance but not necessarily the content of the client's statement. *The therapist avoids generalizations and normalizing comments, and instead focuses on the client's personal experience.*
2. *Reattribution:* In the next phase, the therapist offers a reattribution for the problem behavior, which generally takes one of three forms:
    - An alternative explanation for the problem behavior (e.g., "perhaps he is getting high to avoid actually his feelings of being a failure").
    - A metaphor that implies an alternative construction of the problem (e.g., perhaps his drug use is his way of "self-medicating" his ADHD).
    - Humor to imply that not everything is as it seems (e.g., perhaps his drug use is his way of communing with his ancestors and showing respect for a nonconformist family tradition).

    For reframing to be successful, it must include the concept of responsibility: the problem behavior was intended, but the motivation was not as malevolent as it might appear.
3. *Assess the Impact of the Reframe and Build on It:* Finally, the therapist listens to family's response to assess the "fit": is it meaningful and useful to the family? The therapist often works with the client to modify the reframe so that it better fits with each member of the family's worldview. The goal is to find a mutually agreeable—or at least plausible—alternative explanation for the problem behavior. The reframing process is likely to continue across sessions, as the family more finely tunes the possible explanations.

Building Organizational Themes

An outcome of reframing, organizational themes are used to describe how the problematic behaviors are motivated by *positive but misguided* intentions (Sexton, 2011). These themes help family members reattribute more positive qualities to one another. Themes are most useful when they are mutually developed by the family and therapist. Common themes are:

- Anger implies hurt.
- Anger implies loss.
- Defensive behavior implies emotional bonds.
- Nagging equals importance.
- Pain interferes with listening.
- Differences can be frightening.
- Protection often involves shutting others out (Sexton, 2011, pp. 4–23).

These organizing themes help to describe the origin of the problem without blaming any one person, eliciting greater understanding if not compassion for one another. Ideally, the theme is mutually developed so that it is meaningful to all members of the family while also feeling supportive too. Most often, these themes help focus the family on the big picture—such as feeling loved, safe, and valued—rather than getting lost in the details. In the case study at the end of the chapter, the therapist helps the family explore their relational themes, which includes how their arguing and even the daughter's defiance solidifies their relational bonds.

Interrupting and Diverting to Structure Sessions

When families begin to escalate or start self-defeating patterns, FFT therapists actively intervene to interrupt or divert the conversation to help structure the session (Sexton, 2011). For example, if a parent begins berating the child, the therapist will quickly step in to stop the unhelpful rant: "I can hear that you are very frustrated by your son's behavior. Can you please describe how it is affecting you? Such comments stop potentially harmful interactions and keep the therapy on track. In addition, FFT therapists use coaching ("why don't you try …?), directing (wait, use the problem-solving steps), or modeling to also help redirect families.

Process Comments

A particular form of diverting, FFT therapists use process comments to draw the family's attention to the immediate interactions in the room. These comments serve two functions: (a) they interrupt the problem behavior patterns, and (b) they help the family to become more consciously aware of the patterns. The process comments can focus on the behavioral sequence patterns or functions of the behaviors. For example, "Did you notice that before he finished making the request, you jumped in with an answer based on your assumption about what he was going to say?" or "Are you noticing how the silence and refusal to look up is how you hold on to your sense of power in the relationship?"

Parent Skill Training

Functional family therapists help parents to be more effective based on current scientific literature on parenting behaviors that are associated with risk factors and protective factors for youth behavior problems (Sexton, 2011). For example, supportive yet challenging parenting is predictive of better school performance and social adjustment than authoritarian parenting. Additionally, clear expectations with consistent reinforcement has long been associated with fewer emotional and behavior problems. Whereas parent training may be more effective with younger children, Sexton (2011) and colleagues have found that with adolescents parenting strategies are best learned by helping families alter their family relational sequences in family therapy.

When working with parents, FFT focuses on three areas:

- *Clear Expectations and Rules:* Creating mutually agreed upon and developmentally appropriate behavioral expectations that are concrete and specific. In some cases,

contracts that specify the rules are written down and signed by all parties, which can help adolescents feel actively involved in the process.
- *Active Monitoring and Supervision:* Helping parents take an active role in monitoring their children, which is summed up by answering the following: "Who is the teen with? Where is he or she? What is he or she doing? When will he or she be home?" (Sexton, 2011, pp. 5–14).
- *Consistent and Enforcement of Behavioral Contingencies:* Although most parents of troubled youth hope for a set of consequences that will magically inspire flawless role-model behavior overnight, FFT therapists recognize that consequences are not as effective with teens as with younger children. However, the process of having parents and teens negotiate reasonable terms is more likely to be beneficial (Sexton, 2011). If consequences are to be used, FFT therapists recommend that it be brief, done without anger, and directly linked to the behavior in question (e.g., taking financial responsibility for damages or writing a letter of apology).

Mutual Problem Solving

A subset of parenting skills, problem solving in FFT involves helping parents and children mutually work together to address concerns in a way that strengthens the relationship rather than have the parents dictate how problems will be solved (Sexton, 2011). The typical steps in problem solving include:

- *Identifying the Problem:* Defining the problem in relational terms (everyone shares responsibility) and in concrete, behavioral terms (e.g., yelling, drinking and driving).
- *Identifying the Outcome Desired:* As simple as it sounds, it is especially important in families with troubled youth to identify the desired outcome in specific, behavioral terms; this discussion alone can help resolve some issues (e.g., having a relatively calm discussion about expectations in which no one curses, raises their voice, or insults another and everyone is seeking a reasonable solution).
- *Agreeing on How to Accomplish Goal:* Next, the family identifies each person's role in helping to solve the problem; this may involve written contracts or agreements.
- *Identify Potential Obstacles:* In most cases, it helps to identify potential obstacles and barriers before trying to act on the plan.
- *Reevaluation of Outcomes:* Finally, the goals are evaluated to establish accountability and determine next steps.

Conflict Management

Some families have painful histories and/or rigid patterns that make it difficult to problem solve successfully. In such cases, therapists need to help them move past these past hurts and struggles. Conflict management may not "solve" the painful issue but instead is used by the therapist to help contain difficult interactions. Some commonly used strategies for containing conflict include:

- *Remaining Focused on a Specific Issue:* The therapist keeps the family focused on a specific current issue rather than bringing up past unresolved issues.
- *Adopting a Conciliatory Mindset and Willingness to Talk:* The therapist sets an emotional tone that helps reduce conflict.
- *Staying Oriented to the Present:* The therapist keeps the focus on reducing the conflict rather than trying to directly solve extremely volatile issues or "rehearsing" them.

If one or more family members get stuck in an issue, the therapist may use the following questions to help move the process forward:

- Exactly, what is the issue of concern for you?
- Exactly what would satisfy you?
- How important is that goal to you?
- Have you tried to get what you want through problem solving?
- How much conflict are you willing to risk to get what you want? (Sexton, 2011, pp. 5–18)

Communication Skill Building

In FFT, the family's communication skills are often addressed as part of helping the family resolve their concerns; however, it is not a goal by itself: it is a means to other ends, such as problem solving and parenting. In helping families restructure their interactions, FFT therapists may focus on one or more of the following issues to help improve family communication:

- *Responsibility:* Each person takes responsibility for their words and communications and avoids speaking for others (e.g., not say "in this house" or "kids should").
- *Directness:* Families are encouraged to direct their comments to the intended recipient and avoid third-person comments (e.g., "No one in this house" or "she never").
- *Brevity:* Messages should be kept short to ensure the recipient understands.
- *Concrete and Specific:* Families are taught to avoid generalizations (e.g., "you never") and broad statements (e.g., "make a good decision") and make very specific requests for action.
- *Congruence:* Helping family members to have verbal and nonverbal messages that match and are consistent (e.g., not sounding angry when you say everything is okay).
- *Active Listening:* Therapists help family members learn how to be responsive listeners in ways that are natural for them (e.g., a head nod to signal you have heard the other's message—remember that may be a big step for a teen that hasn't communicated with parents in years).

Matching to Fit the Family

In FFT, therapists maintain realistic goals for their clients. They do not expect their clients to communicate and interact in some idealistic "perfect family" way. Instead, the use ideal forms of communication, parenting, and conflict management to help them develop realistic modifications to the family's current interactions that will help them have better outcomes. For example, many therapists consider yelling "unhealthy" and are quick to target yelling for change. However, in FFT, the therapist first assesses how the yelling functions in the family—how it creates closeness, distance, and determines hierarchy—before determining whether and how it needs to change. The key is to introduce new behavioral skills the *fit* for a particular family.

- *Matching to the Problem Sequence:* When targeting the sequence of problem interactions, the therapist looks for easiest (rather than ideal) places to introduce a new behavior.
- *Matching to the Relational Functions:* Rather than trying to change the fundamental quality of relational functions (interdependence, independence, and hierarchy), the therapist helps the family find more effective ways of maintaining the same relational function (e.g., rather than avoid conflict, use a relatively short, structured approach to problem resolution *or* to express closeness, find mutually enjoyable activities that help the family to feel connected rather than having a parent ask a series of endless questions).
- *Matching to Organization Theme:* Interventions are linked back to the organization themes identified early in treatment to create a sense of coherency and continuity.

Putting It All Together

Case Conceptualization Template

- *Relational Connection:* Describe family's preferred approach to relational connection:
    - High independence
    - High Interdependence
    - Midpointing

- *Relational Hierarchy:* Describe current hierarchy in the family:
  - Parent-up/adolescent-down
  - Adolescent-up/parent-down
  - Symmetrical
- *Relational Function of Symptoms*
  - What function does the symptom serve?
    - To create independence or interdependence?
    - To establish hierarchy or distribute power?
  - How can the family achieve a similar function with more effective relational interactions?
- *Risk and Protective Factors*
  - *Individual Youth and Parent Risk Factors*
    - History of violence or victimization
    - History of early aggressive behavior and general poor behavior control
    - Substance, alcohol, and/or tobacco use/abuse
    - Diagnosis of an emotional or psychological concern, including ADHD, or other deficits in social, cognitive, or information processing
    - Low IQ
    - Antisocial beliefs or attitudes
  - *Family Risk Factors*
    - Lack of mutual attachment and nurturing by parents
    - Ineffective parenting
    - Chaotic home environment
    - Lack of a significant relationship with a caring adult
    - Caregiver who abuses drugs, commits crimes, or is diagnosed with a mental disorder
  - *Peer/School Risk Factors*
    - Associates with other troubled youth, including gang involvement
    - Frequent social rejection by peers
    - Lack of involvement in conventional activities
    - Poor academic performance; little commitment to school
  - *Community Risk Factors*
    - Diminished economic opportunities; high concentration of poor residents
    - High level of transience and low levels of community participation
    - High levels of family disruption
  - *Protective Factors*
    - Strong bond between children and family
    - Parental involvement in child's life
    - Supportive parenting that meets financial, emotional, cognitive, and social needs of child
    - Clear limits and consistent enforcement of discipline
- *Multisystem Assessment:* Describe function of symptomatic behavior in other social systems, such as neighborhood, school, peers, employment, human service agencies, cultural groups, region, etc.
- *Culture and Community*
  - *Cultural Background:* What is the family's cultural background(s)? In which ways do they identify with this background? In which ways do they not?
  - *Culture and Family Norms:* What are the norms for family structure, hierarchy, role, and emotional expression in the family's culture? In which ways does the family embrace these? In which ways do they not? How does this compare with norms from their local and regional community?
  - *Local Community:* How does the family fit within their local community? Are they connected to meaningful and supportive groups within the community? How are their problems viewed within this community? Are there support persons that can be engaged?

- ○ *Socioeconomic Context:* How does the family's socioeconomic status affect their role in the community and contribute to the problems they experience?
- ○ *Adapting Therapy:* How does the therapy process need to be adapted to be respectful of the family's cultural and community contexts, values, and norms? What type of therapeutic relationship, goals, assessments, and interventions would be most useful?
- *Strengths and Resiliency:* Describe forms of individual and family strengths and resiliency.

The following is a treatment plan template that you can use to help develop treatment plans for families with teens having conduct issues.

Treatment Plan Template for Family

Initial Phase of Treatment: FFT

Initial Phase Counseling Tasks

1. Build family *engagement* in therapy and their *motivation*. *Diversity Note: Adapt engagement and motivation strategies to cultural, socioeconomic, legal status, age, gender, and other diversity factors; also address sense of social marginalization due to conduct issues.*
   Relationship building approach/intervention:
   a. Demonstrate *respectful* and *collaborative attitude* with each member of the family to create a context of safety.
   b. Develop motivation by establishing self as a *credible helper*.

2. Assess individual, systemic, and broader cultural dynamics. *Diversity Note: Attend to cultural, socioeconomic, gender, sexual orientation, age, and other diversity factors that may inform family systemic interaction patterns.*
   Assessment strategies:
   a. Assess *relational function*, including *relational connection* (balance of independence and interdependence), and *relational hierarchy*.
   b. Assess *risk and protective factors* as well as *strengths and resiliency*.
   c. Assess how *multiple systems* (school, peer, extended family, community, social service agencies, cultural group, etc.) intersect with family problems and resiliency.
   d. Identify *organizational themes* that characterize family conflict.

Initial Phase Client Goals

Manage crisis; reduce distressing symptoms.

1. Decrease *within-family blame* and increase *within family alliance* to reduce within-family conflict.
   a. *Identify the problem sequence*, including each family member's role in the dynamic, to reduce within-family blame.
   b. *Reframe problem* and interactions *in relational terms* and identify *organization themes* to reduce within-family blame and increase motivation to change and sense of family alliance.

Middle Phase of Treatment: FFT

Middle Phase Counseling Task

1. Monitor quality of the working alliance. *Diversity Note: Monitor client response to interventions to ensure they are culturally appropriate.*
   a. *Assessment Intervention:* Monitor nonverbal responses to interventions and continue to assess for each person's motivation and engagement.

(continued)

Middle Phase Client Goals

1. Increase family's *relational competencies* (e.g., parenting, communication, problem solving) to reduce within-family conflict and improve problem interaction sequence.
   a. *Parent training* and/or *mutual problem solving* matched to the family's problem sequence, relational functioning, and/or organization themes.
   b. *Conflict management* and/or *communication skill training* matched to the family's problem sequence, relational functioning, and/or organization themes.

2. Change *expression of relational functions* to reduce conflict and conduct issues.
   a. *Interrupt self-defeating patterns* and *coach* or *direct* family in more functional interactions.
   b. Use *process comments* to increase family's awareness of their interaction patterns.
   c. *Conflict management, parent training, problem solving* and/or *communication skill training* matched to the family's problem sequence, relational functioning, and/or organization themes.

3. Increase adolescent's and family's *protective factors* to reduce frequency of conduct issues.
   a. Identify and *engage resources* in other systems: peer, school, social service, community, etc.
   b. Build on *organizational themes* to help teen and family access their own strengths and resources.

Closing Phase of Treatment: FFT

Closing-Phase Counseling Task

1. Develop aftercare plan and maintain gains. *Diversity Note: Connect with resources within diverse communities, especially those offered within an ethnic group, religion, sexual orientation support groups, neighborhoods, etc.*
   a. Identify ways to connect to supportive elements within *multiple communities/ systems*.
   b. Identify *strategies for handling future high-risk situations.*

Closing-Phase Client Goals

Determined by theory's definition of health and normalcy.

1. Increase *positive interactions* with [specify *external system*: school, peer, extended family, etc.] to reduce conflict and conduct issues
   a. *Generalize problems solving, conflict management, and communication skills learned* to improve relationships with persons outside the family.
   b. Identify potential *sources of support in community*, including sources of emotional support, information, and practical assistance.

2. Increasing family's *self-efficacy* and within-family alliance to reduce conflict.
   a. *Coaching* family to successfully respond to new challenges as they arise.
   b. *Process comments* to reinforce new, functional patterns.

Tapestry Weaving: Diversity Considerations

Ethnic, Racial, and Cultural Diversity

FFT is an approach that has been widely used but not necessarily widely studied with ethnically diverse populations (Henggeler & Sheidow, 2012; Sexton, 2011). In a large Florida-based study examining the effectiveness of FFT across diverse groups, no significant differences were found in terms of recidivism, post-treatment crime severity, and program completion rates (Durham, 2010). Embedded within the approach's view of relational function is a respect for a wide variety of family structures, allowing it to be applicable to culturally diverse family forms. The therapist respects culture by acknowledging the relational function of problem exchanges and helping the family to find more effective ways to meet those functions, whether it be establishing a more hierarchical or democratic family structure or allowing for more interdependence or independence between family members. Thus, the therapist adapts the goals based on the family's cultural norm, allowing them to honor both collectivist and individualist value systems. Of course, immigrant families are likely to have a significant difference in values between generations, which cannot always be easily bridged, but FFT has many strategies to help create a meaningful common ground.

Research on FFT with diverse populations provides clinically relevant guidance for therapists using FFT with diverse families. For example, one study found that Hispanic adolescents had better outcomes when they worked with a Hispanic therapist; however, ethnic match did not affect outcomes of Anglo adolescents in the same study (Flicker, Waldron, Turner, Brody, & Hops, 2008). This study underscores the importance of cultural competence for Anglo therapists and the need to attend to ethnic match with diverse clients when possible. Future research is likely to refine the implementation recommendations for ethnically diverse families.

Sexual Identity Diversity

Little has been written specifically about FFT and gay, lesbian, bisexual, transgendered, and questioning (GLBTQ) youth or families. However, one case study has considered using FFT principles when working with a gay male (Datchi-Phillips, 2011). Using FFT and general family systems principles, Datchi-Phillips recommends assessing the multiple systems of which a GLBTQ client is a member, including family of origin, current relationship, GLBTQ community, school, work, neighborhood, religious community, etc. Each system is likely to have a unique response to the client's sexuality. Furthermore, the family's organizing theme may be useful for GLBTQ individuals to make sense of and depersonalize some of their own struggles. Datchi-Phillips also emphasizes that often people who identify as GLBTQ are often not fully accepted by their families of origin and therefore must create their own "families." When appropriate these "family members" should also be involved in treatment. Furthermore, an individual's apparent "pathology" should be viewed within the network of family and other social contexts, helping to reframe the youth's problem in ways that are less blaming and more hopeful.

Evidence Base

With its first efficacy trial over 40 years ago, FFT was one of the first evidence-based treatments (Baldwin, Christian, Berkeljon, & Shadish, 2012; Henggeler & Sheidow, 2012). Used internationally, FFT is one of four widely recognized family treatments for adolescent conduct issues that appears to be moderately superior to other forms of treatment or treatment-as-usual (Sexton & Turner, 2010, 2011). Over the years, FFT has been on the vanguard with many "firsts" and landmark trials in the research world. For example, Alexander and Parsons (1973) study was the one of the first clinical trials to show favorable outcomes in the juvenile justice system. In addition, FFT was the first to have an efficacy study conducted by independent investigators, researchers not directly associated with the theory being studied (and therefore less biased; Sexton & Turner, 2011). In addition, they had one of the largest trails to examine the transportability of FFT to real world settings (Sexton & Turner, 2011). Additionally, the rates of adolescent recidivism and relapse have been correlated to the therapist's adherence to the model (Sexton & Turner, 2011). In sum, FFT has a strong evidence base for treating conduct disordered and substance-abusing youth; future research will hopefully refine applications, especially with diverse families.

Online Resources

Emotionally Focused Therapy (Sue Johnson, Canada):
www.eft.ca

Emotionally Focused Therapy (San Diego, California):
www.sdeft.us

Emotionally Focused Therapy (Los Angeles and Houston):
www.theeftzone.com

Emotion-Focused Therapy (Les Greenberg, Canada):
www.emotionfocusedtherapy.org

Functional Family Therapy:
www.fftinc.com

National Juvenile Justice Publication on FFT:
www.ncjrs.gov/pdffiles1/ojjdp/184743.pdf

References

*Asterisk indicates recommended introductory readings.

Alexander, J. F., & Parsons, B. V. (1973). Short-term behavioral intervention with delinquent families: Impact on family process and recidivism. *Journal of Abnormal Psychology, 81*(3), 219–225. doi:10.1037/h0034537

*Alexander, J., & Parsons, B. V. (1982). *Functional family therapy*. Belmont, CA: Brooks/Cole.

Alexander, J., & Sexton, T. L. (2002). Functional Family Therapy (FFT) as an integrative, mature clinical model for treating high risk, acting out youth. In J. Lebow (Ed.), *Comprehensive handbook of psychotherapy, Vol IV: Integrative/Eclectic* (pp. 111–132). New York: Wiley.

Baldwin, S., Christian, S., Berkeljon, A., & Shadish, W. (2012). The effects of family therapies for adolescent delinquency and substance abuse: A meta-analysis. *Journal of Marital and Family Therapy, 38*, 281–304.

Bowlby, J. (1988). *A secure base: Parent-child attachment and healthy human development*. London: Routledge.

Bradley, B., & Furrow, J. L. (2004). Toward a mini-theory of the blamer softening event: Tracking the moment-by-moment process. *Journal of Marital and Family Therapy, 30*, 233–246.

Clotheir, P., Manion, I., Gordon-Walker, J., & Johnson, S. M. (2002). Emotionally focused interventions for couples with chronically ill children: A two year follow-up. *Journal of Marital and Family Therapy, 28*, 391–399.

Datchi-Phillips, C. (2011). Family systems (the relational contexts of individual symptoms). In C. Silverstein (Ed.), *The initial psychotherapy interview: A gay man seeks treatment* (pp. 249–264). Amsterdam Netherlands: Elsevier. doi:10.1016/B978-0-12-385146-8.00012-2

Flicker, S. M., Waldron, H., Turner, C. W., Brody, J. L., & Hops, H. (2008). Ethnic matching and treatment outcome with Hispanic and Anglo substance-abusing adolescents in family therapy. *Journal of Family Psychology, 22*(3), 439–447. doi:10.1037/0893-3200.22.3.439

Furrow, J. L., & Bradley, B. (2011). Emotionally focused couple therapy: Making the case for effective couple therapy. In J. L. Furrow, S. M. Johnson, & B. A. Bradley (Eds.), *The emotionally focused casebook: New directions in treating couples* (pp. 3–29). New York: Routledge/Taylor & Francis Group.

Furrow, J., Ruderman, L., & Woolley, S. (2011). Emotionally focused therapy four day externship. Santa Barbara, CA. September 7–10.

Greenberg, L. S., & Goldman, R. N. (2008). *Emotion-focused couple therapy: The dynamics of emotion, love, and power*. Washington, DC: American Psychological Association.

Henggeler, S., & Sheidow, A. (2012). Empirically supported family-based treatments for conduct disorder and delinquency in adolescents. *Journal of Marital and Family Therapy, 38*, 30–58.

Institute of Medicine (2001). *Crossing the quality chasm: A new health system for the 21st century*. Washington, DC: National Academy Press.

Johnson, S. M. (2000). *Emotionally focused couple therapy with trauma survivors: Strengthening attachment bonds*. New York: Guilford.

*Johnson, S. M. (2004). *The practice of emotionally focused marital therapy: Creating connection* (2nd ed.). New York: Brunner/Routledge.

Johnson, S. (2008). *Hold me tight: Seven conversations for a lifetime of love*. New York: Little, Brown, and Co.

Johnson, S. M., Bradley, B., Furrow, J., Lee, A., Palmer, G., Tilley, D., & Woolley, S. (2005). *Becoming an emotionally focused couples therapist: A workbook*. New York: Brunner/Routledge.

Johnson, S. M., & Greenberg, L. S. (1985). The differential effects of experiential and problem solving interventions in resolving marital conflicts. *Journal of Consulting and Clinical Psychology, 53*, 175–184.

Johnson, S. M., & Greenberg, L. S. (Eds.). (1994). *The heart of the matter: Perspectives on emotion in marital therapy*. New York: Brunner/Mazel.

Johnson, S. M., Hunsley, J., Greenberg, L. S., & Schindler, D. (1999). Emotionally focused couples therapy: Status and challenges. *Clinical Psychology: Science and Practice, 6*, 67–79.

Johnson, S. M., & Wittenborn, A. K. (2012). New research findings on emotionally focused therapy: Introduction to special section. *Journal of Marital and Family Therapy.* doi: 10.1111/j.1752-0606.2012.00292.x

Lebow, J. (2006). *Research for the psychotherapist: From science to practice.* New York: Routledge.

Lebow, J., Chambers, A., Christensen, A., & Johnson, S. (2012). Research on the treatment of couple distress. *Journal of Marital and Family Therapy, 38,* 145–168.

Liu, T., & Wittenborn, A. (2011). Emotionally focused therapy with culturally diverse couples. In J. L. Furrow, S. M. Johnson, & B. A. Bradley (Eds.), *The emotionally focused casebook: New directions in treating couples* (pp. 295–316). New York: Routledge/Taylor & Francis Group.

Mason, O., Platts, H., & Tyson, M. (2005). Early maladaptive schemas and adult attachment in a UK clinical population. *Psychology and Psychotherapy: Theory, Research and Practice, 78*(4), 549–564.

McWilliams, L. A., & Bailey, S. (2010). Associations between adult attachment ratings and health conditions: Evidence from the National Comorbidity Survey Replication. *Health Psychology, 29*(4), 446–453. doi:10.1037/a0020061

Meredith, P. J., Strong, J., & Feeney, J. A. (2006). The relationship of adult attachment to emotion, catastrophizing, control, threshold and tolerance, in experimentally-induced pain. *Pain, 120*(1–2), 44–52. doi:10.1016/j.pain.2005.10.008

Parra-Cardona, J., Córdova, D. R., Holtrop, K., Escobar-Chew, A., & Horsford, S. (2009). Culturally informed emotionally focused therapy with Latino/a immigrant couples. In M. Rastogi & V. Thomas (Eds.), *Multicultural Couple Therapy* (pp. 345–368). Thousand Oaks, CA: Sage Publications.

*Sexton, T. L. (2011). *Functional family therapy in clinical practice: An evidence-based treatment model for working with troubled adolescents.* New York: Routledge.

Sexton, T. L., & Alexander, J. F. (2000, December). Functional family therapy. *Juvenile Justice Bulletin,* U.S. Department of Justice, NCJ 184743.

Sexton, T., & Turner, C. W. (2010). The effectiveness of functional family therapy for youth with behavioral problems in a community practice setting. *Journal of Family Psychology, 24*(3), 339–348. doi:10.1037/a0019406

Sexton, T., & Turner, C. W. (2011). The effectiveness of functional family therapy for youth with behavioral problems in a community practice setting. *Couple and Family Psychology: Research and Practice, 1*(S), 3–15.

Sexton, T. L., & van Dam, A. E. (2010). Creativity within the structure: Clinical expertise and evidence-based treatments. *Journal of Contemporary Psychotherapy, 40*(3), 175–180. doi:10.1007/s10879-010-9144-2

Siegel, D. J. (2010). *The mindful therapist: A clinician's guide to mindsight and neural integration.* New York: Norton.

Slomski, A. (2010). Evidence-based medicine: Burden of proof. *Proto Magazine: Massachusetts General Hospital, 5,* 22–27.

Zuccarini, D., & Karos, L. (2011). Emotionally focused therapy for gay and lesbian couples: Strong identities, strong bonds. In J. L. Furrow, S. M. Johnson, & B. A. Bradley (Eds.), *The emotionally focused casebook: New directions in treating couples* (pp. 317–342). New York: Routledge/Taylor & Francis Group.

FFT Case Study: Mexican American Family with Substance-Abusing Teen

Roberta (AF36), Denise (CF19), and Selena (CF2) have been referred to therapy by the Department of Children and Family Services after Denise recently tested positive for cocaine. Denise was previously in a drug treatment program while she was pregnant with Selena. She has been clean since the birth of her daughter, but recently started using again when her boyfriend was deported to Argentina. Denise's mother Roberta is scared that her granddaughter Selena will be taken away from the family if her daughter is unable to "grow up and take responsibility." Roberta has three other children of her own, ages 16, 13, and 8. Their father left the family shortly after the birth of the last child and has not been heard from in years; he also had problems with drugs. Since then Roberta has struggled to support their four children and has recently had to also support her granddaughter. Her mother, who immigrated from Mexico before Roberta was born, and often helps care for the children, although this has become less frequent as she has gotten older. Roberta has been hospitalized twice in the past year for stress-related symptoms and is fearful of losing Selena to the system. She does not trust her daughter to properly care for Selena and is pushing for Denise to give her custody. Denise wants to keep custody of her daughter and establish independence from her own mother.

FFT Case Conceptualization
Relational Connection

The relationship pattern in this family is one of *high interdependence*, which is consistent with the norms of the family's Mexican heritage. There is a maladaptive pattern of dependency that has emerged between AF36 and her CF19. CF19's addiction has prevented her from becoming independent and capable of financially supporting herself and CF2.

Relational Hierarchy: Adolescent-Up/Parent-Down

AF36 expresses that she has had little control over her daughter's behavior. AF36's parenting style has been historically permissive and through CF19's teenage years she provided little monitoring or supervision, as she worked long hours. CF19 continues to model this parenting style in raising her own daughter. CF19 states that she is now an "adult" and can make her own decisions, yet she continues to be financially dependent on her mother. AF36 enables her daughter's behavior by not setting clear expectations and inconsistent enforcement of consequences. She has threatened to kick her out if she continues to use drugs, but has not followed through. CF19 holds CF2 as a source of bargaining power in her relationship with AF36 by insisting CF2 will be taken away from the family if AF36 does not help her.

Relational Functional Assessment

- *What Function Does the Symptom Serve?*
    - *To Create Independence or Interdependence?* CF19's drug use and AF's permissive parenting both serve to heighten interdependence.
    - *To Establish Hierarchy or Distribute Power?* CF19's behavior and AF's behavior have reversed the hierarchy with CF19 in many ways having a position of more power than AF.
- *How Can the Family Achieve a Similar Function with More Effective Relational Interactions?* The family can find new ways to maintain interdependence and connection while allowing CF19 to assume responsibility for her life and choices, taking on the responsibilities that come with the freedom she desires. Additionally, these strategies will assist AF36 in finding ways to express love for her daughter and satisfy her desire to maintain a close relationship with her without supporting her addiction.

Risk and Protective Factors

Individual Risk Factors

- CF19's history of drug use, early pregnancy, and poor choice of boyfriends.
- History of abuse in AF36 family of origin and in her personal romantic relationships.

Family Risk Factors

- Ineffective parenting by both AF36 and CF19.
- Chaotic home environment with limited space for family to live.
- Lack of father figure in the home.
- Father's history of drug use.

Community Risk Factors

- High concentration of residents in poverty.
- Lack of positive role models within community.
- Lack of positive peer influences and relationships for CF19.
- School has high dropout rate and delinquency.

Protective Factors

- Support of AF36 for CF19.
- Strong bond between all family members.
- Younger siblings generally do well in school and help around the house.
- Support from extended family members, especially grandmother.
- AF has strong religious faith.

Multisystem Assessment

- *School:* CF19 is in an adult education program through which she is trying to earn her GED. This program allows her maximum time to be with her daughter as she finishes school.
- *Peers:* CF19 hangs out with teens who have similar problems, drug use, alcohol use, and conduct issues. She does have one childhood friend who "stays out of trouble" and has been a support to her over the years.

- *Neighborhood:* Neighborhood is generally impoverished with a relatively high level of crime and gang activity.
- *Extended Family:* The family has little contact with father's family of origin. AF's mother sometime cares for the children. AF's sister lives over an hour away and is supportive by phone but has too many of her own problems to be of much help.
- *Church:* AF attends mass weekly and finds great solace there. The family could find additional support here.

Community and Culture

Cultural Background AF's parents emigrated from Mexico before she was born. AF strongly identifies with her cultural heritage; however, her children are less identified with the culture although they are all fluent in Spanish, which is often spoken at home. AF36 has strayed from traditional gender roles by choosing to raise her children without a husband since the father left; she has adopted a more American attitude toward men and her ability to be self-sufficient. CF19 expresses more traditional gender roles and would like to find a man to marry who will take care of her. Typically there is a hierarchy of authority in Hispanic families with respect given to elders. This family does not identify with that norm with the exception that everyone is highly respectful of grandmother.

Socioeconomic Context The Gonzalez family struggles financially and this impacts their lives on a daily basis. The community in which they live is one that is primarily socioeconomically disadvantaged. Many teens in this community do not complete high school so that they may instead work in order to help support their families. There is additionally a sense of hopelessness present in the community, which contributes to a high rate of drug use and crime.

Strengths

- AF36 is supportive and willing to participate in counseling with CF19 to improve their relationship, as well as provide support for CF19 in her sobriety and in caring for CF2.
- CF19 and CF2 have a strong bond and CF19 is motivated by her daughter to change.
- Grandmother is positively involved and a positive influence, one to whom CF19 may respond.
- AF36 has strong religious faith and resources through church.
- AF36's younger children, CF16, AM13, and CF8, are doing relatively well in school and have stayed out of trouble for the most part. CF19's example of poor choices and their consequences have "inspired" the younger children to ban together and try to create fewer hassles for their mother.

FFT Treatment Plan

Initial Phase of Treatment: FFT

Initial Phase Counseling Tasks

1. Build family *engagement* in therapy and their *motivation. Diversity Note: Adapt engagement and motivation strategies to include warmth and personalismo. Be mindful of how socioeconomic status may impact motivation; also address sense of social marginalization due to history of drug use and family's immigration status.*
    a. Demonstrate *respectful* and *collaborative attitude* with AF36 and CF19 to create a context of safety.
    b. Develop motivation by establishing self as a *credible helper*.

(continued)

2. Assess individual, systemic, and broader cultural dynamics. *Diversity Note: Assess and consider degree to which CF19 and AF36 identify with Mexican vs. American culture. Additionally, consider impact of immigration status, low socioeconomic status, and lack of educational opportunity.*
   a. Assess *relational function*, including *relational connection* (balance of independence and interdependence) and *relational hierarchy* between AF36 and CF19 as well as the similarities and differences in how each interacts with CF2.
   b. Assess *risk and protective factors* as well as *strengths and resiliency*. Carefully consider and address risk factors that may impact the safety of CF2 in the home.
   c. Assess how *multiple systems* (school, peer, extended family, community, social service agencies) intersect with CF19's pattern of drug use and the parenting practices of AF36 and CF19. Also consider how systems impact resiliency.
   d. Identify *organizational themes* that characterize family conflict.

Initial Phase Client Goals

1. Decrease *within-family blame* and increase *within-family alliance* to reduce CF19 substance abuse and reduce within-family conflict.
   a. *Identify the problem sequence*, including AF36 and CF19's role in the dynamic, to reduce within-family blame for CF19's addiction and DCFS involvement in family's life.
   b. *Reframe problem* and interactions *in relational terms* and identify *organization themes* to reduce within-family blame and increase CF19's motivation to stop drug use.

Working Phase of Treatment: FFT

Working-Phase Counseling Task

1. Monitor quality of the working alliance. *Diversity Note: Monitor client response to interventions to ensure they are culturally appropriate and honor both AF36 and CF19's degree of acculturation.*
   a. Monitor nonverbal responses to interventions and continue to assess for AF36 and CF19's motivation and engagement.

Working-Phase Client Goals

1. Increase family's *relational competencies* including parenting, communication, and problem solving for AF36 and CF19 to reduce within-family conflict and reduce CF19 substance use.
   a. *Parent training* for both CF19 and AF36 that focuses on setting clear expectations and rules, as well as consistent enforcement of consequences.
   b. *Communication skill training* to address maladaptive patterns and increase *responsibility, directness, congruence, and active listening*.
   c. *Conflict management* to contain difficult interactions between AF36 and CF19. Focus on *staying oriented to the present*.

2. Change *expression of relational functions* to reduce conflict and conduct issues by finding more adaptive strategies for maintaining sense of interdependence respective of the family's culture.
   a. *Interrupt self-defeating patterns* and *coach* or *direct* family in more functional ways to express interdependence.
   b. Use *process comments* to increase family's awareness of their processes and how to achieve a sense of connection that is more functional.
   c. *Conflict management, parent training, and communication skill training* matched to increase trust between AF36 and CF19 and encourage a more healthy balance of interdependence and independence within the family.

(continued)

3. Increase CF19's and family's *protective factors* to reduce relapse by CF19.
   a. Identify and *engage resources* in other systems: peer, social service, community, church, extended family, and drug treatment program.
   b. Build on *organizational themes* to help AF36 and CF19 access their own strengths and resources.

Closing Phase of Treatment: FFT

Closing-Phase Counseling Task

1. Develop aftercare plan and maintain gains. *Diversity Note: Connect CF19 with Mommy and Me group in her community for social support and continued practice with parenting. Support AF36 in accessing support from church and family.*
   a. Identify ways to connect to supportive elements within *multiple communities/systems.*
   b. Identify strategies for *identifying early signs* of potential relapse by CF19.

Closing-Phase Client Goals

1. Increase *positive interactions* with family to reduce conflict and prevent CF19 from continued use of drugs.
   a. *Generalize problems solving, conflict management, and communication skills learned* to improve relationships with persons outside the family.
   b. Identify potential *sources of support in community*, including sources of emotional support, information, and practical assistance.

2. Increasing family's *self-efficacy* and within-family alliance to reduce conflict.
   a. *Coaching* family to successfully respond to new challenges as they arise, paying close attention to CF19's triggers for drug use and AF36's pattern of enabling her daughter.
   b. *Process comments* to reinforce new, functional patterns.
   c. Support CF19 in becoming more self-reliant.

CHAPTER 16

Evidence-Based Group Treatments for Couples and Families

Most parents with two or more children claim that adding a second child doesn't double the work—it quadruples it. The same can be said for clients: adding a second (e.g., a couple) or third (e.g., a family) person to the room exponentially increases the complexity of the situation. If you then add multiple couples and families into a single therapy session, things really get "rockin'." Remember the nursery tale about the old lady in the shoe: so many kids, she didn't know what to do. Well, this chapter provides you with a basic overview of evidence-based group work with couples and families so that you will know what to do and will do what works best.

Lay of the Land

The first section of this chapter provides a brief overview of general group therapy, in case you skimmed most of your group therapy text or have not yet taken or will not be required to take a group therapy class. If you aced your group class, of course, you can go ahead and skip this section. The rest of the chapter focuses on evidence-based group therapy models for working with the following common client concerns:

- Severe mental illness
- Partner abuse and domestic violence
- Marital enrichment
- Parent training

Brief Overview of Types of Groups

Groups can be divided into three general categories based on their relative emphasis on *content* (information provided) versus *process* (emotional processing): psychoeducational groups, process groups, and combined groups.

Psychoeducational Groups or "Classes"

Psychoeducational groups, often referred to as "classes," are content-focused and educational, rooted in the cognitive-behavioral assumption that "if people just knew better, they'd do (i.e., behave or feel) better." In the arena of couple and family relations, research has particularly supported psychoeducational groups for the following problems:

- Families with members diagnosed with severe mental illness (McFarlane, Dixon, Luksted, & Lucksted, 2002; Lucksted, McFarlane, Downing, & Dixon, 2012)
- Domestic violence (Stith, Rosen, & McCollum, 2002; Stith, McCollum, Amanor-Boadu, & Smith, 2012)
- Marital enrichment (i.e., couples who do not present as distressed; Halford, Markman, Stanley, & Kline, 2002; Markman & Rhoades, 2012)
- Parent training (Kaslow, Broth, Smith, & Collins, 2012; Northey, Wells, Silverman, & Bailey, 2002)

Process Groups

As the name suggests, process groups focus on *interpersonal process* rather than content. Although they are typically composed of individuals, increasingly they include couples and families. Process groups more closely address the goals targeted in experiential or psychodynamic forms of individual therapy, such as increasing insight into and confronting problematic interpersonal patterns. The group is often used as a forum for better understanding past or current couple and family relationships. Process groups are frequently used as an alternative to individual therapy or with clients coping with grief, divorce, loss, or abuse.

Yalom (1970/1985) identifies 11 elements that make process groups effective:

- *Instillation of Hope:* Gaining hopefulness for the future from meeting and interacting with others
- *Universality:* Realizing one is not alone in suffering or struggling with life
- *Imparting Information:* Benefiting from information shared by the therapist while facilitating the group
- *Altruism:* Learning how to give and receive from others without needing personal gain
- *The Corrective Recapitulation of the Primary Family Group:* Providing an opportunity to work through family-of-origin issues and respond in new ways
- *Development of Socializing Techniques:* Improving social skills by interacting intimately with others
- *Imitative Behavior:* Modeling self after chosen role models in the group
- *Interpersonal Learning:* Developing insight and working through transference issues in a group context
- *Group Cohesiveness:* Relating to the therapist and group as a whole to work through conflict and differences
- *Catharsis:* Experiencing and expressing emotions of deep significance, which, when combined with cognitive learning, predict positive outcomes
- *Existential Factors:* Recognizing existential truths, such as the inevitability of death and loss, the need for taking responsibility for one's own life, and the inescapability of suffering

Combined Groups

Combined psychoeducational-process groups generally include some psychoeducational instruction but reserve time for group members to share their thoughts, feelings, and internal processes related to the material, allowing members to address issues at a more personal and deeper level. Such groups are particularly common with victims of child abuse and domestic violence, and increasingly with perpetrators of abuse.

General Group Guidelines

Number of People

The number of people allowed into a group is the key to its success. Generally, for process groups, the limit is four children or eight teens and adults. Combined psychoeducational-process groups can run with more, but the process element generally diminishes significantly once the group exceeds 12 people. Psychoeducational groups that do not include a process component can successfully run with 15 to 30 adults.

Number of People in Groups

	Psychoeducational Groups	Combined Groups	Process Groups
Adults and Adolescents	2–30 adults	3–12 adults	3–8 adults
Elementary-Age Children	2–4 children	2–4 children	2–4 children
Middle-School-Age Children	2–6 children	2–6 children	2–6 children
Couples and Families	2–10 couples or families	2–7 couples or families	3–4 couples or families

Open Versus Closed Groups

Open groups are groups that can be joined at any time; closed groups are "closed" to new members once the group begins.

- *Open Groups:* The primary advantage of open groups is that clients do not have to wait to join; once referred, they can immediately start while their motivation is at its peak. Having an open group requires that all the psychoeducational components can be taken in any sequence, which is often difficult if you are trying to build on skills from the prior week.
- *Closed Groups:* Most process groups and court-mandated psychoeducational groups are closed groups. The closed format allows for the intimacy and trust necessary for process groups to function and also enables mandated groups to develop a coherent and in-depth curriculum that allows group facilitators to monitor individual members' progress.

Group Selection: Who's in Your Group?

As any group leader quickly realizes, who is in the group is a critical factor for success, not only for process-oriented groups but also for psychoeducational groups. For process and combined groups, most therapists screen potential group members to ensure that they are appropriate for the group. Group facilitators look at several different factors, depending on their group:

- *Presenting Problem:* When groups are organized around a presenting problem, such as sexual abuse or loss, the group facilitator screens to ensure that potential members are addressing the same general issue.
- *Level of Problem Resolution:* If a group is organized around a particular issue, the facilitator may want to assess the level of problem resolution so that participants are close enough in the recovery and healing process that they can meaningfully interact; this is particularly important when working with trauma victims (e.g., sexual abuse, national disaster recovery, loss, domestic violence) or substance abuse recovery (e.g., using versus sober and clean). Alternatively, some facilitators prefer to have a spectrum of problem resolution in the group so that members can learn from each other, in which case the group should be organized to accommodate these differences.
- *Ability to Meaningfully Contribute:* Most facilitators screen for a person's ability to meaningfully contribute, which is typically defined as being able to actively

participate without dominating or developing hostile interactions with others. Therapists also consider the client's ability to speak and actively engage others in a group setting; clients who are unable to relate with others may not be good candidates for process groups.

Group Rules

Depending on the group situation, therapists may develop rules for how group participants interact in and out of the group session.

- *Confidentiality:* In most cases, as a condition for membership, group participants must agree to maintain the confidentiality of other members in a group, typically in writing. Although group participants cannot be held to the same legal standards as therapists for maintaining confidentiality, having a formal, written process increases members' awareness of the seriousness of the issue.
- *Respectfulness:* Most group leaders moderate discussion to ensure that members speak respectfully to one another and the group leaders, often directly intervening if a conflict gets out of control. However, in more traditional psychodynamic and experiential process groups, the facilitator does *not* interrupt certain conflicts, instead encouraging the members to grow by working through their issues together.
- *Punctuality and Attendance:* Often groups have rules for punctuality and attendance, especially if courts mandate treatment. In some cases, if members arrive after a designated time, they are not allowed to join that group session. Similarly, if the group has a set curriculum, a member may not be allowed to continue if he/she misses too many sessions.

Evidence-Based Couple and Family Groups

Psychoeducational Multifamily Groups for Severe Mental Illness

Family psychoeducation has been used as part of treatment in schizophrenia, bipolar, and childhood mood disorders (Lucksted et al., 2012). Numerous controlled, clinical studies have demonstrated that patients diagnosed with schizophrenia whose families receive psychoeducation have reduced incidents of psychotic relapse and rehospitalization (McFarlane et al., 2002). The relapse rate for these patients is about 15%, less than half the rate of those who receive individual therapy and medication or medication alone (relapse rate 30%–40%). Several comparison studies indicate that multifamily groups are more effective than single-family sessions in which the family learns the same psychoeducational material, making groups not only the more cost-effective treatment but also the more effective overall. Multifamily groups lead to lower relapse rates and higher rates of employment than single-family sessions, probably because of the enhanced support that group settings offer. There is sufficient evidence at this time to consider family psychoeducation for families with a schizophrenic member as an *evidence-based practice* (Lucksted et al., 2012; McFarlane et al., 2002).

The U.S. Substance Abuse and Mental Health Services Administration (SAMHSA) describe multifamily psychoeducational groups as characterized by the following (Lucksted et al., 2012):

- Assumes involved family members of persons diagnosed with schizophrenia need information, assistance, and support
- Assumes the way relatives behave toward a consumer with a mental illness affects the individuals well-being and clinical outcomes
- Combines information, cognitive, behavioral, problem solving, emotional, coping, and consultation elements
- Created and led by mental health professionals
- Offered as part of a clinical treatment plan for the diagnosed individual

- Focuses on improving the diagnosed individual's outcomes but also addressed family member outcomes
- Includes at least the following content:
  - Illness, medication, and treatment management
  - Services coordination
  - Attention to all parties' expectations, emotional experiences, and distress
  - Improving family communication
  - Structured problem-solving instructions
  - Expanding social support networks
  - Explicit crisis planning
- Are generally diagnosis specific, although some cross-diagnosis programs are being developed

Research indicates that psychodynamic and insight-oriented therapies are counterindicated (not recommended) with families who have a member diagnosed with schizophrenia.

The *psychoeducational curriculum* for multifamily groups for families with a member diagnosed with schizophrenia typically includes the following:

- Education about the illness, prognosis, psychological treatment, and medication
- The biological, psychological, and social factors related to the illness
- Education on the role of high emotional expression in the home in increasing the likelihood of relapse (Hooley, Rosen, & Richters, 1995)
- Possible social isolation and stigmatization
- The financial and psychological burden on the family
- The importance of developing a strong social network to support the family
- Problem-solving skills

The *process* aspect of multifamily groups includes the following:

- Creating a forum for mutual aid and support
- Normalizing the family's experience
- Reducing the sense of stigma
- Creating a sense of hope by seeing how other families have coped
- Socializing with each other outside the group context

When families participate in these groups, the patient and the family have fewer social problems, fewer clinical management problems, reduced secondary stress (stress caused by the hospitalization of the patient), increased employment, and decreased relapse and rehospitalization. In the past decade, several studies have been conducted in countries outside the United States, primarily in Asia and Europe, with some showing similar positive outcomes as in the United States and some not showing significant differences; thus, more exploration into cultural adaptation and applicability is needed (Lucksted et al., 2012).

Although most of the research on groups has been done with schizophrenic and schizoaffective disorders, several studies have used multifamily psychoeducation groups as a component of treatment for other severe mental health disorders (McFarlane et al., 2002), including the following:

- Bipolar disorder
- Dual diagnosis (e.g., substance abuse and an Axis I diagnosis)
- Obsessive compulsive disorder
- Depression
- Posttraumatic stress disorder in veterans
- Alzheimer's disease
- Suicidal ideation in adolescents
- Congenital abnormalities
- Intellectual impairment
- Child molesters and pedophilia
- Borderline personality disorder

Groups for Partner Abuse

A widespread and difficult-to-treat phenomenon with numerous physical and emotional effects on children, women, and families, physical abuse of one's partner is increasingly targeted for intervention by state and local authorities. Today, most states mandate therapy in cases of partner abuse, a specific form of domestic violence. Treatment is most frequently delivered in the form of gender-based group therapy (i.e., groups for male batterers or female batterers and groups for male or female victims of abuse; Stith et al., 2012; Stith, Rosen, & McCollum, 2002). However, increasingly, couple-based approaches are being researched as alternative approaches, with initial studies indicating that carefully designed couple-based approaches to partner violence have comparable outcomes to the traditional gender-based groups, with no evidence of an increased potential for violence (Stith, McCollum, Rosen, & Locke, 2002; Stith, Rosen, & McCollum, 2002).

Traditional Gender-Based Groups for Partner Abuse

The traditional gender-based groups for partner abuse are based on the work of Lenore Walker (1979) and the closely related Duluth model (Pence & Paymar, 1993). These models are psychoeducational cognitive-behavioral groups grounded in feminism that are widely used in the criminal justice system. Lenore Walker used the *cycle of violence* to describe the dynamics of the abuse relationship. The cycle has three phases: honeymoon, tension building, and acting out:

- *Honeymoon Phase:* This is the phase in which the batterer pursues the victim, asking forgiveness, promising to stop the violence, and attempting to woo the victim back.
- *Tension-Building Phase:* Victims often describe the tension-building phase as a state of "living on egg-shells." During this period, conflict and tension are rising and the victim tries to avoid behaviors that will "set off" the perpetrator.
- *Acting Out or Explosion:* The final phase is characterized by a physically violent episode; after this episode, the cycle repeats as the batterer tries to repair the relationship in the honeymoon phase.

In the early years of violent relationships, the cycle typically moves slowly, perhaps with only one or two mildly violent episodes per year (e.g., pushing, shoving, throwing objects, one slap). Over time, the frequency and severity of the violent episodes increase.

In addition to the cycle of violence, the Duluth model uses the idea of the *wheel of power and control*, which describes the means by which batterers are believed to maintain control over their partners. The eight spokes of the wheel are as follows:

- Coercion and threats
- Intimidation
- Emotional abuse
- Isolation
- Minimizing, denying, and blaming
- Using children
- Economic abuse
- Male privilege

Critique of the Duluth Model Although the Duluth model is one of the most frequently used models for treating batterers, Dutton and Corvo (2006, 2007) note that the research literature has identified numerous and significant weaknesses with the model:

- *Poor Outcomes:* Studies on the effectiveness of batterer intervention groups indicate a high recidivism rate, with 40% of participants remaining nonviolent. In comparison, untreated batterers have 35% chance of remaining nonviolent, indicating that typical batterer intervention programs made very little impact on recidivism (Stith et al., 2012).
- *De-emphasis on the Therapeutic Alliance:* One of the chief criticisms of the Duluth model is that it fails to promote a supportive therapeutic alliance with clients and instead often involves an adversarial and blaming stance toward clients.

- *Unfounded Gender Bias:* The gender-based, feminist view is not supported by incidence statistics, which indicate that *personality* rather than gender predicts partner violence.
- *Unfounded Emphasis on Power:* The Duluth model describes partner violence as an issue of power and male privilege; however, the research indicates that there is no clear correlation between violence and patriarchal privilege.
- *Unfounded De-emphasis on Anger:* In contrast to the Duluth position that partner violence is not related to anger but to power, research indicates that men who batter have significantly higher levels of anger and hostility than those who do not batter.
- *Ignoring Bidirectional Violence:* The Duluth model is based on the assumption that violence is unilateral, whereas national surveys indicate that most domestic violence is *bidirectional*, with both partners acting violently; nonetheless, men are far more likely than women to severely injure their partners.
- *Unfounded Power and Control Assumptions:* The Duluth model asserts that partner violence is due to the male's desire for power and control, whereas research indicates that both genders vie equally for power and control in relationships.

Thus, although the Duluth model is the most common treatment model, the research literature does not support either the model's theoretical assumptions or its effectiveness.

Conjoint Treatment of Partner Abuse

Because of the limitations of the Duluth model, therapists and researchers have explored other models of treatment of interpersonal violence, primarily conjoint therapy using systemic and cognitive-behavioral approaches (Stith et al., 2012).

Until recently, couples therapy was considered "inappropriate" for the treatment of domestic violence, largely because of the assumption that there is a significant struggle for power, control, and dominance in all battering relationships. However, as already noted, there is little research to support this assumption (Dutton & Corvo, 2006, 2007), and there is increased interest and funding in this new approach (Stith et al., 2012).

Emerging research indicates that there is more than one type of batterer (Holtzworth-Munroe & Stuart, 1994): (a) those who are violent only within their families, (b) those who have more severe pathology, such as depression or borderline personality disorder, and (c) those who are generally violent and have antisocial characteristics. Conjoint couples therapy is designed for only one type of batterer: *family-only batterers without apparent psychopathology*. Arguments for using conjoint couples treatment for mild cases of family-only battering include the following (Stith, Rosen, & McCollum, 2002):

- It is estimated that as many as 67% of all couples presenting for couples therapy have a history of violence. Therapists may begin treating a couple for other issues before discovering violence; stopping therapy because of this revelation can result in the couple choosing to not pursue further treatment.
- Statistics show that 50%–70% of wives who are battered return to the abusive partner after separating or going to a shelter; conjoint couples therapy provides one of the few safe forums for these couples to safely discuss their relationship.
- The majority of violent relationships are *bidirectional*, with both parties committing acts of violence. Because cessation of violence by one partner is highly dependent on the cessation of violence by the other, conjoint treatment, in which both parties are de-escalating violence, has a greater likelihood of reducing bidirectional violence.

Conjoint Couples Treatment Several approaches for working conjointly with couples have been developed; most combine couple-only sessions with multicouple groups (Stith, Rosen, & McCollum, 2002). The common characteristics of conjoint couples treatment of partner violence include the following:

- Each partner is carefully screened to determine if the couple should participate in couple sessions or multicouple group sessions. Couples in which there has been serious injury to one or both partners are excluded. In addition, in private screening

interviews, both partners must state that they prefer couples therapy and that they are not afraid of speaking freely in session with their partner.
- The primary goal is *decreasing violence and emotional abuse,* not saving the marriage or relationship.
- Partners are encouraged to take responsibility for their own violence.
- The skill-building component emphasizes identifying when one is angry, knowing how to de-escalate self and other, and taking time-outs.
- The effectiveness of the program is measured by reduction or elimination of violence.

Effectiveness of Conjoint Couples Treatment Although the research is relatively new, several forms of conjoint couples treatment and multicouple group treatment of partner violence appears to be *more effective* than traditional gender-based models in reducing physical violence (Stith et al., 2012). For men in particular, the multicouple groups had more positive outcomes than single-couple sessions. In the carefully designed conjoint programs studied, there was no evidence that women in conjoint couples treatment for domestic violence were more likely to be abused than when the batterers were treated separately; thus, if thoughtfully approached, it does not appear that conjoint couples treatment increases the incidence of violence (Stith et al., 2002). Furthermore, using conjoint couples treatment when the batterer also had a substance abuse issue was *superior* to individual treatment in reducing violence (Stith et al., 2012).

Relationship Enhancement Programs

One of the most common forms of multicouple group sessions, relationship enhancement programs are preventive interventions for nondistressed couples who want to improve the quality of their relationship (Halford et al., 2002; Markman & Rhoades, 2012). These programs are typically offered as multicouple psychoeducational groups, which are generally grounded in cognitive-behavioral therapies. These groups may include lectures, demonstrations, audiovisual programs, practice, and discussion. Some of the more prominent programs have included the following:

- Relationship Enhancement (RE; Guerney, 1987)
- Prevention and Relationship Enhancement Program (PREP; Markman, Stanley, & Blumberg, 2001)
- Minnesota Couples Communication Program (MCCP; Miller, Nunnally, & Wackman, 1976)
- Practical Application of Intimate Relationship Skills (PAIRS; DeMaria & Hannah, 2002)

All these programs promote positive communication, conflict management, and positive expression of affection. Variations in the curriculum include the following:

- Preventing destructive conflict (emphasized in PREP)
- Developing empathy for one's partner (emphasized in RE)
- Strong emphasis on conflict management (emphasized in MCCP and RE)
- Commitment, respect, love, and friendship (emphasized in PREP)

Effectiveness of Multicouple Relationship Enhancement Programs

In the research literature on these programs, most couples report high satisfaction, rating skill training in communication as the most helpful intervention (Halford et al., 2002; Markman & Rhoades, 2012). Meta-analyses report an effect size of .44 (almost half a standard deviation difference from control groups) for communication at follow-up, which is impressive when you consider that they were not distressed initially, with stronger findings for more rigorously designed studies (Markman & Rhoades, 2012). In addition, skill-based relationship enhancement results in a significant increase in relationship skills upon completion of the program. Studies vary from no improvement to modest improvement in relationship satisfaction immediately following completion of a course.

However, there is limited research on the long-term effectiveness of these programs. Such research has only been conducted on skill-based groups that met for four to eight sessions and that targeted dating, engaged, or recently married couples. These long-term studies indicate the following:

- Relational skills learned in the group are maintained for a period of several years but begin to trail off in 5 to 10 years.
- PREP participants reported enhanced relationship satisfaction and functioning 2 to 5 years after the course.
- PREP participants reported fewer incidents of violence at the 5-year follow-up than those who did not participate.
- Certain effects, such as increased satisfaction and decreased divorce rate, seemed to become evident at the 4- to 5-year mark, meaning that the difference between couples who participated in PREP and those who began marriage without an enrichment program became *detectable only several years later.*
- In one study involving older, more distressed couples, PREP was not effective, indicating that enrichment programs that work early in a relationship might not be equally effective for more established or more distressed couples.

Guidelines

Based on their review of the research literature, Halford et al. (2002) propose the following as best practices in relationship enhancement:

- *Assess Risk and Protective Factors:* Before admitting couples to a relationship enhancement group, therapists should screen them for risk factors associated with higher stress and divorce, such as parental divorce, age, previous marriages, length of time the partners have known each other, cohabitation history, and the presence of stepchildren. Additionally, therapists should consider protective factors and strengths, such as clear communication and realistic relationship expectations.
- *Encourage High-Risk Couples to Attend:* Relationship enhancement is more likely to benefit high-risk couples early in their relationships because they are more likely to need better skills to manage their stressors.
- *Assess and Educate About Relationship Aggression:* Because physical and verbal aggression occur at high rates early in couple relationships, education about aggression and how to handle it can better prepare couples to reduce hostile and dangerous behavior.
- *Offer Relationship Education at Change Points:* In addition to the beginning of a relationship, couples can benefit from relationship enhancement at numerous other life and developmental transition points, such as the birth of a child, relocation, major illness, and unemployment.
- *Promote Early Presentation of Relationship Problems:* Because long-standing distress predicts poor response to couples therapy, couples should be encouraged to address mild distress as it emerges rather than allow the relationship to further deteriorate.
- *Match Content to Couples with Special Needs:* Program enhancement needs to be adapted to fit a couple's special needs, such as a major mental health diagnosis, alcohol or substance abuse, and stepfamily issues.
- *Enhance Accessibility of Evidence-Based Relationship Education Programs:* Programs can be made more accessible by increasing the formats in which they are offered: weekly, on weekends, or online.

Parent Training

Described in the Juice in Chapter 11, parent training is a cognitive-behavioral intervention that has shown effectiveness in treating a wide range of childhood problems. It is the treatment of choice for oppositional defiant disorder (ODD) and the *only treatment that produces normalization of child functioning with attention-deficit, hyperactivity disorder* (ADHD; Northey et al., 2002). Parent training has been shown to be effective with both individual families and multifamily groups (DeRosier & Gilliom, 2007; van den Hoofdakker, van der Veen-Mulders, & Sytema, 2007).

Parent training group curricula include similar elements to parent training with individual families (see Chapter 11):

- Encouraging positive reinforcement to shape desired behaviors
- Using natural and logical consequences whenever possible
- Improving parents' ability to listen and communicate understanding to their child
- Increasing parents' use of "I" statements
- Providing multiple-choice options for children in negotiations when appropriate
- Training parents in emotional coaching to help their children identify, express, and appropriately manage their emotions
- Contingency contracting and point systems to motivate children
- Training in how to use time-outs and other punishments to reduce negative behaviors

Summary

Group therapy with couples and families is gaining recognition as an evidence-based approach for working with a number of common clinical issues, including schizophrenia, bipolar disorder, domestic violence, ADHD, and ODD. Most evidence-based forms of couple and family group psychotherapy are psychoeducational groups that integrate a cognitive-behavioral, family systems approach. New therapists are strongly encouraged to seek advanced training in groups they plan to lead and, whenever possible, cofacilitate groups to "learn the ropes."

Compared with other forms of therapy, psychoeducational groups require a unique set of skills:

- The ability to build rapport with an entire group rather than an individual or a family
- Public speaking skills
- An understanding of curriculum design
- Technical skills to produce overheads and handouts
- Creativity to select and design engaging group activities
- The ability to manage conflict between unrelated persons

Although initially challenging, learning to facilitate couple and family groups adds refreshing diversity to one's clinical workload and provides numerous opportunities for further professional growth.

Online Resources

Minnesota Couples Communication Program (MCCP):
www.couplecommunication.com

Relationship Enhancement Program (PE), National Institute for Relationship Enhancement:
www.nire.org

Practical Application of Intimate Relationship Skills (PAIRS):
www.pairs.com

Prevention and Relationship Enhancement Program (PREP):
www.prepinc.com

SAMHSA Description of PREP, Review of Evidence-Based Program:
www.nrepp.samhsa.gov/programfulldetails.asp?PROGRAM_ID=86

SAMHSA Family Psychoeducation Toolkit:
www.mentalhealth.samhsa.gov/cmhs/communitysupport/toolkits/family

References

DeMaria, R., & Hannah, M. (2002). *Building intimate relationships: Bridging treatment, education and enrichment through the PAIRS Program.* New York: Brunner/Routledge.

DeRosier, M. E., & Gilliom, M. (2007). Effectiveness of a parent training program for improving children's social behavior. *Journal of Child and Family Studies, 16,* 660–670.

Dutton, D. G., & Corvo, C. (2006). Transforming a flawed policy: A call to revive psychology and science in domestic violence research and practice. *Aggression and Violent Behavior, 11,* 457–483.

Dutton, D. G., & Corvo, C. (2007). The Duluth Model: A data-impervious paradigm and a failed strategy. *Aggression and Violent Behavior, 12,* 658–667.

Guerney, B. G. (1987). *Relationship enhancement manual.* Bethesda, MD: Ideal.

Halford, W. K., Markman, H. J., Stanley, S., & Kline, G. H. (2002). Relationship enhancement. In D. Sprenkle (Ed.), *Effectiveness research in marriage and family therapy* (pp. 191–222). Alexandria, VA: American Association for Marriage and Family Therapy.

Holtzworth-Munroe, A., & Stuart, G. L. (1994). Typologies of male batterers: Three subtypes and the differences among them. *Psychological Bulletin, 116,* 476–497.

Hooley, J. M., Rosen, L. R., & Richters, J. E. (1995). Expressed emotion: Toward clarification of a critical construct. In G. A. Miller (Ed.), *The behavioral high-risk paradigm in psychopathology* (pp. 88–120). New York: Springer Verlag.

Kaslow, N., Broth, M., Smith, C., & Collins, M. (2012). Family-based interventions for child and adolescent disorders. *Journal of Marital and Family Therapy, 38,* 82–100.

Lucksted, A., McFarlane, W., Downing, D., & Dixon, L. (2012). Recent developments in family psychoeducation as an evidence-based practice. *Journal of Marital and Family Therapy, 38,* 101–121.

Markman, H., & Rhoades, G. (2012). Relationship education research: Current status and future directions. *Journal of Marital and Family Therapy, 38,* 169–200.

Markman, H., Stanley, S., & Blumberg, S. (2001). *Fighting for your marriage.* San Francisco, CA: Wiley.

McFarlane, W. R., Dixon, L., Lukens, E., & Lucksted, A. (2002). Severe mental illness. In D. Sprenkle (Ed.), *Effectiveness research in marriage and family therapy* (pp. 255–288). Alexandria, VA: American Association for Marriage and Family Therapy.

Miller, S. L., Nunnally, E. W., & Wackman, D. B. (1976). A communication training program for couples. *Social Casework, 57,* 9–18.

Northey, W. F., Wells, K. C., Silverman, W. K., & Bailey, C. E. (2002). Childhood behavioral and emotional disorders. In D. Sprenkle (Ed.), *Effectiveness research in marriage and family therapy* (pp. 89–122). Alexandria, VA: American Association for Marriage and Family Therapy.

Pence, E., & Paymar, M. (1993). *Education groups for men who batter: The Duluth Model.* New York: Springer.

Stith, S., McCollum, E., Amanor-Boadu, Y., & Smith, D. (2012). Systemic perspectives on intimate partner violence treatment. *Journal of Marital and Family Therapy, 38,* 220–240.

Stith, S. M., McCollum, E. E., Rosen, K. H., & Locke, L. D. (2002). Multicouple group treatment for domestic violence. In F. Kaslow (Ed.), *Comprehensive textbook of psychotherapy* (Vol. 4). New York: Wiley.

Stith, S. M., Rosen, K. H., & McCollum, E. E. (2002). Domestic violence. In D. Sprenkle (Ed.), *Effectiveness research in marriage and family therapy* (pp. 223–254). Alexandria, VA: American Association for Marriage and Family Therapy.

van den Hoofdakker, B. J., van der Veen-Mulders, L., & Sytema, S. (2007). Effectiveness of behavioral parent training for children with ADHD in routine clinical practice: A randomized controlled study. *Journal of the American Academy of Child and Adolescent Psychiatry, 46,* 1263–1271.

Walker, Lenore E. (1979). *The battered woman.* New York: Harper & Row.

Yalom, I. D. (1970/1985). *The theory and practice of group psychotherapy* (3rd ed.). New York: Basic Books.

Part III

Cross-Theoretical Case Conceptualization and Integration

CHAPTER 17

Cross-Theoretical Case Conceptualization and Integration

Knowing Where to Look

There are few moments as exciting or neuroses-inducing in a therapist's training as "the first session": seeing your first client and losing your therapeutic "virginity," so to speak. We know the questions that follow: "Was it good for you?" "Did I get it right?" "Was I okay?" Before the session, there are other predictable questions: "What do I say?" "What do I do?" "What if I can't remember X?" Although logical, these questions can quickly get a new therapist lost and off course. That is where knowing where to look makes a big difference. To be effective, therapists need to master the art of *viewing*, which in a talking profession like therapy refers to knowing where to focus your attention while listening.

The heart of therapy—the seeming brilliance of a great therapist—has always been in the viewing. The most useful question for new therapists to ask their supervisors is: *What should I be noticing and listening for when I talk with this client?* Thankfully, this is much easier than trying to memorize what to say. The apparent magic that distinguishes master therapists from average therapists and average therapists from the average person lies in what the person attends to when another is speaking. Essentially, the better you get at knowing how to focus your viewing, the better therapist you will be. I believe it is a skill that great therapists continually develop and refine over the course of their careers—so don't plan on mastering it anytime soon.

Case conceptualization is the technical term for the therapeutic art of viewing. Although it is also sometimes called *assessment, assessment* is a tricky term because it can refer to two different therapeutic tasks: a case conceptualization approach to assessing individual and family dynamics (covered in this chapter) or a diagnostic approach to assessing symptomology (covered in my other text *Mastering Competencies in Family Therapy*). To reduce confusion, I refer to a theory-informed assessment as *case conceptualization* and to a diagnostic assessment as a *clinical assessment*.

Theories provide therapists with unique lenses through which to view clients' problems. Like a detailed map, they place problems in a broader and more comprehensive context that allows therapists to see how pieces fit together and that provides clues to the best path out of a tangled situation.

Case Conceptualization and the Art of Viewing

Case conceptualization enables therapists to generate new perspectives that enable them to be helpful to clients. For example, Ron and Suzie have been arguing for months. They seek out a therapist hoping that somehow, some way, the therapist can help them resolve their differences. When they enter therapy, each is thinking: "I hope the therapist can help my spouse see what he/she is doing wrong and encourage him/her to fix it. I know I have problems too, but I am sure that as soon as my spouse changes, it will be easy for me to change." They have told this and less benevolent versions of their stories to their family and friends, yet no one has been able to help either Ron or Suzie to make meaningful changes. So what makes the therapist different from the family, friends, and hairdressers who listen to this story? Does the therapist have special knowledge that will allow him/her to provide the "answer" to the couple's problem? Or does the therapist simply serve as a socially sanctioned referee to this dispute? Or does the therapist do something more?

Although some therapists define themselves as "educators" or "mediators," most define their roles as promoting change through a transformational interpersonal process that is far more nuanced than education or mediation. Therapists differ from friends, educators, and mediators in their ability to view the situation in new and useful ways. From these new perspectives, possibilities for intervention emerge that did not arise in any of the previous conversations between the partners or with their family and friends.

Theory-Specific vs. Cross-Theory Case Conceptualization

In the preceding chapters, you have learned about theory-specific case conceptualization, which helps you conceptualize a client using a single theory. In this chapter, you will learn a more advanced approach of cross-theoretical case conceptualization, an approach that considers key conceptualization elements from all major schools of family therapy. This is a far more complex assignment that will likely require 10 or more hours to complete your first one; with practice, it goes much faster. Why bother with such a laborious task, especially if you plan to work from only one theoretical position? Let me list the reasons:

Benefits of Cross-Theoretical Conceptualization

- *Clinical Usefulness:* Even when working with a single theory, evaluating your client from multiple perspectives can only deepen and broaden your understanding of your client, and you are likely to get some useful ideas from other sources.
- *Self-Supervision:* Whenever you feel stuck with a client, you can use cross-theoretical case conceptualization to help you identify a dynamic you may not be considering to help get therapy unstuck.
- *Exam Preparation:* Using multiple theories to conceptualize a single case will significantly improve your ability to distinguish one theory from another, helping you to prepare for class and licensing exams.
- *Insight into Theory:* By routinely comparing and contrasting theories, you begin to gain a deeper insight into how therapeutic concepts interrelate, with some typically correlating (such as Satir's communication stances and attachment patterns).

Cross-Theoretical Conceptualization vs. Integration

Cross-theoretical conceptualization is different than theoretical integration. Cross-theoretical conceptualization is a practice of considering a client's situations from multiple theoretical frames without necessarily using every part. In contrast, integration is a process in which a therapist uses multiple theories to intervene. Cross-theoretical conceptualization could be used as part of an integrative approach or to support using a single approach. If you would like to use an integrative approach, you have many options.

Integration Options

The term *integration* is used a lot in the field, but rarely are folks using the same definition. This perhaps accounts for why there is so much debate about the issue. So the next time you finding yourself arguing with a colleague about integration, you might just want to clarify which variety of integration is being discussed:

- Common factors
- Theoretical integration
- Assimilative integration
- Technical integration
- Systematic treatment selection
- Syncretism and sloppy thinking

Common Factors Approach to Integration

As described in Chapter 2, therapists who use the *common factors approach* for integration conceptualize treatment using a common factors model, which identifies the core ingredients of counseling models, such as counseling alliance, client resources, and hope. Therapists who use this approach can then integrate concepts from various approaches using the common factors model. For example, a therapist may use the empathic approach of Satir (Chapter 8) to build the counseling relationship and solution-generating questions from solution-focused counseling (see Chapter 12) to generate hope. These practices are then incorporated and coherently connected using the common factors theory.

Theoretical Integration

As the name implies, therapists who use *theoretical integration* combine two or more theories into a single coherent theoretical model with the intention to improve or expand the potentials of either alone. Most evidence-based treatments (manualized treatments for a specific population; vaguely remember Chapter 2) are examples of theoretical integration. Emotionally focused therapy is an example of how systems theory, humanistic theory, and attachment theory have been combined to treat couples with better than average results (Chapter 15; Johnson, 2004). Similarly, the now well-established cognitive-behavioral approaches (see Chapter 11) are an example of integrating behavioral and cognitive theories. In addition to these more formalized integrative models, individual clinicians often integrate theories with shared philosophical foundations, as described in detail in Chapter 4. This is distinguished from technical eclecticism (see below) by the clear and thought-out combining of two or more theories.

Assimilative Integration or Selective Borrowing

Closely related to theoretical integration, *assimilative integration* and *selective borrowing* refer to having a firm theoretical grounding in one approach while selectively integrating practices from other approaches to allow for a broader range of intervention possibilities (Norcross & Beutler, 2008). In this approach, for example, a therapist may integrate solution-based scaling into a humanistic approach to identify goals and progress. After years of practice, the vast majority of therapists who identify with a single approach most likely also practice some form of assimilative integration.

Technical Eclecticism

Therapists using *technical eclecticism* draw freely from techniques of various theories, without significant concern about the theoretical underpinnings. Instead, the primary focus is identifying interventions that are likely to work for a specific client with a specific concern (Norcross & Beutler, 2008). Eclectics put little emphasis on the philosophical and theoretical principles underlying a technique and instead focus on trying to achieve desired outcomes from a specific theory. Increasingly, therapists are able to include evidence-based practice (using a literature review of research studies to identify potential treatments; see Chapter 2) to more precisely inform the choice of treatment.

Systematic Treatment Section

Systematic treatment selection models, such as Norcross and Beutler's (2008), provide an organized template for integrating theories that incorporates evidence-based practice (see Chapter 2). These approaches are designed to identify appropriate treatments and therapeutic relationships for a specific client. In Norcross and Beutler's approach, they (a) use evidence-based practices to identify what the research has shown to be an effective treatment for a particular client or problem, (b) draw from multiple systems of therapy to address client issues, (c) use five diagnostic and nondiagnostic dimensions to develop a case conceptualization, and (d) adapt treatment methods *and* the counseling relationship to best meet client needs. The five areas used to develop a case conceptualization include:

- *Mental Health Diagnosis:* The client is diagnosed using standard medical diagnoses in the *DSM* (APA, 2000).
- *Stages of Change:* The client's readiness for change is described in one of four stages: contemplation, preparation, action, or maintenance.
- *Coping Style:* The client's coping style is generally described as *internalizing* (e.g., depressed, self-critical, inhibited) or *externalizing* (e.g., aggressive, impulsive, stimulation-seeking).
- *Resistance Level:* A client's resistance level refers to their willingness to take direct intervention from the therapist: if the client's resistance level is low, the therapist can use more directive techniques, whereas if the client's resistance is high, nondirective and paradoxical interventions are more appropriate.
- *Patient Preferences:* Finally, the therapist considers client's preferences for the type of relationship and treatment; this often serves to accommodate to diversity issues as well as the interaction of diversity issues between therapist and client (e.g., gender differences, age differences, ethnicity differences, educational differences).

Syncretism and Sloppy Thinking

As you might imagine, it is quite possible—or even likely—to use the labels "integration" or "eclectic" to simply mask a haphazard approach of doing what you feel like when you feel like it, without much understanding of what you are doing or why or if there is any scientific or theoretical reason to do (a.k.a., "bad therapy"). Norcross and Beutler (2008) refer to this as *syncretism,* which simply refers to combining various philosophies. It can take the form of a less-than-optimal approach to an outright lawyers-will-be-involved harmful approach.

Too often a therapist develops pet theories or techniques because they have not received proper training, often unaware of what is being missed. Unfortunately, this is increasingly common as intensive post-degree in specific treatment approaches or areas of specialty are less common in the United States due to the increasing costs and drop in reimbursement rates over the years. Perhaps the best way to ensure you don't slip into sloppy syncretic thinking is to seek intensive training in at least one approach, including an integrative one. In addition, understanding philosophical differences between schools of counseling will also help prevent the mass confusion that leads to less-than-effective forms of syncretism.

Integration with Integrity

The key to integrating with integrity is understanding the difference between *philosophy* and *theory*. Each theoretical model is built upon basic philosophical assumptions, such as what counts as "truth," what counts as "real," what does it mean to be human, and how do humans change. They also define certain practices based on these assumptions. In most cases, skillful integration involves grounding yourself clearly in *one* set of philosophical assumptions and then *integrating* practices from other models and modifying them as necessary to *fit largely within a single set of philosophical principles*. At this point, you may find it helpful to return to Chapter 4 to review your options.

Overview of Cross-Theoretical Case Conceptualization

As therapists become more experienced, case conceptualization takes place primarily in their heads—while clients are talking. It happens so fast that often they have difficulty tracing their steps. However, new therapists must take things more slowly. Similar to learning a new dance step, the new move needs to be broken down into small pieces with specific instructions for where the hands and feet go; with practice, the dancer is able to put the pieces together more quickly and smoothly until it becomes "natural." That is what we are going to do here with case conceptualization. So, let's start with identifying the components of a systemic cross-theoretical case conceptualization:

1. Introduction to Client
2. Presenting Concern
3. Background Information
4. Client/Family Strengths and Diversity
5. Family Structure
6. Interactional Pattern
7. Intergenerational and Attachment Patterns
8. Solution-Based Assessment
9. Postmodern and Cultural Discourse Conceptualization
10. Client Perspectives

The complete form for case conceptualization is included at the end of this chapter and online (www.cengage.com; www.masteringcompetencies.com), and in each of the following sections you will find detailed instructions and guidelines for completing this conceptualization.

Introduction to Client and Significant Others

I. Introduction to Client and Significant Others

*Denotes clients involved in therapy process

        Age, ethnicity, occupation/grade, other relevant identifiers

AF/AM: _____

AF/AM: _____

CF/CM: _____

CF/CM: _____

Case conceptualization starts by identifying (a) who the client is (individual, couple, or family) and (b) the most salient demographic features that relate to treatment. Common demographic information includes the following.

- Age
- Ethnicity
- Gender
- Sexual orientation/HIV status
- Current occupation/work status or grade in school

I use a combination of abbreviations that make it easy to track family members using confidential notation:

AF = Adult female

AM = Adult male

CF = Child female

CM = Child male

To distinguish members in large families and same-sex couples, I add the age after each abbreviation (AF36, CM8). This can also be particularly helpful to a supervisor or instructor to follow your notes.

Presenting Concern

The presenting concern is a description of how all parties involved are defining the problem: client, family, friends, school, workplace, legal system, and society.

II. Presenting Concern(s)

Client's Description(s) of Problem(s):

AF/AM: _____

AF/AM: _____

CF/CM: _____

CF/CM: _____

Extended Family Description(s) of Problems: _____

_____

_____

Broader System Problem Descriptions: Description of problem from referring party, teachers, relatives, legal system, etc.:

_____: _____

_____: _____

Often new and even experienced therapists assume that this description is a straightforward and clear-cut matter. Though sometimes it is, usually it is surprisingly complex. Anderson and Goolishian (Anderson, 1997; Anderson & Gehart, 2006) developed a unique means of conceptualizing the presenting problem in their collaborative language systems approach, also referred to as collaborative therapy (Chapter 14). This postmodern approach maintains that each person who is talking about the problem is part of the problem-generating system, the set of relationships that produced the perspective or idea that there is a problem. Each person involved has a different definition of the problem; sometimes the difference is slight and sometimes it is stark. For example, when parents bring a child to therapy, the mother, father, siblings, grandparents, teachers, school counselors, doctors, and friends have different ideas of what the problem actually is. The mother may think it is a medical problem, such as ADHD; the father may believe it is related to his wife's permissiveness; the teacher may say it is poor parenting; and the child may think there really is not a problem at all.

Historically, therapists have moved rapidly to define the problem according to their theoretical worldview—either a formal diagnosis (ADHD, depression, etc.) or another mental health category (such as parenting style, defense mechanism, family dynamics, etc.)—with little reflection on the contradictory opinions and descriptions of the various people involved. Although this may be practical at one level, the more therapists can remain open to the family's alternative descriptions of the problem, the more they can remain adaptable and creative. They can maintain stronger rapport with each person involved, honoring each person's perspective and referring to it throughout treatment.

Furthermore, being mindful of the multiple problem definitions gives the therapist greater maneuverability when treatment stagnates or conditions do not improve.

A description of the presenting problem should include the following:

1. The reason(s) each client states he/she is seeking counseling or has been referred
2. Any information from the referring agent (teacher, doctor, psychiatrist, etc.) and his/her description of the problem
3. A brief history of the problem and family (if applicable)
4. Descriptions of the attempted solutions and the outcome of these attempts
5. Any other problem-related information that may be relevant to the situation

Background Information

> **III. Background Information**
>
> *Trauma/Abuse History* (recent and past): _____
>
> *Substance Use/Abuse* (current and past; self, family of origin, significant others): _____
>
> *Precipitating Events* (recent life changes, first symptoms, stressors, etc.): _____
>
> *Related Historical Background* (family history, related issues, previous counseling, medical/mental health history, etc.): _____

Obtaining background information about the problem is the next step. Traditionally, therapists have included information such as the following:

- History of trauma and abuse, including sexual abuse, physical violence, assaults, national disasters, etc.
- Substance use and abuse, current or past
- Precipitating events: recent events that may be related to the problem, occurrence of first symptoms, recent stressors, etc.
- Related historical background, such as health situation and medications, previous counseling, medical issues, etc.

Often, this background information is considered the "facts" of the case. However, as family therapists have historically cautioned, how we describe the facts makes all the difference (Anderson, 1997; O'Hanlon & Weiner-Davis, 1989; Watzlawick, Weakland, & Fisch, 1974). For example, whether you begin by saying that the client "recently won a state-level academic decathlon" or whether you begin with "her mother recently divorced her alcoholic father" paints two very different pictures of the same client for both you, the therapist, and anyone else who reads the assessment. Therefore, although this may seem like the "factual" part of the report in which you as a professional are not imposing any bias; in fact, therapists impose bias by their subtle choice of words, their ordering of information, and their emphasis on particular details.

Based on research about the importance of the therapeutic relationship and of hope (Lambert & Ogles, 2004; Miller, Duncan, & Hubble, 1997), I recommend that therapists write the background section so that they and anyone reading the report, including potentially the client, will have a positive impression of the client and hope for the client's recovery; these two factors affect the outcome of treatment.

Strengths and Diversity Resources

> **IV. Client/Family Strengths and Diversity Resources**
>
> *Strengths and Resources*
>
> *Personal:* _____
>
> *Relational/Social:* _____
>
> *Spiritual:* _____
>
>
> *Diversity: Resources and Limitations*
>
> Identify potential resources and limitations available to clients based on their age, gender, sexual orientation, cultural background, socioeconomic status, religion, regional community, language, family background, family configuration, abilities, etc.
>
> *Unique Resources:* _____
>
> *Potential Limitations:* _____

Client strengths and resources should be the first thing assessed. This is a lesson I learned the hard way. When I began teaching systemic assessment, I put the client strength section at the end because it is more clearly associated with solution-based and postmodern approaches (Chapters 12, 13, and 14; Anderson, 1997; de Shazer, 1988; White & Epston, 1990), which were developed later historically. What I discovered is that after reading about the presenting problem, history, and problematic family dynamics, I was often feeling quite hopeless about the case. However, often upon reading the strengths section at the end, I would immediately perk up and find myself having hope, deep respect, and even excitement about the clients and their future. I have since decided to start by assessing strengths and believe it puts the therapist in a more resourceful mindset, whether working from either a systemic or postmodern perspective.

Emerging research supports the importance of identifying client strengths and resources. Researchers who developed the *common factors model* (discussed in Chapter 2; Lambert & Ogles, 2004; Miller, Duncan, & Hubble, 1997) estimate that 40% of outcomes can be attributed to client factors, such as severity of symptoms, access to resources, and support system; the remaining factors include the therapeutic relationship (30%), therapist interventions (15%), and the client's sense of hope (15%). Assessing for resources leverages client factors (40%), strengthens the therapeutic relationship (30%), and instills hope (15%), thus drawing on three of the four common factors. Thus, the benefit of assessing strengths is hard to overestimate.

To conceptualize a full spectrum of client strengths and resources, therapists can include strengths at several levels:

1. Personal/individual strengths
2. Relational/social strengths and resources
3. Spiritual resources

Personal or Individual Strengths

When assessing for personal or individual strengths and resources, the therapist can begin by reviewing two general categories of strengths: abilities and personal qualities.

- *Abilities:* Where and how are clients functioning in daily life? How do they get to sessions? Are they able to maintain a job, a hobby, or a relationship? Do they have any special talents, either now or in the past? If you look, you will always find a wide range of abilities with even the most "dysfunctional" of clients, especially if you consider the past as well as the present and future.

    Naming their abilities can increase clients' sense of hope and confidence to address the problem at hand. I find this especially helpful with children. If a child is having academic problems at school, the family, teachers, and child may not notice that the child is excelling in an extracurricular activity such as karate, soccer, or piano. Often noticing these areas of accomplishment makes it easier for all involved to find hope for improving the situation.

    Identifying abilities may also give clients or therapists creative ideas about how to solve a current problem. For example, I worked with a recovering alcoholic who hated the idea of writing but often spoke of how music inspired her. By drawing on this strength, we developed the idea of creating a special "sobriety mix" of favorite songs to help maintain sobriety and prevent relapse, an activity that had deep significance and inspiration for her.

- *Personal Qualities:* Ironically, the best place to find personal qualities is within the presenting problem or complaint. What brings clients to see a therapist is usually the flip side of a strength. For example, if a person complains about worrying too much, that person is equally likely to be a diligent and productive worker. Persons who argue with a spouse or child are more likely to speak up for themselves and are generally invested in the relationship in which they are arguing. In virtually all cases, the knife cuts both ways: each liability contains within it a strength in another context. Conversely, a strength in one context is often a problem in another. Here is a list of common problems and their related strengths.

Problems and Related Strengths

Problem	Possible Associated Strength
Depression	• Is aware of what others think and feel • Is connected to others and/or desires connection • Has dreams and hopes • Has had the courage to take action to realize dreams • Has a realistic assessment of self/others (according to recent research; Seligman, 2004)
Anxiety	• Pays attention to details • Desires to perform well • Is careful and thoughtful about actions • Is able to plan for the future and anticipate potential obstacles
Arguing	• Stands up for self and/or beliefs • Fights injustice • Wants the relationship to work • Has hope for better things for others/self
Anger	• Is in touch with feelings and thoughts • Stands up against injustice • Believes in fairness • Is able to sense his/her boundaries and when they are crossed
Overwhelmed	• Is concerned about others' needs • Is thoughtful • Is able to see the big picture • Sets goals and pursues them

Identifying strengths relies heavily on the therapist's viewing skills. A skilled therapist is able to see the strengths that are the flip side of the presenting problem while still remaining aware of the problem.

Relational or Social Strengths and Resources

Family, friends, professionals, teachers, coworkers, bosses, neighbors, church members, salespeople, and numerous others in a person's life can be part of a social support network that helps the client in physical, emotional, and spiritual ways:

- *Physical* support includes people who may help with running errands, picking up the children, or doing tasks around the house.
- *Emotional* support may take the form of listening or helping resolve relational problems.
- *Community* support includes friendships and acceptance provided by any community and is almost always there in some form for a person who may be feeling marginalized because of culture, sexual orientation, language, religion, or similar factors. These communities are critical for coping with the stress of marginalization.

Simply naming, recognizing, and appreciating that there is support can immediately increase a client's sense of hope and reduce feelings of loneliness.

Spiritual Resources

Increasingly, family therapists are becoming aware of how clients' spiritual resources can be used to address their problems (Walsh, 2003). For this reason, therapists should become familiar with the major religious traditions in their community, such as Protestantism, Catholicism, Judaism, Islam, New Age religions, and Native American practices.

Spirituality can be defined as how a person conceptualizes his/her relationship to the universe, life, or God (or however he/she constructs that which is larger than the self). Everyone has some form of spirituality, or a belief in how the universe operates; see Bateson (1972, 1979/2002) and Gergen (1999).

Spirituality

The rules of "how life should go" always inform (a) what the person perceives to be a problem, (b) how a person feels about it, and (c) what that person believes can "realistically" be done about it, all of which a therapist wants to know to develop an effective therapeutic plan.

Questions for Assessing Spirituality A therapist can use some of the following questions to assess a client's spirituality, whether traditional or nontraditional:

- Do you believe there is a god or some form of intelligence that organizes the universe? If so, what types of things does that being or force have control over?
- If there is not a god, by what rules does the universe operate? Why do things happen? Or is life entirely random?
- What is the purpose and/or meaning of human existence? How does this understanding inform how a person should approach life?
- Is there any reason to be kind to others? To one's self?
- What is the ideal versus the realistic way to approach life?
- Why do "bad" things happen to "good" people?
- Do you believe things happen for a reason? If so, what reason?
- Do you belong to a religious community or spiritual circle of friends that provides spiritual support, inspiration, and/or guidance in some way?

With the answers to these questions, therapists can create a map of the client's world that they can use to develop conversations and interventions that are deeply meaningful and a good "fit" for the client. An accurate understanding of a client's "map of life" reveals what logic and actions will motivate the client to make changes, providing therapists with invaluable resources. Often many clients from traditional religious backgrounds and New Age groups believe that "things happen for a reason." These clients can use this one belief to radically and quickly transform how they feel, think, and respond to difficult situations. For example, a recent client of mine who was feeling depressed after being unexpectedly fired experienced a rapid improvement in mood when she began to see the situation as a sign from the divine that she needed to pursue an old career dream she had been putting off for years; this belief allowed her to mobilize her energy and hope to start moving in a positive direction.

Diversity Resources and Limitations

In addition to client and family strengths, all clients bring with them certain resources and limitations from their experiences of diversity, including race, ethnicity, age, gender, sexual orientation, gender orientation, socioeconomic status, educational level, abilities, religion, language, etc. Thus, therapists can assess for these too.

Common resources due to diversity include:

- Strong support network of people who understand client's situation
- Sense of community and connection
- Sense of purpose and direction
- Resources for solving problems
- Beliefs that provide comfort
- Connections with persons outside of immediate network
- Access to social services

Common limitations include:

- Isolation, difficulty meeting others
- Experiences of harassment and discrimination
- Difficulty finding opportunities
- Difficulty communicating with institutions
- Difficulty accessing social services
- Lack of sufficient financial resources, housing, legal representation, etc.

Family Structure

This assessment can be used with individuals, couples, or families. The approach presented here draws from the major theories in family therapy. Consistent with both systemic (earlier family therapies) and postmodern (later forms of family therapy) practices, it uses multiple descriptors to generate a complete, "both/and" perspective (Keeney, 1983) and a rich, multivoiced depiction of the problem (Anderson, 1997).

> **V. Family Structure**
>
> *Family Life Cycle Stage (Check all that apply)*
>
> ❏ Single Adult ❏ Committed Couple ❏ Family With Young Children ❏ Family With Adolescent Children ❏ Divorce ❏ Blended Family ❏ Launching Children ❏ Later Life
>
> Describe struggles with mastering developmental tasks in one or more of these stages:
>
> _____
> _____
> _____
>
> *Typical style for regulating closeness and distance in couple/family:* _____
>
> _____
> _____
> _____

(continued)

Boundaries with/between

Primary couple (A_/A_): ❑ Enmeshed ❑ Clear ❑ Disengaged ❑ NA. Description/example:
_____

Parent A_ & Children: ❑ Enmeshed ❑ Clear ❑ Disengaged ❑ NA. Description/example:
_____

Parent A_ & Children: ❑ Enmeshed ❑ Clear ❑ Disengaged ❑ NA. Description/example:
_____

Siblings: ❑ Enmeshed ❑ Clear ❑ Disengaged ❑ NA. Description/example: _____
_____

Extended Family: ❑ Enmeshed ❑ Clear ❑ Disengaged ❑ NA. Description/example:
_____

Friends/Colleagues/Other: ❑ Enmeshed ❑ Clear ❑ Disengaged ❑ NA. Description/example: _____

Triangles/Coalitions

❑ Cross-generational coalitions: Describe: _____

❑ Coalitions with family of origin: Describe: _____

❑ Other coalitions: _____

Hierarchy Between Parents and Children ❑ NA

AF/AM: ❑ Effective; ❑ Insufficient (permissive); ❑ Excessive (authoritarian)
❑ Inconsistent

AM/AF: ❑ Effective; ❑ Insufficient (permissive); ❑ Excessive (authoritarian)
❑ Inconsistent

Description/Example to Illustrate: _____

Complementary Patterns between _____

and _____:

❑ Pursuer/distancer; ❑ Over-/under-functioner; Emotional/logical; ❑ Good/bad parent;

❑ Other:_____; Ex: _____

Satir Communication Stances: Describe most commonly used stance under stress.

AF or _____ : ❑ Congruent ❑ Placator ❑ Blamer ❑ Superreasonable ❑ Irrelevant

AM or _____ : ❑ Congruent ❑ Placator ❑ Blamer ❑ Superreasonable ❑ Irrelevant

CF or _____ : ❑ Congruent ❑ Placator ❑ Blamer ❑ Superreasonable ❑ Irrelevant

CM or _____ : ❑ Congruent ❑ Placator ❑ Blamer ❑ Superreasonable ❑ Irrelevant

Describe: _____

> *Gottman's Divorce Indicators:* _____
>
> Criticism: ❑ AF/M; ❑ AM/F; Ex: _____
>
> Defensiveness: ❑ AF/M; ❑ AM/F; Ex: _____
>
> Contempt: ❑ AF/M; ❑ AM/F; Ex: _____
>
> Stonewalling: ❑ AF/M; ❑ AM/F; Ex: _____
>
> Failed repair attempts: ❑ AF/M; ❑ AM/F; Ex: _____
>
> Not accept influence: ❑ AF/M; ❑ AM/F; Ex: _____
>
> Harsh startup: ❑ AF/M; ❑ AM/F; Ex: _____

Family Life Cycle Stage

Assessment of the family structure often begins by identifying the client or family's stage in the family life cycle. Each stage of the life cycle involves different developmental tasks, often requiring rebalancing independence and interdependence (Carter & McGoldrick, 1999). For example, the stage of single adult involves high levels of independence, whereas the committed couple stage requires greater degrees of interdependence and the stage of families with young children requires even greater degrees of interdependence. Often the presenting problem is closely linked to the changes required as a person transitions from one stage to the next. In the case of divorce or blended families, a person may actually be experiencing challenges related to more than one stage: such as raising a teen while trying to form a committed partnership, creating even more complex developmental challenges. The stages are:

- *Leaving Home, the Single Adult:* Accepting emotional and financial responsibility for oneself.
- *Committed Relationship:* Committing to a new system; realigning boundaries with family and friends.
- *Families with Young Children:* Adjusting marriage to make space for children; joining in child-rearing tasks; realigning boundaries with parents and grandparents.
- *Families with Adolescent Children:* Adjusting parental boundaries to increase freedom and responsibility for adolescents; refocusing on marriage and career life.
- *Divorce:* Divorce represents an interruption to the family life cycle, typically requiring most members to increase their sense of independence yet, if children are involved, also maintain a new form of interdependence.
- *Blended families:* Blended families, an increasingly common form of families, typically involves a complex balance of independence and interdependence as two or more family systems become entwined, often at different stages of the family life cycle. Often explicit discussion of needs for independence and interdependence is required to navigate this complex transition, that typically requires several years (Visher & Visher, 1979).
- *Launching Children:* Renegotiating the marital subsystem; developing adult-to-adult relationships with children; coping with aging parents.
- *Family in Later Life:* Accepting the shift of generational roles; coping with loss of abilities; middle generation takes more central role; creating space for wisdom of the elderly.

Boundaries: Regulating Closeness and Distance

Most commonly associated with structural family therapy (Chapter 7), *boundaries* are the rules for negotiating interpersonal closeness and distance (Minuchin, 1974). Boundaries exist *internally* within the family and *externally* with those outside the nuclear family.

These rules are generally unspoken and unfold as two people interact over time, each defining when, where, and how he/she prefers to relate to the other. With couples, these rules are often highly complex and difficult to track. Boundaries can be clear, diffuse, or rigid; all boundaries are strongly influenced by culture.

- *Clear Boundaries and Cultural Variance:* Clear boundaries refer to a range of possible ways that couples and families can negotiate a healthy balance between closeness (we-ness) and separation (individuality). Cultural factors shape how much closeness or separation is preferred. Collectivist cultures tend toward greater degrees of closeness, whereas individualistic cultures tend to value greater independence. The best way to determine whether boundaries are clear is to determine whether symptoms have developed in the individual, couple, or family. If they have, boundaries are probably too diffuse or too rigid. Most people who come in for therapy have reached a point where boundaries that may have worked in one context are no longer working, often due to shifting needs in the family life cycle. For relationships to weather the test of time, couples and families must constantly renegotiate their boundaries (rules for relating) to adjust to each person's evolving needs. The more flexible couples are in negotiating these rules, the more successful they will be in adjusting to life transitions and setbacks.
- *Diffuse Boundaries and Enmeshed Relationships:* When couples or families begin to overvalue togetherness at the expense of respecting each other's individuality, their boundaries become *diffuse* and the relationship becomes *enmeshed* (note: technically boundaries are not enmeshed; they are diffuse). In these relationships, one or more parties may feel that they are being suffocated, that they lack freedom, or that they are not cared for enough. Often people in these relationships feel threatened whenever the other disagrees or does not affirm them, resulting in an intense tug-of-war to convince the other to agree with them. Couples with diffuse boundaries may also have diffuse boundaries with their children, families of origin, and/or friends, with the result that these outside others becoming overly involved in one or both of the partners' lives (e.g., parents, friends, or children become involved in couple's arguments).
- *Rigid Boundaries and Disengaged Relationships:* When couples or families privilege independence over togetherness, their boundaries can become *rigid* and the relationship *disengaged*. In these relationships, a person may not allow the other to influence them, often valued a career or external interest more than the relationship, and frequently has minimal emotional connection. Such couples may have a pattern of keeping others at a distance or compensate by having diffuse boundaries with children, friends, family, or an outside love interest (e.g., an emotional or physical affair). Often difficult to accurately assess, the key indicator of rigid boundaries is whether they are creating problems individually or for the partnership/family.

Questions for Assessing Boundaries

Here are some sample questions to think about while working with an individual, couple, or family to assess boundaries:

Assessing Boundaries in Couple Relationships

- Does the couple have clear boundaries that are distinct from their parenting and family-of-origin relationships?
- Does the couple spend time alone not talking about the children?
- Does the couple report an active sex and romantic life?
- Does the couple still feel a sense of connection apart from being parents?

(continued)

> **Assessing Boundaries in Couple, Family, and Social Relationships**
>
> - Does one or more persons experience anxiety or frustration when there is a difference of opinion?
> - Is one hurt or angry if another has a different opinion or perspective on a problem?
> - Do they use "we" or "I" more often when speaking? Is there a balance?
> - Does each person have a set of personal friends and activities separate from the family and partnership?
> - How much energy goes into the coupe/family vs. outside relationships?
> - What gets priority in each person's schedules? Children? Work? Personal activities? Couple time? Friends?

Triangles and Coalitions

Problem systems are identified in most systemic family therapy approaches. Triangles (Kerr & Bowen, 1988), covert coalitions (Minuchin & Fishman, 1981), and cross-generational coalitions (Minuchin & Fishman, 1981) all refer to a similar process: tension between two people is resolved by drawing in a third person (or forth, a "tetrad"; Whitaker & Keith, 1981) to stabilize the original dyad. Many therapists include inanimate objects or other processes as potential "thirds" in the triangulation process. In this situation, some*thing* else is used to manage the tension in the dyad, such as drinking, drug use, work, or hobbies, which are used to help one or both partners soothe their internal stress at the expense of the relationship.

Therapists assess for triangles and problematic subsystems in several ways:

- Clients overtly describe another party as playing a role in their tension; in these cases the clients are aware of the process at some level.
- When clients describe the problem or conflict situation, another person plays the role of confidant or takes the side of one of the partners (e.g., one person has a friend or another family member who takes his/her side against the other).
- After being unable to get a need met in the primary dyad, a person finds what he/she is not getting in another person (e.g., a mother seeks emotional closeness from a child rather than from a husband).
- When therapy is inexplicably "stuck," there is often a triangle at work that distracts one or both parties from resolving critical issues (e.g., an affair, substance abuse, a friend who undermines agreements made in therapy).

Identifying triangles early in the assessment process enables therapists to intervene more successfully and quickly in a complex set of family dynamics.

Hierarchy Between Child and Parents

A key area in assessing parent and child relationships is hierarchy, a structural family concept (see Chapter 7). When assessing parental hierarchy, therapists must ask themselves: Is the parent-child hierarchy developmentally and culturally appropriate? If the hierarchy is appropriate, the child usually has few behavioral problems. If the child is exhibiting symptoms or there are problems in the parent-child relationship, there is usually some problem in the hierarchical structure: either an excessive (authoritarian) or insufficient (permissive) parental hierarchy given the family's current sociocultural context(s). Immigrant families, because they usually have two different sets of cultural norms for parental hierarchy (the traditional and the current cultural context), have difficulty finding a balance between authoritarian and permissive hierarchies.

Assessing hierarchy is critical because it tells the therapist where and how to intervene. If therapists assess only the symptoms, they may make inappropriate interventions. For example, though children diagnosed with ADD have similar symptoms—hyperactivity, defiance, failing to follow through on parents' requests—these same symptoms can occur in two dramatically different family structures: either too much or too little parental hierarchy. When the parental hierarchy is too rigid, the therapist works with the parents to develop a stronger personal relationship with the child and set developmentally and culturally appropriate expectations. If there is not enough parental hierarchy, the therapist helps parents become more consistent with consequences and set limits and rules. Thus the same set of symptoms can require very different interventions.

When conceptualizing parental hierarchy, it can also be helpful to consider the balance of roles within the parental system. Raser (1999) describes the parenting relationship as comprising *business roles* (setting rules, socializing) and *personal roles* (warmth, fun, caring, play), the former correlating most often with an effective hierarchy and the latter with secure attachment (see Attachment below). Typically, a parent is better at one than the other, which often leads to problematic polarization between the parents. Ideally, both parents are able to balance *within themselves* the business and personal sides of parenting, and both parents can then set an effective hierarchy as well as maintain a close emotional bond. Assessing this single dimension alone can provide therapists with a razor-sharp point of focus for treatment, often resulting in rapid improvements.

Complementary Patterns

Complementary patterns characterize most relationships to a certain degree, especially long-term committed relationships. *Complementary* in this case refers to each person taking on opposite or complementary roles, which range from functional to problematic. For example, a complementary relationship of introvert/extrovert can exist in a balanced and well-functioning relationship as well as in an out-of-balance, problematic relationship; the difference is in the rigidity of the pattern. Classic examples of complementary roles that often become problematic include pursuer/distancer, emotional/logical, over-functioner/under-functioner, friendly parent/strict parent, and so forth. Gottman (1999) indicated that the female-pursue (demand) and male-withdraw patterns existed to some extent in the majority of marriages he studied. However, in distressed marriages, this pattern becomes exaggerated and begins to be viewed as innate personality traits. Assessing for these patterns can help therapists intervene in couples' interactions. In most cases, couples readily identify their complementary roles in their complaints about the relationship: "he's too strict with the kids"; "she always emotionally overreacts"; "I have to do it all the time"; "she never wants sex." These broad, sweeping descriptions of the other suggest a likely problematic complementary pattern.

Satir's Communication Stances

Virginia Satir (Chapter 8; Satir et al., 1991) developed an approach for assessing communication patterns based on five communication stances: placating, blaming, superreasonable, irrelevant, and congruent, with the congruent stance being the model for healthy communication. The first four stances are *survival stances,* which people first develop as children when they are confronted with difficult situations, such as an angry parent, peer pressure, or the threat of rejection. Without the emotional and cognitive capacities to skillfully balance their own needs and wants (*self*), the *other*'s needs and wants, and what is appropriate to the *context,* children attend to one or two of these areas and neglect the rest. For example, a child who uses *placating* may quickly learn that by appeasing parents and friends she is able to keep people from being angry at her and get what she wants most: acceptance. Conversely, a child who copes by *blaming* may learn that by always fighting for his rights he is more likely to get adults to give him what he wants. Alternatively, a child who uses a *superreasonable* stance may find that when she always follows the contextual rules—home, school, church, societal rules, etc.—regardless of what she or others may feel, her world stays organized and stable. Lastly, a child who uses the *irrelevant* stance may learn that by always joking around, changing the subject, and being a little "off the wall," he is able to gain approval by being entertaining and/or keep people distracted enough to not be a problem to him. The survival stances, unfortunately, are unbalanced and do not work well in the long term.

The following table shows how the different stances recognize the three elements of self, other, and context.

Satir's Communication Stances

Style	Recognizes	Ignores
Congruent	Self, other, context	None
Placator	Other, context	Self
Blamer	Self, context	Other
Superreasonable	Context	Self, other
Irrelevant	None	Self, other, context

To someone unfamiliar with these communication styles, they may initially seem simplistic and without significant clinical merit. However, the brief process of assessing these styles provides therapists with excellent information about how best to communicate with clients in session and how to design interventions. The following are some clinical implications.

- *Placator:* When working with people who use the placating stance, therapists are more successful when they use less directive therapy methods, such as multiple-choice questions and open-ended reflections, that ask clients to voice their opinion and take a stand, which is often quite painful and scary for placators. With clients who tend toward placating, therapists carefully avoid giving opinions (or even the slightest hint that they may have an opinion) or offering too much personal information. Clients will often use this type of information to know what parts of themselves to hide and what parts to foreground to gain therapist approval. Never underestimate placators; they are skilled in the art of people pleasing and regularly lie to make the therapist feel good about how therapy is progressing (Gehart & Lyle, 2001). Not until the client regularly and openly disagrees with the therapist has rapport been fully established with a client who tends toward the placating stance.
- *Blamer:* When working with those who frequently use the blaming stance, therapists strive to increase clients' awareness of others' thoughts and feelings and help them learn how to communicate their personal perspectives in a way that is respectful of others. Often therapists are more effective when they confront blamers directly and boldly. Counter to what one might expect, this approach typically strengthens the therapeutic relationship with a person who uses the blaming stance. Most people who use blaming lose respect for "wimpy" (think "placating") therapists who do not speak their minds honestly and directly, a skill a blamer has mastered. They generally prefer more upfront and direct communication than is generally tolerated in polite society.
- *Superreasonable:* When working with superreasonable clients, logic and rules reign supreme. To communicate effectively, therapists must identify and refer to the context these clients use to make meaning and inform their action, which may be perceived societal rules, religious norms, professional expectations, or even an idiosyncratic set of personal rules for living. The goal is to help these clients value their own internal, subjective realities and those of others.
- *Irrelevant:* The irrelevant type creates a unique challenge for therapists because there are no prepackaged "grips" of self, other, or context on which the therapist can rely. The therapist needs to find the unique anchors of the client's reality with which to connect. Often the first step is to make the therapeutic relationship a place of

utmost safety for the client so that there is less need to distract. As treatment progresses, the therapist works to increase the client's ability to recognize thoughts and feelings of self and others as well as acknowledge the demands of context. Progress may be slower than with other communication stances.

Gottman's Divorce Indicators

John Gottman (see Chapter 11; Gottman, 1999) has researched couples and their communication patterns for over 30 years. His experience has given him the ability to predict divorce with 97.5% accuracy by assessing five key variables: the Four Horsemen—criticism, defensiveness, contempt, and stonewalling—and effective repair attempts. His extensive research provides therapists with a unique system for assessing couple functioning.

The Four Horsemen of the Apocalypse

In Gottman's research on divorce predictors, he found that negative interactions were not equally corrosive (e.g., expression of anger). Four specific interactions were more significant than others in predicting divorce; he has named these the *Four Horsemen of the Apocalypse*. The presence of these four factors predicts divorce with 85% accuracy.

- *Criticism:* Statement implying that something is globally wrong with the partner rather than stating dissatisfaction with a single, particular event (e.g., "always" or "never" statements; broad characterizations about the other's personality).
- *Defensiveness:* Statement or action to ward off perceived attack or criticism from partner: "I'm innocent."
- *Contempt:* Any statement or gesture that puts oneself on a higher level than one's partner, typically involving mockery, eye rolls, and contemptuous facial expressions. This is the most corrosive of the horsemen and is virtually nonexistent in happy marriages. *Contempt is the best single predictor of divorce.*
- *Stonewalling:* An action in which the listener withdraws from interaction, either physically leaving the room or refusing to continue the conversation.

Therapists assess for the horsemen by considering what the client says, in-session interactions, and through written assessments (Gottman, 1999). The presence of the horsemen should not cause automatic alarm; happily married couples regularly engage in criticism, defensiveness, and stonewalling. However, the frequency of these behaviors is different with happy couples, as is their ability to successfully repair the interactions. Thus, therapists need to assess the frequency as well as the presence. Successful couples maintain a *ratio of 5 positive interactions for every 1 negative interaction* during conflict. Couples heading for divorce will have fewer than 5 positives for every negative and often will be closer to 1:1 positive to negative, or even more negative than positive interactions. The following are some of the interactions that therapists need to assess for frequency:

- *Failed Repair Attempt:* A repair attempt is any verbal or nonverbal attempt on the part of either partner to de-escalate conflict. It can include humor, apologies, touch, or compromise. A "failed" repair attempt means that the other partner rejected the attempt and continued the conflictual interaction. Distressed couples more frequently make repair attempts, and they are rejected more frequently.
- *Rejection of Influence:* Often one partner rejects the other's influence (e.g., suggestions, input). Husbands rejecting their wives' influence predicts divorce with 80% accuracy; the reverse is not true for women.
- *Harsh Startup:* This phrase refers to how problem issues are raised: with positive affect (soft startup) or negative affect (harsh startup). Harsh startup by the wife is associated with greater marital instability and higher rates of divorce; the reverse is not true for men.

When working with couples, therapists can keep Gottman's divorce indicators in mind to identify the severity of distress and determine how best to proceed.

Problem Interaction Patterns

> **VI. Interactional Patterns**
>
> *Problem Interaction Pattern* (A ⇆ B):
>
> Start of tension: _____
>
> Conflict/symptom escalation: _____
>
> Return to "normal"/homeostasis: _____
>
> *Hypothesized homeostatic function of presenting problem: How might the symptom serve to maintain connection, create independence/distance, establish influence, reestablish connection, or otherwise help create a sense of balance in the family?*
>
> _____
>
> _____

One of the hallmarks of family therapy is the ability to assess the family's interaction patterns around the presenting problem. In all honesty, I prefer to do this *before* assessing family structure; but most of my students seem to prefer this order as they find structure a bit easier. You should feel free to do it in whichever order works for you.

This ability is central to the Mental Research Institute (MRI) approach (see Chapter 5; Watzlawick et al., 1974), strategic therapy (Haley, 1976), and the Milan approach (see Chapter 5; Selvini Palazzoli, Boscolo, Cecchin, & Prata, 1978) and is featured prominently in Satir's communication approach (see Chapter 8; Satir, Banmen, Gerber, & Gomori, 1991) and Whitaker's symbolic-experiential therapy (see Chapter 9; Whitaker & Bumberry, 1988). In interactional assessment, the therapist traces reciprocal relational patterns: how person A responds to person B and vice versa. Because more than one person may be involved (persons C, D, E, etc.) in families or larger groups, patterns can become quite complex. Whether the client reports an individual symptom or relational problem, the therapist addresses the fact that the behavior is always embedded within larger systems and that the symptoms help maintain the system's homeostasis or sense of normalcy (even if the behavior is not considered normal by the members of the system). I find it most helpful to think of tracing the problem interaction through three basic phases, which can vary significantly from problem to problem:

> ### Three Phases of Problem Interaction Pattern
>
> - *Start of Tension:* What are the behaviors that signal a rise of tension or the start of the problem? How do things unfold from here? How does each person respond and react to the rise in tension?
> - *Conflict/Symptom Escalation:* What happens when the problem fully emerges, which may be conflict for a family or a depressive episode for an individual? The focus here is on the behavioral actions and responses of each person involved, even in cases of "individual" problems, such as anxiety, depression, or psychosis.
> - *Return to Normal/Homeostasis:* Finally, and often the most enlightening part, is to trace the interaction cycle back to "normal" or homeostasis. What does each person do to get back to the sense of "normal"?

A therapist can assess these patterns using a series of questions, first by identifying the emergence of the problem and then tracing each person's emotional and/or behavioral responses to others until "normalcy" or homeostasis is achieved again. The process looks something like this:

Assessing Interaction Patterns

Client describes how the problem begins:

Example: Mother gets a call from the school saying her son is failing a class.

The therapist inquires about the mother's next actions and the son's response:

Example: Mother lectures son and sets an extensive punishment; the child argues and says she is being unfair. Mother says, "Wait until I tell your dad."

The therapist continues to trace this exchange in terms of how each responded to the other until they return to normal:

Example: Mother responds to son's accusations of her being unfair by adding more consequences and punishments.

The therapist also inquires about how significant others in the family system respond to the problem situation:

Example: How did the father participate? What does the younger sister do while this is going on? What effects does this have in the marital relationship?

The therapist continues assessing the interaction pattern until it is clear that the entire family has returned to a sense of "normalcy."

Systemic Hypothesis

After assessing family structure and interaction patterns, therapists develop a working hypothesis about the problem, a potential role the symptom may be playing in maintaining the family homeostasis (see Chapter 5; Selvini Palazolli et al., 1978). A classic example is that a child's symptom (e.g., tantrums, running away, school failure, eating disorder) (a) creates a common problem that forces the couple to cooperate, work together, and stay together, and/or (b) distracts one or more parents from an ailing marriage. Obviously, a child is rarely thinking, "Gee, I think I'll act out to keep mom and dad together," and a parent isn't thinking, "I'll just obsessively overfocus on the children to distract me from my pathetic marriage." Instead, the symptoms naturally emerge to fill "gaps" or systemic needs to maintain a sense of balance without anyone consciously cooking up a plan.

The following strategies may be used for developing hypotheses:

- *Client Language and Metaphors:* The MRI team often described the entire homeostatic pattern using client language (Watzlawick et al., 1974). For example, if the family were sports enthusiasts, they would describe the interaction pattern using the metaphor of a game in which person A has to play defense (withdraw) when person B plays offense (pursues).
- *Positive Connotation:* The Milan team made it a rule to emphasize the positive and helpful effects of the symptom in the family (Selvini Palazzoli et al., 1978). For example, they would praise a child who was sacrificing personal success to give his mother someone to care for so she felt useful.
- *Love and Power:* Strategic therapists developed hypotheses around love and power. For example, they might hypothesize that the seemingly helpless person's role (e.g., as depressed, compliant, sexually uninterested) had the hidden dimension of giving the person power (through influencing others' behavior) that she could not acquire by other means.

Intergenerational Patterns

> **VII. Intergenerational and Attachment Patterns**
>
> Construct a family genogram and include all relevant information including:
>
> - ages, birth/death dates
> - names
> - relational patterns
> - occupations
> - medical history
> - psychiatric disorders
> - abuse history
>
> Also include a couple of adjectives for persons frequently discussed in session (these should describe personality and/or relational patterns, i.e., quiet, family caretaker, emotionally distant, perfectionist, helpless, etc.). Genogram should be attached to report. Summarize key findings below.
>
> Family Strengths: _____
>
> Substance/Alcohol Abuse: ❑ N/A; ❑ Hx: _____
>
> Sexual/Physical/Emotional Abuse: ❑ N/A; ❑ Hx: _____
>
> Parent/Child Relations: ❑ N/A; ❑ Hx: _____
>
> Physical/Mental Disorders: ❑ N/A; ❑ Hx: _____
>
> Historical Incidents of Presenting Problem: ❑ N/A; ❑ Hx: _____
>
> *Attachment Patterns: Describe most common attachment pattern for each.*
>
> AF/AM: ❑ Secure ❑ Anxious ❑ Avoidant ❑ Anxious/Avoidant. Describe: _____
>
> AF/AM: ❑ Secure ❑ Anxious ❑ Avoidant ❑ Anxious/Avoidant. Describe: _____
>
> CF/CM: ❑ Secure ❑ Anxious ❑ Avoidant ❑ Anxious/Avoidant. Describe: _____
>
> CF/CM: ❑ Secure ❑ Anxious ❑ Avoidant ❑ Anxious/Avoidant. Describe: _____

Assessing for intergenerational patterns is easiest when using a genogram (McGoldrick, Gerson, & Petry, 2008), which provides a visual map of these patterns. Traditionally an intergenerational assessment instrument (Chapter 10; McGoldrick et al., 2008), genograms have also been adapted for use with other models (Hardy & Laszloffy, 1995; Kuehl, 1995; Rubalcava & Waldman, 2004) and are regularly used by all therapists to conceptualize family dynamics.

Therapists can create comprehensive genograms, which map numerous intergenerational patterns, or problem-specific genograms, which focus on patterns related to the presenting problem and how family members have dealt with similar problems across generations (e.g., how other couples have dealt with marital tension). See McGoldrick et al. (2008) for comprehensive instructions on using a genogram. The following figure depicts the symbols commonly used in creating a genogram.

The following patterns are frequently included in genograms:

- Family strengths and resources
- Substance and alcohol abuse and dependence
- Sexual, physical, and emotional abuse
- Personal qualities and/or family roles; complementary roles (e.g., black sheep, rebellious one, overachiever/underachiever)
- Physical and mental health issues (e.g., diabetes, cancer, depression, psychosis)
- Historical incidents of the presenting problem, either with the same people or how other generations and family members have managed this problem

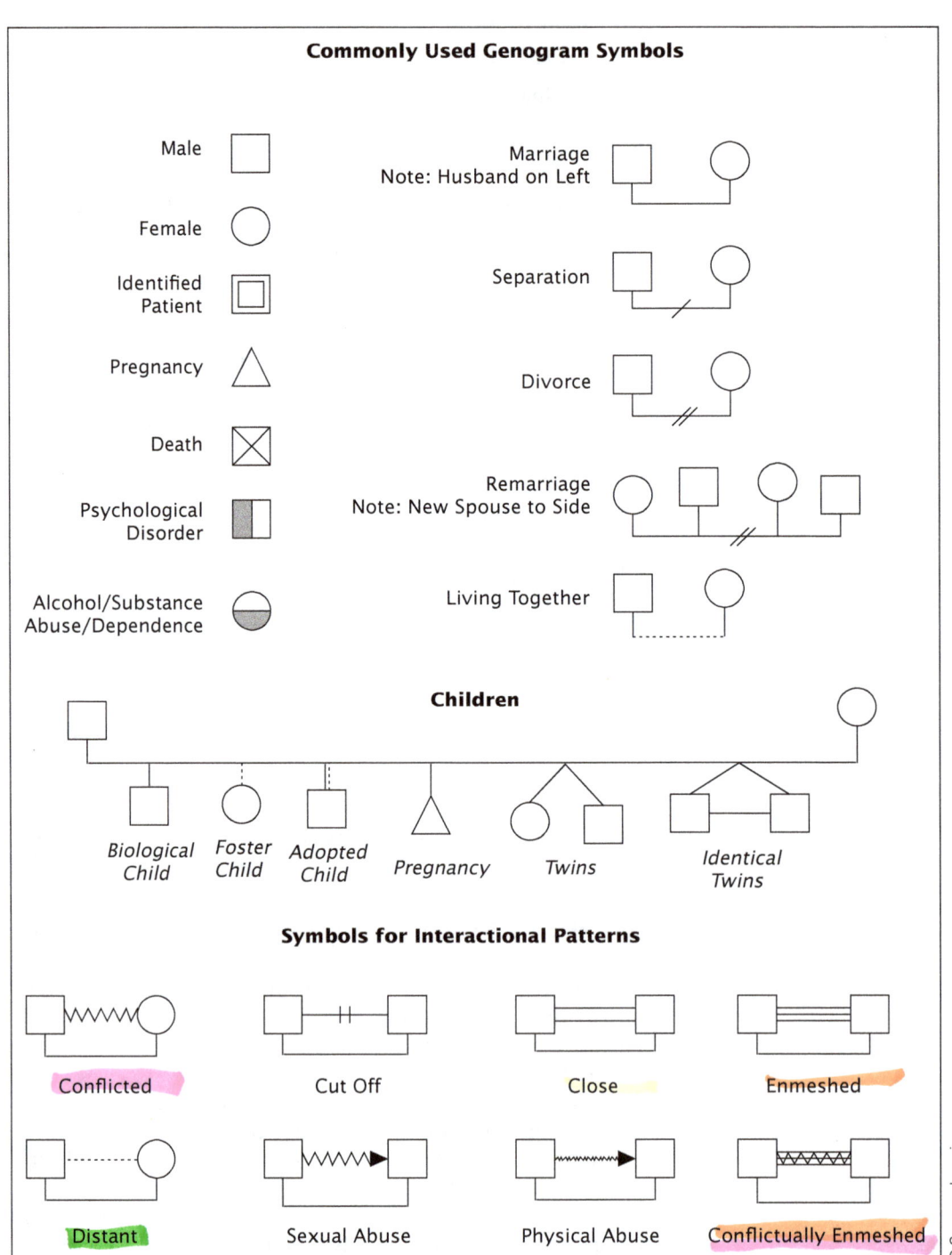

Attachment Patterns

In addition, when considering intergenerational patterns, therapists should also assess for attachment patterns, which are often central in psychodynamic (Chapter 10) and EFT approaches (Chapter 15). Attachment describes a basic human need for connection, warmth, emotional safety, and nurturing from another human being. Although there are several systems for labeling attachment patterns in research settings, Sue Johnson's (2004) research indicates that the following four categories are generally sufficient for couple and family therapy:

- *Secure:* When attachment needs are not met, the person is generally able to still maintain a sufficient level of coping and is able to pursue reconnection in constructive ways.
- *Anxious and Hyperactivated:* When attachment needs are not met, the person becomes anxiously clingy, relentlessly pursues connection, and may become aggressive, blaming, and critical.
- *Avoidance:* When attachment needs are not met, the person suppresses attachment needs and instead focuses on tasks or other distractions.
- *Combination Anxious and Avoidant:* In this style, when attachment needs are not met, the person pursues closeness and then avoids it once offered; this is often associated with a trauma history.

Solution-Based Assessment

> **VIII. Solution-Based Assessment**
>
> *Attempted Solutions That DIDN'T Work:*
>
> 1. _____
>
> 2. _____
>
> 3. _____
>
> *Exceptions and Unique Outcomes (Solutions That DID Work)*: Times, places, relationships, contexts, etc. when problem is less of a problem; behaviors that seem to make things even slightly better:
>
> 1. _____
>
> 2. _____
>
> 3. _____
>
> *Miracle Question Answer*: If the problem were to be resolved overnight, what would client be doing differently the next day? (Describe in terms of doing X rather than not doing Y).
>
> 1. _____
>
> 2. _____
>
> 3. _____

Previous Solutions That *Did Not* Work

When assessing solutions, therapists need to assess two kinds: those that have worked and those that have not. The MRI group (Watzlawick et al., 1974) and cognitive-behavioral therapists (Baucom & Epstein, 1990) are best known for assessing what has not worked, although they use these in different ways when they intervene. With most clients, it is generally easy to assess failed previous solutions.

Questions for Assessing Solutions That Did Not Work
The therapist may begin by asking a straightforward question:

> *What have you tried to solve this problem?*

Most clients respond with a list of things that have not worked. If they need more prompting therapists may ask:

> *I am guessing you have tried to solve this problem (address this issue) on your own and that some things were not as successful as you had hoped. What have you tried that did not work?*

Exceptions and Unique Outcomes: Previous Solutions That *Did* Work

Solution-focused therapists (de Shazer, 1988; O'Hanlon & Weiner-Davis, 1989) assess for previous solutions that *did* work, a process that is similar to identifying *unique outcomes* to the dominant problem story in narrative therapy (Freedman & Combs, 1996; White & Epston, 1990). Assessing previous solutions and unique outcomes is difficult because most clients are less aware of when the problem is not a problem and how they have kept things from getting worse. Some of the questions therapists ask are the following:

Questions for Assessing Solutions and Unique Outcomes
- What keeps this problem from being worse than it is right now?
- Is there any solution you have tried that worked for a while? Or made things slightly better?
- Are there times or places when the problem is less of a problem or not a problem?
- Have you ever been able to respond to the problem so that it is less of a problem or less severe?
- Does this problem occur in all places with all people, or is it better in certain contexts?

These questions generally require more thought and reflection from the client and more follow-up questions from the therapist. Their answers often provide invaluable clues on how best to proceed and intervene in therapy.

Miracle Question Answer

Solution-based therapists use the miracle question (de Shazer, 1988) and similar questions—crystal ball questions (de Shazer, 1988), magic wand questions (Selekman, 1997), and time machine questions (Bertolino & O'Hanlon, 2002)—to assess the client's preferred solution or outcome (see Chapter 12 for a detailed discussion of exactly how to do ask this question successfully—it is harder than you'd think). The answer to these questions are often very helpful in setting goals for treatment.

The Miracle Question

"Imagine that you go home tonight and during the middle of the night a miracle happens: all the problems you came here to resolve are miraculously resolved. However, when you wake up, you have no idea a miracle has occurred. What are some of the first things you would notice that would be different? What are some of the first clues that a miracle has occurred?"

The therapist then helps the client generate a behavioral description of what the client would be *doing*. Often this requires gently asking follow-up questions when clients describe what they or others would *not* be doing or what they would be feeling. Instead, the therapist tries to get a clear videolike picture of what the client(s) are actually doing in their miracle scenario. These are then used to create specific goals for therapy.

Postmodern and Cultural Discourse Conceptualization

> **IX. Postmodern and Cultural Discourse Conceptualization**
>
> *Dominant Discourses* informing definition of problem:
>
> - *Cultural, ethnic, SES, religious etc.: How do key cultural discourses inform what is perceived as a problem and the possible solutions?* _____
> _____
> _____
>
> - *Gender, sexual orientation, etc.: How do the gender/sexual discourses inform what is perceived as a problem and the possible solutions?* _____
> _____
> _____
>
> - *Contextual, family, community, school, and other social discourses: How do other important discourses inform what is perceived as a problem and the possible solutions?* _____
> _____
> _____
>
> *Identity/Self-Narratives: How has the problem shaped each family member's identity?:*
> _____
> _____
>
> *Local or Preferred Discourses: What is the client's preferred identity narrative and/or narrative about the problem? Are there local (alternative) discourses about the problem that are preferred?:* _____
> _____
> _____

Narratives and social discourses outline the broader contexts in which the client's problem occurs. They can be divided into dominant discourses, identity narratives, and local and preferred discourses.

Dominant Discourses

Assessing the *dominant social discourses* in which a client's problems are embedded often creates a new and broader perspective on a client's situation (Freedman & Combs, 1996; White & Epston, 1990). I frequently find that this broader perspective helps me to feel more freedom and possibility, increasing my emotional attunement to clients and allowing me to be more creative in my work. For example, when I view a client's reported "anxiety" as part of a larger discourse in which the client feels powerless, such as being gay or lesbian, I begin to see the anxiety as part of this larger social dance. I also see how it is possible for this person to give less "faith," weight, or credence to the dominant discourse and generate new stories about what is "normal" sexual behavior and what is not. By discussing the difficulty in concretely defining what is "normal" sexual behavior and what is not, the client and I can begin to explore the truths that this person has experienced. We join in an exploratory process that offers new ways for the client to understand the anxiety as well as his/her identity.

Common dominant discourses or broader narratives that inform clients' lives include the following:

- Culture, race, ethnicity, immigration
- Gender, sexual orientation, sexual preferences
- Family-of-origin experiences, such as alcoholism, sexual abuse, adoption, etc.
- Stories of divorce, death, and loss of significant relationships
- Wealth, poverty, power, fame
- Small-town, urban, regional discourses
- Health, illness, body image, etc.

Identity Narratives

When clients come for therapy, the problem discourse has usually become a significant part of their *identity narrative*, the story they tell themselves about who they are. For example, a child having problems in school may begin to think, "I am stupid," and his mother may also be feeling, "I have failed as a mother" because of her child's academic performance. Obviously, these negative sweeping judgments of a person's value or ability need to be addressed in therapy. Assessing them early in the process helps the therapist understand how to engage and motivate each person.

Local and Preferred Discourses

Local discourses are the stories that occur at the local (as opposed to the dominant societal) level (Anderson, 1997). These narratives often are built on personal beliefs and unique interpretations that contradict or significantly modify dominant discourses. Although it is theoretically possible for local discourses to contribute to the problem (e.g., having more oppressive versions of dominant discourses), they are usually a significant source of motivation, energy, and hope in addressing problems. Most often, local discourses are a person's "personal truth" that he/she has been hiding or is ashamed of because of anticipated disapproval from others. They can be common subculture values, such as a preference for certain religious or sexual practices, or they can be unique to the individual, such as wanting to sell everything and travel the world. Local discourses are often the source of *preferred discourses* in narrative therapy (Freedman & Combs, 1996), the versions of one's life and identity that serve as the goal in narrative therapy.

Client Perspectives

Finally, therapists should reflect on areas of client agreement and disagreement with a case conceptualization. Although it is often not clinically appropriate to hand the entire case conceptualization over to a client, discussing the key findings is. Which descriptions seem accurate from the client's perspective? Which do not? Would the client be surprised to hear any of this? If there is a significant disagreement between the client's and therapist's perspective, the therapist needs to consider how he/she will navigate this difference and remain open to the possibility that the assessment is not accurate.

> **X. Client Perspectives**
>
> *Areas of Agreement:* Based on what the client(s) has(ve) said, what parts of the above assessment do they agree with or are likely to agree with? _____
> _____
>
> *Areas of Disagreement:* What parts do they disagree with or are likely to disagree with? Why? _____
> _____
>
> *How do you plan to respectfully work with areas of potential disagreement?* _____
> _____

Consider that the client perspective is particularly important when the client differs from the therapist in age, cultural background, gender, sexual orientation, or socioeconomic status, in which case the therapist may be less likely to fully understand the client's context. Conversely, many therapists have even more trouble with clients who are similar to them because they assume they know more than they do or that they already know the answer to the situation. Whether clients are similar or different, therapists need to carefully consider their perspectives when conceptualizing treatment.

Case Conceptualization, Diversity, and Sameness

Just in case you were beginning to feel that you finally understood something about case conceptualization, let me throw a wrench or two into the mix: diversity and sameness. The problem with case conceptualization and assessments in general is that unfortunately there are no objective standards against which a person can be measured for "clear boundaries," "healthy hierarchies," or "clear communication." Healthy, emotionally engaged boundaries look quite different in a Mexican American family and an Asian American family. In fact, problematic boundaries in a Mexican American family (e.g., cool, disengaged) may look *more like* healthy boundaries (e.g., quietly respectful) than problem boundaries (e.g., overly involved) in Asian American families. Thus, therapists cannot rely simply on objective descriptions of behavior in assessment. They must consider the broader culture norms, which may include *more than one* set of ethnic norms as well as local neighborhood cultures, school contexts, sexual orientation subcultures, religious communities, and so forth. Although you will undoubtedly take a course on cultural issues and will read that professional Codes of Ethics require respecting diversity, it takes working with a diverse range of families and a willingness to learn from them to cultivate a meaningful sense of cultural sensitivity. I believe this to be a lifelong journey.

Ironically, I have found that new therapists in training today sometimes have the most difficulty accepting diversity in clients from within their *same* culture of origin. The more similar clients are to us, the more we expect them to share our values and behavioral norms and the lower our tolerance of difference. For example, middle-class Caucasian therapists often expect middle-class Caucasian clients to have particular values regarding emotional expression, marital arrangements, extended family, and parent-child relationships, and may be quick to encourage particular systems of values—namely, their own. Thus, whether working with someone very similar or very different than yourself, you need to *slowly* assess and evaluate, always considering clients' broader sociocultural context and norms. Therapists who excel in conceptualization and assessment approach these tasks with profound humility and a continual willingness to learn.

Online Resources

Family History Maker (U.S. Department of Health and Human Services): www.hhs.gov/familyhistory/

Genogram Maker (Genogram Software): www.genogram.org

References

American Psychiatric Association. (2000). *Diagnostic and statistical manual for mental disorders* (4th ed., Text Revision). Washington, D.C.: Author.

Anderson, H. (1997). *Conversations, language, and possibilities.* New York: Basic.

Anderson, H., & Gehart, D. R. (Eds.). (2006). *Collaborative therapy: Relationships and conversations that make a difference.* New York: Brunner-Routledge.

Bateson, G. (1972). *Steps to an ecology of mind.* New York: Ballantine.

Bateson, G. (1979/2002). *Mind and nature: A necessary unity.* Cresskill, NJ: Hampton.

Baucom, D. H., & Epstein, N. (1990). *Cognitive-behavioral marital therapy.* New York: Brunner/Mazel.

Bertolino, B., & O'Hanlon, B. (2002). *Collaborative, competency-based counseling and therapy.* New York: Allyn & Bacon.

Carter, B., & McGoldrick, M. (1999). *The expanded family life cycle: Individuals, families, and social perspectives* (3rd ed.). New York: Allyn and Bacon.

de Shazer, S. (1988). *Clues: Investigating solutions in brief therapy.* New York: Norton.

Freedman, J., & Combs, G. (1996). *Narrative therapy: The social construction of preferred realities.* New York: Norton.

Gehart, D. R., & Lyle, R. R. (2001). Client experience of gender in therapeutic relationships: An interpretive ethnography. *Family Process, 40,* 443–458.

Gergen, K. J. (1999). *An invitation to social construction.* Thousand Oaks, CA: Sage.

Gottman, J. M. (1999). *The marriage clinic: A scientifically based marital therapy.* New York: Norton.

Haley, J. (1976). *Problem-solving therapy: New strategies for effective family therapy.* San Francisco, CA: Jossey-Bass.

Hardy, K. V., & Laszloffy, T. A. (1995). The cultural genogram: Key to training culturally competent family therapists. *Journal of Marital and Family Therapy, 21,* 227–237.

Johnson, S. M. (2004). *The practice of emotionally focused marital therapy: Creating connection* (2nd ed.). New York: Brunner/Routledge.

Keeney, B. P. (1983). *Aesthetics of change.* New York: Guilford.

Kerr, M., & Bowen, M. (1988). *Family evaluation.* New York: Norton.

Kuehl, B. P. (1995). The solution-oriented genogram: A collaborative approach. *Journal of Marital and Family Therapy, 21,* 239–250.

Lambert, M. J., & Ogles, B. M. (2004). The efficacy and effectiveness of psychotherapy. In M. J. Lambert (Ed.), *Bergin and Garfield's handbook of psychotherapy and behavior change* (5th ed., pp. 139–193). New York: Wiley.

McGoldrick, M., Gerson, R., & Petry, S. (2008). *Genograms: Assessment and intervention* (3rd ed.). New York: Norton.

Miller, S. D., Duncan, B. L., & Hubble, M. (1997). *Escape from Babel: Toward a unifying language for psychotherapy practice.* New York: Norton.

Minuchin, S. (1974). *Families and family therapy.* Cambridge, MA: Harvard University Press.

Minuchin, S., & Fishman, H. C. (1981). *Family therapy techniques.* Cambridge, MA: Harvard University Press.

Norcross, J. C., & Beutler, L. E. (2008). Integrative psychotherapies. In R. J. Corsini & D. Wedding (Eds.), *Current psychotherapies* (8th ed.) (pp. 481–511). Pacific Grove, CA: Brooks/Cole.

O'Hanlon, W. H., & Weiner-Davis, M. (1989). *In search of solutions: A new direction in psychotherapy.* New York: Norton.

Raser, J. (1999). *Raising children you can live with: A guide for frustrated parents* (2nd ed.). Houston: Bayou.

Rubalcava, L. A., & Waldman, K. M. (2004). Working with intercultural couples: An intersubjective-constructivist perspective. *Progress in Self Psychology, 20,* 127–149.

Satir, V., Banmen, J., Gerber, J., & Gomori, M. (1991). *The Satir model: Family therapy and beyond.* Palo Alto, CA: Science and Behavior Books.

Selekman, M. D. (1997). *Solution-focused therapy with children: Harnessing family strengths for systemic change.* New York: Guilford.

Seligman, M. (2004). *Authentic happiness: Using the new positive psychology to realize your potential for lasting fulfillment.* New York: Free Press.

Selvini Palazzoli, M., Boscolo, L., Cecchin, G., & Prata, G. (1978). *Paradox and counterparadox.* New York: Jason Aronson.

Visher, E., & Visher, J. (1979). *Stepfamilies: A guide to working with stepparents and stepchildren.* New York: Brunner/Mazel.

Walsh, F. (Ed.). (2003). *Spiritual resources in family therapy.* New York: Guilford.

Watzlawick, P., Weakland, J., & Fisch, R. (1974). *Change: Principles of problem formation and problem resolution.* New York: Norton.

Whitaker, C. A., & Bumberry, W. M. (1988). *Dancing with the family: A symbolic experiential approach.* New York: Brunner/Mazel.

Whitaker, C. A., & Keith, D. V. (1981). Symbolic-experiential family therapy. In A. S. Gurman & D. P. Kniskern (Eds.), *Handbook of family therapy* (pp. 187–224). New York: Brunner/Mazel.

White, M., & Epston, D. (1990). *Narrative means to therapeutic ends.* New York: Norton.

Case Conceptualization Form

CASE CONCEPTUALIZATION USING SYSTEMIC THEORIES

For use with individual, couple, or family clients.

Therapist: _____   Client/Case #: _____   Date: _____

> *Symbols*
> AF = Adult Female; AM = Adult Male; CF = Child Female; CM = Child Male
> Ex. = Example; Hx = History; NA = Not Applicable

I. Introduction to Client and Significant Others

*Denotes clients involved in therapy process

    Age, ethnicity, occupation/grade, other relevant identifiers

AF/AM: _____

AF/AM: _____

CF/CM: _____

CF/CM: _____

_____: _____

_____: _____

II. Presenting Concern(s)

Client's Description(s) of Problem(s):

AF/AM: _____

AF/AM: _____

CF/CM: _____

CF/CM: _____

_____: _____

_____: _____

Extended Family Description(s) of Problems: _____

_____

_____

(continued)

Broader System Problem Descriptions: Description of problem from referring party, teachers, relatives, legal system, etc.:

_____: _____

_____: _____

III. Background Information

Trauma/Abuse History (recent and past): _____

Substance Use/Abuse (current and past; self, family of origin, significant others): _____

Precipitating Events (recent life changes, first symptoms, stressors, etc.): _____

Related Historical Background (family history, related issues, previous counseling, medical/mental health history, etc.): _____

IV. Client/Family Strengths and Diversity

Strengths and Resources

   Personal: _____

   Relational/Social: _____

   Spiritual: _____

(continued)

Diversity: Resources and Limitations

Identify potential resources and limitations available to clients based on their age, gender, sexual orientation, cultural background, socioeconomic status, religion, regional community, language, family background, family configuration, abilities, etc.

Unique Resources: _____

_____

Potential Limitations: _____

_____

V. Family Structure

Family Life Cycle Stage (Check all that apply)

❏ Single Adult ❏ Committed Couple ❏ Family With Young Children ❏ Family With Adolescent Children
❏ Divorce ❏ Blended Family ❏ Launching Children ❏ Later Life

Describe struggles with mastering developmental tasks in one or more of these stages: _____

_____

_____

Typical style for regulating closeness and distance in couple/family: _____

_____

_____

Boundaries with/between

Primary couple (A_/A_): ❏ Enmeshed ❏ Clear ❏ Disengaged ❏ NA. Description/example: _____

_____

Parent A_ & Children: ❏ Enmeshed ❏ Clear ❏ Disengaged ❏ NA. Description/example: _____

_____

Parent A_ & Children: ❏ Enmeshed ❏ Clear ❏ Disengaged ❏ NA. Description/example: _____

_____

Siblings: ❏ Enmeshed ❏ Clear ❏ Disengaged ❏ NA. Description/example: _____

_____

Extended Family: ❏ Enmeshed ❏ Clear ❏ Disengaged ❏ NA. Description/example: _____

_____

(continued)

Friends/Colleagues/Other: ❏ Enmeshed ❏ Clear ❏ Disengaged ❏ NA. Description/example: _____
_____

Triangles/Coalitions

❏ Cross-generational coalitions: Describe: _____

❏ Coalitions with family of origin: Describe: _____

❏ Other coalitions: _____

Hierarchy Between Parents and Children ❏ NA

AF/AM: ❏ Effective; ❏ Insufficient (permissive); ❏ Excessive (authoritarian) ❏ Inconsistent

AM/AF: ❏ Effective; ❏ Insufficient (permissive); ❏ Excessive (authoritarian) ❏ Inconsistent

Description/Example to Illustrate: _____
_____
_____

Complementary Patterns between _____ and _____ :

❏ Pursuer/distancer; ❏ Over-/under-functioner; Emotional/logical; ❏ Good/bad parent;

❏ Other: _____; Ex: _____

Satir Communication Stances: Describe most commonly used stance under stress.

AF or _____ : ❏ Congruent ❏ Placator ❏ Blamer ❏ Superreasonable ❏ Irrelevant

AM or _____ : ❏ Congruent ❏ Placator ❏ Blamer ❏ Superreasonable ❏ Irrelevant

CF or _____ : ❏ Congruent ❏ Placator ❏ Blamer ❏ Superreasonable ❏ Irrelevant

CM or _____ : ❏ Congruent ❏ Placator ❏ Blamer ❏ Superreasonable ❏ Irrelevant

Describe: _____
_____

Gottman's Divorce Indicators:

Criticism: ❏ AF/M; ❏ AM/F; Ex: _____

Defensiveness: ❏ AF/M; ❏ AM/F; Ex: _____

Contempt: ❏ AF/M; ❏ AM/F; Ex: _____

Stonewalling: ❏ AF/M; ❏ AM/F; Ex: _____

Failed repair attempts: ❏ AF/M; ❏ AM/F; Ex: _____

(continued)

Not accept Influence: ❑ AF/M; ❑ AM/F; Ex: _____

Harsh startup: ❑ AF/M; ❑ AM/F; Ex: _____

VI. Interactional Patterns

Problem Interaction Pattern (A ⇆ B):

Start of tension: _____

Conflict/symptom escalation: _____

_____

Return to "normal"/homeostasis: _____

_____

Hypothesized homeostatic function of presenting problem: How might the symptom serve to maintain connection, create independence/distance, establish influence, reestablish connection, or otherwise help create a sense of balance in the family? _____

_____

_____

VII. Intergenerational and Attachment Patterns

Construct a family genogram and include all relevant information including:

- ages, birth/death dates
- names
- relational patterns
- occupations
- medical history
- psychiatric disorders
- abuse history

Also include a couple of adjectives for persons frequently discussed in session (these should describe personality and/or relational patterns, i.e., quiet, family caretaker, emotionally distant, perfectionist, helpless, etc.). Genogram should be attached to report. Summarize key findings below.

Family Strengths: _____

Substance/Alcohol Abuse: ❑ N/A; ❑ Hx: _____

Sexual/Physical/Emotional Abuse: ❑ N/A; ❑ Hx: _____

Parent/Child Relations: ❑ N/A; ❑ Hx: _____

Physical/Mental Disorders: ❑ N/A; ❑ Hx: _____

Historical Incidents of Presenting Problem: ❑ N/A; ❑ Hx: _____

(continued)

Attachment Patterns: Describe most common attachment pattern for each.

AF/AM: ❏ Secure ❏ Anxious ❏ Avoidant ❏ Anxious/Avoidant. Describe: _____

AF/AM: ❏ Secure ❏ Anxious ❏ Avoidant ❏ Anxious/Avoidant. Describe: _____

CF/CM: ❏ Secure ❏ Anxious ❏ Avoidant ❏ Anxious/Avoidant. Describe: _____

CF/CM: ❏ Secure ❏ Anxious ❏ Avoidant ❏ Anxious/Avoidant. Describe: _____

VIII. Solution-Based Assessment

Attempted Solutions That DIDN'T Work:

1. _____
2. _____
3. _____

Exceptions and Unique Outcomes (Solutions That DID Work): Times, places, relationships, contexts, etc. when problem is less of a problem; behaviors that seem to make things even slightly better:

1. _____
2. _____
3. _____

Miracle Question Answer: If the problem were to be resolved overnight, what would client be doing differently the next day? (Describe in terms of doing X rather than not doing Y).

1. _____
2. _____
3. _____

IX. Postmodern and Cultural Discourse Conceptualization

Dominant Discourses informing definition of problem:

- *Cultural, ethnic, SES, religious etc.:* How do key cultural discourses inform what is perceived as a problem and the possible solutions? _____

- *Gender, sexual orientation, etc.:* How do the gender/sexual discourses inform what is perceived as a problem and the possible solutions? _____

(continued)

- *Contextual, family, community, school, and other social discourses*: How do other important discourses inform what is perceived as a problem and the possible solutions? _____
_____
_____

Identity/Self-Narratives: How has the problem shaped each family member's identity?: _____
_____
_____

Local or Preferred Discourses: What is the client's preferred identity narrative and/or narrative about the problem? Are there local (alternative) discourses about the problem that are preferred?: _____
_____
_____

X. Client Perspectives

Areas of Agreement: Based on what the client(s) has(ve) said, what parts of the above assessment do they agree with or are likely to agree with? _____
_____
_____

Areas of Disagreement: What parts do they disagree with or are likely to disagree with? Why? _____
_____
_____

How do you plan to respectfully work with areas of potential disagreement? _____
_____
_____

© 2012. Diane R. Gehart, Ph.D. www.masteringcompetencies.com

Name Index

A
Ackerman, Nathan, 217, 219
Adair, C., 255
Alexander, James, 375, 388
Andersen, Tom, 301, 313, 331, 334–335, 338, 340–341, 346
Anderson, Harlene, 301, 329–331, 333, 335–336, 342, 346, 414
Anderson, S., 225
Anger-Diaz, Barbara, 72
Aponte, Harry, 128, 140
Auerswald, Dick, 128, 331
Azpeitia, Lynne, 152

B
Bakhtin, Mikhail, 56, 331
Baldwin, S. A., 22
Bandler, Richard, 292
Banmen, John, 152
Barnes, S., 255
Bateson, Gregory, 48–49, 51–53, 62, 70, 72, 83–84, 86, 88, 103, 418
Baucom, D. H., 255
Bava, Saliha, 332
Beadle, S., 276
Beaudoin, M. N., 348
Beck, Aaron, 235, 239
Beck, Judith, 236
Berg, Insoo Kim, 271, 274
Berman, P. S., 37
Bertolino, Bob, 286–287
Beutler, L. E., 412
Biever, J. L., 348
Block-Lerner, J., 255
Blow, A. J., 18, 21
Bobele, M., 348
Bobrow, Ellen, 104
Bodin, Art, 72
Bohart, A. C., 19
Boscolo, Luigi, 70, 83–85, 88, 331

Boszmormenyi-Nagy, Ivan, 217, 220, 227
Bowen, Murray, 203–204, 206, 208–209, 227
Bowlby, John, 359–360
Brossart, D. F., 226
Brown, K. W., 255
Bruner, Jerome, 311
Bumberry, W. M., 173–174, 179
Butler, C., 96
Buxton, A. P., 294

C
Caldwell, Benjamin E., 15, 24
Campbell, W. K., 255
Carnes, S., 17
Carson, J. W., 255
Carson, K. M., 255
Carter, Betty, 206, 218, 226
Cecchin, Gianfranco, 70, 83–86, 331
Christensen, Andrew, 235
Colapinto, Jorge, 128–129, 132
Combs, Gene, 303, 310, 313
Connell, Gary, 175, 196
Coolhart, D., 118
Cooper, B., 196
Corcoran, J., 294
Cordova, J. V., 254
Corvo, C., 400

D
Datchi-Phillips, C., 388
Dattilio, Frank, 235, 239, 264
de Shazer, Steve, 41, 271, 273–275, 278, 280, 283, 285
Deissler, Klaus, 332
Derrida, Jacques, 310
DeVine, J. L., 96
Dolan, Yvonne, 274–275, 286–287

Doty, N. D., 264
Duncan, Barry, 19, 165, 274
Dutton, D., 400

E
Ehlers, N., 294
Ellis, Albert, 239, 245
Epstein, Norman, 235, 240
Epston, David, 301–303, 315
Erickson, Milton, 48, 71, 103–104, 111–112, 271–275, 285, 291–292

F
Falloon, I. R. H., 233, 238
Fernandez, Elena, 332
Fisch, Richard, 70, 72, 128, 148
Fishman, H. Charles, 125, 128, 132, 134, 140
Fogarty, Thomas, 206
Forgatch, Marion, 234–235, 237, 242
Foucault, Michel, 55, 57, 302
Framo, James, 217–219, 221, 227
Frankfurt, M., 340
Freedman, Jill, 303, 310, 313
Friedman, E. H., 203
Fry, William, 48, 52, 72

G
Garcia-Preto, N., 95
Gehart, D. R., 37–38, 251
Gergen, Kenneth, 55, 331
Gil, K. M., 255
Giordano, J., 95
Gladding, Samuel T., 25
Gomori, Maria, 152
Goolishian, Harry, 301, 330–332, 335, 346, 414
Gottman, John, 233, 235, 255–256, 258–260, 270, 424, 426

Greenberg, Les, 359–361
Grinder, John, 292
Guerin, Philip, 206
Guerney, Bernard G., 128

H
Hageseth, Christian, 30n
Haley, Jay, 48, 52, 69–71, 103–107, 109–110, 117, 128, 148, 274, 291–292, 331
Halford, W. K., 403
Hoffman, Lynn, 85, 331
Holtzworth-Munroe, Amy, 238, 243
Hsu, W., 294
Hubble, Mark, 19, 274

J
Jackson, Don, 48, 52, 70–72, 217
Jacobson, Neil, 235, 238, 243, 260
Jenkins, A., 316
Johnson, Sue, 61, 359–364, 368, 431
Jongsma, A. E., 37

K
Kabat-Zinn, Jon, 251
Karam, E., 16
Karos, L., 373
Keala, D. K., 226
Keeney, Bradford, 48, 72, 85
Keeton, Jennifer, 31
Keim, Jim, 104, 109
Keith, David, 175
Kerr, Michael, 206
King, Charlie, 128
Krasner, B. R., 220
Krusemark, E., 255

L
Lambert, Michael, 18–20, 164
Lane, G., 86
Lankton, Stephen, 275, 291
Lappin, Jay, 128
Lawson, D. M., 226
Levin, Sue, 332
Lindblad-Goldberg, Marion, 128, 139
Liu, T., 373
London, Sylvia, 332

M
Macy, Josiah, 48
Madanes, Cloe, 70, 103–104, 107, 109
Malone, Thomas, 174–175
McCollum, Eric, 251, 275
McGoldrick, Monica, 95, 206, 429
McNamee, Sheila, 55, 331
Mead, Margaret, 48
Mena, M. P., 140
Metcalf, Linda, 274
Miller, R. B., 17, 225
Miller, Scott, 19, 260, 274
Minuchin, Salvador, 69, 103, 125, 127–130, 132, 134, 140, 217, 331
Mitten, Tammy J., 196

Monk, Gerald, 303
Montalvo, Braulio, 128
Mylott, K., 263

N
Napier, Augustus, 175
Nardone, Giorgio, 72, 76
Nelson, Thorana, 275
Nichols, Michael, 128
Norcross, J. C., 412
North, M., 348

O
O'Hanlon, Bill, 271, 273–276, 286–287, 291–292
Orsillo, S. M., 255

P
Palazzoli, Mara Selvini, 70, 84
Palazzoli, Selvini, 70, 84–85, 90
Papp, Peggy, 218, 226
Parsons, Bruce, 375, 388
Patterson, Gerald, 234–235, 237, 242
Patterson, J. E., 17
Pavlov, Ivan, 234, 241
Pearson, V. J., 141
Penn, Peggy, 85, 331–332, 340
Perez, P. J., 349
Perls, Fritz, 61, 185
Plumb, J. C., 255
Prata, Giuliana, 70, 84–85

R
Rabinowitz, Clara, 128
Ray, Wendel, 49, 72, 86
Remley, Theodore P., 25
Rhatigan, D. L., 255
Riskin, Jules, 72
Rodriguez-Jazcilevich, Irma, 332
Rogers, Carl, 61, 153, 276, 359
Rogers, T., 263
Rogge, R. D., 255
Rosman, Bernice L., 128
Rossi, Ernest, 292
Ryan, M. C., 175

S
Safren, S. A., 263
Santisteban, D. A., 140
Satir, Virginia, 61–62, 69–70, 147–157, 159–160, 165–166, 185, 274, 331, 411, 424–425, 427
Scharff, David, 218
Scharff, Jill, 218
Schlanger, Karin, 72
Schnarch, David, 206, 213
Schumer, Florence, 128
Schwartz, Richard, 185–186, 190–191
Segal, Z. V., 251
Seikkula, Jaakko, 332, 346–347
Selekman, Matthew, 275
Sexton, Thomas, 357, 375, 382

Shadish, W. R., 22
Shotter, John, 55, 331
Siegel, D., 347–348
Silverstein, Olga, 218, 226
Skinner, B. F., 234, 241
Sprenkle, D. H., 16, 18, 21
St. George, S., 339
Sullivan, Harry Stack, 128
Szapocznik, Jose, 104, 116

T
Tallman, K., 19
Tarragona, Margarita, 332
Teasdale, J. D., 251
Trepper, Terry, 275, 294
Treyger, S., 294
Tuttle, A. R., 37

V
Von Bertalanffy, Ludwig, 49
Von Foerster, Heinz, 48
Vygotsky, Lev, 312

W
Wachs, K., 254
Walden, Marcia, 31
Walker, Lenore, 400
Walters, Marianne, 218, 226
Wampold, B. E., 19
Wang, C. C., 294
Ward, Julea, 31
Warkentin, John, 174
Watzlawick, Paul, 51–52, 72, 76, 79, 84, 128, 148, 274, 331
Weakland, John, 48, 52, 71–72, 274
Weiner-Davis, Michelle, 271, 275–276
Whitaker, Carl, 61–62, 173–177, 179–180, 195–196, 217, 427
White, Michael, 55, 301–303, 305, 309–310, 312, 314–315
Wilcoxon, Allen P., 25
Wilkins, E. J., 196
Williams, J. M. C., 251
Willoughby, B. L. B., 264
Wilson, S., 17
Winslade, John, 303
Wittenborn, A., 373
Wittgenstein, Ludwig, 55–56, 274, 331
Wulff, D., 339

Y
Yalom, I. D., 396
Yang, L., 141
Yarhouse, M. A., 322
Yutang, Lin, 277

Z
Zajicek, L., 294
Zeig, Jeffrey, 292
Zimmerman, J., 348
Zuccarini, D., 373

Subject Index

A

A-B-C theory, 239, 245
Absurdity, 175
Acceptance, 363
Acceptance and commitment therapy (ACT), 251, 254
Ackerman Institute, 217, 332
Action-based interventions, 70
Adult attachment, 359, 361–362
Affect-infused descriptions, 348
Affective confrontation, 180
African Americans, 95, 116–117
    Bowen intergenerational therapy, 229–230
    cognitive-behavioral family therapy (CBFT), 262
    internal family systems therapy and, 196–197
    structural therapy and, 140–141, 142–143
Agency, 306, 312, 336
Alternative discourses, 306
American Association for Marriage and Family Therapy, 4, 7, 25, 30
American Counseling Association, 7, 25, 30–31
American Psychological Association, 7, 21, 25, 31, 72
Americanism vs. Hispanicism, 118
Anxious attachment, 359
Appropriately unusual comments, 331, 338
Asian Americans
    cognitive-behavioral family therapy (CBFT), 262–263, 266–267
    Satir growth model and, 165
    solution-based therapy, 294
    structural therapy, 140–141
Assessment
    boundaries, 422–423
    Bowen intergenerational therapy, 207–210
    client strengths, 273
    clinical, 409
    cognitive-behavioral family therapy (CBFT), 237
    collaborative therapy, 335
    dominant discourses, 433–434
    emotionally focused therapy (EFT), 364, 366
    family life cycle stage, 421
    family structure, 419–421
    functional family therapy, 377
    Gottman method couples therapy, 256–257
    hypothesizing, 428
    identity narratives, 434
    interactions, 427–428
    intergenerational patterns, 429–430
    internal family systems therapy, 187
    local discourses, 434
    narrative therapy, 305, 308
    object relations family therapy, 219
    Satir growth model, 154–156
    solution-based therapy, 277–278, 431–433
    strategic therapy, 107
    structural therapy, 130
    symbolic-experiential therapy, 176
    theory-informed. *See* Case conceptualization
Assimilative integration, 411
Attachment injury, 365–366
Attachment theory, 359–362, 365, 368, 431
Attempted solutions, 76
Attention deficit, hyperactivity disorder (ADHD), 255, 403
Autonomy, 26, 180, 204
AVANTA Network. *See* Virginia Satir Global Network
Avoidant attachment, 359

B

Bateson Group, 48, 52, 71, 93
Behavior, 54, 79, 241–242, 377–379, 381
    exchanges, 243
    problem, 375–382
Behavioral family therapy, 233–235
    mindfulness-based, 250
Behavioral norms, 261
Behaviorism, 5, 234
    consistency, 234–235
    reinforcement, 234–235
Beneficence, 26
Bidirectional violence, 401
Biracial individuals, 196
Blame-focused definitions, 381
Blamer communication stance, 148, 150, 424–425
Blended families, 98–99
Bottom-up processing, 347
Boundaries, 125–126, 131, 421–422
    assessment, 422–423
    enmeshment, 126, 422
    renegotiating, 132
Boundary making, 133
Bowen intergenerational therapy, 36, 40, 44, 203, 207, 218
    African Americans, 229–230
    assessment, 207–210
    case conceptualization, 207, 213, 229
    detriangulation, 212
    differentiation, 204, 207–211, 213, 226
    diversity/multicultural issues, 226
    emotional systems, 208–211
    genograms, 205–206, 212, 230
    goal setting, 211
    interventions, 206, 211
    multigenerational patterns, 208–209
    nonreactive stance, 207
    research and evidence base, 225–226
    sexual crucible model, 213

448 Subject Index

Bowen intergenerational therapy (*Continued*)
    sexual identity, 227
    sibling position, 210
    treatment plans, 214–216, 231–232
Brief Strategic and Systemic World Network, 72
Brief strategic family therapy (BSFT), 79, 93, 104, 116, 138–140, 291
    African Americans, 95, 116–117
    case conceptualization, 116–117
    drug abuse, 116, 118
    goals, 116
    interactions, 118
    interventions, 117
    Latino/Hispanics, 95, 116–118
    reframing, 118
Brief Therapy Project (MRI), 70–72, 104

C

California Association of Marriage and Family Therapists, ethics codes, 25, 31
Case conceptualization, 35, 41, 409
    Bowen intergenerational therapy, 207, 213, 229
    brief strategic family therapy (BSFT), 116–117
    client perspectives, 434–435
    cognitive-behavioral family therapy (CBFT), 237, 246, 267–269
    collaborative therapy, 335, 342
    coping style, 412
    cross-theoretical, 410
    diversity/multicultural issues, 435
    ecosystemic structural family therapy (ESFT), 139
    emotionally focused therapy (EFT), 364, 369
    forms, 437–443
    functional family therapy, 377, 384–386, 390
    Gottman method couples therapy, 257
    internal family systems therapy, 187, 192
    mental health diagnosis, 412
    Milan systemic therapy, 90–91
    MRI therapy, 80–81
    multisystemic therapy, 94
    narrative therapy, 305, 325–326
    object relations family therapy, 219
    patient preferences, 412
    postmodern therapy, 352–353
    psychoanalytic family therapy, 222
    resistance level, 412
    Satir growth model, 154, 161, 168–169
    solution-based therapy, 288
    state of change, 412
    strategic therapy, 107–109, 112–113, 121–122
    structural therapy, 130, 135, 143–144
    symbolic-experiential therapy, 176, 181–182, 198
    theoretical integration, 412
    theory-specific, 36, 410
Center for Family Studies (Florida), 104, 116
Center for Self Leadership, 186

Centro di Terapia Strategica (Arezzo, Italy), 72
Chabot Emotional Differentiation Scale, 209
Child abuse, reporting, 29
Child sexual abuse therapy, 286–287
Children
    behavioral problems, 375–376
    narrative therapy and, 316
Chronic anxiety, 208, 211
Circular questions, 84, 87, 89
Classical conditioning, 241
Client-centered outcome-informed therapy, 274
Client-centered therapy, 61, 153
Client-generated change, 283
Clients
    agency, 306, 312, 336
    background information, 415
    battle for initiative, 174
    boundaries, 421–422
    characteristics, 19
    diversity resources, 419
    goals, 38, 41–43
    introduction to, 413
    motivation, 278
    personal qualities, 417
    perspectives, 71, 434–435
    presenting problems, 414–415
    resources, 19, 416, 418
    self-actualization, 157, 178–179
    self-efficacy, 286
    solution-focused, 272
    spirituality, 418–419
    strengths, 416, 418
Clinical treatment plans, 37–38, 43
Closed groups, 397
Cognitive restructuring, 381
Cognitive-behavioral family therapy (CBFT), 61, 233, 431
    assessment, baseline, 237
    case conceptualization, 237, 246, 267–269
    common factors, 236
    communication, 243
    contingency contracting, 242
    creating rapport, 236
    diversity/multicultural issues, 261
    empathy, 236
    family schemas, 239–240
    functional analysis, 237–239
    goal setting, 240
    guided discovery, 245–246
    homework tasks, 246
    interventions, 241
    irrational beliefs, 239, 244
    mindfulness-based, 250–251
    operant conditioning, 241–242
    parenting, 234
    point charts, 242
    psychoeducation, 243–244
    research and evidence base, 260
    sexual identity, 263–264
    token economies, 242–243
    treatment plans, 235, 247–250, 269–270
    written contracts, 237
Collaborative language systems, 331, 414
Collaborative reflecting team, 340–342

Collaborative therapy, 55, 57, 86, 301–302, 329–330
    agency, 336
    assessment, 335
    being public, 339–340
    case conceptualization, 335, 342
    client as expert, 333–334
    conversational questions, 337
    dialogues, 332, 334–335, 346
    diversity/multicultural issues, 348–349
    goal setting, 336–337
    "as if" reflecting, 342
    interventions, 337–338
    multiple voices, 340
    mutual puzzling, 329, 338–339
    not knowing/pre-knowing, 330
    presenting problems, 414
    problem-dissolving systems, 335–336
    problem-organizing systems, 335
    professional communication, 339
    reflecting teams, 340–342
    relationships, 332
    research and evidence base, 347
    sexual identity, 349–350
    transformation, 336–337
    treatment plans, 343–345
    "witness," 333
    writing, 332, 340
Command (communication), 51–52
Common factors, 18, 64, 274, 411, 416
    diversity/multicultural, 20
    models, 18–19
    research, 358
    of theories, 17–18
    therapeutic relationship, 20
Communication
    coaching, 159
    metacommunication, 52
    nonverbal, 368, 373
    patterns, 51, 424–425
    relationship enhancement, 402
    relationships, defining, 52
    skills training, 376
    theory, 71
Communication stances, 148–151, 424–425
Community contexts, 379–380
Competencies, 4–7
Complementarity, 132
Complementary patterns, 77, 424
Complementary relationships, 51
Compliments, 135, 285–286
Conditioned stimuli and response, 241
Conduct disorder, 374
Confidentiality, 27
Conflict resolution, 376, 383
Congruent communication stance, 148–149, 157–159, 424–425
Conjoint couples therapy, 401–402
Conjoint Family Therapy (Satir), 152
Consistency, 64–65, 234–235, 242
Constructivism, 56–57, 63–64
Contextual therapy, 203, 217–218, 221
Coping questions, 285
Cotherapy, 175–176
Counterparadox, 89
Countertransference, 7, 187, 218
Couples cognitions, 240

Couples therapy, 221–222, 363
    empirically validated, 374
    ethics, 24
    group therapy, 221–222, 401
    identifying patient, 26–27
    intercultural, 348–349
    legal issues, 24
    marital myths, 256
    mindfulness-based, 254–255
    partner abuse, 29–30, 400–401
    pursue/withdraw patterns, 364, 367
    quid pro quo, 243, 256
    same-sex, 349
Crisis inductions, 134
Cross-generational coalitions, 131, 423
Cross-theoretical case conceptualization, 410–419
Crystal ball questions, 274, 279–280, 432
Cuento therapy, 321–322
Cultural contexts, 126, 128, 379–380
Cultural discourses, 306
Cultural values, 57, 140–141, 167, 261, 348–349, 373, 388, 392, 422, 435
Cybernetic systems theory, 48–49, 53, 62, 70, 78, 152
Cycle of violence, 400

D

Detriangulation, 212, 221
Diagnostic and Statistical Manual (DSM), 12
Dialectic behavioral therapy (DBT), 251, 253–254
Differentiation, 204, 207–211, 213, 226
Diffuse boundaries, 126, 422
Directives, 103–104, 106, 109–111
Disengagement, 126, 422
Diversity, 5–6, 20, 95, 117, 128
    Bowen intergenerational therapy, 226
    case conceptualization, 435
    cognitive-behavioral family therapy (CBFT), 261
    collaborative therapy, 348–349
    emotionally focused therapy (EFT), 373
    experiential therapy, 196
    functional family therapy, 387–388, 392
    narrative therapy, 320–321, 324–325
    postmodern therapy, 58
    Satir growth model, 165
    sexual identity. See Sexual identity diversity
    solution-based therapy, 293–294, 297
    strategic therapy, 117, 120–122
    structural therapy, 128, 140–144
    symbolic-experiential therapy, 196
    systemic therapy, 95
Divorce busting, 275
Divorce predictors, 257–258, 426
Documentation, 26–27
Domestic violence
    cycle of violence, 400
    group therapy, 400–402
    narrative therapy and, 316
    psychoeducational group therapy, 396
    wheel of power and control, 400
Dominant discourses, 302, 306, 322
    assessment, 433–434

Double-bind theory
    counterparadox, 89
    problem-generating, 79–80
    schizophrenia, 48, 52–53
    therapeutic, 79–80
Drug abuse, brief strategic family therapy (BSFT), 116, 118
Duluth model, 400–401
Dulwich Centre (Adelaide, Australia), 303

E

Ecosystemic structural family therapy (ESFT), 93, 128, 139, 379
    case conceptualization, 139
    enactments, 140
    goals, 139
    interventions, 139–140
Education, collaborative practices, 346
Emotional cutoff, 204, 210
Emotional interdependence, 208
Emotional logic, 173
Emotional transactions, 147
Emotional triangles, 209
Emotionally focused therapy (EFT), 61, 139, 147, 357–361, 373
    assessment, 364, 366
    attachment theory, 361–362, 365–366, 368–369
    case conceptualization, 364, 369
    contraindications, 366
    couples, 93, 359
    diversity/multicultural issues, 373
    empathetic conjecture, 368
    enactments, 369
    evidence base, 374
    goal setting, 366
    interactions, 362, 367
    interventions, 367
    primary emotions, 365, 367–368
    pursue/withdraw patterns, 364, 367
    reframing, 368
    relational patterns, 362
    secondary emotions, 365, 367–368
    sexual identity, 373–374
    softening of emotions, 369
    treatment plan, 361, 363–364, 370–372
    validation, 367
Emotions, 159, 165, 365, 367, 369
Empathy, 130, 148, 153–154, 362–363, 368
Empirically supported treatments (ESTs), 21–23
Empirically validated treatments (EVTs), 21
Enactments, 126–127, 132–133, 140, 369
Enmeshment, 126, 422
Epistemology, 53, 60, 88
Ericksonian hypnosis, 271, 274–275, 291–292
Ethics, 5–7, 15, 24–26
    codes, 25–27, 30–32
    family systems of, 217
European Brief Therapy Association, 275
Evidence-based practice (EBP), 17, 21

Evidence-based treatments, 6, 15–17, 21–23, 69, 93
    emotionally focused therapy (EFT), 359, 361
    functional family therapy, 375
    group therapy, 395, 398, 404
    myths, 357–358
    theoretical integration, 412
Exception questions, 272, 277–278
Exiles (internal family systems therapy), 187
Experiential family therapy, 40, 147–148, 153, 165, 173
    diversity/multicultural issues, 196
    sexual identity, 165–166
Externalizing, 307, 309–310

F

Families
    ethical systems, 217
    guidelines, 159
    homeostasis, 50, 63, 70–71, 428
    life cycle, 206, 421
    reconstruction, 160–161
    relational hierarchy, 108, 378, 391
    relationship patterns, 377–378
    roles, 155
    rules, 159
    schema, 239–240
    sculpting, 160
    structural organization, 177
    structure of, 125, 419–422
    subsystems, 130–131
Family games (interactional sequences), 83, 87
Family projection process, 209–210
Family therapy, 15
    community contexts, 379–380
    competencies, 4–5
    cultural contexts, 379–380
    evidence-based, 6, 15–17, 357
    feminist, 218, 226
    identifying patient, 26–27
    interactions, 78, 427
    legal issues, 24–25
    minors and, 28–29
    observational research, 23
    overview, 58–60
    philosophical foundations, 58–60, 64
    technology and, 30
    theories, 3, 10–11, 18, 22–24, 58–59
    training programs, 152
Family Therapy Institute (Washington, D.C.), 104
Family-of-origin therapy, 203, 217, 221–222
Firefighters (internal family systems therapy), 188
First-order change, 50–51, 76–77
First-order cybernetics, 62–63
Four Horsemen of the Apocalypse, 257–258, 426
Functional analysis, 237–239
Functional family therapy, 93, 116, 139, 357, 374
    adaptive behavior, 375
    assessment, 377

Functional family therapy (*Continued*)
behavioral problems, function of, 376
blame-focused definitions, 381
case conceptualization, 377, 384–386, 390
client strengths, 380
coaching, 382
cognitive restructuring, 381
communication, 376, 384
community contexts, 379–380
conflict resolution, 376, 383
consistency, 383
cultural contexts, 379–380
diversity/multicultural issues, 387–388, 392
diverting, 382
engagement, 376–377
evidence base, 388
family-focused definitions, 381
goal setting, 380, 384
interactions, 382
interventions, 381
motivation, 376
organizational themes, 382
parent training, 376
parental expectations, 382
problem-solving, 376, 383
process comments, 382
reframing, 381
relational hierarchy, 378
relational reframing, 381
relationship patterns, 377–378
sexual identity, 388
supervision, 383
treatment plans, 376, 386–387, 392–394
troubled youths, 375, 377, 379, 388, 390

G
Gay-affirmative therapy, 227, 322
Gay/lesbian/bisexual/transgendered/questioning (GLBTQ), 95–96
adolescents, 96
Bowen intergenerational therapy, 227
cognitive-behavioral family therapy (CBFT), 263–264
collaborative therapy, 349–350
coming out, 96, 264, 374
emotionally focused therapy (EFT), 373–374
experiential therapy, 165–166
functional family therapy, 388
internal family systems therapy, 197
narrative therapy, 322
partner adjustment, 294
psychoanalytic family therapy, 227
solution-based therapy, 294
strategic therapy, 118
structural therapy, 141
therapists and, 31
traumatic betrayals, 374
Gender discourses, 306
Gender identity, 118–119
Gender power dynamics, 226
Gender roles, 96, 226
Gender-based group therapy, 400, 402
General systems theory, 48–49, 62, 70
Genograms, 204–206, 212, 429–430

Genuineness, 363
Gestalt therapy, 61
Goal setting, 157, 279
Goal writing process, 41–43
Gottman method couples therapy, 233–234, 255–256, 259–260, 426
assessment, 256–258
case conceptualization, 257
Group therapy, 395–398, 400, 404
multicouples, 402
multifamily, 398–399
participants, 396–398
psychoeducational groups, 395–396
Grupo Campos Elisios (Mexico City), 332
Guided discovery, 245

H
Heightening, 368
Hero/heroine therapy, 322
Hierarchy, 108–109
cultural contexts, 117–118
parental, 108, 131, 141, 181, 378, 423–424
Hispanics. *See* Latino/Hispanics
Homeostasis, 49–50, 63, 70–71, 428
Hope, 20, 154, 277, 305
Houston Galveston Institute, 332
Humanistic therapy, 44, 60–62, 152–153, 164
Humanity, 57–58
Hypnosis, 48, 291. *See also* Ericksonian hypnosis
Hypothesis, 88–89, 428

I
Identified patient (IP), 130
Identity narratives, assessment, 434
Improvisational therapy, 49
Indirect directives, 109
Inductive reasoning, 245
Initial phase goals, 43
Inner dialogues, 331, 334–335
Inner parts, 185–191
Insecure attachment, 359
Insight-oriented therapy, 399
Integrative behavioral couples therapy, 233, 235, 260
Interactional behavior patterns, 72–73, 78, 427–428
Interactional sequences, 74–76
Intercultural couples, 348–349
Intergenerational patterns, 429–431
Intergenerational therapy. *See* Bowen intergenerational therapy
Internal family systems therapy, 147, 173
African Americans, 196–197
assessment, 187
biracial issues, 196
case conceptualization, 187, 192
common factors, 195
countertransference, 187
development burdens, 188
environmental burdens, 188
exiles, 185–187, 191
firefighters, 185–188, 191
inner parts, 185–191
interventions, 189

legacy burdens, 188
managers, 185–188, 191
polarization, 188–189
self, 185–191
sexual identity, 197
in-sight, 191
tangible burdens, 188
transference, 187
traumatic burdens, 188
treatment plans, 192–195
Interpersonal process, 396
Interpersonal violence, 400–401
Interventions, 38, 44, 134, 179
failed, 41
language-based, 83
paradoxical, 111
Intimate partner violence (IPV), 29–30
Invariant prescription, 85, 90
Irrelevant communication stance, 148, 151, 424–426
Irreverence, 86–87

L
Lambert's Common Factors Model, 18–19
Language, 57, 70, 83, 276
Latino Brief Therapy and Training Center (MRI), 72, 95
Latino/Hispanics, 116–117
cognitive-behavioral family therapy (CBFT), 261–262
emotionally focused therapy (EFT), 373
families, 72, 95
functional family therapy, 388
immigration and, 262
narrative therapy, 321–322
structural therapy, 140–141
values, 118
Laws, in therapy, 25
Ledger of entitlement and indebtedness, 217, 220
Ledger of merits, 220
Licensing exams, 12
Linguistics, 87–88, 335
Local discourses, 302, 306, 434
Long-term brief therapy. *See* Milan systemic therapy
Love maps, 259
Loving kindness meditation, 255

M
Macy Conferences, 48–49
Magic wand questions, 279–280, 432
Managers (internal family systems therapy), 187–188
Mantras, 251
Mapping in the landscapes of action and identity/consciousness, 311
Mapping influence, 308
Marburg Institute (Germany), 332
Marital enrichment, 402
Marriage, Couple, Family & Child Counseling standards, 4
Marriage and Family Therapy Core Competencies, 4
Marriages, 256, 258, 260, 396
Meaning, construction of, 56
Measurability, 37

Medical symptoms, 37
Menninger Clinic, 206, 217
Men's discourse, 302
Mental health professionals, 4–7, 16
Mental illness, 395–396, 398–399
Mental Research Institute (Palo Alto, CA), 62, 69–72, 84, 104, 128, 148, 152, 272, 274, 331, 358, 427, 431
Meta-analytically supported treatments (MASTs), 22
Metacommunication, 52, 74, 77
Metaphorical tasks, 111–112
Metaphors, externalizing, 309
Milan Institute, 340
Milan systemic therapy, 55–56, 62–63, 69–70, 72, 74, 83, 85, 427
  case conceptualization, 90
  circular questions, 87, 89
  counterparadox, 89
  curiosity, 86
  family games (interactional sequences), 83, 87–88
  hypothesizing, 88–89
  invariant prescription, 90
  irreverence, 86–87
  neutrality, 85–86
  positive connotations, 89
  rituals, 89–90
  second-order cybernetics, 87
  treatment plans, 90–93
  tyranny of linguistics, 87–88
Milan team, 84–86
Milton H. Erickson Foundation, 292
Milwaukee Brief Family Therapy Center, 271, 274
Mimesis, 129
Mind-body connection, 157
Mindfulness Based Stress Reduction (MBSR), 251
Mindfulness-based therapy, 233, 250–255
Minnesota Couples Communication Program (MCCP), 402
Minors, 28–29
Miracle questions, 274, 279–280, 432
Modernist therapy, 60–61
Monologues, 332, 335
More-of-the-same solutions, 76
Motivation, 376
MRI therapy, 55–56, 62–63, 69–71
  attempted solutions, 73, 76, 78–79
  behavioral goals, 78–79
  case conceptualization, 80–81
  client's language, 79
  dangers of improvement, 80
  interactional sequences, 75–76
  interrupting interventions, 73
  metacommunication, 77
  observation team, 78
  presenting problems, 72–73
  reframing, 79
  therapeutic double bind, 79–80
  treatment plans, 80–83
Multicouple relationship enhancement programs, 402–403
Multicultural competencies, 5
Multidimensional family therapy, 93, 116, 139
Multidirectional partiality, 218–219

Multigenerational patterns, 208–209
Multiple impact therapy, 331
Multiple voices, 340
Multisystemic family therapy, 94, 116, 375, 379, 391–392
Multisystemic therapy (MST), 93–95, 139
Mutual puzzling, 329, 338–339

N
Narrative letters, 314–315
Narrative therapy, 44, 55, 57, 301, 347
  agency, 306, 312
  assessment, 305, 308
  case conceptualization, 305, 316–317, 325–326
  certificates, 315
  children, 316
  definitional ceremony, 314
  discourses, 302–303, 306, 322
  diversity/multicultural issues, 320–321, 324–325
  domestic violence, 316
  externalizing, 307, 309–310
  goal setting, 306–307
  hope, 305
  identity, 306–307
  interventions, 307–308
  leagues, 314
  mapping in the landscapes of action and identity/consciousness, 311
  mapping influence, 308
  native cultures, 321
  optimism, 305
  permission questions, 312–313
  preferred reality, 306–307, 434
  problem deconstruction, 310–311
  reflecting teams, 313
  relative influence questioning, 308
  re-membering conversations, 314
  research and evidence base, 320
  scaffolding conversations, 312
  separating problems and clients, 304–305, 307–308
  sexual identity, 322–325
  situating comments, 313
  sparkling events, 305–306
  statement of position map, 309
  therapist as coauthor, 305
  thickening descriptions, 304
  treatment plans, 303, 317–320, 327–328
  unique outcomes, 305–306
Narratives, 348
National Association of Social Workers, 7, 25, 31
National Institute for Drug Abuse, 116
National Institute of Mental Health (NIMH), 152, 206
Native Americans, narrative therapy, 321
Natural systems, 208
Naturalistic trance. See Ericksonian hypnosis
Negative affect reciprocity, 257
Negative punishment, 242
Negative reinforcement, 242
Neutrality, 85–86
Nonverbal communication, 368, 373
Not-knowing, 330, 347

O
Object relations family therapy, 203, 217–218
  assessment, 219. See also psychoanalytic family therapy
  case conceptualization, 219
  goal setting, 221
  holding environment, 218
  interventions, 221
  parental interjects, 219–220
  project identification, 219
  repression, 219
  self-object relations, 219
  splitting, 219
  transference, 220
Objective reality, 56
Observational research, 23, 78, 93
Office of Juvenile Justice and Delinquency Prevention, 293
Online relationships, 31
Ontology, 60
Open dialogue approach, 332, 346
Open groups, 397
Operant conditioning, 241–242
Oppositional defiant disorder (ODD), 403
Optimism, 277, 305
Ordeal therapy, 112
Organizational themes, 382
Outer dialogues, 331, 334–335

P
Paradox, 76–77, 175
Paradoxical interventions, 111
Parent training, 403
  conflicts, 155
  consistency, 234–235, 242, 383
  expectations, 382
  functional family therapy, 376, 382
  group therapy, 404
  mindfulness-based, 255
  operant conditioning, 241–242
  positive reinforcement, 242
  problem-solving, 383
  psychoeducational group therapy, 396
  reinforcement, 234–235
  supervision, 383
Parental hierarchy, 108, 131, 141, 181, 378, 423–424
Parental interjects, 219–220
Parenting relationship, 424
Partner abuse, 400–402
Permission questions, 312–313
Person-of-the-therapist, 5, 7
Philadelphia Child and Family Therapy Training Center, 128, 139
Philosophical assumptions, 412
Placator communication stance, 148–150, 424–425
Placebo effect, 20
Point charts, 242
Polarization, 188–189
Positive reinforcement, 242
Possibility therapy, 271
Postmodern therapy, 54–56, 60, 63–64, 301
  case conceptualization, 352–353
  diversity/multicultural issues, 58

Postmodern therapy (*Continued*)
  ethics, 55
  influence of, 58
  lived experiences, 347
  not-knowing, 347
  presenting problems, 414
  reflection, 347
  storytelling, 347–348
  treatment plans, 354–355
Postmodernism, 48, 331
Poststructuralists, 57, 64
Power metaphor, 103
Power struggles, 155
Practical Application of Intimate Relationship Skills (PAIRS), 402
Preferred realities, 306–307
Pre-knowing, 330
Presuppositional questions, 284
Pretend techniques, 112
Prevention and Relationship Enhancement Program (PREP), 402
Problem deconstruction, 310–311
Problem systems, 423
Problem-dissolving systems, 335
Problem-generating double bind, 79–80
Problem-organizing systems, 335
Problem-saturated stories, 305
Process comments, 382
Process groups, 396–397
Process-oriented therapy, 207
Projective identification, 219
Psychoanalytic family therapy, 203, 217
  case conceptualization template, 222
  detriangulation, 221
  diversity/multicultural issues, 226–227
  holding environment, 218
  process, 399
  research and evidence base, 225–226
  sexual identity, 227
  treatment plans, 222–225
Psychodynamic therapy, 61, 399
Psychoeducation, 243
Psychoeducational groups/classes, 395–399
Psychoeducational-process groups, 396
Punishment, 242
Pursue/withdraw patterns, 364, 367, 424

Q
Quid pro quo, 243, 256

R
Radical constructivism, 48
Reality construction, 56–57
Reflecting teams, 313, 340–342
Reflection, 347
Reflexivity, 57–58
Reframing, 71, 79, 84, 132, 368, 381
Reinforcement, 234–235, 242
Relational experiments, 212
Relational functions, 375, 377–378, 381, 388, 391
Relational goals, 157
Relational hierarchy, 375, 378, 391
Relational therapists, 27
Relationship Conflict Inventory (RCI), 72
Relationship enhancement programs, 402
Relative influence questioning, 308
Re-membering conversations, 314
Reorientation therapy, 322
Repair attempts, 257
Report (communication), 51–52
Repression, 219
Research, collaborative practices, 346
Research-informed clinician model, 16
Resistance, 41
Resource-focused therapy, 49, 72
Rhizome theory, 331
Rigid boundaries, 126, 422
Rituals, 89–90
Room technique, 191

S
Satir growth model, 69–70, 147–148, 152
  assessment, 154–156
  case conceptualization, 154, 161, 168–169
  chaos, 152
  communication stances, 148–149
  congruent communication, 157, 159
  diversity/multicultural issues, 165
  emotional expression, 159, 165
  family reconstruction, 160–161, 166
  goal setting, 157
  humanistic assumptions, 153
  interactions, 159–160
  interventions, 158
  levels of experience, 156
  mind-body connection, 157
  parts party, 161, 166
  positive feedback loop, 152
  self-esteem, 156
  status quo, 152
  symptoms, 155
  systemic view, 153
  touch, 160
  treatment plans, 162–164, 169–171
Satir Professional Development Institute (Manitoba), 152
Satir's communication approach, 36, 44, 61–62, 424–427
Scaffolding conversations, 312
Scaling questions, 281–284
Schizophrenia, 70, 93
  double-bind theory, 48, 52–53
  psychoeducational group therapy, 398–399
Sculpting, 160
Second-order change, 50–51, 76–77, 152
Second-order cybernetics, 53, 55, 62–63, 85
Secrets policies, 27–28
Secure attachment, 359
Selective borrowing, 411
Self, 185–191
Self-actualization, 61–62, 157, 178–179
Self-compassion, 156–157
Self-correction, 70
Self-disclosure, 363
Self-esteem, 150, 156
Self-leadership, 187, 189, 195–196
Self-mandala, 153
Self-object relations, 219
Self-worth, 156–157
Sequential siding, 219
Session Rating Scale, 20
Sexual crucible model, 206–207, 213
Sexual identity diversity, 95–96, 118, 141, 165, 197, 227, 263, 294, 322, 349–350, 373, 388
Shamanism, 48
Shaping competence, 135
Sibling positions, 210
Situating comments, 313
Social constructionist epistemology, 333, 346
Social constructionist theories, 47, 54–55, 57, 64, 336
Social courtesy, 107
Socratic method, 245
Solution-based therapy, 44, 48, 271–272
  3-D Model: Dissociate, Disown, Devalue, 287
  assessment, 277–278, 431–433
  beginner's mind, 276
  case conceptualization, 277, 288, 297–298
  channeling language, 276
  child sexual abuse, 286–287
  client motivation, 278
  client strengths, 273
  client word choice, 276
  compliments, 285
  constructive questions, 287
  coping questions, 285
  crystal ball questions, 279–280, 432
  diversity/multicultural issues, 293–294, 297
  encouragement, 285
  exceptions, 277–278
  formula first session task, 284
  goal setting, 279
  hope, 277
  interventions, 283, 286
  magic wand questions, 279–280, 432
  miracle questions, 279–280, 432
  one thing different, 283
  optimism, 277
  outcomes, 432
  presuppositional questions, 284
  reflection, 276
  research and evidence base, 292–293
  scaling questions, 281, 284
  sexual identity, 294
  time machine questions, 279–280, 432
  treatment, 275
  treatment plans, 288–291, 299–300
  utilization techniques, 285
  videotalk, 287
Solution-Focused Brief Therapy Association, 274–275
Solution-focused brief therapy (SFBT), 271, 274
Solution-Focused Recovery Scale, 287
Solution-generating questions, 279–281
Solution-oriented couples therapy, 286
Solution-oriented Ericksonian hypnosis, 271
Solution-oriented hypnosis. *See* Ericksonian hypnosis
Solution-oriented therapy, 271, 274
Solutions, 272
Sparkling events, 305–306

Spatial metaphor, 160
Spirituality, 418–419
Splitting, 219
Stage theory, 96
Stage-of-life problems, 108
Standard of care, 26
Standard practice, 358
Standards, professional practice, 21, 25–26
Statement of position map, 309
Storytelling, 347–348
Straightforward directives, 109
Strategic Family Therapy and Training Center (MRI), 104
Strategic therapy, 69–70, 427
  assessment, 107
  case conceptualization, 107–109, 112–113, 121
  cultural diversity, 95
  dangers of improvement, 80
  directives, 103–104, 106, 109–110
  diversity/multicultural issues, 117, 120–122
  evidence-based, 116
  family hierarchy, 108, 117
  humanism, 109
  initial interview, 104
  interactions, 104, 106
  intergenerational conflict, 120–122
  interventions, 109
  metaphorical tasks, 112
  new action, 109
  one-down stance, 107
  ordeals, 112
  phone contact, 105
  pretend techniques, 112
  sexual identity, 118
  stages, 105–109
  transgendered youth, 118–119
  treatment plans, 113–115, 122–123
Strength-based therapy, 271–272, 275
Structural family therapists, 36
Structural therapy, 36, 43–44, 69, 71, 125, 127–129
  assessment, 130
  boundaries, 125–126, 131–133, 140
  case conceptualization, 135
  complementarity, 132
  diversity/multicultural issues, 128, 140–144
  empathy, 130
  enactments, 126–127, 132–133, 140
  families, 132–134
  goals, 132
  interactions, 127
  interventions, 132, 134
  parental hierarchy, 131, 134, 140–141, 423–424
  redirecting alternative transactions, 127
  reframing, 132–133
  role of symptom, 130
  sexual identity, 141
  treatment plans, 135–138, 144–146
  unbalancing, 134
Structuralism, 64
Subcultures, 57
Substance Abuse & Mental Health Services Administration Registry of Evidenced-Based Practices, 260

Subsystems, 130–131
Superreasonable communication stance, 148, 150–151, 424–425
Survival stances, 148–151, 424
Survival triad, 156
Symbolic-experiential therapy, 61–62, 71, 147, 173, 175, 427
  absurd fantasy, 181
  affective confrontation, 180
  assessment, 176
  augumenting despair, 180–181
  authentic encounters, 176
  autonomy, 180
  case conceptualization, 176, 181–182, 198
  common factors, 195
  competencies, 177–178
  diversity/multicultural issues, 196
  emotional process, 177
  family cohesion, 178
  goal setting, 178
  interventions, 179
  metaphors, 181
  mutual growth, 176
  parental hierarchy, 181
  research and evidence base, 196
  responsibility, 176
  self-actualization, 178–179
  spontaneity, 179
  structural organization, 177
  symbolic world, 179
  symptoms, 178–179
  therapist use of self, 175
  treatment plans, 182–185, 200
  trial of labor, 176–177
Symmetrical patterns, 77
Symmetrical relationships, 51
Symptom bearer, 130
Symptom prescription, 111
Symptom-based treatment plans, 37, 43
Symptoms, emotional functions of, 155
Syncretism, 412
Systematic treatment selection models, 412
Systemic assessment, 416
Systemic dynamics, 41, 118
Systemic family therapists, 36, 41, 49, 53–55
Systemic reframing, 132–133
Systemic therapy, 44, 49, 60, 153
  assumptions, 62
  case conceptualization form, 437–443
  complementary patterns, 77
  cultural diversity, 95
  evidence-based treatments, 93
  history of, 69–70, 93
  metacommunication, 77
  problem systems, 423
  process of, 73–75
  sexual identity diversity, 95
  symmetrical patterns, 77
  treatment plans, 99–101
Systems theory, 47–48, 55

T
Taos Institute, 331
Teasing and Bullying Survey (TABS), 72
Technical eclecticism, 411–412

Technology, in therapy, 30–31
Terrible simplifications, 76
Thematic Apperception Tests, 322
Theoretical integration, 410–412
Theoretical language, 43
Theory
  common factors, 18, 20–21
  in therapy, 4, 10–11
  value to therapists, 21
Theory of attachment. See Attachment theory
Theory-based treatment plans, 37
Theory-specific case conceptualization, 36
Therapeutic curiosity, 86
Therapeutic double bind, 79–80
Therapeutic impasse, 333
Therapeutic models, 18, 20–21, 41
Therapeutic presence, 7, 153
Therapeutic relationship, 19–20, 40–41, 164–165
Therapeutic spontaneity, 129
Therapeutic tasks, 38, 40–41
Therapeutic theories, 4–5, 36
Therapists
  acceptance, 363
  adapting to clients, 74
  advocacy stance, 349
  agents of social change, as, 226
  alliances, 363, 376
  being public, 339–340
  challenging irrational beliefs, 244–246
  child abuse reporting, 29
  coaching, 257
  coauthor, 305
  collaboration, 186, 236
  communicating with other systems, 28
  compliments, 135, 285–286
  congruence, 153–154
  consistency, 64–65
  conveying hope, 154
  cotherapists, 175–176
  countertransference, 187
  credibility, 154, 377
  curiosity, 86, 333
  differentiation, 204, 207, 212
  effectiveness, 18, 20
  empathy, 148, 153–154, 236, 362–363, 368
  ethics, 6–7, 24–27, 32, 435
  evocative responses, 367–368
  fearlessness, 87
  genuineness, 363
  inner dialogues, 339
  irreverence, 86–87
  maneuverability, 74
  minors and, 28
  mutual growth, 176
  nonanxious presence, 207
  nonjudgmental, 260
  nonverbal communication, 362, 368, 373
  personal integrity, 176
  personal qualities, 5, 7
  philosophical stance, 333, 335–336
  playfulness, 179–180
  process comments, 382
  process questions, 211
  professional communication, 339

Therapists (*Continued*)
  progress notes, 26
  reflecting emotions, 367
  relationship variables, 20
  respect, 377
  responsibility, 176
  secrets policies, 27–28
  self-disclosure, 363
  shaping competence, 135
  tentative inferences, 105
  transference, 187
  unconditional positive regard, 153
  use of self, 158, 175
  validation, 367
  values, 31
  viewing, 409
  warmth of, 148, 153, 173
  "withness," 333
Therapy of the absurd, 175
Third-party payment
  therapeutic tasks, 40
  treatment plans and, 37, 43
3-D Model: Dissociate, Disown, Devalue, 287
Three-generational emotional process, 203–204
Time machine questions, 279–280, 432
Togetherness, 204
Token economies, 242–243
Top-down process, 347
Touch, 160
Transactions, structural therapy and, 127
Transference, 187, 218, 220
Transformation, 336–337
Transgendered youth, 118–119
Trauma resolution, 347–348
Trauma-informed clinical models, 139
Treatment plans, 35–36, 45
  Bowen intergenerational therapy, 214–216, 231–232
  client goals, 41–43
  clinical, 37–40
  cognitive-behavioral family therapy (CBFT), 247, 269–270
  diversity/multicultural issues, 41–42
  internal family systems therapy, 192–195
  interventions, 44
  medical model, 37
  MRI therapy and couples/families, 82–83
  MRI therapy and individuals, 80–82
  psychoanalytic family therapy, 222–225
  Satir growth model, 162–164, 169–171
  structural therapy, 135–138, 144–146
  symbolic-experiential therapy, 182–185, 200
  symptom-based, 37
  systemic therapy, 99–101
  theory-based, 37
  therapeutic tasks, 40–41
Triangles, 423
Truth, 56, 60–62

U

U. S. Substance Abuse and Mental Health Services Administration (SAMHSA), 398
Unbalancing, 134
Unconditioned stimuli and response, 241
Unique outcomes, 305–306
Utilization techniques, 285
Utopian syndrome, 76

V

Validation, 155, 212
Values, 31
Videoconference therapy, 30
Videotalk, 286–287
Viewing, 409–410
Virginia Satir Global Network, 148, 152

W

Wampold's Common Factors Model, 19
Wheel of power and control, 400
Women, Satir growth model and, 165
Women's discourse, 302–303
The Women's Project, 218, 226
Working-phase goals, 43
Writing, in therapy, 332, 340

Y

Youth
  behavioral problems, 375–383
  functional family therapy, 375–383, 388, 390
  minority, 379
  risk factors, 378–379, 391
  strengths, 380, 392

Z

Zones of proximal development, 312

CPSIA information can be obtained
at www.ICGtesting.com
Printed in the USA
FFHW01n0407300718
47567815-51039FF

9 781285 456430